A Complete Solution Guide to Real and Complex Analysis I

by Kit-Wing Yu, PhD

kitwing@hotmail.com

ISBN: 978-988-78797-8-7 (eBook)
ISBN: 978-988-78797-9-4 (Paperback)

About the author

Dr. Kit-Wing Yu received his B.Sc. (1st Hons), M.Phil. and Ph.D. degrees in Math. at the HKUST, PGDE (Mathematics) at the CUHK. After his graduation, he has joined United Christian College to serve as a mathematics teacher for at least seventeen years. He has also taken the responsibility of the mathematics panel since 2002. Furthermore, he was appointed as a part-time tutor (2002 – 2005) and then a part-time course coordinator (2006 – 2010) of the Department of Mathematics at the OUHK.

Between 2012 and 2014, Dr. Yu was invited to be a Judge Member by the World Olympic Mathematics Competition (China). In the research aspect, he has published over twelve research papers in international mathematical journals, including some well-known journals such as J. Reine Angew. Math., Proc. Roy. Soc. Edinburgh Sect. A and Kodai Math. J.. His research interests are inequalities, special functions and Nevanlinna's value distribution theory.

Preface

Professor Walter Rudin[a] is the author of the classical and famous textbooks: *Principles of Mathematical Analysis*, *Real and Complex Analysis*, and *Functional Analysis*. (People commonly call them *"Baby Rudin"*, *"Papa Rudin"* and *"Grandpa Rudin"* respectively.) Undoubtedly, they have produced important and extensive impacts to the study of mathematical analysis at university level since their publications.

As far as you know, *Real and Complex Analysis* keeps the features of *Principles of Mathematical Analysis* which are well-organized and expositions of theorems are clear, precise and well-written. Therefore, *Real and Complex Analysis* is always one of the main textbooks or references of graduate real analysis course in many universities. Actually, some universities will request their Ph.D. students to study this book for their qualifying examinations.

After the publication of the book *A Complete Solution Guide to Principles of Mathematical Analysis*, some purchasers have suggested me to write a solution book of *Real and Complex Analysis*. (To the best of the author's knowledge, *Papa Rudin* has **no** solution manual.) I was afraid of doing so at the beginning because the exercises are at graduate level and they are much more difficult than those in *Baby Rudin*. Fortunately, I was used to keeping solutions of mathematics exercises done by me in my undergraduate and graduate study. In fact, I have kept at least 25% of the exercises in the first six chapters of *Papa Rudin*. Therefore, after thorough consideration, I decided to start the next project of a solution book to Rudin's *Real and Complex Analysis*.

Since *Papa Rudin* consists of two components, I plan to write the solutions for "Real Analysis" part first. In fact, the present book *A Complete Solution Guide to Real and Complex Analysis I* covers all the exercises of Chapters 1 to 9 and its primary aim is to help every mathematics student and instructor to understand the ideas and applications of the theorems in Rudin's book. To accomplish this goal, I have adopted the way I wrote the book *A Complete Solution Guide to Principles of Mathematical Analysis*. In other words, I intend writing the solutions as comprehensive as I can so that you can understand *every* detailed part of a proof easily. Apart from this, I also keep reminding you what theorems or results I have applied by quoting them *repeatedly* in the proofs. By doing this, I believe that you will become fully aware of the meaning and applications of each theorem.

Before you read this book, I have two gentle reminders for you. Firstly, as a mathematics instructor at a college, I understand that the growth of a mathematics student depends largely on how hard he/she does exercises. When your instructor asks you to do some exercises from Rudin, you are not suggested to read my solutions unless you have tried your best to prove them yourselves. Secondly, when I prepared this book, I found that some exercises require knowledge that Rudin did not cover in his book. To fill this gap, I refer to some other analysis or topology

---

[a]`https://en.wikipedia.org/wiki/Walter_Rudin`.

books such as [9], [10], [22], [42] and [47]. Other useful references are [1], [12], [24], [28], [29], [59], [64], [66] and [67]. Of course, it is not a surprise that we will regard the exercises in *Baby Rudin* as some known facts and if you want to read proofs of them, you are strongly advised to read my book [63].

The features of this book are as follows:

- It covers all the 176 exercises from Chapters 1 to 9 with *detailed* and *complete* solutions. As a matter of fact, my solutions show every detail, every step and every theorem that I applied.

- There are 11 illustrations for explaining the mathematical concepts or ideas used behind the questions or theorems.

- Sections in each chapter are added so as to increase the readability of the exercises.

- Different colors are used frequently in order to highlight or explain problems, lemmas, remarks, main points/formulas involved, or show the steps of manipulation in some complicated proofs. (ebook only)

- Necessary lemmas with proofs are provided because some questions require additional mathematical concepts which are not covered by Rudin.

- Many useful or relevant references are provided to some questions for your future research.

Since the solutions are written solely by me, you may find typos or mistakes. If you really find such a mistake, please send your valuable comments or opinions to

`kitwing@hotmail.com.`

Then I will post the updated errata on my website

`https://sites.google.com/view/yukitwing/`

irregularly.

Kit Wing Yu
April 2019

List of Figures

Contents

CHAPTER **1**

Abstract Integration

1.1 Problems on σ-algebras and Measurable Functions

> **Problem 1.1**
>
> *Rudin Chapter 1 Exercise 1.*

Proof. Let X be a set. Assume that $\mathfrak{M}$ was an infinite σ-algebra in X which has only countably many members. Then we have $\mathfrak{M} = \{\varnothing, X, A_1, A_2, \ldots\}$. Let $x \in X$ be fixed, we define the set

$$S_x = \bigcap_{x \in A} A, \tag{1.1}$$

where $x \in A \in \mathfrak{M}$. (By this definition (1.1), it is clear that $x \in S_x$.) Since $\mathfrak{M}$ is countable, the intersection in (1.1) is actually at most countable. Thus it follows from Comment 1.6(c) that $S_x \in \mathfrak{M}$. Next we define the set

$$\mathfrak{S} = \{S_x \mid x \in X\}$$

so that $\mathfrak{S} \subseteq \mathfrak{M}$. We need a lemma about the set $\mathfrak{S}$:

> **Lemma 1.1**
>
> If $S_x, S_y \in \mathfrak{S}$ and $S_x \cap S_y \neq \varnothing$, then $S_x = S_y$.

Proof of Lemma 1.1. Let $z \in S_x \cap S_y$. We want to show that $S_z = S_x$. On the one hand, since $z \in S_x$, we know from the definition (1.1) that $z \in A$ *for all* A containing x. Now if $w \in S_x$, then $w \in A$ for all A containing x. By the previous observation, such A must contain z, so we get $w \in S_z$ and then

$$S_x \subseteq S_z. \tag{1.2}$$

On the other hand, if $w \in S_z$, then $w \in A$ *for all* A containing z. Thus the hypothesis $z \in S_x \cap S_y$ implies that $w \in S_x \cap S_y$, i.e., $S_z \subseteq S_x \cap S_y$. By this result and the definition (1.1), we obtain

$$S_z \subseteq S_x \cap S_y \subseteq S_x. \tag{1.3}$$

Combining the observations (1.2) and (1.3), we have the desired result that $S_z = S_x$. Similarly, we can show that $S_z = S_y$. Hence we conclude that $S_x = S_y$. ∎

1

Now we return to the proof of the problem. By Lemma 1.1, we may assume that all elements in $\mathfrak{S}$ are distinct. If $B \in \mathfrak{M}$, then the definition (1.1) certainly gives $y \in S_y$ for every $y \in B$ so that

$$B \subseteq \bigcup_{y \in B} S_y. \tag{1.4}$$

Therefore, if the cardinality of $\mathfrak{S}$ is N, then the cardinality of its power set $2^{\mathfrak{S}}$ is 2^N and so there are at most 2^N elements in $\mathfrak{M}$ by the inclusion (1.4), a contradiction. Thus the set $\mathfrak{S}$ must be infinite.

Recall that $\mathfrak{S} \subseteq \mathfrak{M}$ and $\mathfrak{M}$ is countable, $\mathfrak{S}$ is also countable by [49, Theorem 2.8, p. 26]. Since $\mathfrak{M}$ is a σ-algebra, it must contain the power set of $\mathfrak{S}$. However, it is well-known that the power set of an infinite countable set is uncountable [47, Problem 22, p. 16]. We conclude from these facts that $\mathfrak{M}$ is uncountable, a contradiction. This completes the proof of the problem. ∎

> **Problem 1.2**
>
> *Rudin Chapter 1 Exercise 2.*

Proof. Put $\mathbf{f}(x) = (f_1(x), \ldots, f_n(x))$. Since $f_1, \ldots, f_n : X \to \mathbb{R}$, $\mathbf{f}$ maps the measurable space X into $\mathbb{R}^n$. Let Y be a topological space and $\Phi : \mathbb{R}^n \to Y$ be continuous. Define

$$h(x) = \Phi\big((f_1(x), \ldots, f_n(x))\big).$$

Since $h = \Phi \circ \mathbf{f}$, Theorem 1.7(b) shows that it suffices to prove the measurability of $\mathbf{f}$.

To this end, we first consider $\mathbf{f}^{-1}(R)$ for an open rectangle R in $\mathbb{R}^n$. By the definition, $R = I_1 \times \cdots \times I_n$, where I_i is an open interval in $\mathbb{R}$ for $1 \le i \le n$. Now we know from [42, Exercise 3, p. 21] that

$$\begin{aligned}
\mathbf{f}^{-1}(R) &= \{x \in X \mid \mathbf{f}(x) \in R\} \\
&= \{x \in X \mid f_i(x) \in I_i \text{ for } 1 \le i \le n\} \\
&= \{x \in X \mid f_1(x) \in I_1\} \cap \cdots \cap \{x \in X \mid f_n(x) \in I_n\} \\
&= f_1^{-1}(I_1) \cap \cdots \cap f_n^{-1}(I_n).
\end{aligned}$$

Since each f_i is measurable and each I_i is open in $\mathbb{R}$, the set $f_i^{-1}(I_i)$ is measurable in X by Definition 1.3(c). Thus this implies that $\mathbf{f}^{-1}(R)$ is measurable in X by Comment 1.6(c).

Our proof will be complete if we can show that *every* open set V in $\mathbb{R}^n$ can be written as a countable union of open rectangles R_j in $\mathbb{R}^n$. We need some topology. By [42, Exercise 8(a), p. 83], the countable collection

$$\mathscr{B} = \{I = (a,b) \mid a < b \text{ and } a, b \in \mathbb{Q}\}$$

is a *basis* that generates the *standard topology* on $\mathbb{R}$. Then we follow from [42, Theorem 15.1, p. 86] that the collection

$$\mathscr{C} = \{I_1 \times \cdots \times I_n \mid I_i = (a_i, b_i), \, a_i, b_i \in \mathbb{Q}, \, 1 \le i \le n\} \tag{1.5}$$

is a basis for the (product) topology of $\mathbb{R}^n$.[a] Since elements in the collection (1.5) are open rectangles and it is countable by [49, Theorem 2.13, p. 29], every open set V in $\mathbb{R}^n$ is, in fact, a countable union of open rectangles R_j, i.e,

$$\mathbf{f}^{-1}(V) = \mathbf{f}^{-1}\Big(\bigcup_{j=1}^{\infty} R_j\Big) = \bigcup_{j=1}^{\infty} \mathbf{f}^{-1}(R_j).$$

---

[a] For details, please read [42, §13 and §15].

Hence $\mathbf{f}^{-1}(V)$ is also a measurable set in X by Definition 1.3(a)(iii). This completes the proof of the problem. $\blacksquare$

> **Problem 1.3**
>
> *Rudin Chapter 1 Exercise 3.*

Proof. Let $f : X \to \mathbb{R} \subset [-\infty, \infty]$ and $\mathfrak{M}$ be the σ-algebra of X. Let α be real. By [49, Theorem 1.20(b), p. 9], we can find a sequence $\{r_n\}$ of rational numbers such that $\alpha < r_n$ for all $n \in \mathbb{N}$, $r_n \to \alpha$ as $n \to \infty$. In other words, we have

$$(\alpha, \infty] = \bigcup_{n=1}^{\infty} [r_n, \infty]$$

which implies that

$$f^{-1}((\alpha, \infty]) = f^{-1}\left(\bigcup_{n=1}^{\infty} [r_n, \infty] \right) = \bigcup_{n=1}^{\infty} f^{-1}([r_n, \infty]).$$

Recall that

$$f^{-1}([r_n, \infty]) = \{x \mid f(x) \geq r_n\}$$

is assumed to be measurable for each n, so we have $f^{-1}([r_n, \infty]) \in \mathfrak{M}$ and then we follow from Definition 1.3(a)(iii) that $f^{-1}((\alpha, \infty]) \in \mathfrak{M}$. Since α is arbitrary, we conclude from Theorem 1.12(c) that f is measurable. This finishes the proof of the problem. $\blacksquare$

> **Problem 1.4**
>
> *Rudin Chapter 1 Exercise 4.*

Proof.

(a) Let $\alpha_k = \sup\{-a_k, -a_{k+1}, \ldots\}$ for $k = 1, 2, \ldots$. By the definition, we have $\alpha_k \geq -a_n$ for all $n \geq k$ and if $\alpha \geq -a_n$ for all $n \geq k$, then $\alpha \geq \alpha_k$. Note that this is equivalent to the fact that $-\alpha_k \leq a_n$ for all $n \geq k$ and if $-\alpha \leq a_n$ for all $n \geq k$, then $-\alpha \leq -\alpha_k$. In other words, we have

$$-\alpha_k = \inf\{a_k, a_{k+1}, \ldots, \}$$

for $k = 1, 2, \ldots$. Thus this implies that

$$\sup_{n \geq k}(-a_n) = \sup\{-a_k, -a_{k+1}, \ldots\} = -\inf\{a_k, a_{k+1}, \ldots\} = -\inf_{n \geq k}(a_n). \tag{1.6}$$

Similarly, we have

$$\inf_{n \geq k}(-c_n) = -\sup_{n \geq k}(c_n) \tag{1.7}$$

for a sequence $\{c_n\}$ in $[-\infty, \infty]$. By applying the equality (1.6) and then the equality (1.7), we achieve that

$$\limsup_{n \to \infty}(-a_n) = \inf_{k \geq 1}\left\{ \sup_{n \geq k}(-a_n) \right\} = \inf_{k \geq 1}\left\{ \underbrace{-\inf_{n \geq k}(a_n)}_{c_n} \right\} = -\sup_{k \geq 1}\left\{ \inf_{n \geq k}(a_n) \right\} = -\liminf_{n \to \infty}(a_n)$$

which is our desired result.

(b) This part is proven in [63, Problem 3.5, pp. 32, 33].

(c) Since $a_n \leq b_n$ for all $n = 1, 2, \ldots$, we must have

$$\alpha_k = \inf_{n \geq k}(a_n) \leq \inf_{n \geq k}(b_n) = \beta_k \tag{1.8}$$

for all $k = 1, 2, \ldots$. Thus $\{\alpha_n\}$ and $\{\beta_n\}$ are two sequences in $[-\infty, \infty]$ such that $\alpha_n \leq \beta_n$ for all $n = 1, 2, \ldots$. By a similar argument, we also have

$$\sup_{k \geq m} \alpha_k \leq \sup_{k \geq m} \beta_k \tag{1.9}$$

for all $m = 1, 2, \ldots$. Combining the inequalities (1.8) and (1.9), we have

$$\sup_{k \geq m} \left\{ \inf_{n \geq k}(a_n) \right\} \leq \sup_{k \geq m} \left\{ \inf_{n \geq k}(b_n) \right\}$$

for all $m = 1, 2, \ldots$. By Definition 1.13, we have the desired result.

For a counterexample to part (b), we consider $a_n = (-1)^n$ and $b_n = (-1)^{n+1}$ for all $n = 1, 2, \ldots$. On the one hand, we have $a_n + b_n = 0$ for all $n = 1, 2, \ldots$ so that

$$\limsup_{n \to \infty}(a_n + b_n) = 0. \tag{1.10}$$

On the other hand, we have

$$\limsup_{n \to \infty} a_n = \lim_{k \to \infty} a_{2k} = 1 \quad \text{and} \quad \limsup_{n \to \infty} b_n = \lim_{k \to \infty} b_{2k+1} = 1$$

which imply that

$$\limsup_{n \to \infty} a_n + \limsup_{n \to \infty} b_n = 2. \tag{1.11}$$

Hence we obtain from the results (1.10) and (1.11) that the strict inequality can hold in part (b). This completes the proof of the problem. ◾

> **Problem 1.5**
>
> *Rudin Chapter 1 Exercise 5.*

Proof.

(a) Let $\mathfrak{M}_X$ be a σ-algebra of X. Further, suppose that $S_f(\pm\infty) = \{x \in X \mid f(x) = \pm\infty\}$, and $S_g(\pm\infty) = \{x \in X \mid g(x) = \pm\infty\}$. We see that

$$S_f(\infty) = \bigcap_{n=1}^{\infty} \{x \in X \mid f(x) > n\}.$$

Since f is measurable and $(n, \infty]$ is open[b] in $[-\infty, \infty]$, $f^{-1}((n, \infty]) = \{x \mid f(x) > n\} \in \mathfrak{M}_X$ by Theorem 1.12(b) for each $n \in \mathbb{N}$. Thus $S_f(\infty) \in \mathfrak{M}_X$ by Comment 1.6(c). Similarly, all the other sets $S_f(-\infty), S_g(\infty)$ and $S_g(-\infty)$ belong to $\mathfrak{M}_X$ too.

Next, we let $X' = X \backslash (S_f(\infty) \cup S_f(-\infty) \cup S_g(\infty) \cup S_g(-\infty))$. Since $S_f(\infty), S_f(-\infty), S_g(\infty)$ and $S_g(-\infty)$ are measurable, $X' \in \mathfrak{M}$ by Definition 1.3(a)(ii) and Comment 1.6(b). By

---

[b]See the proof of Theorem 1.12(c) in [51, p. 13]

the comment following Proposition 1.24, we know that X' is itself a measure space and if we let $\mathfrak{M}_{X'}$ be a σ-algebra of X', then

$$\mathfrak{M}_{X'} \subseteq \mathfrak{M}_X. \tag{1.12}$$

Furthermore, the restricted mappings $f_{X'} : X' \to \mathbb{R}$ and $g_{X'} : X' \to \mathbb{R}$ are also measurable because of Definition 1.3(c) and the fact that *any* open set V in $\mathbb{R}$ is a countable union of segments of the type (α, β) so that V is also open in the extended number system $[-\infty, \infty]$.[c] In fact, we have

$$f_{X'}^{-1}(V) = f^{-1}(V) \in \mathfrak{M}_{X'} \quad \text{and} \quad g_{X'}^{-1}(V) = g^{-1}(V) \in \mathfrak{M}_{X'}.$$

We need to prove one more thing: the mapping $-g : X \to [-\infty, \infty]$ is measurable. Since g is measurable, we know from [49, Definition 11.13] that $\{x \in X \mid g(x) > -a\} \in \mathfrak{M}_X$ for every real a. It is obvious that $\{x \in X \mid -g(x) < a\} = \{x \in X \mid g(x) > -a\}$ for every real a, so we deduce from [49, Theorem 11.15, p. 311] that $-g$ is measurable.

Now we are ready to prove the desired results. Notice that

$$\{x \in X \mid f(x) = g(x)\} = \{x \in X \mid h(x) = 0\} = h^{-1}(0),$$

where $h = f - g$. Since f and $-g$ are measurable, the new (real) function $h = f - g$ is also measurable by Theorem 1.9(c). Since

$$h^{-1}(0) = \bigcap_{n=1}^{\infty} h^{-1}\left(\left(-\frac{1}{n}, \frac{1}{n}\right)\right)$$

and $h^{-1}((-\frac{1}{n}, \frac{1}{n})) \in \mathfrak{M}_{X'}$ for every $n \in \mathbb{N}$, we yield from this and the relation (1.12) that $h^{-1}(0) \in \mathfrak{M}_{X'} \subseteq \mathfrak{M}_X$. This shows the second assertion. For the first assertion, we note that

$$\begin{aligned}
\{x \in X \mid f(x) < g(x)\} &= \{x \in X \mid h(x) = f(x) - g(x) < 0\} \\
&= \{h^{-1}((-\infty, 0)) \cup [S_f(-\infty) \setminus S_g(-\infty)] \cup [S_g(\infty) \setminus S_f(\infty)]\} \\
&\quad \setminus \{[S_f(\infty) \cap S_g(\infty)] \cup [S_f(-\infty) \cap S_g(-\infty)]\}.
\end{aligned}$$

Recall that $S_f(\infty), S_f(-\infty), S_g(\infty), S_g(-\infty) \in \mathfrak{M}_X$, so we have

$$S_f(-\infty) \setminus S_g(-\infty), \; S_g(\infty) \setminus S_f(\infty), \; S_f(\infty) \cap S_g(\infty), \; S_f(-\infty) \cap S_g(-\infty) \in \mathfrak{M}_X. \tag{1.13}$$

By the measurability of h and the relation (1.12), we have

$$h^{-1}((-\infty, 0)) \in \mathfrak{M}_X. \tag{1.14}$$

Hence the facts (1.13) and (1.14) show that $\{x \mid f(x) < g(x)\} \in \mathfrak{M}_X$.

(b) This part is proven in [63, Problem 11.3, p. 339].

This completes the proof of the problem.   ▨

Problem 1.6

Rudin Chapter 1 Exercise 6.

---

[c] This can be seen, again, from the proof of Theorem 1.12(c) or from the comment following Proposition 1.24.

Proof. We prove the assertions one by one.

- **$\mathfrak{M}$ is a σ-algebra in X.** We check Definition 1.3(a). Since $X^c = \varnothing$, it is at most countable. Thus we have $X \in \mathfrak{M}$. Let $A \in \mathfrak{M}$. If A^c is at most countable, then $A^c \in \mathfrak{M}$. Similarly, if A is at most countable, then since $(A^c)^c = A$, we have $A^c \in \mathfrak{M}$. Suppose that $A_n \in \mathfrak{M}$ for $n = 1, 2, \ldots$. Then we have either A_n or A_n^c is at most countable for $n = 1, 2, \ldots$. If all A_n *are* at most countable, then we deduce from the corollary following [49, Theorem 2.12, p. 29] that the set

$$A = \bigcup_{n=1}^{\infty} A_n \tag{1.15}$$

is also at most countable so that $A \in \mathfrak{M}$. Otherwise, without loss of generality, we suppose that A_1 is uncountable but A_1^c is at most countable. Then we consider

$$A^c = \Big(\bigcup_{n=1}^{\infty} A_n \Big)^c = \bigcap_{n=1}^{\infty} A_n^c \subseteq A_1^c$$

which means that A^c is at most countable too. Hence $A^c \in \mathfrak{M}$ and then $\mathfrak{M}$ is a σ-algebra in X.

- **μ is a measure on $\mathfrak{M}$.** We check Definition 1.18(a). Since $\varnothing$ is at most countable, we have $\mu(\varnothing) = 0$ so that μ is *not* identically ∞. In fact, it is clear that we have $\mu : \mathfrak{M} \to \{0, 1\} \subset [0, \infty]$. Let $\{A_n\}$ be a disjoint countable collection of members of $\mathfrak{M}$.

  Case (i): All A_n are at most countable. Recall that the set A given by (1.15) is at most countable. By the definition of μ, we have $\mu(A) = \mu(A_n) = 0$ for all $n = 1, 2, \ldots$. Thus we have

$$\mu(A) = \sum_{n=1}^{\infty} \mu(A_n) \tag{1.16}$$

  in this case.

  Case (ii): There is *at least* one A_k is uncountable. Since $A_k \in \mathfrak{M}$, A_k^c must be at most countable. Since $A_n \cap A_k = \varnothing$ for all $n \neq k$, we have $A_n \subseteq A_k^c$ for all $n \neq k$. In other words, the measurable sets A_n are at most countable for all $n \neq k$. Therefore, we have $\mu(A_k) = 1$ and $\mu(A_n) = 0$ for all $n \neq k$. Since

$$A^c = \bigcap_{n=1}^{\infty} A_n^c \subseteq A_k^c,$$

  A^c is at most countable and then $\mu(A) = 1$. Hence the equality (1.16) also holds in this case.

  This completes the proof that μ is a measure on $\mathfrak{M}$.

- **The determination of measurable functions and their integrals.** Let $f : X \to \mathbb{R}$ be a measurable function. Since X is uncountable, we have $\mu(X) = 1$ which is the only thing that we know and start with. For every $n \in \mathbb{Z}$, we know that

$$[n, n+1) = \mathbb{R} \setminus \big[(-\infty, n) \cup [n+1, \infty) \big].$$

By the fact that[d] $f^{-1}(A \setminus B) = f^{-1}(A) \setminus f^{-1}(B)$, we obtain

$$f^{-1}\big([n, n+1)\big) = f^{-1}(\mathbb{R}) \setminus \big[f^{-1}\big((-\infty, n)\big) \cup f^{-1}\big([n+1, \infty)\big) \big]. \tag{1.17}$$

---

[d] See [42, Exercise 2(d), p. 20].

By [49, Theorem 11.15, p. 311], each set on the right-hand side in (1.17) is measurable. This implies that

$$f^{-1}\big([n, n+1)\big) \in \mathfrak{M}.$$

Let $E_n = f^{-1}\big([n, n+1)\big)$, where $n \in \mathbb{Z}$. By the definition of $\mathfrak{M}$ and then the definition of μ, we have either $\mu(E_n) = 0$ or $\mu(E_n) = 1$. For every $x \in X$, we must have $f(x) \in [n, n+1)$ for some $n \in \mathbb{Z}$, i.e., $x \in E_n$ for some $n \in \mathbb{Z}$. Therefore, we have

$$\bigcup_{n=-\infty}^{\infty} E_n = X.$$

It is clear that $\mathbb{R} = \bigcup_{n=-\infty}^{\infty} [n, n+1)$, so $\{E_n\}$ is a *disjoint* countable collection of members of $\mathfrak{M}$. By Definition 1.18(a), we see that

$$\mu(X) = \mu\Big(\bigcup_{n=-\infty}^{\infty} E_n \Big) = \sum_{n=-\infty}^{\infty} \mu(E_n). \tag{1.18}$$

The fact $\mu(X) = 1$ and the equality (1.18) force that there exists an integer n_0 such that $\mu(E_{n_0}) = 1$. Without loss of generality, we may assume that $n_0 = 0$, i.e.,

$$\mu(E_0) = \mu\big(f^{-1}([0, 1))\big) = 1. \tag{1.19}$$

Next if we write $[0, 1) = [0, \frac{1}{2}) \cup [\frac{1}{2}, 1)$, then the above argument and the value (1.19) imply that either $\mu\big(f^{-1}([0, \frac{1}{2}))\big) = 1$ or $\mu\big(f^{-1}([\frac{1}{2}, 1))\big) = 1$. This process can be done continuously so that a sequence of intervals $\{[a_n, b_n)\}$ is constructed such that

$$\mu\big(f^{-1}([a_n, b_n))\big) = 1, \quad 0 \le b_n - a_n \le \frac{1}{2^n} \quad \text{and} \quad \lim_{n \to \infty} (b_n - a_n) = 0,$$

where $n = 1, 2, \ldots$. In other words, it means that

$$\mu\big(f^{-1}(a)\big) = 1 \quad \text{and} \quad \mu\big(f^{-1}(b)\big) = 0$$

for some $a \in \mathbb{R}$ and all other real numbers $b \ne a$, but this is equivalent to saying that $f^{-1}(a)$ is uncountable and $f^{-1}(b)$ is at most countable for all $b \ne a$. Now we have completely characterized every measurable function on X.

Finally, it is clear that $X \setminus f^{-1}(a)$ is a set of measure 0. Therefore, we have

$$\int_X f \, d\mu = \int_{f^{-1}(a)} f \, d\mu = a.$$

Hence, this completes the proof of the problem.

1.2   Problems related to the Lebesgue's MCT/DCT

> **Problem 1.7**
>
> *Rudin Chapter 1 Exercise 7.*

Proof. Let $E_k = \{x \in X \mid f_1(x) > k\} = f_1^{-1}((k, \infty])$ and $E = \{x \in X \mid f_1(x) = \infty\}$, where $k = 1, 2, \ldots$. It is clear that each $(k, \infty]$ is a Borel set in $[0, \infty]$, so each E_k is measurable by Theorem 1.12(b). Since

$$E = \bigcap_{k=1}^{\infty} E_k,$$

E is also measurable by Comment 1.6(c). If $\mu(E) > 0$, then we know from the definition that

$$\int_E f_1 \, d\mu = \infty. \tag{1.20}$$

By using the result (1.20), Proposition 1.24(b) and Theorem 1.33, we conclude that

$$\int_X |f_1| \, d\mu \geq \left| \int_X f_1 \, d\mu \right| \geq \left| \int_E f_1 \, d\mu \right| = \infty$$

which contradicts the hypothesis that $f_1 \in L^1(\mu)$. In other words, we must have $\mu(E) = 0$ and thus $f_1 \in L^1(\mu)$ on $X \setminus E$.

Here we may assume that the form of Theorem 1.34 (Lebesgue's Dominated Convergence Theorem) is also valid for measurable functions defined a.e. on X.[e] Now we see that the measurable function f_1 in the problem plays the role of g in the theorem and hence our desired result follows from Theorem 1.34 (Lebesgue's Dominated Convergence Theorem) immediately.

For a counterexample, we consider $X = \mathbb{R}, I_n = (n, \infty), \mu = m$ the Lebesgue measure (see [49, Definition 11.5, pp. 302, 303]) and define for each $n = 1, 2, \ldots$,

$$f_n(x) = \chi_{I_n}(x) = \begin{cases} 1, & \text{if } x \in I_n; \\ 0, & \text{if } x \notin I_n. \end{cases}$$

It is clear that $f_1 \geq f_2 \geq \cdots \geq 0$ on X and

$$\int_{\mathbb{R}} |f_n| \, dm = \int_{\mathbb{R}} f_n \, dm = \int_{I_n} dm = \infty \tag{1.21}$$

for every $n = 1, 2, \ldots$. In particular, the integrals (1.21) show that $f_1 \notin L^1(m)$. Furthermore, we have

$$f(x) = \lim_{n \to \infty} f_n(x) = \lim_{n \to \infty} \chi_{I_n}(x) = 0 \tag{1.22}$$

for every $x \in \mathbb{R}$. Thus we deduce from the expression (1.22) and Proposition 1.24(d) that

$$\int_{\mathbb{R}} f \, dm = 0. \tag{1.23}$$

Hence the inconsistence of the integrals (1.21) and (1.23) show that the condition "$f_1 \in L^1(\mu)$" cannot be omitted. This completes the proof of the problem. ∎

> **Problem 1.8**
>
> *Rudin Chapter 1 Exercise 8.*

---

[e]Actually, Rudin [51, p. 29] assumed this fact in the proof of Theorem 1.38 or the reader may refer to the comment between Theorem 11.32 and its proof in [49, p. 321].

Proof. If $x \in E$, then for all $k \in \mathbb{N}$, we have

$$\lim_{k \to \infty} f_{2k}(x) = \lim_{k \to \infty} \left(1 - \chi_E(x)\right) = 0.$$

Similarly, if $x \notin E$, then for all $k \in \mathbb{N}$, we have

$$\lim_{k \to \infty} f_{2k+1}(x) = \lim_{k \to \infty} \chi_E(x) = 0.$$

Thus we have

$$\liminf_{n \to \infty} f_n(x) = 0$$

for all $x \in X$. However, we have

$$\int_X f_n \, d\mu = \begin{cases} \displaystyle\int_X \chi_E \, d\mu, & \text{if } n \text{ is odd;} \\[2ex] \displaystyle\int_X (1 - \chi_E) \, d\mu, & \text{if } n \text{ is even.} \end{cases}$$

$$= \begin{cases} \mu(E), & \text{if } n \text{ is odd;} \\[2ex] \mu(X \setminus E), & \text{if } n \text{ is even.} \end{cases} \tag{1.24}$$

Therefore we obtain from the results (1.24) that

$$\liminf_{n \to \infty} \int_X f_n \, d\mu = \min(\mu(E), \mu(X \setminus E)) \neq 0$$

if we assume that $\mu(E) > 0$ and $\mu(X \setminus E) > 0$. Hence the example here shows that the inequality in Fatou's lemma can be strict, completing the proof of the problem.[f] ▧

Problem 1.9

Rudin Chapter 1 Exercise 9.

Proof. If $\mu(X) = 0$, then Proposition 1.24(e) implies that $c = \int_X f \, d\mu = 0$, a contradiction. Let $E = \{x \in X \mid f(x) = \infty\}$. We claim that $\mu(E) = 0$. Otherwise, Proposition 1.24(b) implies that

$$c = \int_X f \, d\mu \geq \int_E f \, d\mu = \infty$$

which is a contradiction. Therefore, in the following discussion, we may assume that $x \in X \setminus E$ so that $0 \leq f(x) < \infty$.

For each $n = 1, 2, \ldots$, we define[g] $f_n : X \setminus E \to [0, \infty)$ by

$$f_n(x) = n \log \left[1 + \left(\frac{f(x)}{n}\right)^\alpha\right].$$

Since f is measurable on $X \setminus E$, $g(x) = [1 + (\frac{x}{n})^\alpha]$ and $h(x) = n \log(1 + x)$ are continuous on $[0, \infty)$, Theorem 1.7(b) implies that

$$f_n(x) = n \log \left[1 + \left(\frac{f(x)}{n}\right)^\alpha\right] = h(g(f(x)))$$

---

[f] There is another example in [63, Problem 11.5, p. 340].
[g] $X \setminus E$ is itself a measure space by the remark following Proposition 1.24.

is also measurable on $X \setminus E$.

Now we are going to show that when $\alpha \geq 1$, there exists a function $g \in L^1(\mu)$ such that

$$|f_n(x)| \leq g(x)$$

holds for all $n = 1, 2, \ldots$ and all $x \in X \setminus E$. We first show the following lemma:

> **Lemma 1.2**
>
> For each $n \in \mathbb{N}$ and $\alpha \geq 1$, we have
>
> $$n \log \left[1 + \left(\frac{x}{n} \right)^\alpha \right] \leq \alpha x \tag{1.25}$$
>
> on $[0, \infty)$.

Proof of Lemma 1.2. For $x \in [0, \infty)$, we let

$$F(x) = \alpha x - n \log \left[1 + \left(\frac{x}{n} \right)^\alpha \right].$$

Then basic calculus gives

$$F'(x) = \alpha - \frac{\alpha \left(\dfrac{x}{n} \right)^{\alpha-1}}{1 + \left(\dfrac{x}{n} \right)^\alpha} = \alpha - \frac{\alpha n x^{\alpha-1}}{n^\alpha + x^\alpha} = \alpha \left(1 - \frac{n x^{\alpha-1}}{n^\alpha + x^\alpha} \right). \tag{1.26}$$

Since

$$\frac{n x^{\alpha-1}}{n^\alpha + x^\alpha} \leq \frac{x^\alpha}{n^\alpha + x^\alpha} < 1$$

for $n \leq x$ and since

$$\frac{n x^{\alpha-1}}{n^\alpha + x^\alpha} < \frac{n^\alpha}{n^\alpha + x^\alpha} < 1$$

for $n > x$ and $\alpha \geq 1$, we deduce from the derivative (1.26) that

$$F'(x) < 0$$

for all $x \in [0, \infty)$. By [49, Theorem 5.11, p. 108] and the continuity of F on $[0, \infty)$, we establish that $F(x)$ is decreasing on $[0, \infty)$ which implies the validity of the inequality (1.25). $\blacksquare$

Let's return to the proof of the problem. By Lemma 1.2, it is evident that for all $n \in \mathbb{N}$ and $\alpha \geq 1$, we have

$$f_n = n \log \left[1 + \left(\frac{f}{n} \right)^\alpha \right] \leq \alpha f \tag{1.27}$$

on $X \setminus E$. Furthermore, we apply Proposition 1.24(c) to get

$$\int_X |\alpha f| \, \mathrm{d}\mu = \alpha \int_X f \, \mathrm{d}\mu = \alpha c < \infty.$$

This means that $\alpha f \in L^1(\mu)$ and then $g = \alpha f$ is the desired function.

Next, there are two cases for consideration:

Case (i): $\alpha = 1$. In this case, we have[h]

$$\lim_{n\to\infty} n\log\left[1 + \left(\frac{f(x)}{n}\right)^{\alpha}\right] = \lim_{n\to\infty} \log\left(1 + \frac{f(x)}{n}\right)^{n} = \log e^{f(x)} = f(x)$$

on $X \setminus E$.

Case (ii): $\alpha > 1$. In this case, we apply L'Hospital's rule [49, Theorem 5.13, p. 109] to conclude that

$$\lim_{n\to\infty} n\log\left[1 + \left(\frac{f}{n}\right)^{\alpha}\right] = \lim_{y\to 0^+} \frac{\log(1 + f^{\alpha}y^{\alpha})}{y} = \lim_{y\to 0^+} \frac{\alpha f^{\alpha}y^{\alpha-1}}{1 + f^{\alpha}y^{\alpha}} = \frac{0}{1+0} = 0 \qquad (1.28)$$

on $X \setminus E$.

Thus it follows from Theorem 1.34 (Lebesgue's Dominated Convergence Theorem) that

$$\lim_{n\to\infty} \int_X n\log\left[1 + \left(\frac{f}{n}\right)^{\alpha}\right] d\mu = \begin{cases} \displaystyle\int_X f\,d\mu, & \text{if } \alpha = 1; \\[2ex] \displaystyle\int_X 0\,d\mu, & \text{if } \alpha > 1, \end{cases}$$

$$= \begin{cases} c, & \text{if } \alpha = 1; \\[2ex] 0, & \text{if } \alpha > 1. \end{cases}$$

It remains the case that $0 < \alpha < 1$. In this case, Theorem 1.28 (Fatou's Lemma) can be applied directly to get

$$\int_X \left\{ \liminf_{n\to\infty} \left\{ n\log\left[1 + \left(\frac{f}{n}\right)^{\alpha}\right] \right\} \right\} d\mu \le \liminf_{n\to\infty} \int_X n\log\left[1 + \left(\frac{f}{n}\right)^{\alpha}\right] d\mu. \qquad (1.29)$$

Since $0 < \alpha < 1$, we have $y^{\alpha-1} = \frac{1}{y^{1-\alpha}}$ so that the limit (1.28) becomes

$$\lim_{n\to\infty} n\log\left[1 + \left(\frac{f}{n}\right)^{\alpha}\right] = \lim_{y\to 0^+} \frac{\alpha f^{\alpha}}{y^{1-\alpha}(1 + f^{\alpha}y^{\alpha})} = \infty. \qquad (1.30)$$

Since $\lim_{n\to\infty} x_n = \infty$ if and only if $\limsup_{n\to\infty} x_n = \liminf_{n\to\infty} x_n = \infty$, the inequality (1.29) and the limit (1.30) combine to imply that

$$\liminf_{n\to\infty} \left\{ n\log\left[1 + \left(\frac{f}{n}\right)^{\alpha}\right] \right\} = \infty$$

and then it certainly gives

$$\lim_{n\to\infty} \int_X n\log\left[1 + \left(\frac{f}{n}\right)^{\alpha}\right] d\mu = \infty.$$

This completes the proof of the problem.   ◼

Problem 1.10

Rudin Chapter 1 Exercise 10.

---

[h]We use the definition $e^x = \lim_{n\to\infty}\left(1 + \frac{x}{n}\right)^n$ here.

Proof. Since $f_n \to f$ uniformly on X, there exists a positive integer N such that $n \geq N$ implies

$$|f_n(x) - f(x)| \leq 1$$

for all $x \in X$. By this, we have

$$|f_n(x)| \leq |f_n(x) - f(x)| + |f(x)| \leq |f(x)| + 1 \tag{1.31}$$

and

$$|f(x)| \leq |f(x) - f_N(x)| + |f_N(x)| \leq |f_N(x)| + 1 \tag{1.32}$$

for all $n \geq N$ and $x \in X$. Combining the inequalities (1.31) and (1.32), we see that

$$|f_n(x)| \leq |f_N(x)| + 2 \tag{1.33}$$

for all $n \geq N$ and $x \in X$.

We define $g : X \to [0, \infty)$ by

$$g(x) = \max\{|f_1(x)|, \ldots, |f_{N-1}(x)|, |f_N(x)| + 2\}. \tag{1.34}$$

Then it is easy to see from the inequality (1.33) and the definition (1.34) that

$$|f_n(x)| \leq g(x)$$

for $x \in X$ and $n = 1, 2, \ldots$.

Next we want to show that $g \in L^1(\mu)$. By Theorem 1.9(b), $|f_1(x)|, \ldots, |f_{N-1}(x)|, |f_N(x)|$ are measurable. Thus it follows from the corollaries following Theorem 1.14 that g is also measurable. Furthermore, we know from the hypothesis that $|f_n(x)| \leq M_n$ on X for some constants M_n, where $n = 1, 2, \ldots, N$. Therefore, this and the definition (1.34) certainly imply that

$$g(x) = |g(x)| \leq M$$

on X for some constant M. Since it is obvious that $g \geq 0$ on X, we get from Proposition 1.24(a) that

$$0 \leq \int_X |g| \, d\mu \leq \int_X M \, d\mu = M\mu(X) < \infty$$

so that $g \in L^1(\mu)$. In conclusion, our sequence of functions $\{f_n\}$ satisfies the hypotheses of Theorem 1.34 (Lebesgue's Dominated Convergence Theorem) and hence the desired result follows immediately from this.

For a counterexample, consider $X = \mathbb{R}$ and $\mu = m$ so that $m(\mathbb{R}) = \infty$. For each $n \in \mathbb{N}$, define $f_n : \mathbb{R} \to \mathbb{R}$ by

$$f_n(x) = \frac{1}{n}.$$

Then it is easy to prove that $f_n \to f \equiv 0$ uniformly on $\mathbb{R}$. Since

$$\int_{\mathbb{R}} f \, dm = 0 \quad \text{and} \quad \int_{\mathbb{R}} f_n \, dm = \frac{m(\mathbb{R})}{n} = \infty$$

for $n = 1, 2, \ldots$, we conclude that

$$\lim_{n \to \infty} \int_{\mathbb{R}} f_n \, dm \neq \int_{\mathbb{R}} f \, dm.$$

This finishes the proof of the problem.

> **Problem 1.11**
>
> *Rudin Chapter 1 Exercise 11.*

Proof. For each $n \in \mathbb{N}$, define

$$B_n = \bigcup_{k=n}^{\infty} E_k.$$

Recall that A is the set of all $x \in X$ which lie in infinitely many E_k. On the one hand, if $x \in A$, then $x \in B_n$ for all $n \in \mathbb{N}$ so that $x \in \bigcap_{n=1}^{\infty} B_n$. In other words, we have

$$A \subseteq \bigcap_{n=1}^{\infty} B_n. \tag{1.35}$$

On the other hand, if $x \in \bigcap_{n=1}^{\infty} B_n$, then $x \in B_n$ for each positive integer n and this is equivalent to the condition that x belongs to *infinitely many* E_k, i.e.,

$$\bigcap_{n=1}^{\infty} B_n \subseteq A. \tag{1.36}$$

Hence the set relations (1.35) and (1.36) imply the desired result that

$$A = \bigcap_{n=1}^{\infty} B_n = \bigcap_{n=1}^{\infty} \bigcup_{k=n}^{\infty} E_k.$$

We have to show that $\mu(A) = 0$. By the definition of B_n, we have

$$B_1 \supseteq B_2 \supseteq B_3 \supseteq \cdots .$$

Furthermore, our hypothesis and the *subadditive* property of a measure (see , for example, [54, Corollary 4.6, p. 26]) show that

$$\mu(B_1) = \mu\left(\bigcup_{k=1}^{\infty} E_k \right) \leq \sum_{k=1}^{\infty} \mu(E_k) < \infty.$$

Thus it follows from Theorem 1.19(e) and the subadditive property of μ again that

$$0 \leq \mu(A) = \mu\left(\bigcap_{n=1}^{\infty} B_n \right) = \lim_{n \to \infty} \mu(B_n) = \lim_{n \to \infty} \mu\left(\bigcup_{k=n}^{\infty} E_k \right) \leq \lim_{n \to \infty} \sum_{k=n}^{\infty} \mu(E_k) = 0.$$

Hence we must have $\mu(A) = 0$, completing the proof of the problem. $\blacksquare$

> **Problem 1.12**
>
> *Rudin Chapter 1 Exercise 12.*

Proof. Let n be a positive integer. Define $f_n : X \to [0, \infty]$ by

$$f_n(x) = \min\left(|f(x)|, n\right) \tag{1.37}$$

for every $x \in X$. This definition (1.37) clearly satisfies

$$0 \le f_n(x) \le n \tag{1.38}$$

for every $x \in X$. Since $f \in L^1(\mu)$, it is obviously measurable. By the corollaries following Theorem 1.14, each f_n is also measurable on X. Furthermore, if $x_0 \in X$ such that $|f(x_0)| \le n$, then the definition (1.37) implies that

$$f_n(x_0) = |f(x_0)| \quad \text{and} \quad f_{n+1}(x_0) = |f(x_0)|; \tag{1.39}$$

if $n < |f(x_0)| \le n+1$, then we know again from the definition (1.37) that

$$f_n(x_0) = n \quad \text{and} \quad f_{n+1}(x_0) = |f(x_0)|; \tag{1.40}$$

if $n + 1 < |f(x_0)|$, then we must have

$$f_n(x_0) = n \quad \text{and} \quad f_{n+1}(x_0) = n + 1. \tag{1.41}$$

Thus we can conclude from the computations (1.39), (1.40) and (1.41) that the inequality

$$0 \le f_n(x) \le f_{n+1}(x) \tag{1.42}$$

holds for all $n \in \mathbb{N}$ and $x \in X$. By Theorem 1.26 (Lebesgue's Monotone Convergence Theorem), we have

$$\lim_{n \to \infty} \int_X f_n \, d\mu = \int_X F \, d\mu,$$

where $F = \lim_{n \to \infty} f_n$. By the definition (1.37), we have $F = |f|$ and so

$$\lim_{n \to \infty} \int_X f_n \, d\mu = \int_X |f| \, d\mu$$

or equivalently,

$$\lim_{n \to \infty} \int_X |f_n - f| \, d\mu = 0.$$

Thus for every $\epsilon > 0$, there exists a positive integer N such that

$$\int_X |f_n - f| \, d\mu < \frac{\epsilon}{2} \tag{1.43}$$

for all $n \ge N$.

We *fix* this N. Since $0 \le |f| \le |f - f_N| + |f_N|$ by the triangle inequality[i], we apply Proposition 1.24(a) and then Theorem 1.27 and the property (1.38) to the inequality (1.43) to derive

$$\int_E |f| \, d\mu \le \int_E |f - f_N| \, d\mu + \int_E |f_N| \, d\mu < \frac{\epsilon}{2} + \int_E N \, d\mu = \frac{\epsilon}{2} + N\mu(E) \tag{1.44}$$

for every $E \in \mathfrak{M}$. Therefore, if we take $\delta = \frac{\epsilon}{2N}$ and $\mu(E) < \delta$, then our inequality (1.44) becomes

$$\int_E |f| \, d\mu < \epsilon$$

as desired. Hence we complete the proof of the problem.   ■

---

[i]See [49, Definition 2.15, p. 30].

> **Problem 1.13**
>
> *Rudin Chapter 1 Exercise 13.*

Proof. Let $f : X \to [0, \infty] \subset [-\infty, \infty]$ be a measurable function. For each $n = 1, 2, \ldots$, we define $f_n : X \to [0, \infty] \subset [-\infty, \infty]$ by

$$f_n(x) = nf(x).$$

It is clear that

$$f_n^{-1}\big((\alpha, \infty]\big) = \{x \in X \mid f_n(x) \in (\alpha, \infty]\} = \Big\{x \in X \,\Big|\, f(x) \in \Big(\frac{\alpha}{n}, \infty\Big]\Big\} = f^{-1}\Big(\Big(\frac{\alpha}{n}, \infty\Big]\Big)$$

for all real α. Since f is measurable, we have $f^{-1}\big((\frac{\alpha}{n}, \infty]\big) \in \mathfrak{M}$ for every real α, where $\mathfrak{M}$ is a σ-algebra in X. Thus we obtain from Theorem 1.12(c) that each f_n is measurable for $n = 1, 2, \ldots$. Furthermore, we have

$$0 \le f_1(x) \le f_2(x) \le \cdots \le \infty$$

on X and $f_n(x) \to \infty \cdot f(x)$ as $n \to \infty$. By Proposition 1.24(c), we have

$$\int_X f_n \, \mathrm{d}\mu = \int_X nf \, \mathrm{d}\mu = n \int_X f \, \mathrm{d}\mu \tag{1.45}$$

for $n = 1, 2, \ldots$. Hence we conclude from the equality (1.45) and Theorem 1.26 (Lebesgue's Monotone Convergence Theorem) that

$$\int_X \infty \cdot f \, \mathrm{d}\mu = \lim_{n \to \infty} \int_X f_n \, \mathrm{d}\mu = \lim_{n \to \infty} n \int_X f \, \mathrm{d}\mu = \infty \cdot \int_X f \, \mathrm{d}\mu.$$

Thus Proposition 1.24(c) also holds when $c = \infty$ and so we complete the proof of the problem. $\blacksquare$

CHAPTER 2

Positive Borel Measures

2.1 Properties of Semicontinuity

> **Problem 2.1**
>
> *Rudin Chapter 2 Exercise 1.*

Proof. We have $f_n : \mathbb{R} \to [0, \infty)$ for all $n \in \mathbb{N}$. In Proposition 2.5, we will prove a property which is equivalent to Definition 2.8 and our proof of **Statement (a)** below becomes simpler. However, we choose to apply Definition 2.8 to prove the statements here.

- **Statement (a):** For any real α, β_1 and β_2, let

$$E(\alpha) = \{x \in \mathbb{R} \mid f_1(x) + f_2(x) < \alpha\} \quad \text{and} \quad F_i(\beta_i) = \{x \in \mathbb{R} \mid f_i(x) < \beta_i\}, \tag{2.1}$$

where $i = 1, 2$. If $\alpha \le 0$, then we see that $E(\alpha) = \varnothing$ which is open in $\mathbb{R}$. Similarly, $F_i(\beta_i) = \varnothing$ if $\beta_i \le 0$. So in the following discussion, we assume that $\alpha > 0, \beta_i > 0$ and furthermore, $E(\alpha) \ne \varnothing$ and $F_i(\beta_i) \ne \varnothing$.

Now we claim that

$$E(\alpha) = \bigcup_{\substack{\beta_1 + \beta_2 \le \alpha \\ \beta_1, \beta_2 > 0}} [F_1(\beta_1) \cap F_2(\beta_2)]. \tag{2.2}$$

To prove the claim, on the one hand, if

$$x \in \bigcup_{\substack{\beta_1 + \beta_2 \le \alpha \\ \beta_1, \beta_2 > 0}} [F_1(\beta_1) \cap F_2(\beta_2)],$$

then there exist real β_1 and β_2 with $\beta_1 + \beta_2 \le \alpha$ and $\beta_1, \beta_2 > 0$ so that $x \in F_1(\beta_1) \cap F_2(\beta_2)$. By the definition (2.1), this x satisfies $f_1(x) < \beta_1$ *and* $f_2(x) < \beta_2$ and their sum implies

$$f_1(x) + f_2(x) < \beta_1 + \beta_2 \le \alpha.$$

Thus we have $x \in E(\alpha)$, i.e.,

$$\bigcup_{\substack{\beta_1 + \beta_2 \le \alpha \\ \beta_1, \beta_2 > 0}} [F_1(\beta_1) \cap F_2(\beta_2)] \subseteq E(\alpha).$$

On the other hand, if $x \in E(\alpha)$, then we let $\eta = f_1(x) + f_2(x)$. Define the two numbers β_1 and β_2 by

$$\beta_1 = \frac{\eta + \alpha}{2} - f_2(x) \quad \text{and} \quad \beta_2 = \frac{\eta + \alpha}{2} - f_1(x).$$

By direct computation, we know that

$$\beta_i > \frac{f_1(x) + f_2(x) + f_1(x) + f_2(x)}{2} - f_i(x) = f_1(x) + f_2(x) - f_i(x) \geq 0,$$

where $i = 1, 2$. Since $\eta = f_1(x) + f_2(x)$, it is easy to check that $f_1(x) + f_2(x) < \alpha$ if and only if $[f_1(x) + f_2(x)] + [f_1(x) + f_2(x)] < \alpha + \eta$ if and only if $f_1(x) + f_2(x) < \frac{\eta + \alpha}{2}$ if and only if

$$f_1(x) < \beta_1. \tag{2.3}$$

Similarly, the above argument can be used to show that $f_1(x) + f_2(x) < \alpha$ if and only if

$$f_2(x) < \beta_2. \tag{2.4}$$

Now we deduce from the inequalities (2.3) and (2.4) that $x \in F_1(\beta_1) \cap F_2(\beta_2)$. Furthermore, since $\beta_1 + \beta_2 = \eta + \alpha - f_1(x) - f_2(x) = \alpha$, we have

$$x \in \bigcup_{\substack{\beta_1 + \beta_2 \leq \alpha \\ \beta_1, \beta_2 > 0}} [F_1(\beta_1) \cap F_2(\beta_2)],$$

i.e.,

$$E(\alpha) \subseteq \bigcup_{\substack{\beta_1 + \beta_2 \leq \alpha \\ \beta_1, \beta_2 > 0}} [F_1(\beta_1) \cap F_2(\beta_2)].$$

Hence the claim (2.2) holds.

Since $F_1(\beta_1)$ and $F_2(\beta_2)$ are open in $\mathbb{R}$, $F_1(\beta_1) \cap F_2(\beta_2)$ is also open in $\mathbb{R}$. Since the union of any collection of open sets is open ([49, Theorem 2.24, p. 34]), we follow from this and the equality (2.2) that $E(\alpha)$ is open in $\mathbb{R}$. By Definition 2.8, $f_1 + f_2$ is upper semicontinuous.

- **Statement (b):** By a similar argument as in part (a), we can show that the sum of two lower semicontinuous functions is lower semicontinuous.

- **Statement (c):** This is not true in general. We use Proposition A.7 to give a counterexample. For each $n \in \mathbb{N}$, let $F_n = [\frac{1}{n+1}, \frac{1}{n}]$. Then each χ_{F_n} is upper semicontinuous because F_n is closed in $\mathbb{R}$. Consider the set

$$E = \left\{ x \in \mathbb{R} \;\middle|\; \sum_{n=1}^{\infty} \chi_{F_n}(x) < \frac{1}{2} \right\}.$$

If $x \in F_n$ for some $n \in \mathbb{N}$, then we have

$$1 \leq \sum_{n=1}^{\infty} \chi_{F_n}(x) \leq 2. \tag{2.5}$$

In other words, we have $F_n \not\subseteq E$ for $n = 1, 2, \ldots$. Since $\bigcup_{n=1}^{\infty} F_n = (0, 1]$,[a] we see that

$$(0, 1] \not\subseteq E. \tag{2.6}$$

---

[a] If $x \in (0, 1]$, then there exists a positive integer k such that $\frac{1}{k+1} < x$ which implies that $x \in F_1 \cup F_2 \cup \cdots \cup F_k$.

However, if $x \leq 0$ or $x > 1$, then $x \notin F_n$ for every $n \in \mathbb{N}$ which means that $\chi_{F_n}(x) = 0$, i.e.,

$$(-\infty, 0] \cup (1, \infty) \subseteq E. \tag{2.7}$$

Hence, by combining the set relations (2.6) and (2.7), we conclude that

$$E = (-\infty, 0] \cup (1, \infty)$$

which is *not* open in $\mathbb{R}$. By Definition 2.8, $\displaystyle\sum_{n=1}^{\infty} \chi_{F_n}$ is not upper semicontinuous.

- **Statement (d):** The set

$$F_n(\alpha) = \{x \in \mathbb{R} \mid f_n(x) > \alpha\}$$

is open for every real α. By applying part (b) repeatedly, we know that $\displaystyle\sum_{n=1}^{N} f_n$ is lower semicontinuous for every positive integer N. For every $x \in [0, \infty)$, let

$$f(x) = \lim_{N \to \infty} \sum_{n=1}^{N} f_n(x) = \sum_{n=1}^{\infty} f_n(x),$$

$$E(\alpha) = \{x \in \mathbb{R} \mid f(x) > \alpha\}, \tag{2.8}$$

$$F_N(\alpha) = \left\{x \in \mathbb{R} \;\middle|\; \sum_{n=1}^{N} f_n(x) > \alpha\right\}$$

for real α and $N \in \mathbb{N}$. We claim that

$$E(\alpha) = \bigcup_{N=1}^{\infty} F_N(\alpha) \tag{2.9}$$

for every real α. Similar to the proof of part (a), we suppose that $\alpha > 0$, $E(\alpha) \neq \mathbb{R}$ and $F_N(\alpha) \neq \mathbb{R}$.

Suppose that $x \in E(\alpha)$, i.e., $f(x) > \alpha$. Then there exists a $\epsilon > 0$ such that

$$f(x) > \alpha + \epsilon.$$

Since each f_n is nonnegative, $\left\{\displaystyle\sum_{n=1}^{N} f_n\right\}$ is an increasing sequence. By this and the definition of f in (2.8), there exists a positive integer N' such that $\displaystyle\sum_{n=1}^{N'} f_n(x) > \alpha + \epsilon$ which implies that $x \in F_{N'}(\alpha)$, i.e.,

$$E(\alpha) \subseteq \bigcup_{N=1}^{\infty} F_N(\alpha). \tag{2.10}$$

To prove the other side, if $x \in \displaystyle\bigcup_{N=1}^{\infty} F_N(\alpha)$, then $x \in F_{N'}(\alpha)$ for some positive integer N', i.e., $\displaystyle\sum_{n=1}^{N'} f_n(x) > \alpha$. Again, the fact that $\left\{\displaystyle\sum_{n=1}^{N} f_n\right\}$ is an increasing sequence implies that

$$f(x) \geq \sum_{n=1}^{N'} f_n(x) > \alpha,$$

i.e., $x \in E(\alpha)$ and then

$$\bigcup_{N=1}^{\infty} F_N(\alpha) \subseteq E(\alpha). \tag{2.11}$$

Hence the set relations (2.10) and (2.11) definitely imply the claim (2.9) is true and since each $F_N(\alpha)$ is open in $\mathbb{R}$ by Definition 2.8, $E(\alpha)$ is also open in $\mathbb{R}$. Since α is arbitrary, f is lower semicontinuous by Definition 2.8.

In the proof of **Statements (a)** and **(b)** above, we don't use the property that f_1 and f_2 are nonnegative. Therefore, they remain valid even if the word "nonnegative" is omitted. However, **Statement (c)** cannot hold anymore if the word "nonnegative" is omitted. In fact, we consider the sequence of real functions $\{f_n\}$ defined by

$$f_1 = \chi_{[-1,1]} \quad \text{and} \quad f_n = -\chi_{[\frac{1}{n},\frac{1}{n-1}]} \quad (n = 2, 3, \ldots).$$

Since $[-1,1$ and $[\frac{1}{n},\frac{1}{n-1}]$ are closed in $\mathbb{R}$, we know that $f_1, f_2, \ldots$ are upper semicontinuous. It is clear that $f_1(0) = -1$, so f_1 is *not* a nonnegative function on $\mathbb{R}$. By definition, we have

$$f = \sum_{n=1}^{\infty} f_n = \chi_{[-1,1]} - \sum_{n=2}^{\infty} \chi_{[\frac{1}{n},\frac{1}{n-1}]} = \chi_{[-1,0]} + \chi_{(0,1]} - \sum_{n=1}^{\infty} \chi_{[\frac{1}{n+1},\frac{1}{n}]}. \tag{2.12}$$

By the inequalities (2.5), we see that

$$-\sum_{n=1}^{\infty} \chi_{[\frac{1}{n+1},\frac{1}{n}]}(x) = \begin{cases} -2, & \text{if } x = \frac{1}{2}, \frac{1}{3}, \ldots; \\ -1, & \text{if } x \in (0, 1] \setminus \{\frac{1}{2}, \frac{1}{3}, \ldots\}; \\ 0, & \text{if } x \leq 0 \text{ or } x > 1. \end{cases} \tag{2.13}$$

Thus we follow from the expressions (2.12) and (2.13) that

$$f(x) = \begin{cases} -1, & \text{if } x = \frac{1}{2}, \frac{1}{3}, \ldots; \\ 0, & \text{if } x \in (0, 1] \setminus \{\frac{1}{2}, \frac{1}{3}, \ldots\} \text{ or } x < -1 \text{ or } x > 1; \\ 1, & \text{if } x \in [-1, 0]. \end{cases} \tag{2.14}$$

Therefore, the expression (2.14) of f gives

$$E = \left\{ x \in \mathbb{R} \,\Big|\, f(x) > \frac{1}{2} \right\} = [-1, 0]$$

which is *not* open in $\mathbb{R}$. By Definition 2.8, f is not upper semicontinuous. By Definition 2.8, a function f is lower semicontinuous if and only if $-f$ is upper semicontinuous. This observation indicates that we can deduce a counterexample to **Statement (d)** from the counterexample (2.12).

Finally, the truths of the **Statements (a)**, **(b)** and **(d)** depend *only* on the range of f and the fact that the union of any collection of open sets in a topological space X is open in X (see the set equalities (2.2) and (2.9)). This completes the proof of the problem. ∎

Problem 2.2

Rudin Chapter 2 Exercise 2.

Proof. We formulate and prove the general statement: Let X be a topological space, U an open set in X containing the point x and $f : X \to \mathbb{C}$. Define

$$\varphi(x,U) = \sup\{|f(s) - f(t)| \mid s, t \in U\} \quad \text{and} \quad \varphi(x) = \inf\{\varphi(x,U) \mid U \text{ is open, } x \in U\}. \quad (2.15)$$

We claim that φ is upper semicontinuous, f is continuous at $x \in X$ if and only if $\varphi(x) = 0$ and the set of points of continuity of an arbitrary complex function is a G_δ.

- **φ is upper semicontinuous.** Let $E = \{x \in X \mid \varphi(x) < \alpha\}$, where $\alpha \in \mathbb{R}$. We have to show that E is open in X. If $E = \varnothing$, then there is nothing to prove. Thus we suppose that $E \neq \varnothing$. In this case, we have $p \in E$ so that

$$\varphi(p) < \alpha.$$

This fact shows that *there exists* an open set U containing p and[b]

$$\varphi(p,U) < \alpha.$$

Pick $q \in U \setminus \{p\}$. Since U is open, *there exists* an open set V containing q such that $q \in V \subseteq U$. Since $p, q \in U$, we know from the definition (2.15) that

$$\varphi(p,U) = \varphi(q,U). \quad (2.16)$$

Furthermore, we observe from the definition (2.15) that

$$\varphi(x,U') \le \varphi(x,U) \text{ for every open sets } U, U' \text{ with } U' \subseteq U. \quad (2.17)$$

Now these facts (2.16) and (2.17) imply that $\varphi(q,V) \le \varphi(q,U) = \varphi(p,U) < \alpha$ and then

$$\varphi(q) \le \varphi(q,V) < \alpha.$$

In other words, $q \in E$. Since $q \in U \setminus \{p\}$ is arbitrary, we have shown that $p \in U \subseteq E$, i.e., E is open for every $\alpha \in \mathbb{R}$ and hence φ is upper semicontinuous by Definition 2.8.

- **f is continuous at x if and only if $\varphi(x) = 0$.** Suppose that f is continuous at x. Recall from the definition of continuity ([42, Theorem 18.1, p. 104] or [51, p. 9]) that for every $\epsilon > 0$, the neighborhood

$$B(f(x), \epsilon) = \left\{ z \in \mathbb{C} \,\middle|\, |f(x) - z| < \frac{\epsilon}{2} \right\} \quad (2.18)$$

has a neighborhood U_ϵ of x such that $f(U_\epsilon) \subseteq B(f(x), \epsilon)$, i.e., $f(y) \in B(f(x), \epsilon)$ for all $y \in U_\epsilon$. Now we take this U_ϵ in the definition (2.15):

$$\varphi(x, U_\epsilon) = \sup\{|f(s) - f(t)| \mid s, t \in U_\epsilon\}$$

and it follows from the definition (2.18) that[c]

$$\varphi(x, U_\epsilon) \le \epsilon. \quad (2.19)$$

By the definition (2.15), we have $\varphi(x) \le \varphi(x, U)$ for every open set U containing x. Therefore, we deduce from this and the inequality (2.19) that $\varphi(x) \le \epsilon$. Since ϵ is arbitrary, we have the desired result that $\varphi(x) = 0$.

---

[b]Otherwise, we have $\varphi(p,U) \ge \alpha$ for every open set U containing p and this implies that $\alpha \le \varphi(p)$, a contradiction.

[c]Note that $|f(s) - f(t)| \le |f(s) - f(x)| + |f(x) - f(t)| < \epsilon$ for every $s, t \in U_\epsilon$.

Conversely, suppose that $\varphi(x) = 0$. Thus for every $\epsilon > 0$, there exists a neighborhood U_ϵ of x such that $\varphi(x, U_\epsilon) < \epsilon$. By the definition (2.15), this implies that

$$|f(s) - f(t)| < \epsilon \tag{2.20}$$

for every $s, t \in U_\epsilon$. In particular, if we take $s = x$ to be *fixed* and $t = y$ vary, then the inequality (2.20) can be rewritten as $|f(x) - f(y)| < \epsilon$ for every $y \in U_\epsilon$. By the definition ([42, Theorem 18.1, p. 104] or [51, p. 9]), f is continuous at x.

- **The set of points of continuity of an arbitrary complex function is a G_δ.** By the previous analysis, we establish that

$$G = \{x \in X \mid f \text{ is continuous at } x\} = \{x \in X \mid \varphi(x) = 0\} = \bigcap_{n=1}^{\infty} \left\{ x \in X \,\middle|\, \varphi(x) < \frac{1}{n} \right\}.$$

Since φ is upper semicontinuous, each set $\{x \in X \mid \varphi(x) < \frac{1}{n}\}$ is open in X. By Definition 1.11, we see that G is actually a G_δ.

This completes the proof of the problem. ◼

> **Problem 2.3**
>
> *Rudin Chapter 2 Exercise 3.*

Proof. The first part is proven in [63, Problem 4.20, pp. 73, 74]. The function in the question can be used to prove Urysohn's Lemma for any metric space X directly.

> **Lemma 2.1 (Urysohn's Lemma)**
>
> Suppose that X is a metric space with metric ρ, A and B are disjoint nonempty closed subsets of X. Then there exists a continuous function $f : X \to [0, 1]$ such that $0 \le f(x) \le 1$ for all $x \in X$, $f(x) = 0$ precisely on A and $f(x) = 1$ precisely on B.

Proof of Lemma 2.1. This lemma is proven in [63, Problem 4.22, pp. 75, 76]. Readers are recommended to read [42, Theorem 32.2, p. 202; Theorem 33.1, pp. 207 – 210]. ◼

This ends the proof of the problem. ◼

> **Problem 2.4**
>
> *Rudin Chapter 2 Exercise 4.*

Proof.

(a) Recall from [51, Eqn. (2), p. 42] that

$$\mu(E) = \inf\{\mu(V) \mid E \subseteq V \text{ and } V \text{ is open}\} \tag{2.21}$$

which is *defined* for every subset E of X.[d] By the result [51, Eqn. (4), p. 42], we have

$$\mu(E_1 \cup E_2) \le \mu(E_1) + \mu(E_2). \tag{2.22}$$

---

[d]Here we don't require E to be a measurable set.

We need a lemma:

> **Lemma 2.2**
>
> For disjoint open sets V_1 and V_2, we have
>
> $$\mu(V_1 \cup V_2) = \mu(V_1) + \mu(V_2).$$

Proof of Lemma 2.2. The proof of the inequality $\mu(V_1 \cup V_2) \leq \mu(V_1) + \mu(V_2)$ can be found in [51, **STEP I**, p. 42], so we only prove the other side. Suppose that $g \in C_c(X)$ and $g \prec V_1$. Similarly, suppose that $h \in C_c(X)$ and $h \prec V_2$. By Definition 2.9, since $C_c(X)$ is a vector space, $f = g + h \in C_c(X)$. Furthermore, we know from Definition 2.9(a) that

$$\operatorname{supp}(f) = \operatorname{supp}(g + h) \subseteq \operatorname{supp}(g) + \operatorname{supp}(h) \subseteq V_1 \cup V_2. \tag{2.23}$$

By assumption, we have $V_1 \cap V_2 = \varnothing$ which means that $g(x) = 0$ on $X \setminus V_1$ and $h(x) = 0$ on $X \setminus V_2$. Thus we deduce from these facts that

$$f(x) = \begin{cases} g(x), & \text{if } x \in V_1; \\ h(x), & \text{if } x \in V_2; \\ 0, & \text{if } x \in X \setminus (V_1 \cup V_2). \end{cases}$$

In other words, we have

$$0 \leq f(x) \leq 1 \tag{2.24}$$

on X. Therefore, we can conclude from the set relation (2.23) and the inequalities (2.24) that

$$f \prec V_1 \cup V_2. \tag{2.25}$$

Now, by Definition 2.1, the relation (2.25) and then [51, Eqn. (1), p.41], we obtain that

$$\Lambda(g) + \Lambda(h) = \Lambda(g + h) = \Lambda(f) \leq \mu(V_1 \cup V_2). \tag{2.26}$$

We first *fix* the h in the inequality (2.26) and since it holds for *every* $g \prec V_1$, the definition of supremum gives

$$\mu(V_1) + \Lambda(h) \leq \mu(V_1 \cup V_2). \tag{2.27}$$

Next, the inequality (2.27) holds for *every* $h \prec V_2$, we establish from the definition of supremum that

$$\mu(V_1) + \mu(V_2) \leq \mu(V_1 \cup V_2).$$

Hence we have $\mu(V_1) + \mu(V_2) = \mu(V_1 \cup V_2)$. ∎

Let's go back to the original proof. We want to compute $\mu(E_1 \cup E_2)$. By the definition (2.21), it needs to consider all open sets containing $E_1 \cup E_2$. Since $V_1 \subseteq V_2$ certainly implies $\mu(V_1) \leq \mu(V_2)$, we can restrict our attention to any open set W such that

$$E_1 \cup E_2 \subseteq W \subseteq V_1 \cup V_2.$$

Since $V_1 \cap V_2 = \varnothing$, we have $W = (W \cap V_1) \cup (W \cap V_2)$, where $(W \cap V_1) \cap (W \cap V_2) = \varnothing$.

Furthermore, since $W \cap V_1$ and $W \cap V_2$ are open sets in X, it yields from Lemma 2.2 that

$$\mu(W) = \mu(W \cap V_1) + \mu(W \cap V_2) \geq \mu(E_1) + \mu(E_2). \tag{2.28}$$

Since W is arbitrary, $\mu(E_1) + \mu(E_2)$ is a lower bound of the set in the definition (2.21). Hence we follow from this fact and the inequality (2.28) that

$$\mu(E_1 \cup E_2) \geq \mu(E_1) + \mu(E_2). \tag{2.29}$$

By the inequalities (2.22) and (2.29), we have the desired result that

$$\mu(E_1 \cup E_2) = \mu(E_1) + \mu(E_2).$$

(b) Define $E_1 = E$. By [51, **STEP V**, p. 44], there exists a compact set K_1 and an open set V_1 such that

$$K_1 \subseteq E_1 \subseteq V_1 \quad \text{and} \quad \mu(V_1 \setminus K_1) < \frac{1}{2^2}.$$

By [51, **STEP II**, p. 43], we have $K_1 \in \mathfrak{M}_F$. Then we follow from [51, **STEP VI**, p. 44] that $E_1 \setminus K_1 \in \mathfrak{M}_F$. Define $E_2 = E_1 \setminus K_1$ so that

$$E = E_1 = E_2 \cup K_1,$$

where $\mu(E_2) = \mu(E_1 \setminus K_1) \leq \mu(V_1 \setminus K_1) < \frac{1}{2^2}$. By induction, we can show that there exists a sequence of compact sets $\{K_n\}$ and a sequence of open sets $\{V_n\}$ such that

$$K_n \subseteq E_n \subseteq V_n, \tag{2.30}$$

where $E_{n+1} = E_n \setminus K_n$, $E_{n+1} \in \mathfrak{M}_F$ and $\mu(E_{n+1}) < \frac{1}{2^{n+1}}$. Therefore, the set E satisfies the following relations:

$$E = E_{n+1} \cup K_1 \cup K_2 \cup \cdots \cup K_n \quad \text{and} \quad \mu(E_{n+1}) < \frac{1}{2^{n+1}} \tag{2.31}$$

for every positive integer n. By the set relations (2.30), we deduce that

$$K_{n+1} \subseteq E_{n+1} = E_n \setminus K_n = E_{n-1} \setminus (K_{n-1} \cup K_n) = E \setminus (K_1 \cup K_2 \cup \cdots \cup K_n) \tag{2.32}$$

for all $n \in \mathbb{N}$. In other words, K_{n+1} is disjoint with all $K_1, K_2, \ldots, K_n$ and thus $\{K_1, \ldots, K_n\}$ is a disjoint collection of compact sets in X. Now, by letting $n \to \infty$ in the relations (2.31), we achieve that E is given by

$$E = N \cup K_1 \cup K_2 \cup \cdots,$$

where $\{K_n\}$ is a *disjoint* countable collection of compact sets in X and

$$N = E \setminus \bigcup_{n=1}^{\infty} K_n.$$

By the equalities in (2.32), $N \subseteq E_{n+1}$ holds for all $n \in \mathbb{N}$ and the inequality in (2.31) shows that

$$\mu(N) = \lim_{n \to \infty} \mu(E_{n+1}) = \lim_{n \to \infty} \frac{1}{2^{n+1}} = 0.$$

This completes the proof of the problem.

2.2 Problems on the Lebesgue Measure on $\mathbb{R}$

In Problems 2.5 to 2.8, m stands for the Lebesgue measure on $\mathbb{R}$.

> **Problem 2.5**
>
> *Rudin Chapter 2 Exercise 5.*

Proof. This problem is proven in [49, Remarks 11.11(f), p. 309].   ■

> **Problem 2.6**
>
> *Rudin Chapter 2 Exercise 6.*

Proof. Recall from [49, Theorem 2.47, p. 42] that a subset F of $\mathbb{R}$ is connected if and only if $(x, y) \subseteq F$ for every $x, y \in F$ with $x < y$. Therefore, if F is a connected subset of a totally disconnected set $K \subset \mathbb{R}$ and F consists of more than one point, then K must contain a segment, a contradiction. By this observation, we know that a totally disconnected set $K \subset \mathbb{R}$ contains no segment and this gives a hint to our construction: It is well-known that the Cantor set E is compact and contains no segment. However, Problem 2.5 shows that $m(E) = 0$, so we have to "modify" the construction of E so as to fit our problem.

Let $K_0 = [0, 1]$. The idea of the construction is to remove the "middle $\frac{1}{2^{2n}}$th" segments. In fact, we first remove the segment $\left(\frac{3}{8}, \frac{5}{8}\right)$ of length $\frac{1}{2^2}$ from K_0 and let

$$K_1 = \left[0, \frac{3}{8}\right] \cup \left[\frac{5}{8}, \frac{8}{8}\right].$$

Next, we remove the middle segments $\left(\frac{5}{32}, \frac{7}{32}\right)$ and $\left(\frac{25}{32}, \frac{27}{32}\right)$ each of length $\frac{1}{2^4}$ from each connected component of K_1 to get

$$K_2 = \left[0, \frac{5}{32}\right] \cup \left[\frac{7}{32}, \frac{12}{32}\right] \cup \left[\frac{20}{32}, \frac{25}{32}\right] \cup \left[\frac{27}{32}, \frac{32}{32}\right].$$

Continuing in this way, we obtain a sequence of compact sets K_n such that

$$K_1 \supset K_2 \supset K_3 \supset \cdots .$$

Therefore, we have

$$m(K_n) = 1 - \sum_{k=1}^{n} \frac{1}{2^{2k}} \cdot 2^{k-1} = 1 - \sum_{k=1}^{n} \frac{1}{2^{k+1}}. \tag{2.33}$$

Define the set

$$K = \bigcap_{n=1}^{\infty} K_n.$$

We know that $K \neq \varnothing$ because of the corollary [49, Theorem 2.36, p. 38]. Applying an argument similar to that used in [49, §2.44, pp. 41, 42], we can show that K is a totally disconnected compact set in $\mathbb{R}$. Furthermore, since $m(K_1)$ is finite, Theorem 1.19(e) and the Lebesgue measure (2.33) ensure that

$$m(K) = \lim_{n \to \infty} m(K_n) = \lim_{n \to \infty} \left(1 - \sum_{k=1}^{n} \frac{1}{2^{k+1}}\right) = 1 - \sum_{k=1}^{\infty} \frac{1}{2^{k+1}} = \frac{1}{2} > 0.$$

For the second assertion, let v be lower semicontinuous and $v \le \chi_K$. Assume that there was a $p \in \mathbb{R}$ such that $v(p) > 0$. Since v is lower semicontinuous, Definition 2.8 implies that the set $A = \{x \in \mathbb{R} \mid v(x) > 0\}$ is open in $\mathbb{R}$. By the assumption, we have $p \in A$, so there exists a $\delta > 0$ such that

$$(p - \delta, p + \delta) \subseteq A. \tag{2.34}$$

Now we follow from the hypothesis $v \le \chi_K$ and the relation (2.34) that

$$\chi_K(x) > 0$$

on $(p - \delta, p + \delta)$, therefore we have $(p - \delta, p + \delta) \in K$ which contradicts to the fact that K contains no segment. Hence we have $v \le 0$ and this completes the proof of the problem. ◼

Problem 2.7

Rudin Chapter 2 Exercise 7.

Proof. By the idea used in Problem 2.6, it is not hard to see that the construction of the compact sets K_n and K still work for removing the "middle $\frac{\epsilon}{2^{2n-1}}$th" segments. In this case, instead of the Lebesgue measure (2.33), we have

$$m(K_n) = 1 - \sum_{k=1}^{n} \frac{\epsilon}{2^{2k-1}} \cdot 2^{k-1} = 1 - \epsilon \sum_{k=1}^{n} \frac{1}{2^k} \tag{2.35}$$

and so

$$m(K) = \lim_{n \to \infty} m(K_n) = 1 - \epsilon.$$

Since K is closed in $[0, 1]$, the complement

$$E = [0, 1] \setminus K \tag{2.36}$$

is definitely open in $[0, 1]$.

Now it remains to show that E is dense in $[0, 1]$. We prove an equivalent definition of a dense set: Let $A \subseteq B \subseteq \mathbb{R}$. We say A is dense in B means that for every $x, y \in B$ with $x < y$, we can find $z \in A$ such that $x < z < y$. In particular, suppose that $x, y \in [0, 1]$ with $x < y$. If $(x, y) \cap E = \varnothing$, then the definition (2.36) says that $(x, y) \subseteq K$ which is impossible. Therefore, we have

$$(x, y) \cap E \ne \varnothing$$

and this means that there exists $z \in E$ such that $x < z < y$. Hence E is dense in $[0, 1]$, completing the proof of the problem. ◼

Problem 2.8

Rudin Chapter 2 Exercise 8.

Proof. Let $\{r_n\}$ be the enumeration of all rationals in $\mathbb{R}$. Let F_n be the segment given by

$$F_n = \left(r_n - \frac{1}{2^{2n}}, r_n + \frac{1}{2^{2n}}\right), \tag{2.37}$$

where $n = 1, 2, \ldots$. Define the sets

$$G_n = F_n \setminus (F_{n+1} \cup F_{n+2} \cup \cdots) = F_n \setminus \bigcup_{k=1}^{\infty} F_{n+k}. \tag{2.38}$$

We divide the proof into several steps.

- **Step 1:** $m(G_n) > 0$. By the definition (2.37), we know that $m(F_{n+1}) = \frac{m(F_n)}{2^2}$ for every $n = 1, 2, \ldots$. By the definition (2.38), for every $n \in \mathbb{N}$, we have

$$F_n = G_n \cup \bigcup_{k=1}^{\infty} F_{n+k}$$

so that the subadditive property of a measure (see the proof of Problem 1.11) implies that

$$m(F_n) \le m(G_n) + \sum_{k=1}^{\infty} m(F_{n+k}) = m(G_n) + m(F_n) \sum_{k=1}^{\infty} \frac{1}{2^{2k}} = m(G_n) + \frac{m(F_n)}{3}.$$

Therefore, we have

$$m(G_n) \ge \frac{2m(F_n)}{3} > 0 \tag{2.39}$$

for every positive integer n.

- **Step 2: Existence of a Borel subset $A_n \subset F_n$ with $m(A_n) = \frac{m(F_n)}{2}$.** It is easy to check from the definition (2.37) that $m(F_n) = \frac{1}{2^{2n-1}}$. By Definition 2.19, we know that[e]

$$F_n' = \frac{1}{m(F_n)}\left[F_n - \left(r_n - \frac{1}{2^{2n}}\right)\right] = (0, 1).$$

Thus, by the proof of Problem 2.7[f], there exists a (totally disconnected) compact set $A_n' \subset F_n'$ such that $m(A_n') = \frac{1}{2}$. Then Theorem 2.20(c) implies that the set A_n given by

$$A_n = m(F_n)A_n' + \left(r_n - \frac{1}{2^{2n}}\right)$$

is a (totally disconnected) compact subset of F_n with measure

$$m(A_n) = m\left(m(F_n)A_n' + \left(r_n - \frac{1}{2^{2n}}\right)\right) = m\big(m(F_n)A_n'\big) = m(F_n)m(A_n') = \frac{m(F_n)}{2}. \tag{2.40}$$

Since A_n is compact, it is closed by the Heine–Borel theorem [49, Theorem 2.41, p. 40]. By Definition 1.11, it is also a Borel set, completing the proof of **Step 2**.

Now we can construct a Borel set with the desired properties. The construction is as follows: For every positive integer n, we define

$$E_n = G_n \cap A_n \quad \text{and} \quad E = \bigcup_{n=1}^{\infty} E_n. \tag{2.41}$$

By the definition (2.38), since each F_n is Borel, each G_n is also Borel. Thus each E_n and their countable union E are also Borel sets. It remains to show that the E satisfies the requirements of the problem.

- **Step 3:** $0 < m(E \cap F_n)$. We first show that $m(E_n \cap F_n) > 0$ for every positive integer n. To prove this claim, we note from the definitions (2.38) and (2.41) that $E_n \subseteq G_n \subseteq F_n$, so we have

$$m(E_n \cap F_n) = m(E_n) = m(G_n \cap A_n).$$

Furthermore, since $G_n \subseteq F_n$ and $A_n \subset F_n$ (by **Step 2**), we have

$$m(G_n \cup A_n) \le m(F_n). \tag{2.42}$$

---

[e] It is just a translation and an enlargement.
[f] Take $\epsilon = \frac{1}{2}$ in the measure (2.35).

Combining the inequality (2.39) and the expression (2.40), we can reduce the inequality (2.42) to

$$m(G_n \cup A_n) < m(G_n) + m(A_n). \tag{2.43}$$

Now we apply a property of additive function [49, Eqn. (7), p. 302] to the inequality (2.43) to obtain

$$m(G_n \cup A_n) < m(G_n \cup A_n) + m(G_n \cap A_n)$$

which means that $m(G_n \cap A_n) > 0$ for every positive integer n. Therefore, we follow from this and the fact that $E_n \cap F_n \subseteq E \cap F_n$

$$0 < m(G_n \cap A_n) = m(E_n \cap F_n) \leq m(E \cap F_n)$$

for every positive integer n.

- **Step 4:** $m(E \cap F_n) < m(F_n)$. Fix an integer n. For $m < n$, we apply the identity $A \setminus B = B^c \cap A$ to the definition (2.38) to get

$$G_m^c = [F_m \setminus (F_{m+1} \cup F_{m+2} \cup \cdots)]^c = (F_{m+1} \cup F_{m+2} \cup \cdots) \cup F_m^c$$

which implies that $F_n \subseteq G_m^c$ for $m = 1, 2, \ldots, n-1$. Recall from the definition (2.41) that $E_n^c = G_n^c \cup A_n^c$, so we must have

$$F_n \subseteq E_m^c$$

for $m = 1, 2, \ldots, n-1$. In other words, it says that $F_n \cap E_m = \varnothing$ for $m = 1, 2, \ldots, n-1$. Therefore, this fact implies

$$m(E \cap F_n) = m\Big(\bigcup_{m=1}^{\infty} E_m \cap F_n \Big) = m\Big(\bigcup_{m=n}^{\infty} (E_m \cap F_n) \Big) \leq m\Big(\bigcup_{m=n}^{\infty} (A_m \cap F_n) \Big). \tag{2.44}$$

By the subadditive property of a measure again, we further deduce the inequality (2.44) to

$$m(E \cap F_n) \leq \sum_{m=n}^{\infty} m(A_m \cap F_n) \leq \sum_{m=n}^{\infty} m(A_m). \tag{2.45}$$

Recall the result (2.40) and the fact $m(F_{n+1}) = \frac{m(F_n)}{2^2}$, so we derive from the inequality (2.45) that

$$m(E \cap F_n) \leq \sum_{m=0}^{\infty} \frac{m(F_{n+m})}{2} = \frac{m(F_n)}{2} \sum_{m=0}^{\infty} \frac{1}{2^{2m}} = \frac{2m(F_n)}{3} < m(F_n)$$

for every positive integer n.

- **Step 5:** $0 < m(E \cap I) < m(I)$ **for every nonempty segment** I. By **Step 3** and **Step 4**, the inequalities

$$0 < m(E \cap F_n) < m(F_n) \tag{2.46}$$

hold for every positive integer n, where F_n and E are given by (2.37) and (2.41) respectively.

Lemma 2.3

Let $I = (\alpha, \beta)$ with $\alpha < \beta$. Then there is a positive integer n_0 such that $F_{n_0} \subseteq I$.

Proof of Lemma 2.3. Let $\gamma = \frac{\alpha+\beta}{2}$ and $\delta = \frac{\gamma+\beta}{2}$. By the density of $\mathbb{Q}$ in $\mathbb{R}$, there exists a rational in (γ, δ). Let it be r_n for some positive integer n. If $F_n \subseteq I$, then we are done. Otherwise, we may find another rational in (r_n, δ), namely r_{n+1}. In fact, this process can be repeated m times, i.e., there is a rational r_{n+m} in (r_{n+m-1}, δ). It is clear that

$$r_{n+m} + \frac{1}{2^{2(n+m)}} \le \delta + \frac{1}{2^{2(n+m)}} < \delta + \frac{1}{2^{2m}} \tag{2.47}$$

and

$$r_{n+m} - \frac{1}{2^{2(n+m)}} > \gamma - \frac{1}{2^{2(n+m)}} > \gamma - \frac{1}{2^{2m}}. \tag{2.48}$$

Now we may pick the m large enough such that

$$\frac{1}{2^{2m}} < \beta - \delta \quad \text{and} \quad \frac{1}{2^{2m}} < \frac{\beta - \alpha}{2}$$

simultaneously. Then it follows from the inequalities (2.47) and (2.48) that

$$r_{n+m} + \frac{1}{2^{2(n+m)}} < \beta \tag{2.49}$$

and

$$r_{n+m} - \frac{1}{2^{2(n+m)}} > \gamma - \frac{1}{2^{2m}} > \frac{\alpha + \beta}{2} - \frac{\beta - \alpha}{2} = \alpha. \tag{2.50}$$

Hence the inequalities (2.49) and (2.50) together imply that $F_{n_0} \subseteq I$, where $n_0 = n+m$, completing the proof of the lemma. $\blacksquare$

We return to the proof of the problem. By Lemma 2.3, we have $E \cap F_{n_0} \subseteq E \cap I$ and the left-hand side of the inequalities (2.46) shows that

$$0 < m(E \cap F_{n_0}) \le m(E \cap I). \tag{2.51}$$

Since $I = F_{n_0} \cup (I \setminus F_{n_0})$ and $F_{n_0} \cap (I \setminus F_{n_0}) = \varnothing$, we have

$$E \cap I = (E \cap F_{n_0}) \cup [E \cap (I \setminus F_{n_0})].$$

Hence, by applying the right-hand side of the inequalities (2.46) and Theorem 1.19(b) twice, we obtain

$$m(E \cap I) = m(E \cap F_{n_0}) + m\big(E \cap (I \setminus F_{n_0})\big) < m(F_{n_0}) + m(I \setminus F_{n_0}) = m(I). \tag{2.52}$$

Now our desired result follows immediately if we combine the inequalities (2.51) and (2.52).

- **Step 6:** $m(E) < \infty$. By the subadditive property of the measure m, the expression (2.40) and the definition (2.41), we get

$$m(E) \le \sum_{n=1}^{\infty} m(E_n) \le \sum_{n=1}^{\infty} m(A_n) = \frac{1}{2} \sum_{n=1}^{\infty} m(F_n) = \frac{m(F_1)}{2} \sum_{n=0}^{\infty} \frac{1}{2^{2n}} = \frac{2m(F_1)}{3} < \infty.$$

This completes the proof of the problem.[g] $\blacksquare$

---

[g]Instead of Borel sets of the real number line $\mathbb{R}$, Rudin proved a similar result for measurable set A in $[0, 1]$, see [50]. Furthermore, there are two interesting results ([14] and [35]) related to this problem and some of its applications can be found in [13], [18], [25], [39], [46] and [60].

2.3   Integration of Sequences of Continuous Functions

> **Problem 2.9**
>
> *Rudin Chapter 2 Exercise 9.*

Proof. For every $n \in \mathbb{N}$, we consider the function $g_n : [-1,1] \to [0,1]$ defined by

$$g_n(x) = \begin{cases} 0, & \text{if } x \notin [-\frac{1}{n}, \frac{1}{n}]; \\[2mm] 1 - n|x|, & \text{if } x \in [-\frac{1}{n}, \frac{1}{n}]. \end{cases} \tag{2.53}$$

It is clear that each g_n is continuous on $[-1,1]$ and $0 \le g_n \le 1$ on $[-1,1]$. The graph of g_n is shown in Figure 2.1 below.

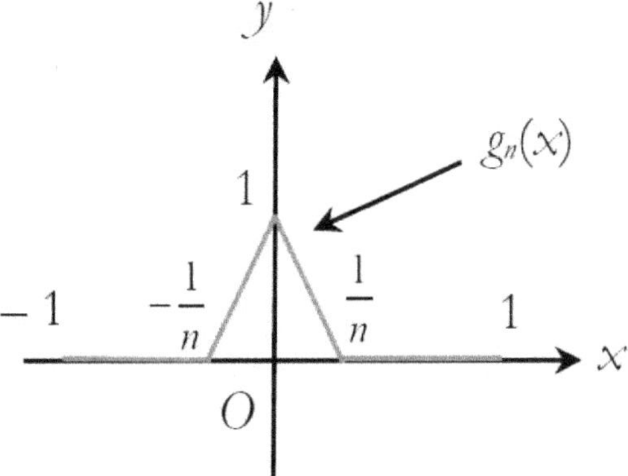

Figure 2.1: The graph of g_n on $[-1,1]$.

Next, we define $g_{n,k} : [0,1] \to [0,1]$ by

$$g_{n,k}(x) = g_n\left(x - \frac{k}{2n}\right), \tag{2.54}$$

where $x \in [0,1]$ and $k = 0, 1, \ldots, 2n$. It is easy to check that $x - \frac{k}{2n} \in [-1,1]$ so that each $g_{n,k}$ is well-defined by (2.54). We claim that if we define the sequence $\{f_1, f_2, \ldots\}$ to be

$$\{g_{1,0}, g_{1,1}, g_{1,2}, g_{2,0}, g_{2,1}, g_{2,2}, g_{2,3}, g_{2,4}, \ldots\},$$

then $\{f_n\}$ satisfies the hypotheses of the problem.

To this end, we first note that since each $g_{n,k}$ is continuous on $[0,1]$ and $0 \le g_{n,k} \le 1$ on $[0,1]$, each f_n is continuous on $[0,1]$ and $0 \le f_n \le 1$. To see why

$$\lim_{n \to \infty} \int_0^1 f_n(x)\,\mathrm{d}x = 0, \tag{2.55}$$

we have to check the behaviour of

$$\int_0^1 g_{n,k}(x)\,\mathrm{d}x$$

for every positive integer n and $k = 0, 1, \ldots, 2n$. In fact, we know from the definitions (2.53) and (2.54) that

$$g_{n,0}(x) = \begin{cases} 1 - n|x|, & \text{if } x \in [0, \tfrac{1}{n}); \\ 0, & \text{if } x \in [\tfrac{1}{n}, 1] \end{cases} \quad \text{and} \quad g_{n,1}(x) = \begin{cases} \tfrac{1}{2} + nx, & \text{if } x \in [0, \tfrac{1}{2n}); \\ \tfrac{3}{2} - nx, & \text{if } x \in [\tfrac{1}{2n}, \tfrac{3}{2n}]; \\ 0, & \text{if } x \in (\tfrac{3}{2n}, 1]. \end{cases} \quad (2.56)$$

Similarly, we have

$$g_{n,2n-1}(x) = \begin{cases} 0, & \text{if } x \in [0, \tfrac{2n-3}{2n}); \\ 1 - n\left|x - \tfrac{2n-1}{2n}\right|, & \text{if } x \in [\tfrac{2n-3}{2n}, 1] \end{cases} \quad (2.57)$$

and

$$g_{n,2n}(x) = \begin{cases} 0, & \text{if } x \in [0, \tfrac{n-1}{n}); \\ 1 + n(x - 1), & \text{if } x \in [\tfrac{n-1}{n}, 1]. \end{cases} \quad (2.58)$$

Finally, for $k = 2, 3, \ldots, 2n - 2$, we have

$$g_{n,k}(x) = \begin{cases} 0, & \text{if } x \in [0, \tfrac{k-2}{2n}); \\ 1 - n\left|x - \tfrac{k}{2n}\right|, & \text{if } x \in [\tfrac{k-2}{2n}, \tfrac{k+2}{2n}]; \\ 0, & \text{if } x \in (\tfrac{k+2}{2n}, 1]. \end{cases} \quad (2.59)$$

The graphs of these $g_{n,k}$ are shown as follows:

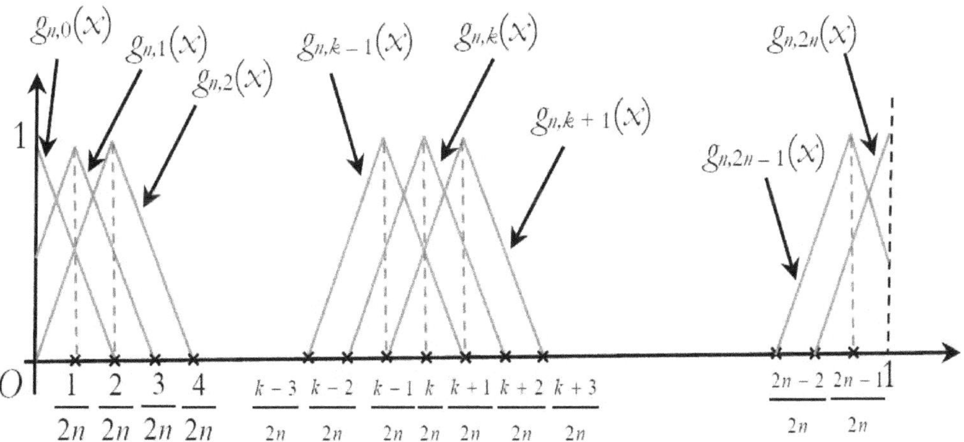

Figure 2.2: The graphs of $g_{n,k}$ on $[0, 1]$.

In other words, it is clear from Figure 2.2 that $\{g_{n,0}, g_{n,1} \ldots, g_{n,2n}\}$ is a family of tent functions centered at $\{\tfrac{0}{2n}, \tfrac{1}{2n}, \ldots, \tfrac{2n}{2n}\}$. Now we have the following cases.[h]

---

[h]We use the fact that each $g_{n,k}$ is a part or the whole of an isosceles triangle with height 1 and base less than or equal to $\tfrac{2}{n}$.

- When $k = 0$, we have

$$\int_0^1 g_{n,0}(x)\,\mathrm{d}x = \int_0^1 g_n(x)\,\mathrm{d}x = \int_0^{\frac{1}{n}} (1 - nx)\,\mathrm{d}x = \frac{1}{2n}. \tag{2.60}$$

- When $k = 1$, we have

$$\int_0^1 g_{n,1}(x)\,\mathrm{d}x = \int_0^{\frac{1}{2n}} \left(\frac{1}{2} + nx\right)\mathrm{d}x + \int_{\frac{1}{2n}}^{\frac{3}{2n}} \left(\frac{3}{2} - nx\right)\mathrm{d}x = \frac{7}{8n}. \tag{2.61}$$

- When $2 \le k \le 2n - 2$, we have

$$\int_0^1 g_{n,k}(x)\,\mathrm{d}x = \frac{1}{2} \times \frac{2}{n} \times 1 = \frac{1}{n}. \tag{2.62}$$

- When $k = 2n - 1$, we have

$$\int_0^1 g_{n,2n-1}(x)\,\mathrm{d}x = \int_{\frac{2n-3}{2n}}^1 \left(1 - n\left|x - \frac{2n-1}{2n}\right|\right)\mathrm{d}x = \frac{7}{8n}. \tag{2.63}$$

- When $k = 2n$, we have

$$\int_0^1 g_{n,2n}(x)\,\mathrm{d}x = \int_{1-\frac{1}{n}}^1 g_n(x-1)\,\mathrm{d}x = \int_{1-\frac{1}{n}}^1 [1 + n(x-1)]\,\mathrm{d}x = \frac{1}{2n}. \tag{2.64}$$

Combining the expressions (2.60) to (2.64), we may conclude that the limit (2.55) holds.

Now it remains to show that $\{f_n(x)\}$ converges for *no* $x \in [0,1]$. If $x = 0$, then we obtain from the definitions (2.56) and (2.57) that

$$g_{n,0}(0) = 1 \quad \text{and} \quad g_{n,2n-1}(0) = 0.$$

In other words, we can find subsequences $\{f_{n_k}(0)\}$ and $\{f_{n_l}(0)\}$ such that $f_{n_k}(0) \to 1$ and $f_{n_l}(0) \to 0$ as $k \to \infty$ and $l \to \infty$ respectively. By Definition 1.13, they imply that

$$\limsup_{n\to\infty} f_n(0) = 1 \quad \text{and} \quad \liminf_{n\to\infty} f_n(0) = 0. \tag{2.65}$$

Similarly, if $x = 1$, then the definitions (2.56) and (2.58) show that

$$g_{n,0}(1) = 0 \quad \text{and} \quad g_{n,2n}(1) = 1$$

which then imply the limits (2.65). If $x \in (0,1) \cap \mathbb{Q}$, then we have $x = \frac{p}{q} = \frac{2p}{2q}$, where $0 < p < q$. Thus it follows from the definitions (2.56) and (2.59) that

$$g_{n,0}\left(\frac{p}{q}\right) = 0 \quad \text{and} \quad g_{nq,2np}\left(\frac{p}{q}\right) = 1 \tag{2.66}$$

for $n \ge q$. Therefore, the limits (2.65) also hold for $x \in (0,1) \cap \mathbb{Q}$. For irrational x in $(0,1)$, the Archimedean Property shows that there exists $N \in \mathbb{N}$ such that $x > \frac{1}{N}$. Thus it implies that

$$g_{n,0}(x) = 0 \tag{2.67}$$

for all $n \ge N$. By the density of rationals, there exists a sequence $\{\frac{p_n}{q_n}\}$ of rationals in $(0,1)$ such that

$$\frac{p_n}{q_n} \to x$$

as $n \to \infty$, where $0 < p_n < q_n$ for all $n \in \mathbb{N}$. Since we always have

$$\frac{p_n}{q_n} \in \left[\frac{np_n - 1}{nq_n}, \frac{np_n + 1}{nq_n} \right],$$

we derive from the second expression (2.66) that

$$g_{nq_n, 2np_n}\left(\frac{p_n}{q_n}\right) = 1 \tag{2.68}$$

for all $n \in \mathbb{N}$. Combining the two results (2.67) and (2.68), we have shown that the limits (2.65) remain valid for irrationals x, completing the proof of the problem. $\blacksquare$

Problem 2.10

Rudin Chapter 2 Exercise 10.

Proof. Given that $\epsilon > 0$. For each positive integer n, we define

$$E_n = \{x \in [0, 1] \mid f_k(x) > \epsilon \text{ for some } k \geq n\}. \tag{2.69}$$

It is clear that each E_n is bounded and

$$E_n \supseteq E_{n+1} \supseteq \cdots .$$

If $E_{n_0} = \varnothing$ for some positive integer n_0, then we know from the definition (2.69) that

$$f_k(x) \leq \epsilon$$

for every $x \in [0, 1]$ and every positive integer $k \geq n_0$. Therefore it implies that

$$0 \leq \int_0^1 f_k(x)\, \mathrm{d}x \leq \epsilon$$

for every positive integer $k \geq n_0$. Since ϵ is arbitrary, we obtain from this that

$$\lim_{n \to \infty} \int_0^1 f_n(x)\, \mathrm{d}x = 0.$$

Without loss of generality, we assume that $E_n \neq \varnothing$ for all $n \in \mathbb{N}$. Let's prove two properties of E_n first:

- **Property 1: E_n is open.** Let $p \in E_n$. Then we have $f_k(p) > \epsilon$ for some $k \geq n$. Define $\epsilon' = \frac{1}{2}(f_k(p) - \epsilon) > 0$. Since f_k is continuous on $[0, 1]$, it is continuous at p. Thus for this particular ϵ', there is a $\delta > 0$ such that

$$|f_k(x) - f_k(p)| < \epsilon' \tag{2.70}$$

 for all points $x \in [0, 1]$ with $|x - p| < \delta$. Now the inequality (2.70) implies that

$$f_k(x) > f_k(p) - \epsilon' = \frac{f_k(p) + \epsilon}{2} > \epsilon.$$

In other words, we have $(p - \delta, p + \delta) \subseteq E_n$ as desired.

- **Property 2:** $\displaystyle\bigcap_{n=1}^{\infty} E_n = \varnothing$. Otherwise, there was a $p \in [0,1]$ such that $p \in E_n$ for all $n \in \mathbb{N}$. Therefore, for each positive integer n, we have

$$f_k(p) > \epsilon \tag{2.71}$$

for some $k \geq n$. If $n \to \infty$, then $k \to \infty$ and the inequality (2.71) implies that

$$0 = \lim_{n\to\infty} f_n(p) = \lim_{k\to\infty} f_k(p) \geq \epsilon > 0,$$

a contradiction.

To finish our proof, we have to study the *lengths* of certain subsets of E_n. Let F be a finite union of bounded (open or closed) intervals of $[0,1]$. Then we may write

$$F = \bigcup_{k=1}^{m} [a_k, b_k] \cup \bigcup_{r=1}^{s} (a_r, b_r) \tag{2.72}$$

where $0 \leq a_k < b_k \leq 1$ and $0 \leq a_r < b_r \leq 1$ for $k = 1, 2, \ldots, m$ and $r = 1, 2, \ldots, s$. Define the length of F, denoted by $\ell(F)$, to be

$$\ell(F) = \sum_{k=1}^{m} \ell\big([a_k, b_k]\big) + \sum_{r=1}^{s} \ell\big((a_r, b_r)\big) = \sum_{k=1}^{m} (b_k - a_k) + \sum_{r=1}^{s} (b_r - a_r).$$

Given a nonempty finite union of bounded interval F expressed in the form (2.72) and assume that F has at least one open interval. In addition, we suppose that $\ell(F) > \epsilon$. Let $\delta = \frac{\epsilon}{4s}$. Then the set

$$G = \bigcup_{k=1}^{m} [a_k, b_k] \cup \bigcup_{r=1}^{s} [a_r + \delta, b_r - \delta]$$

is clearly a nonempty finite union of bounded and *closed* intervals of $[0,1]$. Now it is easy to see that

$$\ell(G) = \sum_{k=1}^{m} (b_k - a_k) + \sum_{r=1}^{s} (b_r - a_r - 2\delta) = \ell(F) - 2s\delta = \ell(F) - \frac{\epsilon}{2}$$

implying that

$$\ell(G) > \ell(F) - \epsilon.$$

Here we need a lemma about the sequence of bounded subsets (2.69):

Lemma 2.4

Let $S_n = \{F \mid F \text{ is a finite union of bounded intervals of } E_n\}$ for every positive integer n and

$$L_n = \sup\{\ell(F) \mid F \in S_n\}. \tag{2.73}$$

Then we have

$$\lim_{n\to\infty} L_n = 0.$$

Proof of Lemma 2.4. By **Property 1**, we know that each S_n is nonempty. It is obvious that $\{L_n\}$ is a bounded decreasing sequence. By the Monotone Convergence Theorem [49, Theorem 3.14, p. 55], $\{L_n\}$ converges in $\mathbb{R}$. Assume that this limit was nonzero.

- **Step 1: The construction of a compact set K_n.** There exists a $\delta > 0$ such that $L_n \geq \delta$ for every positive integer n. By the definition (2.73), $L_n - \frac{\delta}{2^n}$ is not an upper bound of $\{\ell(F) \mid F \in S_n\}$. In other words, *there exists* a $F_n \in S_n$ such that

$$\ell(F_n) > L_n - \delta \cdot 2^{-n} \tag{2.74}$$

  for every positive integer n. By the observation preceding Lemma 2.4, we may assume further that each F_n is closed. Let $K_n = \bigcap_{k=1}^{n} F_k \subseteq F_n$. Since each F_k is closed in $\mathbb{R}$, K_n is closed in $\mathbb{R}$ too. By the Heine-Borel Theorem, the set K_n must be compact. Furthermore, we have $K_n \supseteq K_{n+1}$ for each $n = 1, 2, \ldots$.

- **Step 2: Each K_n is nonempty.** *There exists* a $F \in S_n$ such that

$$\ell(F) \geq \delta \tag{2.75}$$

  for each $n \in \mathbb{N}$. Otherwise, there is a $N \in \mathbb{N}$ such that $\ell(F) < \delta$ for all $F \in S_N$ which means that δ is an upper bound of L_N, but this is a contradiction. Now we are going to show that if $K_n = \varnothing$, then it is impossible to have the inequality (2.75). This contradiction bases on the following two facts:

  - **Fact 1:** Suppose that G is a finite union of bounded intervals of $E_n \setminus F_n$ for every $n \in \mathbb{N}$, where F_n are those sets considered in **Step 1**. Then we have $G \cap F_n = \varnothing$ and $G \cup F_n \in S_n$ so that $\ell(G) + \ell(F_n) = \ell(G \cup F_n) \leq L_n$. This and the inequality (2.74) give, for every $n \in \mathbb{N}$,

$$\ell(G) < \delta \cdot 2^{-n}. \tag{2.76}$$

  - **Fact 2:** Suppose that F is a finite union of bounded intervals of $E_n \setminus K_n$ for every $n \in \mathbb{N}$. By De Morgan's law (see [42, p. 11]), we have

$$(F \setminus F_1) \cup \cdots \cup (F \setminus F_n) = F \setminus (F_1 \cap \cdots \cap F_n) = F \setminus K_n = F. \tag{2.77}$$

  Note that $F \setminus F_k$ is also a finite union of bounded intervals of E_k (and hence of $E_k \setminus F_k$) for $k = 1, 2, \ldots, n$, so it follows from the inequality (2.76) that

$$\ell(F \setminus F_k) < \delta \cdot 2^{-k} \tag{2.78}$$

  for $k = 1, 2, \ldots, n$. Hence it follows from the expression (2.77) and the inequality (2.78) that

$$\ell(F) < \delta. \tag{2.79}$$

  If $K_m = \varnothing$ *for some* $m \in \mathbb{N}$, then the F considered in **Fact 2** is a subset of E_m, but the inequality (2.79) definitely contradicts the inequality (2.75).

- **Step 3: A contradiction to Property 2.** By **Step 1** and **Step 2**, we deduce from [49, Theorem 2.36, p. 38] that $\bigcap_{n=1}^{\infty} K_n \neq \varnothing$. Since $F_k \subseteq E_k$ for $k = 1, 2, \ldots, n$, we must have $K_n \subseteq E_n$. However, these two facts contradict **Property 2**.

Hence the limit must be 0 which completes the proof of the lemma.   ∎

We return to the proof of the problem. By Lemma 2.4, for $\epsilon > 0$, there is a positive integer N such that for $n \geq N$, we have

$$\ell(F) \leq L_n < \epsilon, \tag{2.80}$$

where F is *any* finite union of bounded intervals of E_n. Now one may think that we can derive

$$\int_0^1 f_n(x)\,\mathrm{d}x < 2\epsilon$$

for $n \geq N$ directly from the uniform boundedness of f_n, the definition (2.69) and the inequality (2.80). However, it fails because we have no way to estimate the integral

$$\int_{E_n \setminus F} f_n(x)\,\mathrm{d}x.$$

To overcome this problem, we play the trick that since f_n is Riemann integrable on $[0,1]$, the integral must be equal to its *lower Riemann integral*, see $[2, \S1.17, \text{p. }74]$. Thus we have to find

$$\sup\left\{ \int_0^1 s_n(x)\,\mathrm{d}x \,\Big|\, 0 \leq s_n(x) \leq f_n(x) \text{ and } s_n \text{ is a step function}\right\},$$

where the sup is taken over all step functions s_n below f_n on $[0,1]$.

For every $n \geq N$, we define

$$\mathcal{E}_n = \{x \in [0,1] \,|\, s_n(x) > \epsilon\} \quad \text{and} \quad \mathcal{F}_n = [0,1] \setminus \mathcal{E}_n.$$

Since s_n is a step function, it is clear that $\mathcal{E}_n$ and $\mathcal{F}_n$ are finite unions of bounded intervals. Furthermore, since $0 \leq s_n(x) \leq f_n(x)$, we have $\mathcal{E}_n \subseteq E_n$ and $\mathcal{E}_n \in S_n$. Then we deduce from the inequality (2.80) that

$$\ell(\mathcal{E}_n) < \epsilon$$

for all $n \geq N$. Hence we obtain from this and repeated uses of $[49, \text{Theorem } 6.12(\text{c}), \text{ p. } 128]$ that for all $n \geq N$, we have

$$0 \leq \int_0^1 s_n(x)\,\mathrm{d}x = \int_{\mathcal{E}_n} s_n(x)\,\mathrm{d}x + \int_{\mathcal{F}_n} s_n(x)\,\mathrm{d}x \leq \int_{\mathcal{E}_n} \mathrm{d}x + \int_{\mathcal{F}_n} \epsilon\,\mathrm{d}x \leq \ell(\mathcal{E}_n) + \epsilon < 2\epsilon.$$

Since ϵ is arbitrary, we have shown that

$$\lim_{n \to \infty} \int_0^1 f_n(x)\,\mathrm{d}x = 0.$$

This completes the proof of the problem. ∎

Remark 2.1

There are many mathematicians who have provided different proofs of Problem 2.10. See, for examples, [37], [38] and [53].

2.4   Problems on Borel Measures and Lebesgue Measures

Problem 2.11

Rudin Chapter 2 Exercise 11.

Proof. We follow the given hint. Suppose that Ω is the family of all compact subsets K_α of X with $\mu(K_\alpha) = 1$. Since X is compact and $\mu(X) = 1$, Ω is not empty. Now we define

$$K = \bigcap_{K_\alpha \in \Omega} K_\alpha. \tag{2.81}$$

Since X is Hausdorff, each K_α is closed in X by the corollaries following Theorem 2.5. Thus K is closed in X and then K is Borel (see Definition 1.11), i.e., $K \in \mathfrak{M}$. Since $K \subseteq X$, Theorem 2.4 says that K is also compact.

Suppose that $K \subseteq V$ and V is open in X. Then V^c is closed in X and so it is compact in X by Theorem 2.4. As we have mentioned in the previous paragraph that each K_α is closed in X, so each K_α^c is open in X. By the definition (2.81) and the fact that $K \subseteq V$, we have

$$V^c \subseteq K^c = \bigcup_{K_\alpha \in \Omega} K_\alpha^c.$$

In other words, $\{K_\alpha^c\}$ forms an open cover of the compact set V^c. Thus we have

$$V^c \subseteq K_{\alpha_1}^c \cup K_{\alpha_2}^c \cup \cdots \cup K_{\alpha_n}^c \tag{2.82}$$

for some positive integer n. Since $\mu(K_{\alpha_m}) = \mu(X) = 1$ for $m = 1, 2, \ldots, n$, we have $\mu(K_{\alpha_m}^c) = 0$ for $m = 1, 2, \ldots, n$. Therefore, it follows from these and the set relation (2.82) that

$$\mu(V^c) \le \sum_{m=1}^{n} \mu(K_{\alpha_m}^c) = 0,$$

i.e., $\mu(V^c) = 0$ or equivalent $\mu(V) = 1$. Since μ is regular and $K \in \mathfrak{M}$, Definition 2.15 implies that

$$\mu(K) = \inf\{\mu(V) \mid K \subseteq V \text{ and } V \text{ is open}\} = 1$$

which proves the first assertion.

For the second assertion, let H be a proper compact subset of K, i.e., $H \subset K$. Assume that $\mu(H) = 1$. Then it means that $H \in \Omega$ and the definition (2.81) shows that $K \subseteq H$, a contradiction. Hence we must have $\mu(H) < 1$. This completes the proof of the problem. ■

Problem 2.12

Rudin Chapter 2 Exercise 12.

Proof. Let $\mathscr{B}$ be the σ-algebra of all Borel sets in $\mathbb{R}$ and K be a nonempty compact subset of $\mathbb{R}$. We have to show that there exists a measure μ on $\mathscr{B}$ such that K is the smallest closed subset with the property $\mu(K^c) = 0$.

Since every compact set K has a countable base (see [49, Exercise 25, Chap. 2, p. 45]), there exists a countable sequence $F = \{x_1, x_2, \ldots\} \subseteq K$ such that

$$K = \overline{F}.$$

Next we define $\mu : F \to [0, \infty]$ by

$$\mu(x_n) = \frac{1}{2^n}$$

and for any $E \in \mathscr{B}$, we define $\mu : \mathscr{B} \to [0, \infty]$ by

$$\mu(E) = \sum_{n=1}^{\infty} \frac{1}{2^n} \chi_{\{x_n\}}(E), \tag{2.83}$$

where $\chi_{\{x_n\}}$ is the characteristic function of the set $\{x_n\}$. Suppose that $\{E_i\}$ is a disjoint countable collections of members of $\mathscr{B}$ and

$$E = \bigcup_{i=1}^{\infty} E_i.$$

If $x_n \in E_i$ for some i, then $x_n \notin E_j$ for all $j \neq i$. In other words, each element of F belongs to *at most* one element of $\{E_i\}$.[i] Thus it follows from the definition (2.83) that

$$\mu(E) = \sum_n \frac{1}{2^n} \chi_{\{x_n\}}(E), \tag{2.84}$$

where the summation runs through all n such that $x_n \in E_i$ for some i. Since the series

$$\sum_{n=1}^{\infty} \frac{1}{2^n}$$

converges absolutely, its rearrangement converges to the same limit (see [49, Theorem 3.55, p. 78]). Therefore, we can rewrite the expression (2.84) as

$$\mu(E) = \sum_{i=1}^{\infty} \sum_{i_k} \frac{1}{2^{i_k}} \chi_{\{x_{i_k}\}}(E_i), \tag{2.85}$$

where the inner summation runs through all i_k such that $x_{i_k} \in E_i$. It may happen that the set $F \cap E_i$ is finite for some $i = 1, 2, \dots$. Since the inner summation in the expression (2.85) is exactly $\mu(E_i)$, we deduce from the expression (2.85) that

$$\mu\left(\bigcup_{i=1}^{\infty} E_i\right) = \mu(E) = \sum_{i=1}^{\infty} \mu(E_i).$$

By Definitions 1.18(a) and 2.15, μ is a Borel measure on $\mathbb{R}$.

By the definition (2.83), it is easy to see that for every $E \in \mathscr{B}$, we have

$$\mu(E) > 0 \quad \text{if and only if} \quad K \cap E \neq \varnothing \tag{2.86}$$

so that $\mu(K^c) = 0$. Assume that there was a proper closed subset $H \subset K$ in X such that $\mu(H^c) = 0$. Since $H \subset K$, we have $K^c \subset H^c$. If $H^c \cap K = \varnothing$, then $H^c \subseteq K^c$, a contradiction. Thus $H^c \cap K \neq \varnothing$. Since H^c is open in X, we have $H^c \in \mathscr{B}$. Therefore, we can conclude from the condition (2.86) that

$$\mu(H^c) > 0$$

which contradicts to our hypothesis. In other words, K is the *smallest* closed subset in $\mathbb{R}$ such that $\mu(K^c) = 0$, or equivalently,

$$K = \operatorname{supp}(\mu).$$

This completes the analysis of the problem.

> **Problem 2.13**
>
> *Rudin Chapter 2 Exercise 13.*

---

[i] However, E_i may contain *more than* one element of F.

Proof. It is clear that the point set $\{0\}$ is a compact subset of $\mathbb{R}$. Assume $f : \mathbb{R} \to \mathbb{R}$ was a continuous function such that $\{0\} = \operatorname{supp}(f)$. Let $V = (-\infty, 0) \cup (0, \infty)$. We know from Definition 2.9 that $\operatorname{supp}(f) = \overline{f^{-1}(V)}$, so

$$\{0\} = \overline{f^{-1}(V)} \tag{2.87}$$

so that $f^{-1}(V) \neq \varnothing$. Since V is open in $\mathbb{R}$ and f is continuous on $\mathbb{R}$, $f^{-1}(V)$ is open in $\mathbb{R}$ by Definition 1.2(c). Thus, if $p \in f^{-1}(V)$, then there exists a $\delta > 0$ such that

$$(p - \delta, p + \delta) \subseteq f^{-1}(V). \tag{2.88}$$

Thus we deduce from the set relations (2.87) and (2.88) that

$$[p - \delta, p + \delta] \subseteq \overline{f^{-1}(V)} = \{0\},$$

a contradiction. Hence there is no continuous function f such that $\{0\} = \operatorname{supp}(f)$.

For the second assertion, we suppose that K is a compact subset of $\mathbb{R}$. We claim that K is the support of a continuous function if and only if K is the closure of an open set V in X. If K is the support of a continuous function $f : \mathbb{R} \to \mathbb{R}$, then Definition 2.9 shows that

$$K = \overline{\{x \in \mathbb{R} \mid f(x) \neq 0\}} = \overline{f^{-1}\big((-\infty, 0) \cup (0, \infty)\big)} = \overline{f^{-1}\big((-\infty, 0)\big) \cup f^{-1}\big((0, \infty)\big)}.$$

Since f is continuous on $\mathbb{R}$, $f^{-1}\big((-\infty, 0)\big)$ and $f^{-1}\big((0, \infty)\big)$ are open in $\mathbb{R}$. Thus this proves one direction. Conversely, if we have

$$K = \overline{V} \tag{2.89}$$

for some open set V, then we consider $F = V^c$ which is closed in $\mathbb{R}$. Define $\rho_F : \mathbb{R} \to \mathbb{R}$ by

$$\rho_F(x) = \inf\{|x - y| \mid y \in F\}.$$

By Problem 2.3, we know that ρ_F is uniformly continuous on $\mathbb{R}$ and $\rho_F(x) = 0$ if and only if $x \in F$. Therefore, we must have

$$\rho_F(x) > 0 \quad \text{if and only if} \quad x \in V. \tag{2.90}$$

Applying the relation (2.90) to the expression (2.89), we have

$$K = \overline{\{x \in \mathbb{R} \mid \rho_F(x) \neq 0\}},$$

i.e., K is the support of the continuous function ρ_F.

We note that the third assertion is *not* valid in other topological spaces. We consider the three-element set $X = \{a, b, c\}$. By [42, Example 1, p. 76], we see that $\{\varnothing, \{b\}, X\}$ is a topology of X. Let $K = \{a, b\}$. Then it is easy to check from the definition that K is compact. Now if $K = \overline{V}$ for some open set V, then K must be closed in X (see [42, p. 95]). However, since $X \setminus K = \{c\}$ which is *not* open in X, K is *not* closed in X. This gives a counter-example to explain that the description "K is the closure of an open set V in X" *cannot* guarantee that K is the support of a continuous function in an arbitrary topological space and hence we complete the proof of the problem. ∎

Problem 2.14

Rudin Chapter 2 Exercise 14.

Proof. Let $f : \mathbb{R}^k \to \mathbb{R}$. We follow the proof of Theorem 2.24 (Lusin's Theorem). Suppose that $0 \le f < 1$. Attach a sequence $\{s_n\}$ of simple measurable functions to f, as in the proof of Theorem 1.17 (The Simple Function Approximation Theorem). Put $t_1 = s_1$ and $t_n = s_n - s_{n-1}$ for $n = 2, 3, \ldots$. Then $2^n t_n$ is the characteristic function of a measurable set $T_n \subseteq \mathbb{R}^k$ and

$$f(\mathbf{x}) = \sum_{n=1}^{\infty} t_n(\mathbf{x}) = \sum_{n=1}^{\infty} 2^{-n} \chi_{T_n}(\mathbf{x}) \tag{2.91}$$

on $\mathbb{R}^k$. By Theorem 2.20(b), there exist $A_n, B_n \in \mathscr{B}$ in $\mathbb{R}^k$ such that

$$A_n \subseteq T_n \subseteq B_n \quad \text{and} \quad m(B_n \setminus A_n) = 0 \tag{2.92}$$

for every positive integer n.[j]

We define $g : \mathbb{R}^k \to \mathbb{R}$ and $h : \mathbb{R}^k \to \mathbb{R}$ by

$$g(\mathbf{x}) = \sum_{n=1}^{\infty} 2^{-n} \chi_{A_n}(\mathbf{x}) \quad \text{and} \quad h(\mathbf{x}) = \sum_{n=1}^{\infty} 2^{-n} \chi_{B_n}(\mathbf{x}). \tag{2.93}$$

We claim that g and h are Borel measurable functions on $\mathbb{R}^k$. It is easy to check that

$$\{\mathbf{x} \in \mathbb{R}^k \,|\, \chi_{A_n}(\mathbf{x}) < \alpha\} = \begin{cases} \varnothing, & \text{if } \alpha \le 0; \\[2mm] A_n^c, & \text{if } 0 < \alpha \le 1; \\[2mm] \mathbb{R}, & \text{if } \alpha > 1. \end{cases}$$

Thus each χ_{A_n} is a Borel measurable function by Definition 1.11. Similarly, each χ_{B_n} is also a Borel measurable function. Since the partial sums g_n and h_n of g and h are just simple functions and $A_n, B_n \in \mathscr{B}$, they are Borel measurable functions by the comment following Definition 1.16. Furthermore, for each $n = 1, 2, \ldots$, we have

$$|2^{-n} \chi_{A_n}(\mathbf{x})| \le 2^{-n} \quad \text{and} \quad |2^{-n} \chi_{B_n}(\mathbf{x})| \le 2^{-n}$$

on $\mathbb{R}^k$. Since $\displaystyle\sum_{n=1}^{\infty} 2^{-n}$ converges, it follows from Weierstrass M-test (see [49, Theorem 7.10, p. 147]) that $\sum 2^{-n} \chi_{A_n}$ and $\sum 2^{-n} \chi_{B_n}$ converge uniformly on $\mathbb{R}^k$ to g and h respectively. By the Corollary (a) following Theorem 1.14, g and h are Borel measurable. This proves our claim.

By the measure (2.92), we know that $g = h$ a.e. $[m]$ on $\mathbb{R}^k$. Obviously, if $F \subseteq E$, then we must have $\chi_F(\mathbf{x}) \le \chi_E(\mathbf{x})$. Therefore, we observe from the set relations (2.92) that

$$\chi_{A_n}(\mathbf{x}) \le \chi_{T_n}(\mathbf{x}) \le \chi_{B_n}(\mathbf{x}) \tag{2.94}$$

on $\mathbb{R}^k$. Hence we apply the inequalities (2.94) to functions (2.91) and (2.93), we obtain

$$g(\mathbf{x}) \le f(\mathbf{x}) \le h(\mathbf{x}) \tag{2.95}$$

on $\mathbb{R}^k$.

Next, if $0 \le f < M$ for some positive constant M on $\mathbb{R}^k$, then the above argument can be applied to $\frac{f}{M}$ so that the inequalities (2.95) hold with g and h replaced by Mg and Mh respectively.

---

[j]In fact, each A_n is an F_σ and each B_n is a G_δ, see Definition 1.11.

In the general case, we consider the measurable sets $E_N = \{\mathbf{x} \in \mathbb{R}^k \mid -N \le f(\mathbf{x}) \le N\}$, where $N \in \mathbb{N}$. Now we have $\chi_{E_N} f \to f$ as $N \to \infty$ on $\mathbb{R}^k$. In fact, the above argument shows that we can find Boreal functions g_N and h_N such that $g_N(\mathbf{x}) = h_N(\mathbf{x})$ a.e. $[m]$ on $\mathbb{R}^k$ and

$$g_N(\mathbf{x}) \le \chi_{E_N} f(\mathbf{x}) \le h_N(\mathbf{x})$$

on $\mathbb{R}^k$. By Theorem 1.14, we see that the functions g and h defined by

$$g(\mathbf{x}) = \limsup_{N \to \infty} g_N(\mathbf{x}) \quad \text{and} \quad h(\mathbf{x}) = \limsup_{N \to \infty} h_N(\mathbf{x}) \tag{2.96}$$

are Borel measurable functions on $\mathbb{R}^k$. Since $g_N(\mathbf{x}) = h_N(\mathbf{x})$ a.e. $[m]$ on $\mathbb{R}^k$, we have $g(\mathbf{x}) = h(\mathbf{x})$ a.e. $[m]$ on $\mathbb{R}^k$. Furthermore, we note that

$$g(\mathbf{x}) = \limsup_{N \to \infty} g_N(\mathbf{x}) \le \limsup_{N \to \infty} \chi_{E_N} f(\mathbf{x}) \le \limsup_{N \to \infty} h_N(\mathbf{x}) = h(\mathbf{x}) \tag{2.97}$$

on $\mathbb{R}^k$. Since

$$\lim_{N \to \infty} \chi_{E_N} f(\mathbf{x}) = \limsup_{N \to \infty} \chi_{E_N} f(\mathbf{x}) = f(\mathbf{x})$$

on $\mathbb{R}^k$, the inequalities (2.95) follow immediately from the inequalities (2.96) and the fact (2.97). Hence our desired results also hold in this general case, completing the proof of the problem. $\blacksquare$

Problem 2.15

Rudin Chapter 2 Exercise 15.

Proof. For each positive integer n, we define $f_n : [0, \infty) \to [0, \infty]$ by

$$f_n(x) = \begin{cases} \left(1 - \dfrac{x}{n}\right)^n e^{\frac{x}{2}}, & \text{if } x \in [0, n]; \\[2mm] 0, & \text{if } x > n. \end{cases}$$

It is easy to see that each f_n is continuous on $[0, \infty)$ and so Definition 1.11 shows that it is Borel measurable. Fix a $x \in \mathbb{R}$. Then the Archimedean Property ensures the existence of a positive integer N such that $N \ge x$. By this fact and the fact that $(1 + \frac{x}{n})^n \to e^x$ as $n \to \infty$, we have

$$\lim_{n \to \infty} f_n(x) = \lim_{n \to \infty} \left(1 - \frac{x}{n}\right)^n e^{\frac{x}{2}} = e^{-x} \cdot e^{\frac{x}{2}} = e^{-\frac{x}{2}}.$$

Since $e^{-\frac{x}{2}} \in L^1(\mathbb{R})$ and

$$|f_n(x)| \le e^{-\frac{x}{2}} \tag{2.98}$$

for all $n = 1, 2, \ldots$ and $x \in [0, \infty)$, it follows from Theorem 1.34 (Lebesgue's Dominated Convergence Theorem) that

$$\lim_{n \to \infty} \int_0^\infty f_n \, dx = \int_0^\infty e^{-\frac{x}{2}} \, dx = 2. \tag{2.99}$$

By the inequality (2.98), we get

$$\left| \int_0^\infty f_n \, dx - \int_0^n f_n \, dx \right| = \left| \int_n^\infty f_n \, dx \right| \le \int_n^\infty |f_n| \, dx \le \int_n^\infty e^{-\frac{x}{2}} \, dx = 2e^{-\frac{n}{2}}$$

which implies that

$$\lim_{n \to \infty} \int_0^\infty f_n \, dx = \lim_{n \to \infty} \int_0^n f_n \, dx. \tag{2.100}$$

Combining the results (2.99) and (2.100), we obtain

$$\lim_{n \to \infty} \int_0^n f_n \, \mathrm{d}x = 2.$$

For the second integral, we define $g_n : [0, \infty) \to [0, \infty]$ by

$$g_n(x) = \begin{cases} \left(1 + \dfrac{x}{n}\right)^n \mathrm{e}^{-2x}, & \text{if } x \in [0, n]; \\[2ex] \dfrac{2^x}{\mathrm{e}^{2x}}, & \text{if } x > n. \end{cases}$$

Therefore each g_n is continuous on $[0, \infty)$. Furthermore, by using a similar argument as above, we have for $x \in \mathbb{R}$,

$$\lim_{n \to \infty} g_n(x) = \lim_{n \to \infty} \left(1 + \frac{x}{n}\right)^n \mathrm{e}^{-2x} = \mathrm{e}^x \cdot \mathrm{e}^{-2x} = \mathrm{e}^{-x}.$$

Since $\mathrm{e}^{-x} \in L^1(\mathbb{R})$ and $|g_n(x)| \le \mathrm{e}^{-x}$ for all $n = 1, 2, \ldots$ and $x \in [0, \infty)$, it follows from Theorem 1.34 (Lebesgue's Dominated Convergence Theorem) that

$$\lim_{n \to \infty} \int_0^\infty g_n \, \mathrm{d}x = \int_0^\infty \mathrm{e}^{-x} \, \mathrm{d}x = 1.$$

Now we apply the same trick as obtaining the result (2.100), we are able to show that

$$\lim_{n \to \infty} \int_0^n g_n \, \mathrm{d}x = 1.$$

We have completed the proof of the problem. ∎

Problem 2.16

Rudin Chapter 2 Exercise 16.

Proof. Let $T : \mathbb{R}^k \to \mathbb{R}^k$ be linear, $Y = T(\mathbb{R}^k)$ with $\dim Y = r < k$. Let $\mathcal{B} = \{\mathbf{b}_1, \ldots, \mathbf{b}_r\}$ be a basis of Y. Since $\mathcal{B}$ is a finite set of linearly independent vectors of $\mathbb{R}^k$, it can be enlarged to a basis $\{\mathbf{b}_1, \ldots, \mathbf{b}_r, \mathbf{b}_{r+1}, \ldots, \mathbf{b}_k\}$ for $\mathbb{R}^k$.[k] Define the linear transformation $T : \mathbb{R}^k \to \mathbb{R}^k$ by

$$T(\alpha_1 \mathbf{b}_1 + \cdots + \alpha_k \mathbf{b}_k) = (\alpha_1, \ldots, \alpha_k).$$

Then it is easy to see that T is one-to-one and so $\det T \neq 0$. Furthermore, we have

$$T(Y) = \{(\alpha_1, \ldots, \alpha_r, 0, \ldots, 0) \,|\, \alpha_1, \ldots, \alpha_r \in \mathbb{R}\}$$

so that

$$T(Y) = \bigcup_{n=1}^{\infty} W_n, \tag{2.101}$$

where $W_n = \{(\xi_1, \ldots, \xi_r, 0, \ldots, 0) \,|\, -n \le \xi_i \le n \text{ and } i = 1, 2, \ldots, r\}$. By the definition, we have

$$W_n \subseteq W_{n+1}$$

---

[k]See, for example, [23, Theorem 30.19, p. 279]. Sometimes, this result is called the **Basis Extension Theorem**.

for every $n \in \mathbb{N}$. It is trivial to see from Theorem 2.20(a) that

$$m(W_n) = \text{vol}\,(W_n) = 0 \tag{2.102}$$

for each $n = 1, 2, \ldots$. In addition, since $\mathbb{R}^k$ is a Borel set, Y is also a Borel set. Hence we follow from this fact, the expressions (2.101), (2.102) and Theorem 1.19(d) that

$$m\big(T(Y)\big) = \lim_{n \to \infty} m(W_n) = \lim_{n \to \infty} \text{vol}\,(W_n) = 0. \tag{2.103}$$

When we read the proof of Theorem 2.20(e) and §2.23 carefully, we see that the validity of the formula

$$m\big(T(E)\big) = |\det T| m(E) \tag{2.104}$$

when T is an one-to-one map of $\mathbb{R}^k$ onto $\mathbb{R}^k$ is *independent* of whether $T(\mathbb{R}^k)$ is a subspace of lower dimension or not. As a consequence, we may apply the formula (2.104) here. Hence we can conclude from $m(Y) = |\det T|^{-1} m\big(T(Y)\big)$ and the result (2.103) that $m(Y) = 0$, completing the proof of the problem. $\blacksquare$

2.5   Problems on Regularity of Borel Measures

> **Problem 2.17**
>
> *Rudin Chapter 2 Exercise 17.*

Proof. Let d be the mentioned distance in the problem, $\mathbf{p} = (x_1, y_1)$ and $\mathbf{q} = (x_2, y_2)$. We are going to prove the assertions one by one.

- **X is a metric space with metric d.** We check the definition of a metric space, see [49, Definition 2.15, p. 30]:

  - If $\mathbf{p} = \mathbf{q}$, then $d(\mathbf{p}, \mathbf{q}) = |y_1 - y_2| = 0$. Otherwise, we have

    $$d(\mathbf{p}, \mathbf{q}) = \begin{cases} |y_1 - y_2|, & \text{if } x_1 = x_2 \text{ and } y_1 \neq y_2; \\ 1 + |y_1 - y_2|, & \text{if } x_1 \neq x_2, \end{cases}$$
    $$\neq 0.$$

  - It is clear that $d(\mathbf{p}, \mathbf{q}) = d(\mathbf{q}, \mathbf{p})$ holds because $|y_1 - y_2| = |y_2 - y_1|$.
  - Let $\mathbf{r} = (x_3, y_3)$. Suppose that $\mathbf{r} \neq \mathbf{p}$ and $\mathbf{r} \neq \mathbf{q}$. Otherwise, the triangle inequality is trivial. There are two cases:
    * **Case (i):** $x_1 = x_2$. Then we have $d(\mathbf{p}, \mathbf{q}) = |y_1 - y_2|$ and

      $$d(\mathbf{p}, \mathbf{r}) = \begin{cases} |y_1 - y_3|, & \text{if } x_3 = x_1 \text{ and } y_3 \neq y_1; \\ 1 + |y_1 - y_3|, & \text{if } x_3 \neq x_1. \end{cases} \tag{2.105}$$

      Similarly, we have

      $$d(\mathbf{q}, \mathbf{r}) = \begin{cases} |y_2 - y_3|, & \text{if } x_3 = x_2 \text{ and } y_3 \neq y_2; \\ 1 + |y_2 - y_3|, & \text{if } x_3 \neq x_2. \end{cases} \tag{2.106}$$

      Since $|y_1 - y_2| \leq |y_1 - y_3| + |y_3 - y_2|$, any combination of the distances (2.105) and (2.106) imply that

      $$d(\mathbf{p}, \mathbf{q}) \leq d(\mathbf{p}, \mathbf{r}) + d(\mathbf{r}, \mathbf{q}). \tag{2.107}$$

    * **Case (ii):** $x_1 \neq x_2$. In this case, we have $d(\mathbf{p}, \mathbf{q}) = 1 + |y_1 - y_2|$. If $x_3 = x_1$, then $x_3 \neq x_2$ so that $d(\mathbf{p}, \mathbf{r}) = |y_1 - y_3|$ and $d(\mathbf{q}, \mathbf{r}) = 1 + |y_2 - y_3|$. The situation for $x_3 = x_1$ is similar. If $x_3 \neq x_1$ and $x_3 \neq x_2$, then we see from the distances (2.105) and (2.106) again that $d(\mathbf{p}, \mathbf{r}) = 1 + |y_1 - y_3|$ and $d(\mathbf{q}, \mathbf{r}) = 1 + |y_2 - y_3|$. Hence the triangle inequality (2.107) remains true.

By the above analysis, we conclude that d is a metric.

- **X is locally compact.** Let $X = (\mathbb{R}^2, d)$ and $\mathbf{p} = (x, y), \mathbf{q} = (u, v) \in \mathbb{R}^2$. Fix $\mathbf{p}$ first. By the definition, $d(\mathbf{p}, \mathbf{q}) < 1$ implies that $x = u$ and $|y - v| < 1$. Thus the neighborhood $B(\mathbf{p}, 1)$ of $\mathbf{p}$ is given by

$$
\begin{aligned}
B(\mathbf{p}, 1) &= \{\mathbf{q} \in X \mid d(\mathbf{p}, \mathbf{q}) < 1\} \\
&= \{\mathbf{q} \in X \mid x = u \text{ and } |y - v| < 1\} \\
&= \{x\} \times (y - 1, y + 1).
\end{aligned} \tag{2.108}
$$

Geometrically, $B(\mathbf{p}, 1)$ is the vertical line segment with half length less than 1 and (x, y) as its midpoint. In addition, this line segment is open in $\mathbb{R}$ with the usual metric. Furthermore, it is clear from the definition (2.108) that

$$
\overline{B(\mathbf{p}, 1)} = \{x\} \times [y - 1, y + 1]. \tag{2.109}
$$

We want to show that $\overline{B(\mathbf{p}, 1)}$ is compact with respect to the metric d. To this end, let $\{V_\alpha\}$ be an open cover of $\overline{B(\mathbf{p}, 1)}$, i.e.,

$$
\overline{B(\mathbf{p}, 1)} \subseteq \bigcup_\alpha V_\alpha.
$$

Let $\mathbf{a} \in \overline{B(\mathbf{p}, 1)}$, where $\mathbf{a} = (s, t)$ with $s = x$ and $t \in [y - 1, y + 1]$. Then $(x, t) \in V_\alpha$ for some α. Since V_α is open in X, there exists a $\delta_t \in (0, 1)$ such that

$$
B(\mathbf{a}, \delta_t) = \{\mathbf{q} = (u, v) \mid u = s = x \text{ and } |v - t| < \delta_t\} = \{x\} \times (t - \delta_t, t + \delta_t) \subseteq V_\alpha \tag{2.110}
$$

so that

$$
\overline{B(\mathbf{p}, 1)} \subseteq \bigcup_{\substack{\mathbf{a} \in \overline{B(\mathbf{p},1)} \\ \mathbf{a} = (s,t)}} B(\mathbf{a}, \delta_t) \subseteq \bigcup_\alpha V_\alpha. \tag{2.111}
$$

Applying the expressions (2.109) and (2.110) to the set relation (2.111), we see that

$$
\overline{B(\mathbf{p}, 1)} = \{x\} \times [y - 1, y + 1] \subseteq \bigcup_{t \in [y-1, y+1]} \{x\} \times (t - \delta_t, t + \delta_t) \subseteq \bigcup_\alpha V_\alpha \tag{2.112}
$$

which means that $\{U_t\}$, where $U_t = (t - \delta_t, t + \delta_t)$ with $t \in [y - 1, y + 1]$, is an open cover of $[y - 1, y + 1]$. Since $[y - 1, y + 1]$ is compact with respect to the *usual metric* $|\cdot|$ by the Heine-Borel Theorem, there are finitely many indices $t_1, \ldots, t_k$ such that

$$
[y - 1, y + 1] \subseteq (t_1 - \delta_{t_1}, t_1 + \delta_{t_1}) \cup (t_2 - \delta_{t_2}, t_2 + \delta_{t_2}) \cup \cdots \cup (t_k - \delta_{t_k}, t_k + \delta_{t_k}).
$$

Suppose that $(x, t_i) \in V_{\alpha_i}$ for $i = 1, 2, \ldots, k$. Hence we follow immediately from the set relation (2.112) that

$$
\overline{B(\mathbf{p}, 1)} \subseteq V_{t_1} \cup V_{t_2} \cup \cdots \cup V_{t_k}.
$$

In other words, the closure $\overline{B(\mathbf{p}, 1)}$ is compact with respect to the metric d and since $\mathbf{p}$ is an arbitrary point in X, Definition 2.3(f) shows that X is locally compact.[1]

---

[1] We remark that we cannot use the Heine-Borel Theorem directly to the closure (2.109) because we are working in the space X with metric d, not $\mathbb{R}^2$ with the usual metric.

- **Construction of a positive linear functional.** Let $f \in C_c(X)$, i.e., $f : (\mathbb{R}^2, d) \to \mathbb{C}$ is a continuous function and $\operatorname{supp}(f)$ is compact. First of all, we notice that the set

$$E = \{x \in \mathbb{R} \mid f(x, y) \neq 0 \text{ for some } y\}$$

is finite. To see this, let $K = \operatorname{supp}(f)$ and $0 < \delta < 1$. Since

$$K \subseteq \bigcup_{\mathbf{p} \in K} B(\mathbf{p}, \delta),$$

where $B(\mathbf{p}, \delta) = \{x\} \times (y - \delta, y + \delta)$. Now the compactness of K ensures that there exists some positive integer n such that

$$K \subseteq \bigcup_{i=1}^{n} B(\mathbf{p}_i, \delta) = \bigcup_{i=1}^{n} \{\mathbf{q} \in X \mid u = x_i \text{ and } |y_i - v| < \delta\}$$

which forces $E = \{x_1, x_2, \ldots, x_n\}$, as claimed.

Next, we claim that the mapping $\Lambda : C_c(X) \to \mathbb{R}$ defined by

$$\Lambda(f) = \sum_{j=1}^{n} \int_{-\infty}^{\infty} f(x_j, y) \, dy$$

is a positive linear functional. To this end, let $f, g \in C_c(X)$. Since $C_c(X)$ is a vector space, we have $\alpha f + \beta g \in C_c(X)$ for any $\alpha, \beta \in \mathbb{C}$. Furthermore, let $f(x_i, y) \neq 0$ and $g(x_j, y) \neq 0$ for at least one y, where $1 \le i \le n$ and $n + 1 \le j \le m$, and $f(z_k, y)g(z_k, y) \neq 0$ for at least one y, where $m + 1 \le k \le r$. By the definition, we have

$$f(x_i, y) = g(x_j, y) = 0,$$

where $n + 1 \le i \le m$, $1 \le j \le n$ and all y, so we get

$$\Lambda(\alpha f + \beta g) = \sum_{s=1}^{r} \int_{-\infty}^{\infty} [\alpha f(x_s, y) + \beta g(x_s, y)] \, dy$$

$$= \alpha \left[\sum_{s=1}^{n} \int_{-\infty}^{\infty} f(x_s, y) \, dy + \sum_{s=m+1}^{r} \int_{-\infty}^{\infty} f(z_s, y) \, dy \right]$$

$$+ \beta \left[\sum_{s=n+1}^{m} \int_{-\infty}^{\infty} g(x_s, y) \, dy + \sum_{s=m+1}^{r} \int_{-\infty}^{\infty} g(z_s, y) \, dy \right]$$

$$= \alpha \Lambda(f) + \beta \Lambda(g).$$

By Definition 2.1, Λ is a linear functional on $C_c(X)$. If $f \ge 0$, then $\Lambda(f) \ge 0$. Thus Λ is positive. This proves the claim.

- **The measures of the x-axis and its compact subsets.** In order to apply Theorem 2.14 (The Riesz Representation Theorem), we have to show that X is Hausdorff. Given $\mathbf{p}, \mathbf{q} \in X$ and $\mathbf{p} \neq \mathbf{q}$. Let $\mathbf{p} = (x, y)$ and $\mathbf{q} = (u, v)$. There are two cases:

  - **Case (i):** $x = u$ but $y \neq v$. Then we define $\delta = |y - x|$ and it obtains from the definition (2.108) that

$$B\left(\mathbf{p}, \frac{\delta}{2}\right) \cap B\left(\mathbf{q}, \frac{\delta}{2}\right) = \left[\{x\} \times \left(y - \frac{\delta}{2}, y + \frac{\delta}{2} \right) \right] \cap \left[\{x\} \times \left(v - \frac{\delta}{2}, v + \frac{\delta}{2} \right) \right] = \varnothing.$$

- **Case (ii):** $x \neq u$ **but** $y = v$. Then the geometric property of the neighborhoods $B(\mathbf{p}, 1)$ and $B(\mathbf{q}, 1)$ ensures that their intersection must be empty.

In conclusion, we have shown that X is Hausdorff.

By Theorem 2.14 (The Riesz Representation Theorem), there exists a unique positive measure μ on $\mathfrak{M}$ associated with Λ. Let

$$E = \{(x, 0) \mid x \in \mathbb{R}\}.$$

Let $\delta \in (0, 1)$ and

$$U_\delta = \bigcup_{\mathbf{p} \in E} B(\mathbf{p}, \delta) = \bigcup_{x \in \mathbb{R}} \{x\} \times (-\delta, \delta).$$

Then U_δ is clearly an open set in X and $E \subseteq U_\delta$. Take $(x_1, 0), \ldots, (x_n, 0) \in E$ and construct

$$K = \overline{B\big((x_1, 0), \delta\big)} \cup \overline{B\big((x_2, 0), \delta\big)} \cup \cdots \cup \overline{B\big((x_n, 0), \delta\big)}$$
$$= \big(\{x_1\} \times [-\delta, \delta]\big) \cup \big(\{x_2\} \times [-\delta, \delta]\big) \cup \cdots \cup \big(\{x_n\} \times [-\delta, \delta]\big).$$

Since each $\overline{B\big((x_i, 0), \delta\big)}$ is compact, K is also compact and $K \subseteq U_\delta$. By Theorem 2.12 (Urysohn's Lemma), there exists an $f \in C_c(X)$ such that

$$K \prec f \prec U_\delta.$$

Since $\mu(U_\delta) = \sup\{\Lambda(f) \mid f \prec U_\delta\}$ (see [51, Eqn. (1), p. 41]), we have

$$\mu(U_\delta) \geq \Lambda(f) = \sum_{j=1}^{m} \int_{-\infty}^{\infty} f(x'_j, y) \, dy, \tag{2.113}$$

where $x'_1, \ldots, x'_m$ are those values of x for which $f(x, y) \neq 0$ for at least one y and for some positive integer m. Since $K \prec f$, we have $f(x_i, y) = 1$ for all $i = 1, 2, \ldots, n$ and $y \in (-\delta, \delta)$. Thus we have $\{x_1, x_2, \ldots, x_n\} \subseteq \{x'_1, x'_2, \ldots, x'_m\}$ and then the inequality (2.113) reduces to

$$\mu(U_\delta) \geq \sum_{i=1}^{n} \int_{-\delta}^{\delta} dy = n\delta$$

so that $\mu(U_\delta) \to \infty$ as $n \to \infty$. If V is an open set containing E, then V must contain the open set U_δ *for some* small δ and thus $\mu(V) = \infty$ for every open set V containing E. By Theorem 2.14(c), we have

$$\mu(E) = \inf\{\mu(V) \mid E \subseteq V, V \text{ is open}\} = \infty.$$

Next, let $K \subseteq E$ be compact. Choose $\delta \in (0, 1)$. Since

$$K \subseteq W = \bigcup_{(a,0) \in K} B\big((a, 0), \delta\big), \tag{2.114}$$

the compactness of K implies that

$$K \subseteq \bigcup_{i=1}^{m} B((a_i, 0), \delta) = \bigcup_{i=1}^{m} (\{a_i\} \times (-\delta, \delta))$$

for some positive integer m. In other words, we know from the definition (2.114) that

$$K = \{(a_1, 0), (a_2, 0), \ldots, (a_m, 0)\}.$$

Besides, since W is open in X, Theorem 2.12 (Urysohn's Lemma) again ensures there is a $g \in C_c(X)$ such that

$$K \prec g \prec W.$$

Recall from [51, Eqn. (7), p. 43] that

$$\mu(K) = \inf\{\Lambda(f) \mid K \prec f\} \leq \Lambda(g) = \sum_{i=1}^{m} \int_{-\infty}^{\infty} g(a_i, y)\,\mathrm{d}y = 2m\delta. \tag{2.115}$$

Since m is fixed and δ is *arbitrary*, it follows from the inequality (2.115) that $\mu(K) = 0$.

This completes the proof of the problem.   ◼

Problem 2.18

Rudin Chapter 2 Exercise 18.

Proof. Recall that X has an order relation '$<$' and every nonempty subset A of X has a first/smallest element a, i.e., $a \leq x$ for all $x \in A$.[m] If $\omega_0 = \min X$ (ω_0 is a smallest element in X), then we have

$$X = [\omega_0, \omega_1].$$

Next, we characteristic the topology on X as follows:

> **The order topology of X:** For $\alpha \in X$, we define
>
> $$P_\alpha = \{x \in X \mid x < \alpha\} = [\omega_0, \alpha) \quad \text{and} \quad S_\alpha = \{x \in X \mid \alpha < x\} = (\alpha, \omega_1].$$
>
> A subset E of X is open in X if it is in one of the following classes
>
> $$[\omega_0, \alpha), \quad (\beta, \omega_1], \quad [\omega_0, \alpha) \cap (\beta, \omega_1] = (\beta, \alpha) \quad \text{or} \quad \text{a union of such sets.} \tag{2.116}$$
>
> See [42, §14, pp. 84 – 86].

Now we prove the assertions one by one.

(a) X **is a compact Hausdorff space.** Suppose that $\mathcal{V}$ is an open cover of $[\omega_0, \omega_1]$. Let $\alpha_0 = \omega_1$. Then there is an open set $V_0 \in \mathcal{V}$ such that $\alpha_0 \in V_0$. If $V_0 = X$, then X is compact. Therefore, in the following discussion, we may assume that $V_0 \neq X$.

Since V_0 is open in X, it follows from the uncountability of X and the order topology (2.116) that there exists an $\alpha_1 \in X$ such that $(\alpha_1, \alpha_0] \subseteq V_0$. We note that $\alpha_1 < \alpha_0$. Now this α_1 can be chosen such that $\alpha_1 \notin V_0$. Otherwise, $\alpha_1 \in V_0$ and we may choose another $\alpha_2 \in X$ such that $(\alpha_2, \alpha_1] \subseteq V_0$. Here we have $\alpha_2 < \alpha_1$. Similarly, this α_2 can be chosen such that $\alpha_2 \notin V_0$. If not, then we continue this process in the same manner, but this must stop in a *finite* number of steps. Otherwise, we can find an infinite decreasing sequence

$$\alpha_0 > \alpha_1 > \alpha_2 > \cdots. \tag{2.117}$$

in X, which is a contradiction to the hint. Therefore, we may assume that

$$\alpha_1 \notin V_0 \quad \text{and} \quad (\alpha_1, \alpha_0] \subseteq V_0. \tag{2.118}$$

---

[m] See [42, p. 63].

Since $\mathcal{V}$ is a cover of $[\omega_0, \omega_1]$, there exists a $V_1 \in \mathcal{V}$ containing α_1. By the above argument, we may find an $\alpha_2 \in X$ such that

$$\alpha_2 \notin V_1 \quad \text{and} \quad (\alpha_2, \alpha_1] \subseteq V_1. \tag{2.119}$$

If $\alpha_2 \in V_0$, then since $[\alpha_2, \alpha_0] \subseteq V_0$ and $\alpha_2 < \alpha_1 < \alpha_0$, we have $\alpha_1 \in V_0$ which contradicts the result (2.118). Thus $\alpha_2 \notin V_0$ and then we deduce from this fact, the results (2.118) and (2.119) that

$$\alpha_2 \notin V_0 \cup V_1 \quad \text{and} \quad (\alpha_2, \alpha_0] \subseteq V_0 \cup V_1.$$

Inductively, we can find $\alpha_{n+1} \in X$ such that

$$\alpha_{n+1} \notin V_0 \cup V_1 \cup \cdots \cup V_n \quad \text{and} \quad (\alpha_{n+1}, \alpha_0] \subseteq V_0 \cup V_1 \cup \cdots \cup V_n.$$

If this process continues infinitely, then we can find an infinite decreasing sequence (2.117) again, a contradiction. Therefore, the sequence *must* stop eventually at the Nth step and it must stop at $\alpha_{n+1} = \omega_0$ for all $n \geq N$, i.e.,

$$\omega_0 \notin V_0 \cup V_1 \cup \cdots \cup V_N \quad \text{and} \quad (\omega_0, \alpha_0] \subseteq V_0 \cup V_1 \cup \cdots \cup V_N.$$

Since $\omega_0 \in V$ for some $V \in \mathcal{V}$, we conclude that

$$[\omega_0, \omega_1] = [\omega_0, \alpha_0] \subseteq V_0 \cup V_1 \cup \cdots \cup V_N \cup V.$$

In other words, X is compact.

Next, we prove that X is Hausdorff and we need the help of the following lemma:

Lemma 2.5

Define the mapping $\gamma : [\omega_0, \omega_1] \to [\omega_0, \omega_1]$ by

$$\gamma(\alpha) = \begin{cases} \min(S_\alpha), & \text{if } \alpha \in [\omega_0, \omega_1); \\ \omega_1, & \text{if } \alpha = \omega_1. \end{cases}$$

Then we have $X \setminus S_\alpha = P_{\gamma(\alpha)} = [\omega_0, \alpha]$.

Proof of Lemma 2.5. For every $\alpha \in [\omega_0, \omega_1)$, $S_\alpha \neq \varnothing$ so that the mapping γ is well-defined. By the definition, we have

$$S_\alpha = [\gamma(\alpha), \omega_1]$$

with $\alpha < \gamma(\alpha)$. Furthermore, there is *no* element between α and $\gamma(\alpha)$. Otherwise, $\gamma(\alpha)$ is not a smallest element of S_α anymore. Hence we deduce from this that

$$X \setminus S_\alpha = [\omega_0, \alpha] = [\omega_0, \gamma(\alpha)) = P_{\gamma(\alpha)}.$$

If $\alpha = \omega_1$, then $S_{\omega_1} = \varnothing$ and $P_{\gamma(\omega_1)} = [\omega_0, \omega_1] = X$ so that $X \setminus S_{\omega_1} = P_{\gamma(\omega_1)}$ in this case, completing the proof of the lemma. $\blacksquare$

Let $\alpha, \beta \in X$ and we may assume that $\alpha < \beta$. Then we see that $\alpha \in P_{\gamma(\alpha)} = [\omega_0, \alpha]$ and $\beta \in S_\alpha$. Thus $P_{\gamma(\alpha)}$ and S_α are neighborhoods of α and β respectively. By Lemma 2.5, they are disjoint and this shows that X is Hausdorff, as required.

(b) $X \setminus \{\omega_1\}$ **is open but not σ-compact.** Let $X \setminus \{\omega_1\} = [\omega_0, \omega_1)$ and $\alpha \in [\omega_0, \omega_1)$. As we have proven in part (a) that $P_{\gamma(\alpha)} = [\omega_0, \gamma(\alpha))$ is a neighborhood of α. Hence $X \setminus \{\omega_1\}$ is open in X.

Assume that $X \setminus \{\omega_1\} = [\omega_0, \omega_1)$ was σ-compact. Let K be a compact subset of $[\omega_0, \omega_1)$. We know that the family

$$\{P_x = [\omega_0, x) \mid x < \omega_1\}$$

is an open cover of K so that there are finitely many indices $x_1, x_2, \ldots, x_k$ such that

$$K \subseteq P_{x_1} \cup P_{x_1} \cup \cdots \cup P_{x_k}.$$

Without loss of generality, we may assume that $x_1 < x_2 < \cdots < x_k$. Thus we have

$$K \subseteq [\omega_0, x_k) \subset [\omega_0, x_k].$$

Since $x_k < \omega_1$, our hypothesis shows that K is at most countable. Since $X \setminus \{\omega_1\}$ is σ-compact, it is a countable union of compact sets so that $X \setminus \{\omega_1\}$ is countable. However, we have

$$X = \left(X \setminus \{\omega_1\}\right) \cup \{\omega_1\}$$

which means that X is countable. Evidently, this result contradicts our hypothesis. Hence $X \setminus \omega_1$ is not σ-compact.

(c) **Every $f \in C(X)$ is constant on S_α for some $\alpha \neq \omega_1$.** Suppose that $x = f(\omega_1)$ and $B(x, \frac{1}{n}) = \{z \in \mathbb{C} \mid |z - x| < \frac{1}{n}\}$, where n is a positive integer. Since f is continuous on X and $B(x, \frac{1}{n})$ is open in $\mathbb{C}$, $f^{-1}\left(B(x, \frac{1}{n})\right)$ is an open subset of X and $\omega_1 \in f^{-1}\left(B(x, \frac{1}{n})\right)$. By the topology (2.116), *there exists* an $\alpha_1 \in X \setminus \{\omega_1\}$ such that

$$(\alpha_1, \omega_1] \subseteq f^{-1}\left(B\left(x, \frac{1}{n}\right)\right). \tag{2.120}$$

By Lemma 2.5, we note that $(\alpha_1, \omega_1]$ is uncountable, so a sequence $\{\alpha_n\}$ exists such that

$$\alpha_1 < \alpha_2 < \cdots < \omega_1. \tag{2.121}$$

Now the results (2.120) and (2.121) together give

$$S_{\alpha_n} \subseteq S_{\alpha_{n+1}} \subseteq f^{-1}\left(B\left(x, \frac{1}{n}\right)\right)$$

for all $n \in \mathbb{N}$. Before we proceed further, we need the following result:[n]

> **Lemma 2.6**
>
> Every well-ordered set A has the least upper bound property.

Proof of Lemma 2.6. Let B be a nonempty subset of A having an upper bound in A. Therefore, the set U of upper bounds of B is nonempty. Since $U \subseteq A$ and A is well-ordered, U has a least element, completing the proof of Lemma 2.6. $\blacksquare$

By the relation (2.121), we see that the nonempty subset $\{\alpha_n\}$ of X is bounded above by ω_1. Therefore, it follows from Lemma 2.6 that $\sup_{n \in \mathbb{N}}\{\alpha_n\}$ exists in X. Call it α. It is clear that $\alpha \leq \omega_1$ and

$$S_\alpha \subseteq S_{\alpha_n} \subseteq f^{-1}\left(B\left(x, \frac{1}{n}\right)\right) \tag{2.122}$$

---

[n] See [42, Exercise 1, p. 66]

for all $n = 1, 2, \ldots$. In fact, it is impossible to have $\alpha = \omega_1$ because the set

$$S_{\omega_1} = \{x \in Y \mid \omega_1 < x\}$$

is *uncountable*, where Y is the well-ordered set as described in the "**Construction**" in the question. As $\omega_1 = \alpha$, we have $S_{\omega_1} = S_\alpha \subseteq S_{\alpha_n}$ so that S_{α_n} is uncountable too, but it contradicts to the fact that α_n is a predecessor of ω_1. Hence we obtain $\alpha < \omega_1$ and we deduce from the set relations (2.122) that $f(S_\alpha) \subseteq B(x, \frac{1}{n})$ for all $n \in \mathbb{N}$ and this means that

$$f(S_\alpha) \subseteq \bigcap_{n=1}^{\infty} B\left(x, \frac{1}{n}\right) = \{x\}.$$

Now this is exactly our desired result.

(d) **The intersection of $\{K_n\}$ of uncountable compact subsets of X is also uncountable compact.** We first prove a lemma which indicates a relationship between the cardinality of a compact set K and the topological property of ω_1 in K.

> **Lemma 2.7**
>
> Suppose that K is a nonempty compact subset of X. Then K is uncountable if and only if ω_1 is a *limit point* of K.

Proof of Lemma 2.7. Let K be uncountable. By the proof of part (b), we know that K cannot lie inside $[\omega_0, \alpha]$ for every predecessor $\alpha < \omega_1$. Otherwise, K will be at most countable, a contradiction. Hence this implies that

$$K \cap (\alpha, \omega_1] \neq \varnothing$$

for every $\alpha < \omega_1$. By the definition (see [42, p. 97]), it follows that ω_1 is a limit point of K.

Conversely, suppose that ω_1 is a limit point of K. By [42, Theorem 17.9, p. 99], K must be infinite. Assume that K was countable. Then K has the following representation

$$K = \{\alpha_n\},$$

where $\alpha_1 \le \alpha_2 \le \cdots$. Since K is compact and X is Hausdorff, K is closed in X by Corollary (a) following Theorem 2.5. Thus it must be $\omega_1 \in K$, i.e., every neighborhood S_α of ω_1 satisfies

$$K \cap (S_\alpha \setminus \{\omega_1\}) \neq \varnothing.$$

Let $\alpha_N \in K \cap (S_\alpha \setminus \{\omega_1\})$ for some positive integer N, i.e., $\alpha < \alpha_N < \omega_1$. Recall that $\{\alpha_n\}$ is increasing, so we must have

$$\alpha_n \in S_\alpha$$

for all $n \ge N$. By the definition (see [42, p. 98]), $\{\alpha_n\}$ converges to ω_1. However, by an argument similar to the proof of part (c), this fact will imply the contradiction that S_{α_n} is uncountable. Hence K is uncountable and we complete the proof of the lemma. $\blacksquare$

Let's go back to the proof of part (d). Denote the intersection of $\{K_n\}$ by K. Since X is Hausdorff, each K_n is closed in X. Since $K \subseteq K_n$, Theorem 2.4 implies that K is compact. For each $n \in \mathbb{N}$, let $K'_n = K_1 \cap K_2 \cap \cdots \cap K_n$. It is easy to check that

$$\bigcap_{i=1}^{n} K'_i = \bigcap_{i=1}^{n} (K_1 \cap \cdots \cap K_i) = \bigcap_{i=1}^{n} K_i,$$

so we may assume that

$$K_1 \supseteq K_2 \supseteq \cdots . \tag{2.123}$$

Therefore, any finite subcollection of $\{K_n\}$ is nonempty and Theorem 2.6 shows that $K \neq \varnothing$. By Lemma 2.7, ω_1 is a limit point of every K_n. Then, using [42, Theorem 17.9, p. 99] again, it means that every neighborhood $(\alpha, \omega_1]$ of ω_1 contains infinitely many points of every K_n. Combining this fact and the sequence (2.123), every neighborhood $(\alpha, \omega_1]$ of ω_1 must contain infinitely many points of the compact set K. Hence ω_1 is a limit point of K and we finally conclude from Lemma 2.7 that K is uncountable.

(e) $\mathfrak{M}$ **is a σ-algebra containing all Borel sets in X.** Suppose that

$$\mathfrak{M} = \{E \subseteq X \,|\, \text{either } E \cup \{\omega_1\} \text{ or } E^c \cup \{\omega_1\} \text{ contains an uncountable compact set}\}.$$

- $\mathfrak{M}$ **is a σ-algebra.** We check Definition 1.3(a). In fact, it is obvious that $X \in \mathfrak{M}$ because X is itself compact uncountable. Let $E \in \mathfrak{M}$. Then either $E \cup \{\omega_1\}$ or $E^c \cup \{\omega_1\}$ contains an uncountable compact set. Since $(E^c)^c = E$, it must be true that $E^c \in \mathfrak{M}$.

  To verify Definition 1.3(a)(iii), let $E_n \in \mathfrak{M}$ and $E = \bigcup_{n=1}^{\infty} E_n$. If *there exists* an $n_0 \in \mathbb{N}$ such that $E_{n_0} \cup \{\omega_1\}$ contains an uncountable compact set K, then

$$K \subseteq E_{n_0} \cup \{\omega_1\} \subseteq \left(\bigcup_{n=1}^{\infty} E_n \right) \cup \{\omega_1\} = E \cup \{\omega_1\}$$

so that $E \in \mathfrak{M}$. Otherwise, *all* $E_n \cup \{\omega_1\}$ contain no uncountable compact sets. This forces that every $E_n^c \cup \{\omega_1\}$ contains an uncountable compact set K_n. We note that

$$E^c \cup \{\omega_1\} = \left(\bigcap_{n=1}^{\infty} E_n^c \right) \cup \{\omega_1\} = \bigcap_{n=1}^{\infty} \left(E_n^c \cup \{\omega_1\} \right),$$

so it is true that

$$K = \bigcap_{n=1}^{\infty} K_n \subseteq \bigcap_{n=1}^{\infty} \left(E_n^c \cup \{\omega_1\} \right) = E^c \cup \{\omega_1\}.$$

  Now the result of part (d) illustrates that K is uncountable compact, thus $E^c \in \mathfrak{M}$ which implies that $E \in \mathfrak{M}$.

- $\mathfrak{M}$ **contains all Borel sets in X.** If $\alpha < \omega_1$, then $(\alpha, \omega_1) \neq \varnothing$. Let $\beta \in (\alpha, \omega_1)$. Since $X = P_\beta \cup [\beta, \omega_1]$ and P_β is countable, $[\beta, \omega_1]$ must be uncountable. Since $[\beta, \omega_1] = X \setminus P_\beta$, $[\beta, \omega_1]$ is closed in X and Theorem 2.4 verifies that $[\beta, \omega_1]$ is compact. Furthermore, we note that

$$[\beta, \omega_1] \subseteq S_\alpha, \tag{2.124}$$

so these facts show that $S_\alpha \in \mathfrak{M}$.

  Similarly, for every $\alpha > \omega_0$, we have $X \setminus P_\alpha = [\alpha, \omega_1]$ and then $X \setminus P_\alpha \in \mathfrak{M}$. Since $\mathfrak{M}$ is a σ-algebra, $\mathfrak{M}$ also contains P_α. Hence $\mathfrak{M}$ contains the order topology τ of X which proves that $\mathscr{B}(X) \subseteq \mathfrak{M}$ as desired.

(f) **λ is a measure on $\mathfrak{M}$ but not regular.** Suppose that $\lambda : \mathfrak{M} \to [0, \infty]$ is defined by

$$
\lambda(E) = \begin{cases} 1, & \text{if } E \cup \{\omega_1\} \text{ contains an uncountable compact set;} \\[2mm] 0, & \text{if } E^c \cup \{\omega_1\} \text{ contains an uncountable compact set.} \end{cases} \tag{2.125}
$$

If $E \in \mathfrak{M}$, then we know from the definition of $\mathfrak{M}$ in part (e) that it is impossible for both $E \cup \{\omega_1\}$ and $E^c \cup \{\omega_1\}$ containing uncountable compact sets, so this function is well-defined.

- λ **is a measure on $\mathfrak{M}$.** Suppose that $\{E_i\}$ is a disjoint countable collection of members of $\mathfrak{M}$. We claim that *at most* one $E_i \cup \{\omega_1\}$ contains an uncountable compact set. To see this, assume that both $E_1 \cup \{\omega_1\}$ and $E_2 \cup \{\omega_1\}$ contained uncountable compact sets K_1 and K_2 respectively. Since $E_1 \cap E_2 = \varnothing$, we have

$$
\begin{aligned}
K_1 \cap K_2 &\subseteq \big(E_1 \cup \{\omega_1\}\big) \cap \big(E_2 \cup \{\omega_1\}\big) \\
&= \big(E_1 \cap \{\omega_1\}\big) \cup \big(E_2 \cap \{\omega_1\}\big) \cup \{\omega_1\} \\
&= \{\omega_1\}. \tag{2.126}
\end{aligned}
$$

However, part (d) tells us that $K_1 \cap K_2$ is uncountable so that the result (2.126) is impossible. This proves the claim.

By the previous analysis, if $E_1 \cup \{\omega_1\}$ is the *only* set containing an uncountable compact set K, then we get

$$
K \subseteq \bigcup_{i=1}^{\infty} \big(E_i \cup \{\omega_1\}\big) = \Big(\bigcup_{i=1}^{\infty} E_i \Big) \cup \{\omega_1\},
$$

so we follow from the definition (2.125) that

$$
\lambda \Big(\bigcup_{i=1}^{\infty} E_i \Big) = 1.
$$

On the other hand, we know from the facts $\lambda(E_1) = 1$ and $\lambda(E_i) = 0$ for $i = 2, 3, \ldots$ that

$$
\sum_{i=1}^{\infty} \lambda(E_i) = 1.
$$

Therefore, we have

$$
\lambda \Big(\bigcup_{i=1}^{\infty} E_i \Big) = \sum_{i=1}^{\infty} \lambda(E_i) \tag{2.127}
$$

in this case. Similarly, if there is *no* $E_i \cup \{\omega_1\}$ contains an uncountable compact set, then each $E_i^c \cup \{\omega_1\}$ contains an uncountable compact set K_i so that

$$
\lambda(E_i) = 0
$$

for every $i = 1, 2, \ldots$. Furthermore, we have

$$
K = \bigcap_{i=1}^{\infty} K_i \subseteq \bigcap_{i=1}^{\infty} \big(E_i^c \cup \{\omega_1\}\big) = \Big(\bigcap_{i=1}^{\infty} E_i^c \Big) \cup \{\omega_1\} = \Big(\bigcup_{i=1}^{\infty} E_i \Big)^c \cup \{\omega_1\}
$$

and part (d) ensures that K is uncountable compact. Therefore, we obtain

$$
\lambda \Big(\bigcup_{i=1}^{\infty} E_i \Big) = 0
$$

and then the formula (2.127) still holds in this case. Finally, since $\lambda(X) = 1 < \infty$. By Definition 1.18(a), λ is a positive measure.

- λ **is not regular.** For every $\alpha < \omega_1$, we know from the set relation (2.124) that $S_\alpha \cup \{\omega_1\}$ must contain an uncountable compact set, so the definition (2.125) gives that $\lambda(S_\alpha) = 1$. If $E = \{\omega_1\}$, then $E^c \cup \{\omega_1\} = \{\omega_1\}^c \cup \{\omega_1\} = X$ so that $\{\omega_1\} \in \mathfrak{M}$. Therefore, we derive from the definition (2.125) that

$$\lambda(\{\omega_1\}) = 0.$$

However, these facts give

$$\lambda(\{\omega_1\}) \neq \inf\{\lambda(S_\alpha) \,|\, \{\omega_1\} \subseteq S_\alpha\}.$$

By Definition 2.15, λ is *not* regular.

(g) **The validity of the integral.** Let $f \in C_c(X)$. By part (c), there exists an $\alpha_0 < \omega_1$ such that $f(x) = f(\omega_1)$ on S_{α_0}. Recall from part (e) that $S_{\alpha_0} \in \mathfrak{M}$. By the set relation (2.124), every $S_\alpha \cup \{\omega_1\}$ contains an uncountable compact set and thus, in particular, $\lambda(S_{\alpha_0}^c) = 0$. By this, we gain

$$\int_X f \,\mathrm{d}\lambda = \int_{S_{\alpha_0}^c} f \,\mathrm{d}\lambda + \int_{S_{\alpha_0}} f \,\mathrm{d}\lambda = \int_{S_{\alpha_0}} f \,\mathrm{d}\lambda = f(\omega_1)\lambda(S_{\alpha_0}) = f(\omega_1)$$

which is the required result.

(h) **The regular μ associates with the linear functional in part (f).** It is clear that $\Lambda : C_c(X) \to \mathbb{C}$ defined by

$$\Lambda(f) = f(\omega_1) \tag{2.128}$$

is a linear functional on $C_c(X)$. By Theorem 2.14 (The Riesz Representation Theorem) and Theorem 2.17(b), we have a unique regular positive Borel measure μ on X. By the result (2.128), we know that

$$\int_X f \,\mathrm{d}\mu = f(\omega_1).$$

By [51, Eqn. (1), p. 41], we have $\mu(V) = \sup\{\Lambda(f) \,|\, f \prec V\} = f(\omega_1)$ for every open set V in X. Thus this implies that

$$\mu(E) = \inf\{\mu(V) \,|\, E \subseteq V \text{ and } V \text{ is open}\} = f(\omega_1)$$

for every $E \in \mathfrak{M}$.

We have completed the proof of the problem.   ■

Remark 2.2

The measure considered in Problem 2.18 is called the **Dieudonné's measure**, see [10, Exampl 7.1.3, pp. 68, 69] and [17].

Problem 2.19

Rudin Chapter 2 Exercise 19.

Proof. Suppose that X is a compact metric space with metric ρ and Λ is a positive linear functional on $C_c(X)$, the space of all continuous complex functions on X with compact support. To begin with the construction of the class μ, we use Λ to define a set function μ^* on every open set V in X by

$$\begin{aligned}
\mu^*(V) &= \sup\{\Lambda(f) \mid f \prec V\} \\
&= \sup\{\Lambda(f) \mid f \in C_c(X),\, 0 \le f \le 1 \text{ on } X \text{ and } \operatorname{supp}(f) \subseteq V\}
\end{aligned} \qquad (2.129)$$

and for every *subset* $E \subseteq X$ that

$$\mu^*(E) = \inf\{\mu^*(V) \mid V \text{ is open in } X \text{ and } E \subseteq V\}. \qquad (2.130)$$

Now we are going to present the proof by quoting several facts (some are with proofs and some are not). In fact, the idea of the following proof is stimulated by Feldman's online article [20].

- **Fact 1: The set function μ^* is an outer measure.** We check the definition [47, p. 346].

  - Since $\operatorname{supp}(f) \subseteq \varnothing$ if and only if $\operatorname{supp}(f) = \varnothing$, we get from Definition 2.9 that $f \equiv 0$ on X which implies that

  $$\mu^*(\varnothing) = \sup\{\Lambda f \mid f \in C_c(X),\, 0 \le f \le 1 \text{ on } X \text{ and } \operatorname{supp}(f) \subseteq \varnothing\} = \Lambda(0) = 0.$$

  - Suppose that $E, F \subseteq X$ and $E \subseteq F$. Since any open set V containing F must also contain E, we obtain from the definition (2.130) that

  $$\begin{aligned}
  \mu^*(E) &= \inf\{\mu^*(W) \mid W \text{ is open in } X \text{ and } E \subseteq W\} \\
  &\le \inf\{\mu^*(V) \mid V \text{ is open in } X \text{ and } F \subseteq V\} \\
  &= \mu^*(F).
  \end{aligned}$$

  - The proof of the subadditivity

  $$\mu^*\left(\bigcup_{i=1}^{\infty} E_i\right) \le \sum_{i=1}^{\infty} \mu^*(E_i) \qquad (2.131)$$

  follows exactly the same as the proof of **Step I**.

  Hence μ^* is an outer measure on X.

- **Fact 2: $\mu^*(K) < \infty$ for every compact set $K \subseteq X$.** Since X is open in X, we follow from the definition (2.129) that $\mu^*(X) \le \Lambda(1)$. By **Fact 1**, we have $\mu^*(K) \le \mu^*(X) \le \Lambda(1)$ for every compact set $K \subseteq X$.

- **Fact 3: A topological result in metric spaces.** Now we need the following topological result about metric spaces which will be used in **Fact 4**:

  > **Lemma 2.8**
  >
  > Let X be a metric space with metric ρ and E a nonempty proper subset of X. Then the set $V = \{x \in X \mid \rho_E(x) < \epsilon\}$ is open in X, where $\epsilon > 0$ and $\rho_E(x)$ is defined in Problem 2.3.

Proof of Lemma 2.8. If $V = \varnothing$, then there is nothing to prove. Thus we assume that $V \neq \varnothing$ and pick $x_0 \in V$ so that $\rho_E(x_0) < \epsilon$. Then there exists a $\delta > 0$ such that $\epsilon - \delta > 0$ and $\rho_E(x_0) < \epsilon - \delta$. We consider the ball

$$B(x_0, \delta) = \{x \in X \mid \rho(x, x_0) < \delta\}$$

and we claim that $B(x_0, \delta) \subseteq V$. To see this, let $x \in B(x_0, \delta)$. For every $y \in E$, we have

$$\rho(x, y) \leq \rho(x, x_0) + \rho(x_0, y) < \delta + \rho(x_0, y)$$

and therefore we obtain

$$\begin{aligned}
\rho_E(x) &= \inf\{\rho(x, y) \mid y \in E\} \\
&\leq \inf\{\rho(x, x_0) + \rho(x_0, y) \mid y \in E\} \\
&< \delta + \inf\{\rho(x_0, y) \mid y \in E\} \\
&= \delta + \rho_E(x_0) \\
&< \epsilon.
\end{aligned}$$

In other words, $x \in V$. Since x is arbitrary, $B(x_0, \delta) \subseteq V$ and then V is open in X, proving Lemma 2.8.   ■

- **Fact 4: If V is open in X, then V is measurable with respect to μ^*.** We recall from [47, p. 347] that $V \subseteq X$ is measurable with respect to the outer measure μ^* if for every subset E of X, we have

$$\mu^*(E) = \mu^*(E \cap V) + \mu^*(E \cap V^c). \tag{2.132}$$

It suffices to show the inequality

$$\mu^*(E) \geq \mu^*(E \cap V) + \mu^*(E \cap V^c) - \epsilon \tag{2.133}$$

holds for every $\epsilon > 0$ because the other side follows directly from the subadditivity (2.131). By the definition (2.130), there is an open set E' such that $E \subseteq E'$ and

$$\mu^*(E) \geq \mu^*(E') - \frac{\epsilon}{2}. \tag{2.134}$$

By the property in **Fact 1**, we have

$$\mu^*(E \cap V) \leq \mu^*(E' \cap V) \quad \text{and} \quad \mu^*(E \cap V^c) \leq \mu^*(E' \cap V^c). \tag{2.135}$$

Thus it is easy to see from the inequalities (2.134) and (2.135) that one can obtain the inequality (2.133) if we can show the following result holds

$$\mu^*(E') \geq \mu^*(E' \cap V) + \mu^*(E' \cap V^c) - \frac{\epsilon}{2}. \tag{2.136}$$

To this end, we first find bounds of $\mu^*(E' \cap V)$ and $\mu^*(E' \cap V^c)$. Since $E' \cap V$ is open in X, the definition (2.129) implies the existence of a continuous function $f_1 : X \to [0, 1]$ such that $\operatorname{supp}(f_1) \subseteq E' \cap V$ and

$$\mu^*(E' \cap V) \leq \Lambda(f_1) + \frac{\epsilon}{8}. \tag{2.137}$$

Let $\epsilon \in \big(0, 8\Lambda(1)\big)$ and $\delta > 0$ be a constant such that $\delta \leq \epsilon[8\Lambda(1) - \epsilon]^{-1}$. Then we have

$$\frac{\delta}{1+\delta}\Lambda(1) \leq \frac{\epsilon}{8} \quad \text{and} \quad F_1 = \frac{f_1}{1+\delta}. \tag{2.138}$$

Recall the fact that the positivity of Λ implies the monotonicity of Λ. In particular, $\Lambda(f) \leq \Lambda(1)$ for every $f \prec E' \cap V$. Thus we apply the results (2.138) and the monotonicity of Λ to the inequality (2.137) to get

$$\mu^*(E' \cap V) \leq \frac{1}{1+\delta}\Lambda(f_1) + \frac{\delta}{1+\delta}\Lambda(f_1) + \frac{\epsilon}{8} \leq \Lambda(F_1) + \frac{\delta}{1+\delta}\Lambda(1) + \frac{\epsilon}{8} = \Lambda(F_1) + \frac{\epsilon}{4}. \quad (2.139)$$

We replace E by the closed set $(E' \cap V)^c$ in Lemma 2.8 to get the open set U. Given $\delta > 0$. Since f_1 is uniformly continuous on X, there exists a $\eta > 0$ such that $|f_1(x) - f_1(y)| \leq \delta$ for all $x, y \in X$ with $\rho(x, y) < \eta$. Recall that f_1 vanishes on the closed set $(E' \cap V)^c$, so if we take $y \in (E' \cap V)^c$, then we have

$$f_1(x) \leq \delta$$

for all $x \in X$ with $\rho(x, y) < \eta$. Since $\rho_{(E' \cap V)^c}(x) < \eta$, Lemma 2.8 ensures that the set

$$U = \{x \in X \mid \rho_{(E' \cap V)^c}(x) < \eta\}$$

is open in X. In other words, we have established from **Fact 3** that there exists an open set U such that

$$U \supset (E' \cap V)^c \supseteq V^c \quad \text{and} \quad f_1(x) \leq \delta \text{ on } U. \quad (2.140)$$

Since $E' \cap U$ is open in X, the definition (2.129) again shows that we can find a continuous function $f_2 : X \to [0, 1]$ such that $\operatorname{supp}(f_2) \subseteq E' \cap U$ and furthermore, by using similar argument as in the proof of the inequality (2.139), we obtain that

$$\mu^*(E' \cap V^c) \leq \mu^*(E' \cap U) \leq \Lambda(f_2) + \frac{\epsilon}{4} \leq \Lambda(F_2) + \frac{\epsilon}{4}, \quad (2.141)$$

where $F_2 = \frac{f_2}{1+\delta}$.

Now we have found the bounds (2.139) and (2.141) of $\mu^*(E' \cap V)$ and $\mu^*(E' \cap V^c)$ respectively. Next, we want to show that $F_3 \in C_c(X)$, where $F_3 = F_1 + F_2$. To this end, we recall the facts that $F_1, F_2 : X \to [0, \frac{1}{1+\delta}]$ because $0 \leq f_1 \leq 1$ and $0 \leq f_2 \leq 1$. Furthermore, we note from the set relation (2.140) that

$$V \cap U \neq \varnothing.$$

Therefore, we have the following facts:

- Since $\operatorname{supp}(f_1) \subseteq E' \cap V$, $\operatorname{supp}(F_1) \subseteq E' \cap V$, i.e., $F_1(x) = 0$ on $(E' \cap V)^c$. Therefore, we conclude that

$$F_3 = F_2 \leq \frac{1}{1+\delta} \leq 1$$

  on $(E' \cap V)^c$. Similarly, since $\operatorname{supp}(F_2) \subseteq E' \cap U$, we have $F_2(x) = 0$ on $(E' \cap U)^c$ and then

$$F_3 = F_1 \leq \frac{1}{1+\delta} \leq 1$$

  on $(E' \cap U)^c$.

- On $(E' \cap V) \cap U$, we immediately follow from the inequality in (2.140) that $f_1(x) \leq \delta$ on $(E' \cap V) \cap U \subseteq U$. It is clear that $f_2(x) \leq 1$ on $(E' \cap V) \cap U$, so these facts imply that

$$F_3 = F_1 + F_2 = \frac{f_1}{1+\delta} + \frac{f_2}{1+\delta} \leq \frac{\delta}{1+\delta} + \frac{1}{1+\delta} = 1.$$

- On $(E' \cap V) \setminus U$, we have $F_1 \leq 1$ and $F_2 = 0$ so that $F_3 \leq 1$.

    – On $U \setminus (E' \cap V)$, we have $F_1 = 0$ and $F_2 \leq 1$ so that $F_3 \leq 1$.

To have a better understanding of the above facts, we can draw some pictures. For examples, Figure 2.3(a) shows the sets V, E' and $E' \cap V$, the shaded blue part in Figure 2.3(b) indicates the closed set $(E' \cap V)^c$ and the part inside the circle in Figure 2.4 is the set $(E' \cap V)^c \setminus U$.

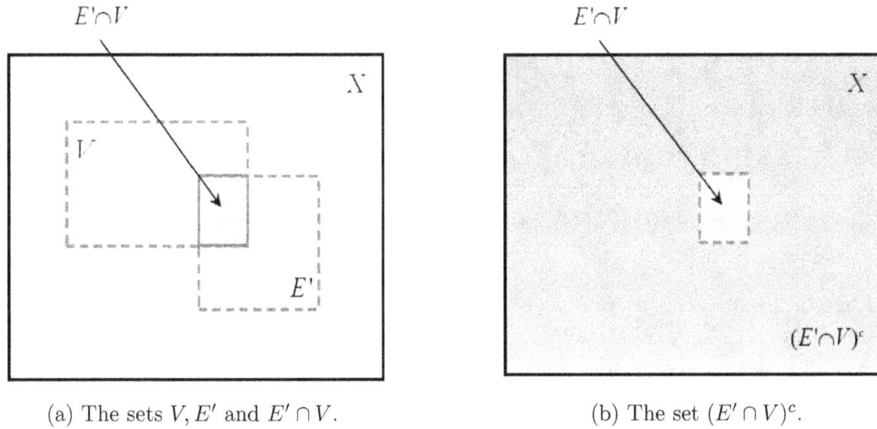

(a) The sets V, E' and $E' \cap V$.   (b) The set $(E' \cap V)^c$.

Figure 2.3: The pictures of $V, E', E' \cap V$ and $(E' \cap V)^c$.

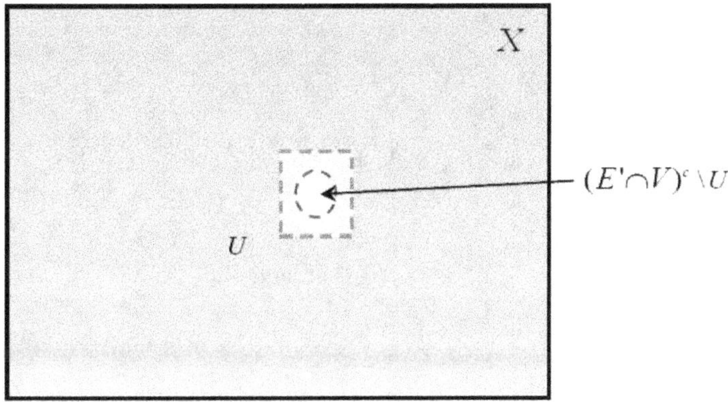

Figure 2.4: The set $(E' \cap V)^c \setminus U$.

    Thus we have shown that $0 \leq F_3(x) \leq 1$ on X. Next, since F_1 and F_2 are continuous on X, F_3 is also continuous on X. By the definition, we know that $x_0 \in \mathrm{supp}\,(F_3)$ if and only if $x_0 \in \mathrm{supp}\,(F_1) \subseteq E' \cap V$ or $x_0 \in \mathrm{supp}\,(F_2) \subseteq E' \cap U$. In addition, we note that $(E' \cap V) \cup (E' \cap U) \subseteq E'$. On the other hand, if $y \in E'$ but $y \notin E' \cap V$, then we have $y \in (E' \cap V)^c$ and we follow from the set relation (2.140) that $y \in U$. Thus this implies that $y \in E' \cap U$ and so
$$E' = (E' \cap V) \cup (E' \cap U).$$

In other words, it gives $x_0 \in \mathrm{supp}\,(F_3)$ if and only if $x_0 \in \mathrm{supp}\,(F_1) \cup \mathrm{supp}\,(F_2) \subseteq E'$, i.e., $\mathrm{supp}\,(F_3) \subseteq E'$.

    Finally, it yields from the inequalities (2.139) and (2.141) as well as the definition (2.129) that

$$\mu^*(E' \cap V) + \mu^*(E' \cap V^c) \leq \Lambda(F_1) + \Lambda(F_2) + \frac{\epsilon}{2} = \Lambda(F_3) + \frac{\epsilon}{2} \leq \mu^*(E') + \frac{\epsilon}{2}$$

which is exactly the inequality (2.136). Hence we obtain the desired result that (2.132).

- **Fact 5: The set $\mathfrak{M}^*$ of all measurable sets (with respect to μ^*) is a σ-algebra and $\mu = \mu^*|_{\mathscr{B}(X)}$ is a (complete) measure.** The first assertion follows directly from Carathéodory's Theorem [47, Theorem 8, p. 349]. By **Fact 4**, we have

$$\mathscr{B}(X) \subseteq \mathfrak{M}^*$$

which is exactly **Step VII**, where $\mathscr{B}(X)$ denotes the set of all Borel sets in X. By Carathéodory's Theorem again, it shows that μ is a (complete) measure on $\mathscr{B}(X)$ and this is the same as **Step IX**. Consequently, **Steps III, IV, V, VI** and **VIII** can all be skipped.[o]

- **Fact 6: Deduction of parts (b) and (d).** Let K be a compact set of X. Since X is a metric space, K is also Hausdorff. By Corollary (a) following Theorem 2.5, we know that K is closed in X and so $K \in \mathscr{B}(X)$. Therefore, we deduce from **Facts 2** and **5** that

$$\mu(K) = \mu^*(K) \leq \mu^*(X) \leq \Lambda(1)$$

which is exactly part (b) of Theorem 2.14 (The Riesz Representation Theorem).

If E is open in X, then $E \in \mathscr{B}(X)$. Thus we observe from the equation (2.132) and then the definition (2.130) that

$$
\begin{aligned}
\mu(E) &= \mu^*(E) \\
&= \mu^*(X) - \mu^*(E^c) \\
&= \mu^*(X) - \inf\{\mu^*(V) \,|\, V \text{ is open in } X \text{ and } E^c \subseteq V\} \\
&= \mu^*(X) - \inf\{\mu^*(X) - \mu^*(V^c) \,|\, V^c \text{ is closed in } X \text{ and } V^c \subseteq E\} \\
&= \sup\{\mu^*(V^c) \,|\, V^c \text{ is closed in } X \text{ and } V^c \subseteq E\} \\
&= \sup\{\mu(V^c) \,|\, V^c \text{ is closed in } X \text{ and } V^c \subseteq E\}.
\end{aligned}
\tag{2.142}
$$

Since X is compact, V^c is compact by Theorem 2.4 so that the expression (2.142) can be expressed as

$$\mu(E) = \sup\{\mu(K) \,|\, K \text{ compact and } K \subseteq E\}$$

which is the same as part (d) of Theorem 2.14 (The Riesz Representation Theorem).

- **Fact 7: Deduction of part (a).** As Rudin pointed out in **Step X** that it suffices to prove

$$\Lambda(f) \leq \int_X f \, d\mu \tag{2.143}$$

holds for $f \in C_c(X)$ because the opposite inequality follows by changing the sign of f. In fact, the proof of our inequality (2.143) for compact metric X is essentially the same as that of **Step X**, except that Rudin applied **Step II** to show that

$$\mu(K) \leq \sum \Lambda(h_i),$$

---

[o]We cannot say at this stage that **Step II** can be omitted because the proof of **Step X** still needs it, but we will see very soon in **Fact 7** that it is eventually allowable to do so.

where $h_i \prec V_i$ and $\sum h_i = 1$ on K.[P] However, this kind of argument can be replaced completely by using **Fact 6** that $\mu(K) \leq \Lambda(1)$ holds for every compact set $K \subseteq X$. Therefore, we finally arrive at the expected inequality (2.143) and furthermore, **Step II** can also be skipped.

Hence we conclude that **Steps II** to **IX** can be replaced by a simpler argument (outer measure in the sense of Carathéodory). In other words, only **Steps I** and **X** should be kept. Of course, the proof of the uniqueness of μ cannot be omitted. This completes the proof of the problem. ■

Remark 2.3

(a) We remark that a Radon measure is a measure μ defined on the σ-algebra $\mathfrak{M}$ containing all Borel sets $\mathscr{B}(X)$ of the Hausdforff space X and it satisfies $\mu(K) < \infty$ for all compact subsets K, outer regular on $\mathfrak{M}$ and inner regular on open sets in X, see [22, p. 212] or [47, p. 455]. Hence the unique positive measure μ in Theorem 2.14 (The Riesz Representation Theorem) is in fact a Radon measure when X is Hausdforff.

(b) You are also advised to read the paper [52] for a similar proof of Problem 2.19.

2.6 Miscellaneous Problems on L^1 and Other Properties

Problem 2.20

Rudin Chapter 2 Exercise 20.

Proof. Let $f = \sup_{n} f_n : [0,1] \to [0,\infty]$. By Definition 1.30, $f \notin L^1$ on $[0,1]$ is equivalent to

$$\int_0^1 |f(x)|\,\mathrm{d}x = \infty.$$

In fact, the inspiration of the construction of $\{f_n\}$ comes from the idea of Problem 2.9. For each positive integer n, we consider the continuous functions $g_n : [-1,1] \to [0,\infty)$ given by

$$g_n(x) = \begin{cases} 0, & \text{if } x \notin [-\frac{1}{n^2}, \frac{1}{n^2}]; \\ n - n^3|x|, & \text{if } x \in [-\frac{1}{n^2}, \frac{1}{n^2}]. \end{cases}$$

Next, we define $f_n : [0,1] \to [0,\infty)$ by

$$f_n(x) = g_n\left(x - \frac{1}{n}\right) = \begin{cases} 0, & \text{if } x \notin [\frac{1}{n} - \frac{1}{n^2}, \frac{1}{n} + \frac{1}{n^2}]; \\ n - n^2|nx - 1|, & \text{if } x \in [\frac{1}{n} - \frac{1}{n^2}, \frac{1}{n} + \frac{1}{n^2}]. \end{cases} \tag{2.144}$$

Recall that $f_n(x)$ is the tent function centered at $\frac{1}{n}$ and the graph of it is an isosceles triangle with height n and base $\frac{2}{n^2}$ so that each f_n is continuous on $[0,1]$ and its area is $\frac{1}{n}$ which implies that

$$\int_0^1 f_n(x)\,\mathrm{d}x = \frac{1}{n} \to 0$$

[P]See **Step X** for the meanings of the notations used here.

as $n \to \infty$. Furthermore, if $x = 0$, then $x \notin [\frac{1}{n} - \frac{1}{n^2}, \frac{1}{n} + \frac{1}{n^2}]$ for all $n \in \mathbb{N}$. and thus

$$f_n(0) = 0$$

for all $n \in \mathbb{N}$ by the definition (2.144). If $x \in (0, 1]$, then there is a positive integer N such that $\frac{1}{n} + \frac{1}{n^2} < x$ for all $n \geq N$. In other words, we have $x \notin [\frac{1}{n} - \frac{1}{n^2}, \frac{1}{n} + \frac{1}{n^2}]$ for all $n \geq N$. Therefore, the definition (2.144) certainly shows that

$$f_n(x) \to 0$$

as $n \to \infty$ for all $x \in (0, 1]$.

Now it remains to show that $f \notin L^1$ on $[0, 1]$. To this end, we study the behaviour of f as $x \to 0$. For every $n \in \mathbb{N}$, we consider every x in $[\frac{1}{n} - \frac{1}{n^2}, \frac{1}{n} + \frac{1}{n^2}]$. If n is sufficiently large enough, we have

$$x = \frac{1}{n} + o\left(\frac{1}{n}\right), \tag{2.145}$$

where $o(x)$ is the little-o notation.[q]   Therefore, it follows from the definition (2.144) and the relation (2.145) that

$$f_n(x) = n + o(1) = \frac{1}{x} + o(1) \tag{2.146}$$

for large enough n and small positive x. By the definition and the expression (5.102), we have

$$f(x) = \sup_n f_n(x) \geq f_n(x) = \frac{1}{x} + o(1) \tag{2.147}$$

for small positive x. Obviously, $\frac{1}{x} \notin L^1$ on $[0, 1]$, so we conclude from the estimate that (2.147) that $f \notin L^1$ on $[0, 1]$. This completes the proof of the problem. ∎

Problem 2.21

Rudin Chapter 2 Exercise 21.

Proof. Let $\alpha \in f(X) \subseteq \mathbb{R}$ and

$$E_\alpha = \{x \in X \mid f(x) \geq \alpha\} = f^{-1}\big([\alpha, \infty)\big).$$

By the definition, it is clear that E_α is nonempty. Furthermore, since

$$E_\alpha = X \setminus f^{-1}\big((-\infty, \alpha)\big) \subseteq X,$$

the set E_α is closed in X. Now we consider the collection $\{E_\alpha\}$ of subsets of X. Suppose that $\alpha_1, \alpha_2, \ldots, \alpha_n \in f(X)$ and we define $\alpha = \max(\alpha_1, \alpha_2, \ldots, \alpha_n)$. Then we have

$$\bigcap_{k=1}^{n} E_{\alpha_k} = \bigcap_{k=1}^{n} \{x \in X \mid f(x) \geq \alpha_k\} = E_\alpha \neq \varnothing.$$

Therefore, $\{E_\alpha\}$ has the *finite intersection property* and then since X is compact, we conclude that[r]

$$\bigcap_{\alpha} E_\alpha \neq \varnothing.$$

Thus suppose that $x_0 \in \bigcap_{\alpha} E_\alpha$ which means that $f(x_0) \geq \alpha$ for all $\alpha \in f(X)$, i.e., $f(x_0) \geq f(x)$ for all $x \in X$. Hence this shows that f attains its maximum at some point of X, completing the proof of the problem. ∎

---

[q] Recall that $g(n) = o(f(n))$ means $\frac{g(n)}{f(n)} \to 0$ as $n \to \infty$.

[r] See, for instances, [42, Theorem 26.9, p. 169] or [64, Problem 4.23, p. 43].

Remark 2.4

Problem 2.21 can be treated as the **Extreme Value Theorem for upper semicontinuous functions**. Similarly, we can show that if $f : X \to \mathbb{R}$ is lower semicontinuous and X is compact, then f attains its minimum at some point of X.

Problem 2.22

Rudin Chapter 2 Exercise 22.

Proof. By the definition and the triangle inequality, for all $p \in X$, we have

$$g_n(x) \le f(p) + nd(x,p) \le [f(p) + nd(y,p)] + nd(x,y).$$

In other words, $g_n(x) - nd(x,y)$ is a lower bound of $\{f(p) + nd(y,p) \,|\, p \in X\}$. Thus, by the definition of infimum, we must have $g_n(x) - nd(x,y) \le g_n(y)$ or equivalently,

$$g_n(x) - g_n(y) \le nd(x,y). \tag{2.148}$$

Likewise, we have

$$g_n(y) - g_n(x) \le nd(x,y). \tag{2.149}$$

Therefore, we conclude from the inequalities (2.148) and (2.149) that

$$|g_n(x) - g_n(y)| \le nd(x,y).$$

Since $f(x) \ge 0$ on X and $d(x,y) \ge 0$ for every $x, y \in X$, we have $g_n(x) \ge 0$ for all positive integers n. Fix $x \in X$, we observe that

$$f(p) + nd(x,p) \ge f(p) + (n-1)d(x,p) \ge g_{n-1}(x)$$

for every $p \in X$ and $n = 2, 3, \ldots$. In other words, this implies that

$$g_n(x) \ge g_{n-1}(x)$$

for every $x \in X$ and $n = 2, 3, \ldots$. Besides, it is trivial that $g_n(x) \le f(x)$ holds for all $n \in \mathbb{N}$ if $f(x) = \infty$. Without loss of generality, we may assume that $f(x) < \infty$. In this case, since $f(p) \le f(p) + nd(x,p)$ for all $p \in X$, if we take $p = x$ in the definition of g_n, then we have

$$g_n(x) \le f(x)$$

for every $n = 1, 2, \ldots$. This proves property (ii).

To prove property (iii), we need the following lemma:

Lemma 2.9

A function $f : X \to \mathbb{R}$ is lower semicontinuous on X if and only if given $x \in X$, for every $\{x_n\} \subseteq X \setminus \{x\}$ converging to x and for every $\epsilon > 0$, there exists a positive integer N such that $n \ge N$ implies

$$f(x) < f(x_n) + \epsilon. \tag{2.150}$$

Proof of Lemma 2.9. For every $\alpha \in \mathbb{R}$, denote $E_\alpha = \{x \in X \mid f(x) \leq \alpha\}$. Recall that $f^{-1}((\alpha, \infty))$ is open in X if and only if $X \setminus f^{-1}((\alpha, \infty)) = E_\alpha$ is closed in X. In other words, f is lower semicontinuous on X if and only if E_α is closed in X.

Let E_α be closed in X. Assume that the inequality (2.150) was invalid. Then it means that there exists a $x_0 \in X$ and a sequence $\{x_n\}$ in X converging to x_0 such that for some $\epsilon > 0$, the inequality

$$f(x_0) - \epsilon \geq f(x_{n_k}) \tag{2.151}$$

holds for *infinitely many* k. Take $\alpha \in (f(x_0) - \epsilon, f(x_0))$. On the one hand, since $f(x_0) > \alpha$, $x_0 \notin E_\alpha$. On the other hand, we gain from the inequality (2.151) that $f(x_{n_k}) < \alpha$ for infinitely many k and this implies that

$$\{x_{n_k}\} \subseteq E_\alpha.$$

Since the set E_α is closed and $x_{n_k} \to x_0$ as $k \to \infty$, we have $x_0 \in E_\alpha$, a contradiction. Hence the inequality (2.150) must hold for a lower semicontinuous function f.

Conversely, we prove that the inequality (2.150) implies E_α is closed in X by showing that E_α contains all its limit points. To this end, pick $\alpha \in \mathbb{R}$ such that $E_\alpha \neq \varnothing$. Let $\{x_n\} \subseteq E_\alpha \setminus \{x\}$ and $x_n \to x$ as $n \to \infty$. Given $\epsilon > 0$. By the inequality (2.150), there exists a positive integer N such that

$$f(x) - \epsilon < f(x_n) \tag{2.152}$$

for all $n \geq N$. Since $f(x_n) \leq \alpha$, we deduce immediately from the inequality (2.152) that

$$f(x) < \alpha + \epsilon.$$

Since ϵ is arbitrary, we must have $f(x) \leq \alpha$ or equivalently $x \in E_\alpha$. This ends the proof of the lemma. $\blacksquare$

(a) Similarly, a function $f : X \to \mathbb{R}$ is upper semicontinuous on X if and only if given $x \in X$, for every $\{x_n\} \subseteq X \setminus \{x\}$ converging to x and for every $\epsilon > 0$, there exists a positive integer N such that $n \geq N$ implies $f(x) > f(x_n) - \epsilon$.

(b) Except the equivalent definition stated in Lemma 2.9, a lower semicontinuous function f on X can also be reformulated in terms of the lower limit of f: $\liminf\limits_{y \to x} f(y) \geq f(x)$ for every $y \to x$ in $X \setminus \{x\}$, see [61, pp. 69, 70] and [65, Definition 3.62, p. 64].

Now it is time to go back to the proof of the problem. On the one hand, since f is lower semicontinuous on X, Lemma 2.9 implies that for every $\epsilon > 0$, there exists a $\delta > 0$ such that

$$f(y) > f(x) - \epsilon \tag{2.153}$$

for all $y \in B(x, \delta)$. In this case, we always have

$$f(y) + nd(x, y) > f(x) - \epsilon \tag{2.154}$$

for every positive integer n. Since $\epsilon > 0$ is arbitrary, the inequality (2.154) becomes

$$f(y) + nd(x, y) \geq f(x). \tag{2.155}$$

On the other hand, we consider $y \notin B(x, \delta)$ so that $d(x, y) \geq \delta$. It is clear that $n\delta - \epsilon > 0$ holds for large enough n,[s] thus we get from the inequality (2.153) that

$$f(y) + nd(x, y) \geq f(y) + n\delta > f(x) - \epsilon + n\delta > f(x)$$

for large enough n. Therefore, we conclude that the inequality (2.155) always holds for all $y \in X$ and for all large enough n. By the definition of limit inferior, we have

$$\liminf_{n \to \infty} g_n(x) \geq f(x). \tag{2.156}$$

By property (ii), we see that

$$\limsup_{n \to \infty} g_n(x) \leq f(x) \tag{2.157}$$

holds for every $x \in X$. Hence it deduces from the inequalities (2.156) and (2.157) that

$$\lim_{n \to \infty} g_n(x) = f(x)$$

on X.

By property (i), each g_n is continuous on X. Next, property (ii) says that $\{g_n\}$ is increasing and bounded by f. Finally, property (iii) ensures that $\{g_n\}$ converges to f on X pointwisely. This ends the proof of the problem. $\blacksquare$

Remark 2.6

With the aid of Lemma 2.9, we can prove Problem 2.1(b) easily. In fact, given $x \in X$, for every $\{x_n\} \subseteq X \setminus \{x\}$ converging to x and for every $\epsilon > 0$, there exists a positive integer N_1 such that $n \geq N_1$ implies that

$$f_1(x) < f_1(x_n) + \frac{\epsilon}{2}. \tag{2.158}$$

Similarly, there exists a positive integer N_2 such that $n \geq N_2$ implies

$$f_2(x) < f_2(x_n) + \frac{\epsilon}{2}. \tag{2.159}$$

Take $N = \max(N_1, N_2)$. If $n \geq N$, then we derive from the inequalities (2.158) and (2.159) that

$$f_1(x) + f_2(x) < f_1(x_n) + f_2(x_n) + \epsilon.$$

Hence it follows from Lemma 2.9 that f is lower semicontinuous on X.

Problem 2.23

Rudin Chapter 2 Exercise 23.

Proof. Let $\mathbf{x} \in \mathbb{R}^k$ and $f : \mathbb{R}^k \to \mathbb{R}$ be defined by

$$f(\mathbf{x}) = \mu(V + \mathbf{x}).$$

Since $V + \mathbf{x}$ is just a translation of V, it is also open in $\mathbb{R}^k$, i.e., $V + \mathbf{x} \in \mathscr{B}$.

---

[s]Obviously, the n depends on δ and ϵ.

We first construct a counter example which is not upper semicontinuous or continuous. By Example 1.20(b), we consider the unit mass concentrated at $\mathbf{0}$, i.e.,

$$\mu(E) = \begin{cases} 1, & \text{if } \mathbf{0} \in E; \\ 0, & \text{if } \mathbf{0} \notin E. \end{cases}$$

If $V = B(\mathbf{0}, 1)$, then we have $V + \mathbf{x} = B(\mathbf{x}, 1)$ and

$$\mu(V + \mathbf{x}) = \begin{cases} 1, & \text{if } \mathbf{0} \in B(\mathbf{x}, 1); \\ 0, & \text{if } \mathbf{0} \notin B(\mathbf{x}, 1) \end{cases} = \begin{cases} 1, & \text{if } |\mathbf{x}| < 1; \\ 0, & \text{if } |\mathbf{x}| \geq 1. \end{cases} \tag{2.160}$$

Thus it establishes from the result (2.160) that

$$f^{-1}\big((-\infty, 1)\big) = \{\mathbf{x} \in \mathbb{R}^k \,|\, f(\mathbf{x}) < 1\} = \{\mathbf{x} \in \mathbb{R}^k \,|\, \mu(V + \mathbf{x}) < 1\} = \mathbb{R}^k \setminus B(\mathbf{x}, 1)$$

which is closed in $\mathbb{R}^k$. By Definition 2.8, f is *not* upper semicontinuous or continuous.

However, we claim that f is always lower semicontinous. Let $\{\mathbf{x}_n\} \subseteq \mathbb{R}^k \setminus \{\mathbf{x}\}$ be such that $\mathbf{x}_n \to \mathbf{x}$ as $n \to \infty$. Recall that both $V + \mathbf{x}$ and $V + \mathbf{x}_n$ are open in $\mathbb{R}^k$. We claim that $\mathbf{y} \in V + \mathbf{x}$ implies that $\mathbf{y} \in V + \mathbf{x}_n$ for all but finitely many n. Let $\mathbf{y} = \mathbf{a} + \mathbf{x}$ for some $\mathbf{a} \in V$. Since V is open in $\mathbb{R}^k$, there exists a $\epsilon > 0$ such that

$$B(\mathbf{a}, \epsilon) \subseteq V.$$

By the hypothesis, there is a positive integer N such that $n \geq N$ implies

$$\mathbf{x}_n - \mathbf{x} = \frac{\epsilon}{2} \quad \text{or} \quad \mathbf{x}_n - \mathbf{x} = -\frac{\epsilon}{2}.$$

In the first case, we have

$$\mathbf{y} = \mathbf{a} + \mathbf{x} = \mathbf{a} - \frac{\epsilon}{2} + \mathbf{x} + \frac{\epsilon}{2} = \mathbf{a} - \frac{\epsilon}{2} + \mathbf{x}_n. \tag{2.161}$$

Since $\mathbf{a} - \frac{\epsilon}{2} \in V$, the expression (2.161) guarantees that $\mathbf{y} \in V + \mathbf{x}_n$ for all $n \geq N$. The other case can be done similarly, so we omit the details here. This proves our claim.

Now if $\chi_{V+\mathbf{x}}(\mathbf{y}) = 1$, then we have $\chi_{V+\mathbf{x}_n}(\mathbf{y}) = 1$ for all but finitely many n. In other words, it means that

$$\liminf_{n \to \infty} \chi_{V+\mathbf{x}_n}(\mathbf{y}) \geq \chi_{V+\mathbf{x}}(\mathbf{y}) \tag{2.162}$$

for every $\mathbf{y} \in \mathbb{R}^k$. By Proposition 1.9(d), each $\chi_{V+\mathbf{x}_n} : \mathbb{R}^k \to [0, \infty]$ is measurable, so we may apply Theorem 1.28 (Fatou's Lemma) to conclude that

$$\liminf_{n \to \infty} \left(\int_{\mathbb{R}^k} \chi_{V+\mathbf{x}_n} \, d\mu \right) \geq \int_{\mathbb{R}^k} \left(\liminf_{n \to \infty} \chi_{V+\mathbf{x}_n} \right) d\mu. \tag{2.163}$$

By Proposition 1.24(f), we have

$$\int_{\mathbb{R}^k} \chi_{V+\mathbf{x}_n} \, d\mu = \int_{V+\mathbf{x}_n} d\mu = \mu(V + \mathbf{x}_n). \tag{2.164}$$

In addition, it follows from the inequality (2.162) and the application of Proposition 1.24(f) again that

$$\int_{\mathbb{R}^k} \left(\liminf_{n \to \infty} \chi_{V+\mathbf{x}_n} \right) \mathrm{d}\mu \geq \int_{\mathbb{R}^k} \chi_{V+\mathbf{x}}(\mathbf{y}) \, \mathrm{d}\mu = \int_{V+\mathbf{x}} \mathrm{d}\mu = \mu(V + \mathbf{x}). \tag{2.165}$$

Thus, by substituting the results (2.164) and (2.165) into the inequality (2.163), we achieve that

$$\liminf_{n \to \infty} \mu(V + \mathbf{x}_n) \geq \mu(V + \mathbf{x})$$

or equivalently,

$$\liminf_{\mathbf{y} \to \mathbf{x}} \mu(V + \mathbf{y}) \geq \mu(V + \mathbf{x})$$

for every $\mathbf{y} \to \mathbf{x}$ in $\mathbb{R}^k \backslash \{\mathbf{x}\}$. Hence we conclude from Remark 2.5(b) that f is lower semicontinuos on $\mathbb{R}^k$, completing the proof of the problem. ∎

Problem 2.24

Rudin Chapter 2 Exercise 24.

Proof. Suppose that S_1 and S_2 are the sets of simple functions and step functions on $\mathbb{R}$ respectively. We divide the proof into several steps.

- **Step 1: S_1 is dense in $L^1(\mathbb{R})$.** Let $f = f^+ - f^1$. By the Corollary (b) following Theorem 1.14, both f^+ and f^- are measurable. Since $f^+ : \mathbb{R} \to [0, \infty]$, we follow from Theorem 1.17 (The Simple Function Approximation Theorem) that there is a sequence $\{s_n\}$ of nonnegative increasing simple functions such that for every $x \in \mathbb{R}$, $s_n(x) \to f^+(x)$ as $n \to \infty$. Thus Theorem 1.26 (Lebesgue's Monotone Convergence Theorem) implies that

$$\int_{-\infty}^{\infty} s_n \, \mathrm{d}x \to \int_{-\infty}^{\infty} f^+ \, \mathrm{d}x \tag{2.166}$$

  as $n \to \infty$. It is clear that $|f^+(x) - s_n(x)| = f^+(x) - s_n(x)$ for every $x \in \mathbb{R}$, so we have[t]

$$\int_{-\infty}^{\infty} |f^+(x) - s_n(x)| \, \mathrm{d}x = \int_{-\infty}^{\infty} [f^+(x) - s_n(x)] \, \mathrm{d}x$$
$$= \int_{-\infty}^{\infty} f^+(x) \, \mathrm{d}x - \int_{-\infty}^{\infty} s_n(x) \, \mathrm{d}x. \tag{2.167}$$

  If we apply the limit (2.166) to the expression (2.167), then we gain

$$\lim_{n \to \infty} \int_{-\infty}^{\infty} |f^+(x) - s_n(x)| \, \mathrm{d}x = 0.$$

  Likewise, the same is true for f^-, so we have

$$\lim_{n \to \infty} \int_{-\infty}^{\infty} |f^-(x) - t_n(x)| \, \mathrm{d}x = 0$$

  where $\{t_n\}$ is a sequence of nonnegative increasing simple functions such that for every $x \in \mathbb{R}$, $t_n(x) \to f^-(x)$ as $n \to \infty$. Hence $\{g_n = s_n - t_n\}$ is a sequence of simple functions such that

$$\int_{-\infty}^{\infty} |f - g_n| \, \mathrm{d}x = \int_{-\infty}^{\infty} |f^+ - f^- - s_n + t_n| \, \mathrm{d}x$$

---

[t] Since $|f^+(x)| \leq |f(x)|$ on $\mathbb{R}$ and $f \in L^1(\mathbb{R})$, we have $f^+ \in L^1(\mathbb{R})$. Thus this makes the second equality holds because the right-hand side is *not* in the form $\infty - \infty$.

$$\le \int_{-\infty}^{\infty} |f^+ - s_n|\,\mathrm{d}x + \int_{-\infty}^{\infty} |f^- - t_n|\,\mathrm{d}x$$

$$\to 0$$

as $n \to \infty$.

- **Step 2: S_2 is dense in S_1.** To this end, it suffices to prove that for every measurable set $E \subset \mathbb{R}$ with $m(E) < \infty$ and every $\epsilon > 0$, there exists a step function g on $\mathbb{R}$ such that

$$\int_{-\infty}^{\infty} |\chi_E - g|\,\mathrm{d}x < \epsilon. \tag{2.168}$$

By Theorem 2.20(b), there exists an open set V in $\mathbb{R}$ such that $E \subseteq V$ and $m(V \setminus E) < \frac{\epsilon}{2}$. By [49, Exercise 29, Chap. 2, p. 45], we have

$$V = \bigcup_{n=1}^{\infty} (a_n, b_n), \tag{2.169}$$

where $(a_i, b_i) \cap (a_j, b_j) = \varnothing$ if $i \ne j$. Assume that there was an unbounded segment in the expression (2.169). Then we have $m(V) = \infty$. However, since $m(V) = m(V \setminus E) + m(E)$, the facts $m(V) = \infty$ and $m(V \setminus E) < \frac{\epsilon}{2}$ will force that $m(E) = \infty$, a contradiction. Furthermore, the expression (2.169) implies that

$$m(V) = \sum_{n=1}^{\infty} m\big((a_n, b_n)\big),$$

so there exists a positive integer N such that

$$m(V) - \frac{\epsilon}{2} < \sum_{n=1}^{N} m\big((a_n, b_n)\big). \tag{2.170}$$

Define $g : \mathbb{R} \to \mathbb{R}$ by

$$g(x) = \sum_{n=1}^{N} \chi_{(a_n, b_n)}(x)$$

and $V_N = (a_1, b_1) \cup \cdots \cup (a_N, b_N)$. Let $x \in (E \setminus V_N) \cup (V_N \setminus E)$. If $x \in E \setminus V_N$, then we have $\chi_E(x) = 1$ and $g(x) = 0$. Similarly, if $x \in V_N \setminus E$, then we have $\chi_E(x) = 0$ and $g(x) = 1$. Therefore, we have

$$|g - \chi_E| = 1 \tag{2.171}$$

on $(E \setminus V_N) \cup (V_N \setminus E)$. On the other hand, we let

$$x \in (E \setminus V_N)^c \cap (V_N \setminus E)^c = (V_N \cup E^c) \cap (E \cup V_N^c) = (V_N \cap E) \cup (V_N^c \cap E^c).$$

If $x \in V_N \cap E$, then we have $\chi_E(x) = g(x) = 1$. Similarly, $x \in V_N^c \cap E^c$ implies that $\chi_E(x) = g(x) = 0$. Both cases give

$$|g - \chi_E| = 0 \tag{2.172}$$

on $(E \setminus V_N)^c \cap (V_N \setminus E)^c$. Now the facts $E, V_N \subseteq V$ and (2.170) definitely imply that

$$m(V_N \setminus E) \le m(V \setminus E) < \frac{\epsilon}{2} \quad \text{and} \quad m(E \setminus V_N) \le m(V \setminus V_N) < \frac{\epsilon}{2}. \tag{2.173}$$

Hence we deduce from the results (2.171), (2.172) and the estimates (2.173) that

$$\int_{-\infty}^{\infty} |g - \chi_E|\,\mathrm{d}x = \int_{E \setminus V_N} \mathrm{d}x + \int_{V_N \setminus E} \mathrm{d}x = m(V_N \setminus E) + m(E \setminus V_N) < \epsilon.$$

This is the desired result (2.168).

- **Step 3:** S_2 **is dense in** $L^1(\mathbb{R})$. We need the following lemma:

> **Lemma 2.10**
>
> Suppose that A, B and C are subsets of a metric space X. If $A \subseteq B \subseteq C$, A is dense in B and B is dense in C, then A is dense in C.

Proof of Lemma 2.10. Recall the definition that A is dense in S if $A \subseteq S \subseteq \overline{A}$, where $\overline{A}$ denotes the closure of A in X. Then we have

$$A \subseteq B \subseteq \overline{A} \quad \text{and} \quad B \subseteq C \subseteq \overline{B}$$

which show immediately that $A \subseteq C$. Let $p \in \overline{B}$. We claim that $p \in \overline{A}$. To this end, we know from the assumption that

$$B(p,\delta) \cap (B \setminus \{p\}) \neq \varnothing$$

for every $\delta > 0$. Let $q \in B(p,\delta) \cap (B \setminus \{p\})$. Since $B \subseteq \overline{A}$, $\overline{A}$ also contains q. If $q \in A$, then we have

$$B(p,\delta) \cap (A \setminus \{p\}) \neq \varnothing \tag{2.174}$$

for every $\delta > 0$. Therefore, the definition implies that $p \in \overline{A}$. Next, suppose that $q \in A'$. Since $B(p,\delta)$ is open in X, there exists a $\epsilon > 0$ such that $B(q,\epsilon) \subseteq B(p,\delta)$. Since q is a limit point of A, the definition shows that $B(q,\epsilon) \cap A \neq \varnothing$ and thus the set relation (2.174) still holds in this case. Therefore, we have proven our claim that $p \in \overline{A}$ and this means that $\overline{B} \subseteq \overline{A}$. Hence we establish the set relations

$$A \subseteq C \subseteq \overline{A}$$

which implies that A is dense in C, completing the proof of Lemma 2.10.

Let's return to the original proof. By applying Lemma 2.10 to the results in **Step 1** and **Step 2**, we conclude easily that S_2 is dense in $L^1(\mathbb{R})$ which is our desired result.

Hence we have completed the proof of the problem.

> **Remark 2.7**
>
> By Definition 1.16, we remark that a simple function s is a finite linear combination of characteristic functions of an arbitrary set A_i and when all A_i are intervals, then s becomes a step function. In fact, the class of step functions is one of the building blocks for the theory of Riemann integration, see [49, Chap. 6].

> **Problem 2.25**
>
> *Rudin Chapter 2 Exercise 25.*

Proof.

(a) We note that $e^t > 0$ and $\log(1 + e^t) > 0$ for every $t > 0$. Furthermore, we know that the functions e^x and $\log x$ are increasing in their corresponding domains respectively. Thus the inequality $\log(1 + e^t) < c + t$ is equivalent to $1 + e^t < e^{c+t}$ and finally, it is equivalent to

$$0 < 1 + e^{-t} < e^c. \tag{2.175}$$

Apply log to both sides of the inequality (2.175), we have

$$\log(1 + e^{-t}) < c \tag{2.176}$$

for every $t > 0$. Since the left-hand side of the inequality (2.176) is a decreasing function of t, the smallest value of c for the validity of the inequality (2.176) on $(0, \infty)$ is found by

$$\lim_{\substack{t \to 0 \\ t > 0}} \log(1 + e^{-t}) = \log 2.$$

(b) Let $f \in L^1$ on $[0, 1]$. Define $E = \{x \in [0, 1] \mid f(x) > 0\} = f^{-1}((0, \infty))$. By Theorem 1.12(b), E is a measurable set in $[0, 1]$. Thus we follow from part (a) that, on E,

$$nf(x) = \log\left(e^{nf(x)}\right) < \log\left(1 + e^{nf(x)}\right) < \log 2 + nf(x)$$

on $(0, \infty)$ and this implies that

$$\int_E f \, d\mu < \frac{1}{n} \int_E \log(1 + e^{nf}) \, d\mu < \frac{\log 2}{n} + \int_E f \, d\mu. \tag{2.177}$$

Next, we have $1 \le 1 + e^{nf(x)} \le 2$ on $[0, 1] \setminus E$. Thus we have

$$0 \le \log(1 + e^{nf}) \le \log 2 \tag{2.178}$$

on $[0, 1] \setminus E$. Since $[0, 1] \setminus E$ is also a measurable set in $[0, 1]$, we deduce from the inequalities (2.178) that

$$0 \le \frac{1}{n} \int_{[0,1] \setminus E} \log(1 + e^{nf}) \, d\mu \le \frac{\log 2}{n}.$$

Now the sum of the inequalities (2.177) and (2.178) give

$$\int_E f \, d\mu < \frac{1}{n} \int_0^1 \log(1 + e^{nf}) \, d\mu < \frac{2\log 2}{n} + \int_E f \, d\mu. \tag{2.179}$$

Hence, by taking $n \to \infty$ in the inequalities (2.179), we get that

$$\lim_{n \to \infty} \frac{1}{n} \int_0^1 \log(1 + e^{nf}) \, d\mu = \int_E f \, d\mu.$$

This completes the proof of the problem. $\blacksquare$

CHAPTER **3**

L^p-Spaces

3.1 Properties of Convex Functions

> **Problem 3.1**
>
> *Rudin Chapter 3 Exercise 1.*

Proof. Let $\{\varphi_\alpha\}$ be a collection of convex functions on (a,b). Define

$$\varphi = \sup\{\varphi_\alpha\} \tag{3.1}$$

and assume that it is finite. Suppose that $x, y \in (a,b)$ and $0 \le \lambda \le 1$. By the definition (3.1) and the convexity of φ_α, we have

$$\varphi_\alpha\big((1-\lambda)x + \lambda y\big) \le (1-\lambda)\varphi_\alpha(x) + \lambda\varphi_\alpha(y) \le (1-\lambda)\varphi(x) + \lambda\varphi(y)$$

for all α. In other words, we have

$$\varphi\big((1-\lambda)x + \lambda y\big) \le (1-\lambda)\varphi(x) + \lambda\varphi(y).$$

By Definition 3.1, φ is convex on (a,b).

Suppose that $\{\varphi_n\}$ is a sequence of convex functions on (a,b). For $x \in (a,b)$, we define $\varphi : (a,b) \to \mathbb{R}$ to be

$$\varphi(x) = \lim_{n\to\infty} \varphi_n(x). \tag{3.2}$$

Now for $x, y \in (a,b)$ and $0 \le \lambda \le 1$, we follow from the limit (3.2) that

$$\lim_{n\to\infty} [(1-\lambda)\varphi_n(x) + \lambda\varphi_n(y)] = (1-\lambda)\varphi(x) + \lambda\varphi(y) \tag{3.3}$$

and

$$\lim_{n\to\infty} \varphi_n\big((1-\lambda)x + \lambda y\big) = \varphi\big((1-\lambda)x + \lambda y\big). \tag{3.4}$$

Since $\varphi_n\big((1-\lambda)x + \lambda y\big) \le (1-\lambda)\varphi_n(x) + \lambda\varphi_n(y)$ for every $n = 1, 2, \ldots$, we take limits to both sides and apply the results (3.3) and (3.4) to conclude that

$$\varphi\big((1-\lambda)x + \lambda y\big) \le (1-\lambda)\varphi(x) + \lambda\varphi(y).$$

Hence φ is convex on (a,b) by Definition 3.1.

69

For each positive integer n, suppose that $f_n : (a, b) \to \mathbb{R}$ is convex. Define $f : (a, b) \to \mathbb{R}$ by

$$f(x) = \limsup_{n \to \infty} f_n(x),$$

where $x \in (a, b)$. By Definition 1.13, we have

$$f(x) = \lim_{k \to \infty} g_k(x), \tag{3.5}$$

where $g_k = \sup\{f_k, f_{k+1}, \ldots\}$ for $k = 1, 2, \ldots$. Thus if each f_n is convex on (a, b), then the first assertion shows that each g_k is convex on (a, b). By the definition (3.5), since f is the pointwise limit of the sequence of convex functions $\{g_1, g_2, \ldots\}$ defined on (a, b), we deduce from the second assertion that f is also convex on (a, b). However, the lower limit of a sequence of convex functions *may not* be convex. For example, consider the functions $f_n : \mathbb{R} \to \mathbb{R}$ defined by

$$f_n(x) = (-1)^n x.$$

It is clear that each f_n is convex on $\mathbb{R}$. However, for each $x \in \mathbb{R}$, we have

$$f(x) = \liminf_{n \to \infty} f_n(x) = -|x|.$$

This shows that f is not convex on $\mathbb{R}$ and we complete the proof of the problem. ■

> **Problem 3.2**
>
> *Rudin Chapter 3 Exercise 2.*

Proof. For every $x, y \in (a, b)$ and $0 \le \lambda \le 1$, since φ is convex on (a, b), we have

$$\varphi\big((1 - \lambda)x + \lambda y\big) \le (1 - \lambda)\varphi(x) + \lambda\varphi(y). \tag{3.6}$$

Since ψ is convex and nondecreasing on $\varphi\big((a, b)\big)$, we obtain from the inequality (3.6) that

$$\psi\Big(\varphi\big((1 - \lambda)x + \lambda y\big)\Big) \le \psi\Big((1 - \lambda)\varphi(x) + \lambda\varphi(y)\Big) \le (1 - \lambda)\psi\big(\varphi(x)\big) + \lambda\psi\big(\varphi(y)\big).$$

By Definition 3.1, $\psi \circ \varphi$ is convex on (a, b).

Let $\varphi > 0$ on (a, b). By [51, Theorem (c), p. 2], the function exp is a monotonically increasing positive function on $\mathbb{R}$. Since $(e^x)' = e^x$, it is convex on $\mathbb{R}$ by the paragraph following Definition 3.1. Hence, if $\log \varphi$ is convex on (a, b), then we conclude from the first assertion that $\varphi = e^{\log \varphi}$ is also convex on (a, b).

However, the converse of the second assertion is false. For instance, we know that

$$\varphi(x) = x$$

is convex on $(0, \infty)$, but $\log x$ is not convex on $(0, \infty)$ because $(\log x)'' = -x^{-2} < 0$ on $(0, \infty)$, i.e., $(\log x)'$ is *not* a monotonically increasing function on $(0, \infty)$. We have completed the proof of the problem. ■

> **Problem 3.3**
>
> *Rudin Chapter 3 Exercise 3.*

Proof. This problem is proven in [63, Problem 4.24, pp. 79 – 81]. ■

3.2 Relations among L^p-Spaces and some Consequences

> **Problem 3.4**
>
> *Rudin Chapter 3 Exercise 4.*

Proof.

(a) Since $0 < r < p < s$, we can find $\lambda \in (0,1)$ such that $p = \lambda r + (1-\lambda)s$. By Theorem 3.5 (Hölder's Inequality), we have

$$
\begin{aligned}
\varphi(p) &= \int_X |f|^p \, d\mu \\
&= \int_X |f|^{\lambda r} \times |f|^{(1-\lambda)s} \, d\mu \\
&\le \left\{ \int_X (|f|^{\lambda r})^{\frac{1}{\lambda}} \, d\mu \right\}^{\lambda} \times \left\{ \int_X [|f|^{(1-\lambda)s}]^{\frac{1}{1-\lambda}} \, d\mu \right\}^{1-\lambda} \\
&= [\varphi(r)]^{\lambda} \times [\varphi(s)]^{1-\lambda}.
\end{aligned}
\tag{3.7}
$$

Since $r, s \in E$, $\varphi(r)$ and $\varphi(s)$ are finite. Hence the inequality (3.7) ensures that $\varphi(p)$ is also finite, i.e., $p \in E$.

(b) We prove the assertion one by one.

- **Case (i): $\ln \varphi$ is convex in E°.** If $E^\circ = \varnothing$, then there is nothing to prove. Assume that $E^\circ \ne \varnothing$. By [49, Theorem 2.47, p. 42] and part (a), the set E is connected. Since $E \subseteq (0, \infty)$, E is either an interval in one of the forms $[a, b]$, $[a, b)$, $(a, b]$ or (a, b) for some positive a and b with $a < b$. In each of the cases, we always have $E^\circ = (a, b)$.

  Let $x, y \in (a, b)$ and $\lambda \in [0, 1]$. Since (a, b) is a convex set, $\lambda x + (1-\lambda)y \in (a, b)$. Thus it follows from the inequality (3.7) that

$$
\varphi\big(\lambda x + (1-\lambda)y\big) \le [\varphi(x)]^{\lambda} \times [\varphi(y)]^{1-\lambda}.
\tag{3.8}
$$

  If $\varphi(p) = 0$ for some $p \in (0, \infty)$, then we have

$$
\int_X |f|^p \, d\mu = 0
$$

  so that $|f(x)| = 0$ almost everywhere on X. By the remark following Definition 3.7, it implies that $\|f\|_\infty = 0$, a contradiction. Hence we have $\varphi(p) > 0$ and we are allowed to take the logarithm to both sides of the inequality (3.8) to get

$$
\ln \varphi\big(\lambda x + (1-\lambda)y\big) \le \ln \left\{ [\varphi(x)]^{\lambda} \times [\varphi(y)]^{1-\lambda} \right\} \le \lambda \ln \varphi(x) + (1-\lambda) \ln \varphi(y).
$$

  By Definition 3.1, $\ln \varphi$ is convex in (a, b).

- **Case (ii): φ is continuous on E.** By Theorem 3.2, $\ln \varphi$ is continuous on (a, b). Thus φ is also continuous on (a, b), so it remains to show that φ is continuous at the endpoints. Let $a \in E$. Then E is either $[a, b)$ or $[a, b]$, but no matter which case E is, a is a limit point of E so that we can find a decreasing sequence $\{p_n\} \subseteq (a, b)$ such that $p_n \to a$ as $n \to \infty$. By this, there exists a $\epsilon > 0$ and a positive integer N such that $p_n \in (a, a + \epsilon) \subset E$ for all $n \ge N$. Fix this ϵ and then $n \ge N$ implies that

$$
|f(x)|^{p_n} \le
\begin{cases}
|f(x)|^{a+\epsilon}, & \text{if } |f(x)| \ge 1; \\[2mm]
|f(x)|^{a}, & \text{if } |f(x)| < 1.
\end{cases}
$$

Hence we obtain that

$$|f(x)|^{p_n} \le |f(x)|^{a+\epsilon} + |f(x)|^a$$

on X.

We recall that $a, a + \epsilon \in E$, so the definition implies that $\varphi(a)$ and $\varphi(a + \epsilon)$ are finite. Since f is measurable, $|f|^a$ and $|f|^{a+\epsilon}$ are measurable by Proposition 1.9(b) and Theorem 1.7(b). Thus these two facts show that $|f|^a, |f|^{a+\epsilon} \in L^1(\mu)$. By Theorem 1.32, we have $|f|^{a+\epsilon} + |f|^a \in L^1(\mu)$. For each $x \in X$, $|f(x)| \ge 0$. Since an exponential function with nonnegative base is continuous on its domain, we have

$$\lim_{n\to\infty} |f(x)|^{p_n} = |f(x)|^a$$

on X. In conclusion, we have shown that the sequence $\{|f|^{p_n}\}$ satisfies the hypotheses of Theorem 1.34 (Lebesgue's Dominated Convergence Theorem) and then

$$\lim_{n\to\infty} \varphi(p_n) = \lim_{n\to\infty} \int_X |f|^{p_n} \, d\mu = \int_X |f|^a \, d\mu = \varphi(a).$$

By definition, φ is continuous at a. Similarly, φ is continuous at b if $b \in E$. Consequently, φ is continuous on E.

(c) We claim that E can be any open or closed connected subset of $(0, \infty)$. We consider $X = (0, \infty)$, $\mu = m$ and $0 < a \le b$. We need the following result (see [2, Examples 1 & 5, pp. 417, 419]):

Lemma 3.1

Let $x > 0$ and $b > 1$. Then we have

$$\int_x^1 t^{-\alpha} \, dt = \begin{cases} \dfrac{1 - x^{1-\alpha}}{1 - \alpha}, & \text{if } \alpha \ne 1; \\[2ex] -\ln x & \text{if } \alpha = 1 \end{cases}$$

and

$$\int_1^b t^{-\beta} \, dt = \begin{cases} \dfrac{b^{1-\beta} - 1}{1 - \beta}, & \text{if } \beta \ne 1; \\[2ex] \ln b, & \text{if } \beta = 1. \end{cases}$$

Furthermore, the improper integrals

$$\int_0^1 x^{-\alpha} \, dx \quad \text{and} \quad \int_1^\infty x^{-\beta} \, dx$$

are convergent if and only if $\alpha < 1$ and $\beta > 1$ respectively.

– **Case (i):** $E = (a, b)$. For $x \in (0, \infty)$, let

$$f(x) = \begin{cases} x^{-\frac{1}{b}}, & \text{if } x \in (0, 1); \\[2ex] x^{-\frac{1}{a}}, & \text{if } x \in [1, \infty). \end{cases}$$

We have to find $p \in (0, \infty)$ such that

$$\varphi(p) = \int_0^\infty |f(x)|^p \, dx < \infty.$$

We assume that the integral can be split as follows:

$$\varphi(p) = \int_0^1 x^{-\frac{p}{b}}\,\mathrm{d}x + \int_1^\infty x^{-\frac{p}{a}}\,\mathrm{d}x. \tag{3.9}$$

Apply Lemma 3.1 to the two improper integrals in the expression (3.9), we know that $\varphi(p) < \infty$ if and only if $\frac{p}{b} < 1$ and $\frac{p}{a} > 1$ if and only if $a < p < b$. In other words, we have $E = (a,b)$ and such split is allowable.

- **Case (ii): $E = [a,b]$.** In this case, we need another lemma:[a]

> **Lemma 3.2**
>
> The improper integral
> $$\int_0^{e^{-1}} \frac{\mathrm{d}x}{x^\alpha |\ln x|^\beta}$$
> converges if and only if (i) $\alpha < 1$ or (ii) $\alpha = 1$ and $\beta > 1$. The improper integral
> $$\int_e^\infty \frac{\mathrm{d}x}{x^\alpha (\ln x)^\beta}$$
> converges if and only if (i) $\alpha > 1$ or (ii) $\alpha = 1$ and $\beta > 1$.

We consider

$$f(x) = \begin{cases} \dfrac{1}{x^{\frac{1}{b}} |\ln x|^{1+\frac{1}{b}}}, & \text{if } x \in (0, e^{-1}); \\[2ex] e^{\frac{1}{b}} + \dfrac{e^{-\frac{1}{a}} - e^{\frac{1}{b}}}{e - e^{-1}}(x - e^{-1}), & \text{if } x \in (e^{-1}, e); \\[2ex] \dfrac{1}{x^{\frac{1}{a}} (\ln x)^{1+\frac{1}{a}}}, & \text{if } x \in [e, \infty). \end{cases} \tag{3.10}$$

It is clear that f is continuous on $(0, \infty)$ and so it is measurable. Similar to **Case (i)**, we assume that the integral can be split as follows:

$$\begin{aligned}
\varphi(p) &= \int_0^\infty |f(x)|^p\,\mathrm{d}x \\
&= \int_0^{e^{-1}} \frac{\mathrm{d}x}{x^{\frac{p}{b}} |\ln x|^{p(1+\frac{1}{b})}} + \int_{e^{-1}}^e \left[e^{\frac{1}{b}} + \frac{e^{-\frac{1}{a}} - e^{\frac{1}{b}}}{e - e^{-1}}(x - e^{-1}) \right]^p \mathrm{d}x \\
&\quad + \int_e^\infty \frac{\mathrm{d}x}{x^{\frac{p}{a}} (\ln x)^{p(1+\frac{1}{a})}}. \tag{3.11}
\end{aligned}$$

By Lemma 3.2, the first integral in the expression (3.11) converges if and only if (i) $\frac{p}{b} < 1$ or (ii) $\frac{p}{b} = 1$ and $p(1 + \frac{1}{b}) > 1$. The first condition gives $p < b$. If $p = b$, then $p(1 + \frac{1}{b}) = 1 + b > 1$ so that the second condition is actually $p = b$. In this case, we have $p \le b$.

Similarly, we apply Lemma 3.2 to the third integral in the expression (3.11), so it converges if and only if (i) $\frac{p}{a} > 1$ or (ii) $\frac{p}{a} = 1$ and $p(1 + \frac{1}{a}) > 1$. The first condition

---

[a] The integrals in Lemma 3.2 are called Bertrand's integrals, see the following webpage
`https://fr.wikipedia.org/wiki/Int%C3%A9grale_impropre`.

shows $p > a$. If $p = a$, then $p(1 + \frac{1}{a}) = 1 + a > 1$ so that the second condition implies that $p = a$. In this case, we have $p \geq a$.

Since the second integral in the expression (3.11) is finite, we combine these observations to conclude that $\varphi(p) < \infty$ if and only if $p \in [a, b]$.

- **Case (iii):** $E = \{a\}$. If we take $b = a$ in the definition (3.10), then we obtain $p \geq a$ and $p \leq a$. Hence we establish $E = \{a\}$.

- **Case (iv):** $E = \varnothing$. Consider $f \equiv 1$ on $(0, \infty)$. Since the integral

$$\int_0^\infty |1|^p \, \mathrm{d}x$$

is obviously divergent for every $p \in (0, \infty)$, we conclude that $E = \varnothing$ in this case.

(d) Let $r < p < s$. By the proof of part (a), we see that $p = \lambda r + (1 - \lambda)s$ for some $\lambda \in (0, 1)$ so that the inequality (3.7) holds. In other words, we have

$$\|f\|_p^p \leq \left(\|f\|_r^r\right)^\lambda \times \left(\|f\|_s^s\right)^{1-\lambda}. \tag{3.12}$$

It is clear that

$$\left(\|f\|_r^r\right)^\lambda \times \left(\|f\|_s^s\right)^{1-\lambda} \leq \begin{cases} \left(\|f\|_s^r\right)^\lambda \times \left(\|f\|_s^s\right)^{1-\lambda}, & \text{if } \|f\|_r \leq \|f\|_s; \\[2mm] \left(\|f\|_r^r\right)^\lambda \times \left(\|f\|_r^s\right)^{1-\lambda}, & \text{if } \|f\|_s \leq \|f\|_r \end{cases}$$

$$= \begin{cases} \|f\|_s^p, & \text{if } \|f\|_r \leq \|f\|_s; \\[2mm] \|f\|_r^p, & \text{if } \|f\|_s \leq \|f\|_r. \end{cases} \tag{3.13}$$

Hence, by combining the inequalities (3.12) and (3.13), we get

$$\|f\|_p \leq \max(\|f\|_r, \|f\|_s). \tag{3.14}$$

Let $f \in L^r(\mu) \cap L^s(\mu)$. By the definition, we have

$$\|f\|_r < \infty \quad \text{and} \quad \|f\|_s < \infty.$$

Hence we follow from these and the inequality (3.14) that $\|f\|_p < \infty$, i.e., $f \in L^p(\mu)$ and then $L^r(\mu) \cap L^s(\mu) \subseteq L^p(\mu)$.

(e) Recall that $\|f\|_\infty > 0$.

- **Case (i):** $\|f\|_\infty < \infty$. Let $E_\alpha = \{x \in X \mid |f(x)| \geq \alpha\}$, where $\alpha \in (0, \|f\|_\infty)$. It is clear that E_α is measurable because $|f|$ is measurable. By Definition 3.7, $\|f\|_\infty$ is the smallest number such that

$$\mu(\{x \in X \mid |f(x)| > \|f\|_\infty\}) = 0.$$

Thus we have $\mu(E_\alpha) > 0$ for every $\alpha \in (0, \|f\|_\infty)$. Assume that $\mu(E_\alpha) = \infty$. Then we have

$$\|f\|_r^r = \int_X |f|^r \, \mathrm{d}\mu = \int_{E_\alpha} |f|^r \, \mathrm{d}\mu + \int_{X \setminus E_\alpha} |f|^r \, \mathrm{d}\mu \geq \alpha^r \mu(E_\alpha) = \infty$$

which contradicts to our hypothesis. Thus we have $0 < \mu(E_\alpha) < \infty$ and then Proposition 1.24(b) implies that

$$\|f\|_p^p = \int_X |f|^p \, \mathrm{d}\mu \geq \int_{E_\alpha} |f|^p \, \mathrm{d}\mu \geq \alpha^p \mu(E_\alpha)$$

which shows that

$$\|f\|_p \geq \alpha[\mu(E_\alpha)]^{\frac{1}{p}}. \tag{3.15}$$

Since $\lim\limits_{p \to \infty} [\mu(E_\alpha)]^{\frac{1}{p}} = 1$, we obtain from the inequality (3.15) that

$$\liminf_{p \to \infty} \|f\|_p \geq \alpha.$$

In particular, since α is arbitrary in $(0, \|f\|_\infty)$, we have

$$\liminf_{p \to \infty} \|f\|_p \geq \|f\|_\infty. \tag{3.16}$$

Next, if $p > r$, then we have

$$\|f\|_p^p = \int_X |f|^p \, d\mu = \int_X |f|^{p-r} \times |f|^r \, d\mu. \tag{3.17}$$

By Definition 3.7, since $|f(x)| \leq \|f\|_\infty$ for *almost all* $x \in X$, we get from the expression (3.17) and Proposition 1.24(b) that

$$\|f\|_p^p \leq \int_{X \setminus E} \|f\|_\infty^{p-r} \times |f|^r \, d\mu = \|f\|_\infty^{p-r} \times \int_{X \setminus E} |f|^r \, d\mu = \|f\|_\infty^{p-r} \times \|f\|_r^r, \tag{3.18}$$

where $E = \{x \in X \mid |f(x)| > \|f\|_\infty\}$ is of measure 0. Rewrite the inequality (3.18) in the form

$$\|f\|_p \leq \|f\|_\infty^{1-\frac{r}{p}} \|f\|_r^{\frac{r}{p}}$$

which, by using the fact that $\|f\|_r < \infty$, implies

$$\limsup_{p \to \infty} \|f\|_p \leq \|f\|_\infty. \tag{3.19}$$

Hence our desired result follows immediately from the inequalities (3.16) and (3.19).

– **Case (ii):** $\|f\|_\infty = \infty$. Then we deduce immediately from the inequality (3.16) that

$$\lim_{p \to \infty} \|f\|_p = \liminf_{p \to \infty} \|f\|_p = \infty = \|f\|_\infty.$$

Hence we have completed the proof of the problem. ∎

Problem 3.5

Rudin Chapter 3 Exercise 5.

Proof.

(a) Suppose that $s = \infty$. Since $|f(x)| \leq \|f\|_\infty$ holds for almost all $x \in X$, we have

$$\mu(E) = \mu(\{x \in X \mid |f(x)| > \|f\|_\infty\}) = 0$$

so that

$$\|f\|_r^r = \int_{X \setminus E} |f|^r \, d\mu \leq \int_{X \setminus E} \|f\|_\infty^r \, d\mu = \|f\|_\infty^r \cdot \mu(X \setminus E) = \|f\|_\infty^r. \tag{3.20}$$

Hence we have $\|f\|_r \leq \|f\|_\infty$ in this case.

Next, we suppose that $s < \infty$. Since f is measurable, $|f|^r$ is also measurable on X by Proposition 1.9(b) and Theorem 1.7(b) for every $r > 0$. We apply Theorem 3.5 (Hölder's Inequality) to the measurable functions $|f|^r$ and 1 to obtain that

$$
\int_X |f|^r \, d\mu \le \left\{ \int_X \left(|f|^r\right)^{\frac{s}{r}} d\mu \right\}^{\frac{r}{s}} \times \left\{ \int_X d\mu \right\}^{1-\frac{r}{s}}
$$
$$
= \left(\int_X |f|^s \, d\mu \right)^{\frac{r}{s}}
$$
$$
= \|f\|_s^r. \tag{3.21}
$$

Hence we conclude that $\|f\|_r \le \|f\|_s$.

(b) Suppose that $0 < r < s < \infty$. Since $\|f\|_r = \|f\|_s$, it means that the equality holds in the inequality (3.21). Recall that the inequality (3.21) is derived from Theorem 3.5 (Hölder's Inequality), so there are constants α and β, not both 0, such that

$$
\alpha(|f|^r)^{\frac{s}{r}} = \beta \quad \text{a.e. on } X.
$$

If $\alpha = 0$, then it is trivial that $\beta = 0$, a contradiction. Thus $\alpha \ne 0$ and then we have

$$
|f| = \left(\frac{\beta}{\alpha} \right)^{\frac{1}{s}} \quad \text{a.e.}
$$

In other words, $|f|$ is a constant a.e. on X.

Next, suppose that $s = \infty$ and $\|f\|_r = \|f\|_\infty < \infty$, where $0 < r < \infty$. Then the equality holds in the inequality (3.20) so that $|f|^r = \|f\|_\infty^r < \infty$ a.e. on X, i.e.,

$$
|f| = \|f\|_\infty \quad \text{a.e. on } X.
$$

In other words, we achieve that $|f|$ is a constant a.e. on X in this case.

Consequently, we conclude that $0 < r < s \le \infty$ and the conditions $\|f\|_r = \|f\|_s < \infty$ hold if and only if $|f|$ is a constant a.e. on X

(c) If $f \in L^s(\mu)$, then $\|f\|_s < \infty$ and we know from part (a) that $\|f\|_r < \infty$, i.e., $f \in L^r(\mu)$. Thus it is true that $L^s(\mu) \subseteq L^r(\mu)$. Now the other inclusion $L^r(\mu) \subseteq L^s(\mu)$ is a consequence of the following lemma[b]:

> **Lemma 3.3**
>
> Suppose that $(X, \mathfrak{M}, \mu)$ is a measure space and $\mathfrak{M}_0 = \{E \in \mathfrak{M} \,|\, \mu(E) > 0\}$. Then the following conditions are equivalent:
>
> - **Condition (1).** $\inf\{\mu(E) \,|\, E \in \mathfrak{M}_0\} > 0$,
>
> - **Condition (2).** $L^r(\mu) \subseteq L^s(\mu)$ for all $0 < r < s \le \infty$.

---

[b]This lemma is part of the statements in [62, Theorem 1] and our proof follows part of the argument there.

Proof of Lemma 3.3. Suppose that **Condition (1)** is true. Suppose that $f \in L^r(\mu)$ and $E_n = \{x \in X \mid |f(x)| \geq n\}$ for every $n \in \mathbb{N}$. We claim that there exists an $N \in \mathbb{N}$ such that $\mu(E_N) = 0$. Otherwise, there exists $\epsilon > 0$ such that $\mu(E_n) \geq \epsilon$ for each $n \in \mathbb{N}$. Let $E = \bigcap_{n=1}^{\infty} E_n$. By the construction of E_n, we have $E_1 \supseteq E_2 \supseteq \cdots$. Since $\mu(E_1) \leq \mu(X) < \infty$, it yields from Theorem 1.19(e) that

$$\mu(E) = \lim_{n \to \infty} \mu(E_n) \geq \epsilon > 0. \tag{3.22}$$

However, we deduce from the inequality (3.22) that

$$\|f\|_r = \left(\int_X |f|^r \, \mathrm{d}\mu \right)^{\frac{1}{r}} \geq \left(\int_E |f|^r \, \mathrm{d}\mu \right)^{\frac{1}{r}} = \infty \cdot \mu(E)^{\frac{1}{r}} = \infty,$$

a contradiction. Therefore, we have our claim that $\mu(E_N) = 0$ for some $N \in \mathbb{N}$. In other words, $|f(x)| \leq N$ holds for *almost all* $x \in X$ and we obtain from Definition 3.7 that $\|f\|_\infty \leq N$, i.e., $f \in L^\infty(\mu)$ and so $f \in L^s(\mu)$. Hence **Condition (2)** holds.

Assume that **Condition (1)** was false. Thus there exists a sequence of $\{E_n\} \subseteq \mathfrak{M}_0$ such that

$$0 < \mu(E_n) < \frac{1}{2^n}. \tag{3.23}$$

Put $F_1 = E_1$ and $F_n = E_n \setminus (E_{n-1} \cup \cdots \cup E_1)$, where $n = 2, 3, \ldots$. Then it is easy to check that $F_i \cap F_j = \varnothing$ for $i \neq j$ and

$$0 < \mu(F_n) = \mu\big(E_n \setminus (E_{n-1} \cup \cdots \cup E_1)\big) \leq \mu(E_n) < \frac{1}{2^n}$$

for every $n = 1, 2, \ldots$. Therefore, we may assume that $\{E_n\}$ is a sequence of *disjoint* measurable sets.

Next, we define $\alpha_n = \mu(E_n)$ for $n = 1, 2, \ldots$ and $E = \bigcup_{n=1}^{\infty} E_n$. Suppose first that $s < \infty$. Then we define $f : X \to \mathbb{R}$ by

$$f(x) = \sum_{n=1}^{\infty} \alpha_n^{-\frac{1}{s}} \chi_{E_n}(x). \tag{3.24}$$

Since $r < s$, we have $1 - \frac{r}{s} < 1$. Note that $f(x) = 0$ if $x \in X \setminus E$. Thus it reduces from the bounds (3.23) and Theorem 1.29 that

$$\|f\|_r^r = \int_{X \setminus E} |f|^r \, \mathrm{d}\mu + \int_E |f|^r \, \mathrm{d}\mu = \sum_{n=1}^{\infty} \int_{E_n} \alpha_n^{-\frac{r}{s}} \, \mathrm{d}\mu = \sum_{n=1}^{\infty} \alpha_n^{1-\frac{r}{s}} < \frac{2^{1-\frac{r}{s}}}{1 - 2^{1-\frac{r}{s}}} < \infty$$

which means $f \in L^r(\mu)$. Similarly, it is easy to derive that

$$\|f\|_s^s = \int_E |f|^s \, \mathrm{d}\mu = \sum_{n=1}^{\infty} \int_{E_n} \alpha_n^{-1} \, \mathrm{d}\mu = \sum_{n=1}^{\infty} 1 = \infty$$

which implies that $f \notin L^s(\mu)$. Hence we have shown that

$$L^r(\mu) \nsubseteq L^s(\mu).$$

If $s = \infty$, then we replace the coefficients $\alpha_n^{-\frac{1}{s}}$ by $\alpha_n^{-\frac{1}{2r}}$ in the expression (3.24). By similar argument, since $\alpha_n^{-\frac{1}{2r}} > 2^{\frac{n}{2r}}$ for all $n \in \mathbb{N}$, we conclude $\|f\|_\infty = \infty$ and then $L^r(\mu) \nsubseteq L^\infty(\mu)$. $\blacksquare$

(d) If $0 < p < q$, then part (a) says that $\|f\|_p \le \|f\|_q < \infty$. Let $E_0 = \{x \in X \,|\, |f(x)| = 0\}$. By Problem 1.5, E_0 is measurable.

- **Case (i):** $\mu(E_0) > 0$. Now the set $E_\infty = \{x \in X \,|\, |f(x)| = \infty\}$ is measurable by Problem 1.5. Pick $p \in (0, r)$, so part (a) indicates that $\|f\|_p \le \|f\|_r < \infty$. If $\mu(E_\infty) > 0$, then we have

$$\|f\|_p = \left\{ \int_X |f|^p \, d\mu \right\}^{\frac{1}{p}} \ge \left\{ \int_{E_\infty} |f|^p \, d\mu \right\}^{\frac{1}{p}} = [\infty \cdot \mu(E_\infty)]^{\frac{1}{p}} = \infty,$$

a contradiction. Thus we have $\mu(E_\infty) = 0$, i.e., $|f(x)| < \infty$ a.e. on X. Consequently, we have

$$\int_X \log|f| \, d\mu = \int_{X \setminus E_\infty} \log|f| \, d\mu = \int_{X \setminus (E_\infty \cup E_0)} \log|f| \, d\mu + \int_{E_0} \log|f| \, d\mu \quad (3.25)$$

Since $\mu(X) = 1$ and $|f(x)|$ is nonzero finite on $X \setminus (E_\infty \cup E_0)$, the first integral on the right-hand side of the equation (3.25) is finite. However, the second integral on the right-hand side of the equation (3.25) is in the form $-\infty$ so that the right-hand side of our desired result is in the form $\exp\{-\infty\}$ which is defined to be 0. On the other hand, we notice that

$$\|f\|_p = \left\{ \int_X |f|^p \, d\mu \right\}^{\frac{1}{p}} = \left\{ \int_X |f|^p \chi_{X \setminus E_0} \, d\mu \right\}^{\frac{1}{p}}. \quad (3.26)$$

Apply Theorem 3.5 (Hölder's Inequality) with $u > 1$ and $v > 1$ as fixed conjugate exponents to the right-hand side of the expression (3.26), we see that

$$\left\{ \int_X |f|^p \chi_{X \setminus E_0} \, d\mu \right\}^{\frac{1}{p}} \le \left\{ \int_X (|f|^p)^u \, d\mu \right\}^{\frac{1}{pu}} \left\{ \int_X (\chi_{X \setminus E_0})^v \, d\mu \right\}^{\frac{1}{pv}}$$

$$= \mu(X \setminus E_0)^{\frac{1}{pv}} \|f\|_{pu}. \quad (3.27)$$

Combining the expression (3.26) and the inequality (3.27), we gain

$$\|f\|_p \le \mu(X \setminus E_0)^{\frac{1}{pv}} \|f\|_{pu}.$$

Since $\mu(X) = 1$ and $\mu(E_0) > 0$, we have $\mu(X \setminus E_0) < 1$ and then

$$\lim_{p \to 0} \mu(X \setminus E_0)^{\frac{1}{pv}} = 0. \quad (3.28)$$

In addition, when p is chosen to be small enough such that $pu < r$, part (a) implies that $\|f\|_{pu} \le \|f\|_r < \infty$. Hence we observe from the limit (3.28) and this fact that

$$\lim_{p \to 0} \|f\|_p \le \lim_{p \to 0} \mu(X \setminus E_0)^{\frac{1}{pv}} \|f\|_{pu} \le \|f\|_r \lim_{p \to 0} \mu(X \setminus E_0)^{\frac{1}{pv}} = 0.$$

In other words, the desired result holds in this case.

- **Case (ii):** $\mu(E_0) = 0$. Then $|f(x)| > 0$ a.e. on X. By the fact used in **Case (i)**, we may assume that

$$0 < |f(x)| < \infty \quad \text{a.e. on } X. \quad (3.29)$$

This fact allows us to apply [51, Eqn. (7), p. 63] to the *almost* positive function $|f|^p$ to obtain

$$\|f\|_p = \left\{ \int_X |f|^p \, d\mu \right\}^{\frac{1}{p}} \ge \exp\left\{ \frac{1}{p} \int_X \log|f|^p \, d\mu \right\} = \exp\left\{ \int_X \log|f| \, d\mu \right\}$$

for every $p \in (0, r)$. In particular,

$$\lim_{p \to 0^+} \|f\|_p \geq \exp\left\{\int_X \log|f|\,\mathrm{d}\mu\right\}. \tag{3.30}$$

For the reverse direction, we consider the function $\phi : (0, \infty) \to \mathbb{R}$ defined by

$$\phi(x) = x - 1 - \log x.$$

Since $\phi'(x) = 1 - \frac{1}{x} = 0$ and $\phi''(x) = \frac{1}{x^2}$, ϕ has the absolute minimum at $x = 1$, i.e., $\phi(x) \geq \phi(1)$ on $(0, \infty)$ or equivalently

$$\log x \leq x - 1 \tag{3.31}$$

on $(0, \infty)$. Recall the fact (3.29), so we replace x by $\|f\|_p^p$ in the inequality (3.31) and then using the fact that $\mu(X) = 1$ to obtain

$$\log\|f\|_p \leq \frac{\|f\|_p^p - 1}{p} = \frac{1}{p}\left(\int_X |f|^p\,\mathrm{d}\mu - \int_X \mathrm{d}\mu\right) = \int_X \frac{|f|^p - 1}{p}\,\mathrm{d}\mu. \tag{3.32}$$

Let $E = \{x \in X \mid |f(x)| = 1\}$ and $F = \{x \in X \mid |f(x)| \neq 1\}$. Then we get

$$\int_X \frac{|f|^p - 1}{p}\,\mathrm{d}\mu = \int_E \frac{|f|^p - 1}{p}\,\mathrm{d}\mu + \int_F \frac{|f|^p - 1}{p}\,\mathrm{d}\mu. \tag{3.33}$$

It is clear from Proposition 1.24(d) that

$$\int_E \frac{|f|^p - 1}{p}\,\mathrm{d}\mu = 0,$$

so it suffices to evaluate

$$\lim_{p \to 0+} \int_F \frac{|f|^p - 1}{p}\,\mathrm{d}\mu.$$

To this end, we need the following lemma:

Lemma 3.4

Let $|f| > 0$ and $\|f\|_r < \infty$ for some $r > 0$. Define $\psi : \mathbb{R} \to \mathbb{R}$ by $\psi(p) = |f|^p$. Then we have

$$\frac{|f|^p - 1}{p} \in L^1(\mu)$$

for all $p \in (0, r)$.

Proof of Lemma 3.4. Since $\psi'' = |f|^p (\log|f|)^2 > 0$ on $\mathbb{R}$, ψ is convex on $\mathbb{R}$. Thus we see from [49, Exercise 23, p. 101] that, for $0 < p < r$,

$$\frac{\psi(p) - \psi(0)}{p - 0} \leq \frac{\psi(r) - \psi(0)}{r - 0}$$

which reduces to

$$\frac{|f|^p - 1}{p} \leq \frac{|f|^r - 1}{r}$$

and so

$$\int_X \frac{|f|^p - 1}{p}\,\mathrm{d}\mu \leq \int_X \frac{|f|^r - 1}{r}\,\mathrm{d}\mu = \frac{1}{r}\|f\|_r^r - \frac{1}{r} < \infty.$$

In other words, we have $\frac{|f|^p - 1}{p} \in L^1(\mu)$ for all $p \in (0, r)$, completing the proof of the lemma.

Let's return to the proof of the problem. By Theorem 1.34 (Lebesgue's Dominated Convergence Theorem), we obtain that

$$\lim_{p\to 0+} \int_F \frac{|f|^p - 1}{p}\, d\mu = \int_F \lim_{p\to 0+} \left(\frac{|f|^p - 1}{p}\right) d\mu. \tag{3.34}$$

By L' Hôspital Rule, we derive

$$\lim_{p\to 0+} \frac{|f|^p - 1}{p} = \lim_{p\to 0+} |f|^p \log |f| = \log |f| \tag{3.35}$$

on X. Therefore, when we combine the results (3.32), (3.33), (3.34) and (3.35), we establish that

$$\lim_{p\to 0+} \log \|f\|_p \le \int_F \lim_{p\to 0+} \left(\frac{|f|^p - 1}{p}\right) d\mu = \int_F \log |f|\, d\mu = \int_X \log |f|\, d\mu. \tag{3.36}$$

Since $\log x$ is continuous for $x > 0$, we get from the inequality (3.36) that

$$\lim_{p\to 0+} \|f\|_p \le \exp\left\{ \int_X \log |f|\, d\mu \right\}. \tag{3.37}$$

Hence the required result follows immediately from the inequalities (3.30) and (3.37). This ends the proof of the problem. ∎

Problem 3.6

Rudin Chapter 3 Exercise 6.

Proof. Let $x > 0$ and $0 \le c \le 1$. We claim that the function $\Phi : [0, \infty) \to \mathbb{R}$ satisfies the relation

$$c\Phi(x) + (1 - c)\Phi(1) = \Phi(x^c). \tag{3.38}$$

Obviously, the relation (3.38) holds for $c = 0$ and $c = 1$. Without loss of generality, we may assume that $0 < c < 1$ in the following discussion.

Since f is bounded measurable and positive on $[0, 1]$, there exists a positive constant M such that

$$\|f\|_1 = \int_0^1 |f|\, dx \le M < \infty.$$

Thus we get from Problem 3.5(d) that

$$\lim_{p\to 0} \|f\|_p = \exp\left\{ \int_0^1 \log |f(t)|\, dt \right\}. \tag{3.39}$$

By putting the expression (3.39) into the relation in question, we obtain

$$\Phi\left(\exp\left\{ \int_0^1 \log |f(t)|\, dt \right\} \right) = \int_0^1 \Phi(f(t))\, dt. \tag{3.40}$$

We define $f : [0, 1] \to \mathbb{R}$ by

$$f(t) = \begin{cases} x, & \text{if } x \in [0, c]; \\ 1, & \text{if } x \in (c, 1]. \end{cases} \tag{3.41}$$

Since $x > 0$, the f is bounded measurable and positive on $[0, 1]$. Therefore, we substitute the function (3.41) into the relation (3.40) to yield

$$
\begin{aligned}
\Phi(x^c) &= \Phi\big(\exp(c \log x)\big) \\
&= \Phi\left(\exp\left(\int_0^c \log x \, dt\right)\right) \\
&= \int_0^c \Phi(x) \, dt + \int_c^1 \Phi(1) \, dt \\
&= c\Phi(x) + (1 - c)\Phi(1).
\end{aligned}
$$

This proves the claim (3.38).

Next, we prove a lemma which is useful of determining the explicit form of the function Φ:

> **Lemma 3.5**
>
> The relation (3.38) holds for all $x > 0$ and $c \geq 0$.

Proof of Lemma 3.5. The case $0 \leq c \leq 1$ is clear by the above analysis. Let $c > 1$ so that $0 < \frac{1}{c} < 1$. Then the relation (3.38) becomes

$$
\Phi\left(x^{\frac{1}{c}}\right) = \frac{1}{c}\Phi(x) + \left(1 - \frac{1}{c}\right)\Phi(1). \tag{3.42}
$$

Take $y = x^{\frac{1}{c}}$. Since $\frac{1}{c} > 0$, y will take all positive real numbers. Therefore, it follows from the expression (3.42) that

$$
\Phi(y) = \frac{1}{c}\Phi(y^c) + \left(1 - \frac{1}{c}\right)\Phi(1)
$$

and after simplification, we have

$$
\Phi(y^c) = c\Phi(y) + (1 - c)\Phi(1).
$$

This proves the lemma. $\blacksquare$

We return to the proof of the problem. Define $\Psi : [0, \infty) \to \mathbb{R}$ by $\Psi(x) = \Phi(x) - \Phi(1)$, then the relation (3.38) can be rewritten as

$$
\Psi(x^c) = c\Psi(x), \tag{3.43}
$$

where $x > 0$ and $c \geq 0$. Take $x = e$ and $c = \ln y$, where $y \geq 1$. Then the expression (3.43) becomes

$$
\Psi(y) = \Psi(e) \ln y, \tag{3.44}
$$

where $y \geq 1$. For $0 < y < 1$, we put $x = e^{-1}$ and $c = \ln(y^{-1}) > 0$ into the expression (3.43) so that

$$
\Psi(y) = \Psi\big(e^{-\ln(y^{-1})}\big) = -\Psi(e^{-1}) \ln y. \tag{3.45}
$$

Combining the results (3.44) and (3.45), we establish that

$$
\Psi(x) = \begin{cases} \Psi(e) \ln x, & \text{if } x \geq 1; \\[2mm] -\Psi(e^{-1}) \ln x, & \text{if } 0 < x < 1 \end{cases}
$$

which imply that

$$\Phi(x) = \begin{cases} \Phi(1) + [\Phi(e) - \Phi(1)]\ln x, & \text{if } x \geq 1; \\[2mm] \Phi(1) - [\Phi(e^{-1}) - \Phi(1)]\ln x, & \text{if } 0 < x < 1, \end{cases}$$

completing the proof of the problem. $\blacksquare$

Problem 3.7

Rudin Chapter 3 Exercise 7.

Proof. In this problem, we assume that μ is a positive measure. We give examples one by one:

Example 3.1. Let $X = \mathbb{N}$ with the counting measure μ. By Definition 3.6, we have

$$\ell^p(\mathbb{N}) = \Big\{ x = \{\xi_n\} \,\Big|\, \|x\|_p = \Big(\sum_{n=1}^{\infty} |\xi_n|^p \Big)^{\frac{1}{p}} < \infty \Big\}.$$

Let $r < s$. Thus if $\{\xi_n\} \in \ell^r(\mathbb{N})$, then $\{|\xi_n|^r\}$ is a convergent real sequence so that there exists a positive integer N such that $|\xi_n| \leq 1$ for all $n \geq N$, see [49, Theorem 3.23, p. 60]. Therefore, we must have $|\xi_n|^s \leq |\xi_n|^r$ for all $n \geq N$ and this implies that

$$\sum_{n=N}^{\infty} |\xi_n|^s \leq \sum_{n=N}^{\infty} |\xi_n|^r < \infty.$$

In other words, we prove that $\{\xi_n\} \in \ell^s(\mathbb{N})$ and then $\ell^r(\mathbb{N}) \subseteq \ell^s(\mathbb{N})$.

Example 3.2. We consider the Lebesgue measure m on $X = [0,1]$. Since $m(X) = 1$, Problem 3.5(c) shows that $L^s([0,1]) \subseteq L^r([0,1])$ if $0 < r < s$.

Example 3.3. Consider $X = (0, \infty)$ and the Lebesgue measure m. Let $1 \leq r < s \leq \infty$. Take $\frac{1}{s} < \alpha < \frac{1}{r}$ and define $f : (0, \infty) \to \mathbb{R}$ by

$$f(x) = x^{-\alpha} \chi_{(0,1)}(x).$$

Then we have

$$\Big\{ \int_0^{\infty} |f|^r \, dx \Big\}^{\frac{1}{r}} = \Big\{ \int_0^1 x^{-\alpha r} \, dx \Big\}^{\frac{1}{r}} \quad \text{and} \quad \Big\{ \int_0^{\infty} |f|^s \, dx \Big\}^{\frac{1}{s}} = \Big\{ \int_0^1 x^{-\alpha s} \, dx \Big\}^{\frac{1}{s}}. \tag{3.46}$$

By Lemma 3.1, the integrals on the right-hand sides of the two expressions (3.46) converge if and only if $\alpha r < 1$ and $\alpha s < 1$ respectively. Since $\frac{1}{s} < \alpha < \frac{1}{r}$, we conclude that $f \in L^r((0, \infty))$ but $f \notin L^s((0, \infty))$, i.e., $L^r((0, \infty)) \not\subseteq L^s((0, \infty))$.

Similarly, we define $g : (0, \infty) \to \mathbb{R}$ by

$$g(x) = x^{-\alpha} \chi_{(1, \infty)}(x).$$

Then we have

$$\Big\{ \int_0^{\infty} |f|^r \, dx \Big\}^{\frac{1}{r}} = \Big\{ \int_1^{\infty} x^{-\alpha r} \, dx \Big\}^{\frac{1}{r}} \quad \text{and} \quad \Big\{ \int_0^{\infty} |f|^s \, dx \Big\}^{\frac{1}{s}} = \Big\{ \int_1^{\infty} x^{-\alpha s} \, dx \Big\}^{\frac{1}{s}}.$$

Thus Lemma 3.1 again shows that $g \in L^s((0, \infty))$ but $g \notin L^r((0, \infty))$. In other words, it means that $L^s((0, \infty)) \not\subseteq L^r((0, \infty))$.

Next, we are going to find conditions on μ under which these situations will occur:

- $L^r(\mu) \subseteq L^s(\mu)$ **for** $0 < r < s \le \infty$. Recall from Lemma 3.3 that it is equivalent to the condition that $\inf\{\mu(E) \mid E \in \mathfrak{M}_0\} > 0$, where $\mathfrak{M}_0 = \{E \in \mathfrak{M} \mid \mu(E) > 0\}$.

- $L^s(\mu) \subseteq L^r(\mu)$ **for** $0 < r < s < \infty$. Similar to the proof of Lemma 3.3, the following result is part of the statements in [62, Theorem 2]:

> **Lemma 3.6**
>
> Suppose that $(X, \mathfrak{M}, \mu)$ is a measure space and $\mathfrak{M}_\infty = \{E \in \mathfrak{M} \mid \mu(E) \text{ is finite}\}$. Then the following conditions are equivalent:
>
> - **Condition (1).** $\sup\{\mu(E) \mid E \in \mathfrak{M}_\infty\} < \infty$,
>
> - **Condition (2).** $L^s(\mu) \subseteq L^r(\mu)$ for all $0 < r < s < \infty$.

Proof of Lemma 3.6. Suppose that **Condition (1)** holds. To begin with, let $f \in L^s(\mu)$ and $E_n = \{x \in X \mid \frac{1}{n+1} \le |f(x)| < \frac{1}{n}\}$ for each $n = 1, 2, \ldots$. Then it is clear that $E_n \in \mathfrak{M}$. Now for each *fixed* n, we see that

$$\int_X |f|^s \, \mathrm{d}\mu \ge \int_{E_n} |f|^s \, \mathrm{d}\mu \ge \frac{1}{(n+1)^s} \int_{E_n} \mathrm{d}\mu = \frac{\mu(E_n)}{(n+1)^s}$$

which implies that

$$\mu(E_n) \le (n+1)^s \int_X |f|^s \, \mathrm{d}\mu = (n+1)^s \|f\|_s^s < \infty.$$

Therefore, we conclude that $E_n \in \mathfrak{M}_\infty$. Note that $E_i \cap E_j = \varnothing$ if $i \ne j$ and μ is a positive measure, if $F_n = E_1 \cup E_2 \cup \cdots \cup E_n$, then we deduce from Theorem 1.19(b) that

$$\mu(F_n) = \mu\left(\bigcup_{k=1}^{n} E_k\right) = \sum_{k=1}^{n} \mu(E_k) < \infty.$$

In other words, each F_n is an element of $\mathfrak{M}_\infty$. Next, we know that $F_1 \subseteq F_2 \subseteq \cdots$ and $\bigcup_{k=1}^{n} F_k = \bigcup_{k=1}^{n} E_k$, so Theorem 1.19(d) implies that

$$\lim_{n \to \infty} \mu(F_n) = \mu\left(\bigcup_{n=1}^{\infty} F_n\right) = \mu\left(\bigcup_{n=1}^{\infty} E_n\right) = \sum_{n=1}^{\infty} \mu(E_n). \tag{3.47}$$

Given a positive integer n, it is clear that

$$\mu(F_n) \le \sup\{\mu(F_k) \mid k \in \mathbb{N}\} \le \sup\{\mu(E) \mid E \in \mathfrak{M}_\infty\}. \tag{3.48}$$

Furthermore, by combining **Condition (1)**, the expression (3.47) and the inequality (3.48), it yields that

$$\sum_{n=1}^{\infty} \mu(E_n) = \lim_{n \to \infty} \mu(F_n) \le \sup\{\mu(E) \mid E \in \mathfrak{M}_\infty\} < \infty.$$

On $X \setminus \bigcup_{n=1}^{\infty} E_n$, we claim that $|f(x)| \geq 1$. Otherwise, there is a $x_0 \in X \setminus \bigcup_{n=1}^{\infty} E_n$ such that $|f(x_0)| < 1$. By the definition, $x_0 \in E_N$ for some $N \in \mathbb{N}$, a contradiction. This observation implies that

$$|f(x)|^r \leq |f(x)|^s$$

on $X \setminus \bigcup_{n=1}^{\infty} E_n$. Finally, we have

$$\int_X |f|^r \, d\mu = \int_{X \setminus \bigcup_{n=1}^{\infty} E_n} |f|^r \, d\mu + \sum_{n=1}^{\infty} \int_{E_n} |f|^r \, d\mu \leq \int_X |f|^s \, d\mu + \sum_{n=1}^{\infty} \frac{\mu(E_n)}{n^r} < \infty.$$

This proves that **Condition (2)** holds.

Conversely, since $L^s(\mu) \subseteq L^r(\mu)$ implies that $L^{\alpha s}(\mu) \subseteq L^{\alpha r}(\mu)$ for every $\alpha \in (0, \infty)$, we may assume that $r \geq 1$. It is known that if $r, s \in [1, \infty]$ and $L^s(\mu) \subseteq L^r(\mu)$, then the mapping

$$T : L^s(\mu) \to L^r(\mu)$$

is continuous, see [62, Lemma 1]. This result guarantees the existence of a positive constant k such that $\|f\|_r \leq k\|f\|_s$ for all $f \in L^s(\mu)$, so

$$\mu(E) \leq k^{\frac{rs}{s-r}}$$

for every $E \in \mathfrak{M}_\infty$ and this implies **Condition (1)**.

We have completed the proof of the problem.

> ### Problem 3.8
>
> *Rudin Chapter 3 Exercise 8.*

Proof. For every $n \in \mathbb{N}$, since $g(x) \to \infty$ as $x \to 0$, there exists a sequence $\{\epsilon_n\}$ of positive numbers such that

$$g(x) \geq n \tag{3.49}$$

for every $x \in (0, \epsilon_n)$. Without loss of generality, we may assume that $\{\epsilon_n\}$ is decreasing and

$$\epsilon_{n+1} < \frac{\epsilon_n}{2} \tag{3.50}$$

for every $n = 1, 2, \ldots$. Now we define $h_n : (0, 1) \to \mathbb{R}$ and $h : (0, 1) \to \mathbb{R}$ by

$$h_n(x) = n\left(1 - \frac{x}{\epsilon_n}\right)\chi_{(0,\epsilon_n)}(x) \quad \text{and} \quad h(x) = \sup_n h_n(x)$$

respectively.

Let $x \in (0, 1)$ be *fixed* but arbitrary. Since $\epsilon_n \to 0$ as $n \to \infty$, there exists a positive integer N_x such that $x \leq \epsilon_n$ for all $1 \leq n \leq N_x$ and $x \geq \epsilon_n$ for all $n > N_x$. Thus we have $x \in (0, \epsilon_n)$ for all $1 \leq n \leq N_x$ and $x \notin (0, \epsilon_n)$ for all $n > N_x$. See Figure 3.1 for an illustration of the distribution of x and ϵ_n.

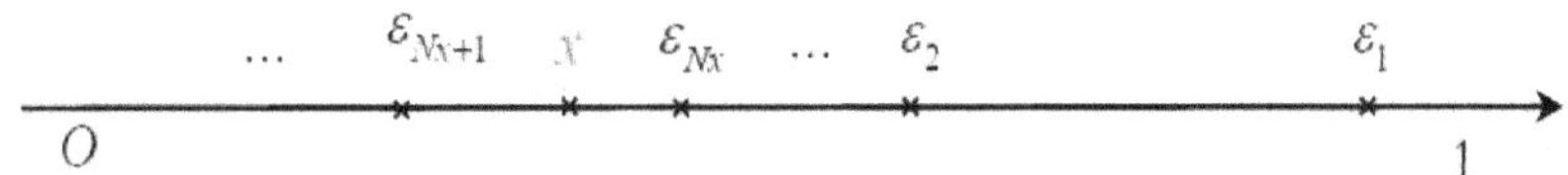

Figure 3.1: The distribution of x and ϵ_n.

These facts imply that

$$h_n(x) = \begin{cases} n\left(1 - \dfrac{x}{\epsilon_n}\right), & \text{if } 1 \leq n \leq N_x; \\[2ex] 0, & \text{if } n > N_x \end{cases} \tag{3.51}$$

so that

$$h(x) = \sup_n h_n(x) = \sup\{h_1(x), h_2(x), \ldots\} = \max_{1 \leq n \leq N_x} n\left(1 - \frac{x}{\epsilon_n}\right) < \infty,$$

i.e., h is finite. Since each h_n is convex on $(0,1)$, Problem 3.1 shows that the h is also convex on $(0,1)$.

Next, we want to compare the magnitudes of $h_n(x)$ and $g(x)$ at the fixed point $x \in (0,1)$. On the one hand, for $1 \leq n \leq N_x$, we have $x \in (0, \epsilon_n)$, so $n(1 - \frac{x}{\epsilon_n}) \leq n$ and we deduce from the inequality (3.49) and the definition (3.51) that

$$h_n(x) \leq g(x) \tag{3.52}$$

for $1 \leq n \leq N_x$. On the other hand, if $n > N_x$, then $x \notin (0, \epsilon_n)$ and we have $h_n(x) = 0$, but the inequality (3.49) still holds for $n = N_x$ (i.e., $x \in (0, \epsilon_{N_x})$) so that the inequality (3.52) remains valid in this case. Hence we have established that

$$h(x) \leq g(x) \tag{3.53}$$

for the fixed point $x \in (0,1)$. Since x is arbitrary, the inequality (3.53) holds on $(0,1)$.

Take $n = N_x - 1$ in the definition (3.51) to get

$$h_{N_x-1}(x) = (N_x - 1)\left(1 - \frac{x}{\epsilon_{N_x-1}}\right). \tag{3.54}$$

By the hypothesis (3.50), we have $2\epsilon_{N_x} < \epsilon_{N_x-1}$. Then since $x < \epsilon_{N_x}$, we have

$$\frac{x}{\epsilon_{N_x-1}} < \frac{x}{2\epsilon_{N_x}} < \frac{1}{2}. \tag{3.55}$$

By substituting the estimate (3.55) into the expression (3.54), we obtain

$$h_{N_x-1}(x) > \frac{N_x - 1}{2}.$$

Since $x \to 0$ if and only if $N_x \to \infty$, we conclude that $h_{N_x}(x) \to \infty$ as $x \to 0$ and the definition implies that $h(x) \to \infty$ as $x \to 0$ as required.

The second assertion is false. To see this, consider $g(x) = x^{\frac{1}{3}}$ which satisfies $g(x) \to \infty$ as $x \to \infty$. Assume that there was a convex function $h : (0, \infty) \to \mathbb{R}$ such that

$$h(x) \leq x^{\frac{1}{3}} \tag{3.56}$$

on $(0, \infty)$ and $h(x) \to \infty$ as $x \to \infty$. We know from $[66, \text{Exercise } 6(c), \text{p. } 262]^c$ that if $h : (0, \infty) \to \mathbb{R}$ is convex on $(0, \infty)$, then the ratio

$$\frac{h(x)}{x}$$

tends to a finite limit or to infinity as $x \to \infty$. If this limit is finite, since $h(x) \to \infty$ as $x \to \infty$, this limit must be positive. Thus it must be true that

$$h(x) \geq kx \tag{3.57}$$

as $x \to \infty$ for some positive constant k. However, the inequality (3.57) definitely contradicts the inequality (3.56). This completes the proof of the problem. ■

Problem 3.9

Rudin Chapter 3 Exercise 9.

Proof. Let $\Phi : (0, \infty) \to (0, \infty)$ be such that $\Phi(p) \to \infty$ as $p \to \infty$. Suppose that $E_1 = (0, \frac{1}{2})$ and

$$E_n = \left[\frac{1}{2} \mid \frac{1}{2^2} \mid \cdots \mid \frac{1}{2^{n-1}}, \frac{1}{2} + \frac{1}{2^2} + \cdots + \frac{1}{2^n} \right) = \left[1 - \frac{1}{2^{n-1}}, 1 - \frac{1}{2^n} \right),$$

where $n \geq 2$. Obviously, we have $E_n \subset (0, 1)$ for every $n \in \mathbb{N}$ and $E_i \cap E_j = \varnothing$ for $i \neq j$. Furthermore, each E_n is measurable and

$$m(E_n) = \frac{1}{2^n} > 0.$$

Suppose that $x \in (0, 1)$. If $x < \frac{1}{2}$, then $x \in E_1$. Otherwise, $\frac{1}{2} \leq x < 1$ and there exists a $\delta > 0$ such that $x < \delta < 1$. Take N to be *least* positive integer such that

$$N > -\frac{\log(1 - \delta)}{\log 2} > 0.$$

Then we have $x < 1 - \frac{1}{2^N} < 1$ so that $x \in E_2 \cup E_3 \cup \cdots \cup E_N$. In other words, we get

$$(0, 1) = \bigcup_{n=1}^{\infty} E_n.$$

Finally, we define $f_n, f : (0, 1) \to (0, \infty)$ by

$$f_n(x) = \sum_{k=1}^{n} k\chi_{E_k}(x) \quad \text{and} \quad f(x) = \lim_{n \to \infty} f_n(x) = \sum_{n=1}^{\infty} n\chi_{E_n}(x) \tag{3.58}$$

respectively. Since each E_n is measurable, each χ_{E_n} is measurable by Proposition 1.9(d). Thus each function f_n is measurable and then Theorem 1.14 implies that f is also measurable. Now it remains to show that the function f satisfies the required properties of the problem.

- **Property 1:** $\|f\|_p \to \infty$. Assume that $f \in L^{\infty}((0, 1))$. Then there exists a positive constant M such that $\|f\|_{\infty} < M$. By Definition 3.7, we see that

$$|f(x)| < M \quad \text{a.e. on } (0, 1). \tag{3.59}$$

---

^{c}See also $[31, \text{Theorem } 126, \text{p. } 99]$ and $[51, \text{Eqn. } (2), \text{p. } 62]$.

Since $E_i \cap E_j = \varnothing$ for $i \neq j$, we have

$$|f(x)| = f(x) \geq f_n(x) \geq n \tag{3.60}$$

for every $x \in E_n$. Since $n \to \infty$, there exists a positive integer N such that $N \geq M$. Recall that $m(E_N) = \frac{1}{2^N} > 0$, so we obtain from the inequality (3.60) that

$$|f(x)| \geq M$$

on E_N, but this contradicts our assumption (3.59). Therefore, we must have $f \notin L^\infty((0,1))$ and in fact $\|f\|_\infty = \infty$. Furthermore, it follows from the Ratio Test (see [49, Theorem 3.34]) that

$$\|f\|_1 = \int_0^1 f(x)\,\mathrm{d}x = \sum_{n=1}^{\infty} n \times m(E_n) = \sum_{n=1}^{\infty} \frac{n}{2^n} < \infty,$$

so we have $\|f\|_1 < \infty$. Hence we deduce from Problem 3.4(e) that

$$\|f\|_p \to \|f\|_\infty = \infty$$

as $p \to \infty$.

- **Property 2: $\|f\|_p \leq \Phi(p)$ for sufficiently large p.** Fix a positive integer n, we consider

$$\left[\frac{n}{\Phi(p)}\right]^p$$

for $p > 0$. Since $\Phi(p) \to \infty$ as $p \to \infty$, when p is sufficiently large, we have $\frac{n}{\Phi(p)} \leq 1$ so that

$$\left[\frac{n}{\Phi(p)}\right]^p \times m(E_n) \leq \frac{1}{2^n}. \tag{3.61}$$

Therefore, we see that

$$\|f\|_p^p = \int_0^1 \left[\sum_{n=1}^{\infty} n\chi_{E_n}(x)\right]^p \mathrm{d}x = \sum_{n=1}^{\infty} n^p \times m(E_n). \tag{3.62}$$

Consequently, we follow from the inequality (3.61) and the expression (3.62) that

$$\frac{\|f\|_p^p}{[\Phi(p)]^p} = \frac{1}{[\Phi(p)]^p} \sum_{n=1}^{\infty} n^p \times m(E_n) = \sum_{n=1}^{\infty} \left[\frac{n}{\Phi(p)}\right]^p \times m(E_n) \leq \sum_{n=1}^{\infty} \frac{1}{2^n} = 1$$

for sufficiently large p, or equivalently,

$$\|f\|_p \leq \Phi(p)$$

for sufficiently large p.

We end the proof of the problem. $\blacksquare$

3.3   Applications of Theorems 3.3, 3.5, 3.8, 3.9 and 3.12

Problem 3.10

Rudin Chapter 3 Exercise 10.

Proof. First of all, we have to show that $f \in L^p(\mu)$. Recall that $L^p(\mu)$ is a metric space with metric $\|\cdot\|$. By the triangle inequality

$$\|f_n - f_m\|_p \leq \|f_n - f\|_p + \|f - f_m\|_p$$

and the hypothesis $\|f_n - f\|_p \to 0$ as $n \to \infty$ together imply that $\{f_n\}$ is a Cauchy sequence in $L^p(\mu)$. Now the completeness of $L^p(\mu)$ ensures that $\{f_n\}$ converges to an element of $L^p(\mu)$. By the uniqueness of limits, this element *must be* f so that $f \in L^p(\mu)$. By Theorem 3.12, $\{f_n\}$ has a subsequence $\{f_{n_k}\}$ such that $f_{n_k} \to f$ a.e. on X. Since $f_n \to g$ a.e. on X as $n \to \infty$, every subsequence of $\{f_n\}$ converges to g a.e. on X. In particular, we take this subsequence $\{f_{n_k}\}$ and the uniqueness of limits again imply that

$$f = g$$

a.e. on X. We have completed the proof of the problem. ■

> **Problem 3.11**
>
> *Rudin Chapter 3 Exercise 11.*

Proof. Since $f, g : \Omega \to [0, \infty]$, the hypothesis $fg \geq 1$ implies that $\sqrt{fg} \geq 1$. By Theorem 3.5 (Hölder's Inequality), we see that

$$\left(\int_\Omega \sqrt{fg}\,\mathrm{d}\mu \right)^2 \leq \int_\Omega f\,\mathrm{d}\mu \times \int_\Omega g\,\mathrm{d}\mu. \tag{3.63}$$

By Proposition 1.24(a), we have

$$\int_\Omega \sqrt{fg}\,\mathrm{d}\mu \geq \int_\Omega \mathrm{d}\mu = \mu(\Omega) = 1. \tag{3.64}$$

Now our desired result follows immediately by combining the inequalities (3.63) and (3.64). This completes the analysis of the problem. ■

> **Problem 3.12**
>
> *Rudin Chapter 3 Exercise 12.*

Proof. Suppose that $A = \infty$. Since $h \leq \sqrt{1 + h^2}$ on Ω, we have

$$\int_\Omega \sqrt{1 + h^2}\,\mathrm{d}\mu = \infty,$$

so the equalities definitely hold in this case. Next, suppose that $A = 0$. By the fact that $1 \leq \sqrt{1 + h^2} \leq \sqrt{1 + 2h + h^2} = 1 + h$, we achieve that

$$\int_\Omega \mathrm{d}\mu \leq \int_\Omega \sqrt{1 + h^2}\,\mathrm{d}\mu \leq \int_\Omega (1 + h)\,\mathrm{d}\mu = \int_\Omega \mathrm{d}\mu. \tag{3.65}$$

Since $\mu(\Omega) = 1$, we conclude from the inequalities (3.65) that

$$\int_\Omega \sqrt{1 + h^2}\,\mathrm{d}\mu = 1.$$

Thus the equalities also hold in this case.

Suppose that $0 < A < \infty$ which means $h(x) \in (0,\infty)$ a.e. on Ω. We consider the function $\varphi : \mathbb{R} \to (0,\infty)$ defined by

$$\varphi(x) = \sqrt{1 + x^2}$$

which implies that

$$\varphi'(x) = \frac{x}{\sqrt{1 + x^2}} \quad \text{and} \quad \varphi''(x) = \frac{1}{(1 + x^2)^{\frac{3}{2}}}. \tag{3.66}$$

Since $\varphi''(x) \geq 0$ on $\mathbb{R}$, we know from [49, Exercise 14, p. 115] that φ is convex on $\mathbb{R}$. In particular, φ is convex on $(0,\infty)$. By Theorem 3.3 (Jensen's Inequality), we obtain that

$$\sqrt{1 + A^2} = \sqrt{1 + \int_\Omega h \, d\mu} \leq \int_\Omega \sqrt{1 + h^2} \, d\mu \tag{3.67}$$

which is exactly the left-hand side inequality. The right-hand side inequality is easy because we always have $\sqrt{1 + h^2} \leq 1 + h$. Hence we have verified the validity of the inequalities.

For the second assertion, we suppose that $\mu = m$ on $\Omega = [0,1]$, $h = f'$ and h is continuous on Ω. Then we have

$$A = \int_0^1 f'(x) \, dx = f(1) - f(0).$$

In addition, we notice that

$$\int_\Omega \sqrt{1 + h^2} \, d\mu = \int_0^1 \sqrt{1 + [f'(x)]^2} \, dx$$

which is the *arc length* of the curve from $(0, f(0))$ to $(1, f(1))$.[d] Thus the inequalities give bounds of such arc length. Geometrically, the upper bound consists of the length from $(0, f(0))$ to $(1, f(0))$ (which is 1) and the length from $(1, f(0))$ to $(1, f(1))$ (which is A). The lower bound is just the hypotenuse of the right triangle with vertices $(0, f(0))$, $(0, f(1))$ and $(1, f(1))$, see Figure 3.2 below:

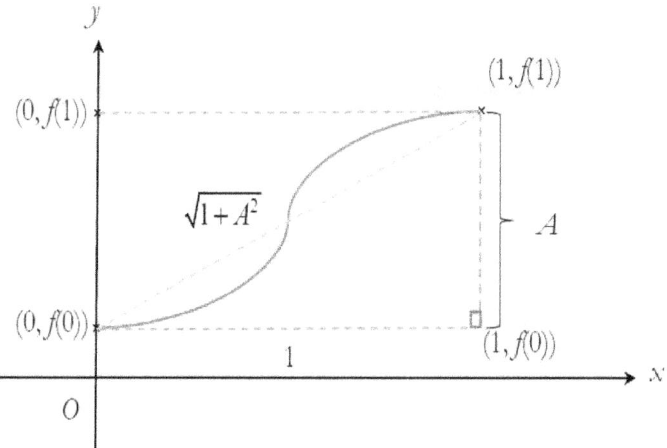

Figure 3.2: The geometric interpretation of a special case.

---

[d]See, for examples, [2, p. 535] and [49, Theorem 6.27, p. 137].

Our first inequality (3.67) comes from the application of Theorem 3.3 (Jensen's Inequality), so the equality of the first inequality (3.67) holds if and only if the equality in [51, Eqn. (2), p. 62] holds if and only if

$$\varphi(A) = \varphi\big(h(x)\big). \tag{3.68}$$

We see from the first derivative in (3.66) that $\varphi'(x) > 0$ in $(0, \infty)$ so that it is *injective* in $(0, \infty)$. Recall that $A \in (0, \infty)$ and $h(x) \in (0, \infty)$ a.e. on Ω, so we deduce that the equality (3.68) holds if and only if

$$h = A \quad \text{a.e. on } \Omega.$$

Next we recall the second inequality is derived from the fact that $\sqrt{1 + h^2} \le 1 + h$ and some simple computations show that the equality holds if and only if

$$h \equiv 0 \quad \text{a.e. on } \Omega.$$

Hence we have ended the proof of the problem. ◼

Problem 3.13

Rudin Chapter 3 Exercise 13.

Proof. There are two cases.

- **Case (i):** $1 < p < \infty$. Then the inequality in Theorem 3.8 is just Hölder's inequality. If $\|f\|_p = 0$ or $\|g\|_p = 0$, then we have $f = 0$ a.e. on X or $g = 0$ a.e. on X by Theorem 1.39(a). In either case, the equality holds trivially. Therefore, we may assume that both $\|f\|_p > 0$ and $\|g\|_p > 0$. Furthermore, by the remark following the proof of Theorem 3.5 (Hölder's Inequality), we may further assume that

$$0 < \|f\|_p < \infty \quad \text{and} \quad 0 < \|g\|_p < \infty.$$

In this case, we see that the equality in Theorem 3.8 holds if and only if there are positive constants α and β such that

$$\alpha f^p = \beta g^q$$

a.e. on X.

In Theorem 3.9, the inequality is in fact Minkowski's inequality. By simple observation, it is clear that the equality holds if $f = 0$ a.e. on X or $g = 0$ a.e. on X. Therefore, we may assume that $f \ne 0$ and $g \ne 0$ on measurable sets E and F with $\mu(E) > 0$ and $\mu(F) > 0$ respectively. By examining the proof of [51, Eqn. (2), Theorem 3.5, pp. 64, 65], we know that it applies Hölder's inequality to the functions f and $(f+g)^{p-1}$ as well as the functions g and $(f+g)^{p-1}$. Thus we deduce from the previous paragraph that the equality in

$$\int_X f \cdot (f+g)^{p-1} \, d\mu \le \left\{ \int_X f^p \, d\mu \right\}^{\frac{1}{p}} \times \left\{ \int_X (f+g)^{(p-1)q} \, d\mu \right\}^{\frac{1}{q}}$$

holds if and only if there are positive constants α and β such that

$$\alpha f^p = \beta(f+g)^q \quad \text{a.e. on } X. \tag{3.69}$$

Similarly, the equality in

$$\int_X g \cdot (f+g)^{p-1} \, d\mu \le \left\{ \int_X g^p \, d\mu \right\}^{\frac{1}{p}} \times \left\{ \int_X (f+g)^{(p-1)q} \, d\mu \right\}^{\frac{1}{q}}$$

holds if and only if there are positive constants λ and ν such that

$$\lambda g^p = \nu (f + g)^q \quad \text{a.e. on } X. \tag{3.70}$$

By combining the equations (3.69) and (3.70), we conclude that

$$f^p = \frac{\beta}{\alpha}(f + g)^q = \frac{\beta\lambda}{\alpha\nu} g^p \quad \text{a.e.}$$

which means that $f = Kg$ a.e. for some positive constant K.

- **Case (ii)**: $p = \infty$. The inequality in Theorem 3.8 comes from the inequality[e]

$$|f(x)g(x)| \le \|f\|_\infty \cdot |g(x)|$$

for almost all x. Therefore, it is easy to see that the equality in Theorem 3.8 holds if and only if $|f(x)g(x)| = \|f\|_\infty \cdot |g(x)|$ for almost all x if and only if $g(x) = 0$ for almost x or $f(x) = \|f\|_\infty$ for almost x.[f]

  Now the triangle inequality
  $$|f + g| \le |f| + |g|$$

  implies the inequality in Theorem 3.9. Thus the equality in Theorem 3.9 holds if and only if f and g are either nonnegative functions or nonpositive functions for almost all x.

This completes the proof of the problem.   ◼

3.4   Hardy's Inequality and Egoroff's Theorem

> **Problem 3.14**
>
> *Rudin Chapter 3 Exercise 14.*

Proof. We follow the suggestions given by Rudin.

(a) We divide the proof into two cases:

- **Special Case:** $f \ge 0$ **and** $f \in C_c\big((0, \infty)\big)$. By the First Fundamental Theorem of Calculus, the function $F(x)$ is differentiable on $(0, \infty)$ and

$$xF'(x) = f(x) - F(x) \tag{3.71}$$

for every $x \in (0, \infty)$. By Integration by Parts[g] and the expression (3.71), we have

$$\int_0^\infty F^p(x)\,dx = [xF^p(x)]_0^\infty - \int_0^\infty px F^{p-1}(x)F'(x)\,dx. \tag{3.72}$$

Now we have to evaluate the limits:

$$\lim_{x \to 0} x F^p(x) \quad \text{and} \quad \lim_{x \to \infty} x F^p(x).$$

---

[e]This is [51, Eqn. (2), Theorem 3.8, p. 66].

[f]The latter case means that f is a nonnegative constant for almost all x.

[g]Here we assume that the formula for integration by parts is also valid for improper integrals, see [3, p. 278].

Since $\operatorname{supp}(f) = \overline{\{x \in (0,\infty) \mid f(x) \neq 0\}}$ is compact, the Heine-Borel Theorem guarantees that $\operatorname{supp}(f) = [a, b] \subset (0, \infty)$. By this and the continuity of f, we conclude that f is bounded on $(0, \infty)$, i.e., $|f(x)| \leq M$ on $(0, \infty)$ for some positive constant M. Thus we follow from this that

$$|F(x)| \leq \frac{1}{x} \int_0^x |f(t)| \, \mathrm{d}t \leq M$$

for every $x \in (0, \infty)$, so we see that

$$\lim_{x \to 0} x F^p(x) = 0. \tag{3.73}$$

Again, the boundedness of f implies that

$$|x F(x)| \leq \int_0^x |f(t)| \, \mathrm{d}t \leq \int_0^\infty |f(t)| \, \mathrm{d}t = \int_a^b |f(t)| \, \mathrm{d}t \leq M(b - a) < \infty \tag{3.74}$$

on $(0, \infty)$. Since $x F^p = \frac{(xF)^p}{x^{p-1}}$ and $p > 1$, it reduces from these and the bound (3.74) that

$$\lim_{x \to \infty} x F^p(x) = 0. \tag{3.75}$$

Therefore, we deduce from the formula (3.72) and the two limits (3.73) and (3.75) that

$$\int_0^\infty F^p(x) \, \mathrm{d}x = -p \int_0^\infty F^{p-1}(x)[f(x) - F(x)] \, \mathrm{d}x$$
$$= -p \int_0^\infty F^{p-1}(x) f(x) \, \mathrm{d}x + p \int_0^\infty F^p(x) \, \mathrm{d}x$$

so that

$$(p - 1) \int_0^\infty F^p(x) \, \mathrm{d}x = p \int_0^\infty F^{p-1}(x) f(x) \, \mathrm{d}x. \tag{3.76}$$

Note that $1 < p < \infty$, let q be its conjugate exponent. Obviously, our hypotheses make sure that f and F^{p-1} have range $[0, \infty]$, so we may apply Theorem 3.5 (Hölder's Inequality) to them. In fact, it yields from the fact $(p - 1)q = p$ and the expression (3.76) that

$$(p - 1)\|F\|_p^p = (p - 1) \int_0^\infty F^p(x) \, \mathrm{d}x$$
$$\leq p \left\{ \int_0^\infty f^p(x) \, \mathrm{d}x \right\}^{\frac{1}{p}} \left\{ \int_0^\infty F^{(p-1)q}(x) \, \mathrm{d}x \right\}^{\frac{1}{q}}$$
$$= p\|f\|_p \cdot \|F\|_p^{\frac{p}{q}}$$

which is exactly

$$\|F\|_p = \|F\|_p^{p - \frac{p}{q}} \leq \frac{p}{p - 1} \|f\|_p. \tag{3.77}$$

– **General Case:** $f \in L^p((0, \infty))$. By Definition 3.6, we always have

$$\|f\|_p = \big\| |f| \big\|_p,$$

so we may assume that $f \geq 0$ on $(0, \infty)$. Since $1 < p < \infty$, Theorem 3.14 says that $C_c((0, \infty))$ is dense in $L^p((0, \infty))$. Thus there exists a sequence $\{f_n\} \subseteq C_c((0, \infty))$ such that

$$\|f_n - f\|_p \to 0 \tag{3.78}$$

as $n \to \infty$. Notice that f_n may *not* be nonnegative, but since $\big||f_n| - |f|\big| \leq |f_n - f|$ and $f \geq 0$, we have

$$\big||f_n(x)| - f(x)\big| \leq |f_n(x) - f(x)| \tag{3.79}$$

for every $x \in (0, \infty)$. Certainly, the inequality (3.79) implies that

$$\big\||f_n| - f\big\|_p \leq \|f_n - f\|_p \tag{3.80}$$

for every $n = 1, 2, \ldots$. Now the result (3.78) and the inequality (3.80) ensure that

$$\big\||f_n| - f\big\|_p \to 0$$

as $n \to \infty$. In other words, we may also assume that $f_n \geq 0$ for each $n = 1, 2, \ldots$. Recall that $f_n \in C_c\big((0, \infty)\big)$, so we derive from the **Special Case** that

$$\|F_n\|_p \leq \frac{p}{p-1}\|f_n\|_p, \tag{3.81}$$

where

$$F_n(x) = \frac{1}{x}\int_0^x f_n(t)\, \mathrm{d}t$$

and $n = 1, 2, \ldots$. Suppose that

$$F(x) = \frac{1}{x}\int_0^x f(t)\, \mathrm{d}t.$$

Since $f \geq 0$ and $f_n \geq 0$, it is easily seen from the definition that $F \geq 0$ and $F_n \geq 0$.

We claim that $\{F_n\}$ *converges pointwise* to F on $(0, \infty)$. To see this, fix $x \in (0, \infty)$, then their definitions give

$$|F_n(x) - F(x)| \leq \frac{1}{x}\int_0^x |f_n(t) - f(t)|\, \mathrm{d}t. \tag{3.82}$$

By the result (3.78), we know that for every $\epsilon > 0$, there exists a positive integer N such that $n \geq N$ implies

$$\|f_n - f\|_p < \epsilon.$$

Apply Theorem 3.5 (Hölder's Inequality) with $\frac{1}{p} + \frac{1}{q} = 1$ to the inequality (3.82), we get immediately that

$$\begin{aligned}
|F_n(x) - F(x)| &\leq \frac{1}{x}\left\{\int_0^x |f_n(t) - f(t)|^p\, \mathrm{d}t\right\}^{\frac{1}{p}} \times \left\{\int_0^x 1^q\, \mathrm{d}t\right\}^{\frac{1}{q}} \\
&\leq \frac{1}{x}\|f_n - f\|_p \times x^{\frac{1}{q}} \\
&= x^{-\frac{1}{p}}\|f_n - f\|_p \\
&< \epsilon x^{-\frac{1}{p}} \tag{3.83}
\end{aligned}$$

for all $n \geq N$. Since x is *fixed* and *independent of* ϵ, the estimate (3.83) verifies the truth of the claim.

Next, recall that each f_n is continuous on $[a, x]$ for every $a > 0$, so the First Fundamental Theorem of Calculus shows that each F_n is continuous on $[a, x]$. As a continuous function, every F_n is measurable on $[a, x]$.[h] Hence, it follows from the

---

[h]See §1.11 or [49, Example 11.14].

pointwise convergence of $\{F_n\}$, Theorem 1.28 (Fatou's Lemma), the inequality (3.81) and then the result (3.78) that

$$
\begin{aligned}
\|F\|_p^p &= \int_0^\infty F^p(x)\,\mathrm{d}x \\
&= \int_0^\infty \liminf_{n\to\infty} F_n^p(x)\,\mathrm{d}x \\
&\le \liminf_{n\to\infty} \int_0^\infty F_n^p(x)\,\mathrm{d}x \\
&\le \lim_{n\to\infty} \int_0^\infty F_n^p(x)\,\mathrm{d}x \\
&= \lim_{n\to\infty} \|F_n\|_p^p \\
&\le \left(\frac{p}{p-1}\right)^p \lim_{n\to\infty} \|f_n\|_p^p \\
&\le \left(\frac{p}{p-1}\right)^p \|f\|_p^p
\end{aligned}
$$

which is equivalent to the desired result

$$
\|F\|_p \le \frac{p}{p-1}\|f\|_p.
$$

Hence the mapping $T : L^p\big((0,\infty)\big) \to L^p\big((0,\infty)\big)$ given by $T(f) = F$ is continuous.

(b) Suppose that $f \in L^p\big((0,\infty)\big)$ and

$$
\|F\|_p = \frac{p}{p-1}\|f\|_p < \infty. \tag{3.84}
$$

Recall that $\|f\|_p = \big\|\,|f|\,\big\|_p$, so we may suppose further that $f \ge 0$ on $(0,\infty)$. By the **General Case** of the proof in part (a), there is a nonnegative sequence $\{f_n\} \subseteq C_c\big((0,\infty)\big)$ such that $\|f_n - f\|_p \to 0$ as $n \to \infty$. Now we replace F and f by F_n and f_n in the expression (3.76) respectively, we gain

$$
\|F_n\|_p^p = \int_0^\infty F_n^p(x)\,\mathrm{d}x = \frac{p}{p-1}\int_0^\infty F_n^{p-1}(x)f_n(x)\,\mathrm{d}x \tag{3.85}
$$

for each $n = 1, 2, \ldots$. Since $f_n \in C_c\big((0,\infty)\big)$, $T(f_n) = F_n \in L^p\big((0,\infty)\big)$, where T is the continuous mapping considered in part (a). By Theorem 3.5 (Hölder's Inequality), it is easy to see that $f_n F_n^{p-1} \in L^p\big((0,\infty)\big)$ for each $n = 1, 2, \ldots$. Since the mapping T is continuous, it follows from the expression (3.85) that

$$
\int_0^\infty F^p(x)\,\mathrm{d}x = \frac{p}{p-1}\int_0^\infty F^{p-1}(x)f(x)\,\mathrm{d}x. \tag{3.86}
$$

In other words, the formula (3.76) holds for $f \in L^p\big((0,\infty)\big)$. We apply Theorem 3.5 (Hölder's Inequality) to the right-hand side of the inequality (3.86) and then using our hypothesis (3.84) to conclude that

$$
\begin{aligned}
\|F\|_p^p &= \frac{p}{p-1}\int_0^\infty F^{p-1}(x)f(x)\,\mathrm{d}x \\
&\le \frac{p}{p-1}\left\{\int_0^\infty f^p(x)\,\mathrm{d}x\right\}^{\frac{1}{p}} \times \left\{\int_0^\infty F^p(x)\,\mathrm{d}x\right\}^{\frac{1}{q}}
\end{aligned} \tag{3.87}
$$

$$= \frac{p}{p-1}\|f\|_p \times \|F\|_p^{\frac{p}{q}}$$

$$= \|F\|_p^p.$$

Hence the equality in the inequality (3.87) must hold. Consequently, there are constants α and β, *not* both 0, such that $\alpha f^p = \beta(F^{p-1})^q = \beta F^p$ a.e. on $(0,\infty)$ and so

$$\alpha^{\frac{1}{p}} f = \beta^{\frac{1}{p}} F \quad \text{a.e. on } (0,\infty). \tag{3.88}$$

Here we have several cases:

- **Case (i):** $\beta = 0$. Then we have $\alpha \neq 0$ and $f = 0$ a.e. on $(0,\infty)$. Thus we are done.
- **Case (ii):** $\alpha = 0$. Then we have $\beta \neq 0$ and $F = 0$ a.e. on $(0,\infty)$. By the hypothesis (3.84), we see that $\|f\|_p = 0$ which gives $f = 0$ a.e. on $(0,\infty)$ and we are done again.
- **Case (iii):** $\alpha\beta \neq 0$. By Theorem 3.8, we see that $f \in L^1([a,b])$, where $[a,b]$ is any bounded interval of $(0,\infty)$. By the comment following the proof of Theorem 11.3 on [49, p. 324], we know that F is differentiable a.e. on $[a,b]$ and

$$\big(xF(x)\big)' = f(x) \tag{3.89}$$

a.e. on $[a,b]$. Since $[a,b]$ is arbitrary, the formula (3.89) holds a.e. on $(0,\infty)$. Combining the equations (3.88) and (3.89), we see immediately that $xF' + xF = f$ which implies

$$xf' = (c-1)f \quad \text{a.e. on } (0,\infty) \tag{3.90}$$

for some nonzero constant c. By solving the differential equation (3.90), we obtain that

$$f(x) = \gamma x^{c-1} \quad \text{a.e. on } (0,\infty)$$

for some constant γ. Since $\gamma x^{c-1} \in L^p((0,\infty))$, it forces that $\gamma = 0$. Hence we have shown that the equality holds only if $f = 0$ a.e. on $(0,\infty)$.

(c) Take $f(x) = x^{-\frac{1}{p}}\chi_{[1,A]}(x)$ for large A. Then we have

$$\|f\|_p = \left\{ \int_0^\infty |x^{-\frac{1}{p}}\chi_{[1,A]}(x)|^p \, dx \right\}^{\frac{1}{p}} = \left\{ \int_1^A x^{-1} \, dx \right\}^{\frac{1}{p}} = (\log A)^{\frac{1}{p}}. \tag{3.91}$$

Next, we know from the definition that

$$F(x) = \frac{1}{x}\int_0^x t^{-\frac{1}{p}}\chi_{[1,A]}(t)\, dt$$

$$= \begin{cases} 0, & \text{if } x \in (0,1); \\[2mm] \dfrac{1}{x}\displaystyle\int_1^x t^{-\frac{1}{p}}\, dt, & \text{if } x \in [1,A]; \\[2mm] \dfrac{1}{x}\displaystyle\int_1^A t^{-\frac{1}{p}}\, dt, & \text{if } x \in (A,\infty); \end{cases}$$

$$= \begin{cases} 0, & \text{if } x \in (0,1); \\[2mm] \dfrac{p}{p-1}(x^{-\frac{1}{p}} - x^{-1}), & \text{if } x \in [1,A]; \\[2mm] \dfrac{p}{p-1}\cdot\dfrac{A^{1-\frac{1}{p}}-1}{x}, & \text{if } x \in (A,\infty). \end{cases} \tag{3.92}$$

Note that A is large and $1 - \frac{1}{p} > 0$, so we have

$$A^{1-\frac{1}{p}} - 1 > 0. \tag{3.93}$$

Therefore, we deduce from the expressions (3.92) and the fact (3.93) that

$$\begin{aligned}
\|F\|_p^p &= \int_0^\infty |F(x)|^p \, \mathrm{d}x \\
&= \int_1^A |F(x)|^p \, \mathrm{d}x + \int_A^\infty |F(x)|^p \, \mathrm{d}x \\
&= \left(\frac{p}{p-1}\right)^p \int_1^A |x^{-\frac{1}{p}} - x^{-1}|^p \, \mathrm{d}x + \left(\frac{p}{p-1}\right)^p \int_A^\infty \left|\frac{A^{1-\frac{1}{p}} - 1}{x}\right|^p \, \mathrm{d}x \\
&= \left(\frac{p}{p-1}\right)^p \int_1^A (x^{-\frac{1}{p}} - x^{-1})^p \, \mathrm{d}x + \frac{p^p}{(p-1)^{p+1}} \left(A^{1-\frac{1}{p}} - 1\right)^p A^{1-p} \\
&> \left(\frac{p}{p-1}\right)^p \int_1^A (x^{-\frac{1}{p}} - x^{-1})^p \, \mathrm{d}x. \tag{3.94}
\end{aligned}$$

Fix p first. Assume that the constant $\frac{p}{p-1}$ could be replaced by a smaller number $\frac{\delta p}{p-1}$ for some $\delta < 1$, i.e.,

$$\|F\|_p \le \frac{\delta p}{p-1} \|f\|_p.$$

Take $\epsilon > 0$ such that $\delta < \epsilon < 1$. Then it is clear that for large A, we have $x^{\frac{1}{p}-1} < 1 - \epsilon$ for $x \ge \frac{A}{2}$ or equivalently,

$$x^{-\frac{1}{p}} - x^{-1} > \epsilon x^{-\frac{1}{p}} > 0 \tag{3.95}$$

for $x \ge \frac{A}{2}$. By this estimate (3.95), we obtain

$$\int_1^A (x^{-\frac{1}{p}} - x^{-1})^p \, \mathrm{d}x > \int_{\frac{A}{2}}^A \epsilon^p x^{-1} \, \mathrm{d}x = \epsilon^p \left(\log A - \log \frac{A}{2}\right) > \epsilon^p \log A. \tag{3.96}$$

By substituting the inequality (3.96) into the inequality (3.94) and then using the fact (3.91), we can show that

$$\|F\|_p^p > \left(\frac{\epsilon p}{p-1}\right)^p \log A > \left(\frac{\delta p}{p-1}\right)^p \|f\|_p^p \ge 0,$$

but this implies

$$\|F\|_p > \frac{\delta p}{p-1} \|f\|_p,$$

a contradiction.

(d) Since $f(x) > 0$ on $(0, \infty)$, there exists a $\delta > 0$ such that we can find an $\alpha \in (0, \infty)$ with

$$\int_0^\alpha f(t) \, \mathrm{d}t > \delta > 0.$$

By the definition, we have $F(x) > 0$ on $(0, \infty)$ so that

$$\|F\|_1 = \int_0^\infty F(x) \, \mathrm{d}x \ge \int_\alpha^\infty F(x) \, \mathrm{d}x = \int_\alpha^\infty \frac{1}{x} \left(\int_0^x f(t) \, \mathrm{d}t\right) \mathrm{d}x > \int_\alpha^\infty \frac{\delta}{x} \, \mathrm{d}x = \infty.$$

Hence $F \notin L^1$. In other words, this shows that Hardy's inequality does not hold when $p = 1$.

Hence we have completed the proof of the problem.   ◼

> ### Problem 3.15
> *Rudin Chapter 3 Exercise 15.*

Proof. We follow the hint. Suppose that $\{a_n\}$ is decreasing. Let $f : (0, \infty) \to (0, \infty)$ be given by

$$f(x) = \sum_{n=1}^{\infty} a_n \chi_{(n-1,n]}(x). \tag{3.97}$$

Firstly, we note that

$$\int_0^{\infty} f^p(x)\, \mathrm{d}x = \sum_{n=1}^{\infty} \int_{n-1}^{n} a_n^p\, \mathrm{d}x = \sum_{n=1}^{\infty} a_n^p \tag{3.98}$$

which implies that $f \in L^p\big((0, \infty)\big)$ if and only if $\sum_{n=1}^{\infty} a_n^p < \infty$. This shows that the desired inequality holds if $f \notin L^p\big((0, \infty)\big)$. Without loss of generality, we assume that $\sum_{n=1}^{\infty} a_n^p < \infty$ in the following discussion.

Secondly, let N be a positive integer and $x \in (N-1, N]$. Then we deduce from the definition (3.97) that

$$
\begin{aligned}
F(x) &= \frac{1}{x} \int_0^x f(t)\, \mathrm{d}t \\
&= \frac{1}{x} \Big[\sum_{n=1}^{N-1} \int_{n-1}^{n} f(t)\, \mathrm{d}t + \int_{N-1}^{x} f(t)\, \mathrm{d}t \Big] \\
&= \frac{1}{x} \Big[\sum_{n=1}^{N-1} \int_{n-1}^{n} a_n\, \mathrm{d}t + \int_{N-1}^{x} a_N\, \mathrm{d}t \Big] \\
&= \frac{1}{x} [a_1 + a_2 + \cdots + a_{N-1} + a_N(x - N + 1)].
\end{aligned}
\tag{3.99}
$$

It is clear from the expression (3.99) and the fact $\{a_n\}$ is decreasing that

$$
\begin{aligned}
\frac{1}{x} \Big[\sum_{n=1}^{N-1} a_n + a_N(x - N + 1) \Big] &= \frac{1}{x} \Big(\sum_{n=1}^{N} a_n - N a_N + x a_N \Big) \\
&= \frac{1}{x} \Big(\sum_{n=1}^{N} a_n - N a_N \Big) + a_N \\
&\geq \frac{1}{N} \Big(\sum_{n=1}^{N} a_n - N a_N \Big) + a_N \\
&= \frac{1}{N} \sum_{n=1}^{N} a_n.
\end{aligned}
\tag{3.100}
$$

Thus we obtain from the expression (3.99) and the inequality (3.100) that

$$\|F\|_p = \Big\{ \int_0^{\infty} F^p(x)\, \mathrm{d}x \Big\}^{\frac{1}{p}} = \Big\{ \sum_{N=1}^{\infty} \int_{N-1}^{N} F^p(x)\, \mathrm{d}x \Big\}^{\frac{1}{p}} \geq \Big\{ \sum_{N=1}^{\infty} \Big(\frac{1}{N} \sum_{n=1}^{N} a_n \Big)^p \Big\}^{\frac{1}{p}}. \tag{3.101}$$

Furthermore, by Problem 3.14(a) and the expression (3.98), we derive that

$$\|F\|_p \le \frac{p}{p-1}\|f\|_p = \frac{p}{p-1}\Big\{\int_0^\infty f^p(x)\,dx\Big\}^{\frac{1}{p}} = \frac{p}{p-1}\Big\{\sum_{n=1}^\infty a_n^p\Big\}^{\frac{1}{p}}. \tag{3.102}$$

Hence our desired inequality follows immediately from the inequalities (3.101) and (3.102).

For the general case, recall the assumption that $a_n > 0$ for all $n \in \mathbb{N}$ and $\sum_{n=1}^\infty a_n^p < \infty$, so the sequence $\{a_n^p\}$ converges absolutely and every rearrangement converges to the same sum (see [49, Theorem 3.55, p. 78]). On the other hand, if we fix a positive integer N, then

$$\sum_{k=1}^N \Big(\frac{1}{k}\sum_{n=1}^k a_n\Big)^p = \Big(\frac{a_1}{1}\Big)^p + \Big(\frac{a_1+a_2}{2}\Big)^p + \Big(\frac{a_1+a_2+a_3}{3}\Big)^p$$
$$+ \cdots + \Big(\frac{a_1+a_2+\cdots+a_N}{N}\Big)^p,$$

so it follows from this form that the sum attains its maximum if and only if $a_1 \ge a_2 \ge \cdots \ge a_N$. Therefore, we have

$$\sum_{N=1}^\infty \Big(\frac{1}{N}\sum_{n=1}^N a_n\Big)^p \le \underbrace{\sum_{N=1}^\infty \Big(\frac{1}{N}\sum_{n=1}^N \alpha_n\Big)^p \le \Big(\frac{p}{p-1}\Big)^p \sum_{n=1}^\infty \alpha_n^p}_{\text{By the special case.}} = \Big(\frac{p}{p-1}\Big)^p \sum_{n=1}^\infty a_n^p,$$

where $\{\alpha_n\}$ is the decreasing rearrangement of $\{a_n\}$. This completes the proof of the problem.
$\blacksquare$

Problem 3.16

Rudin Chapter 3 Exercise 16.

Proof. Let's prove the assertions one by one.

- **A proof of Egoroff's Theorem.** Put

$$S(n,k) = \bigcap_{i,j>n} \Big\{x \in X \,\Big|\, |f_i(x) - f_j(x)| < \frac{1}{k}\Big\}. \tag{3.103}$$

For each $k \in \mathbb{N}$, we follow from the definition (3.103) that

$$S(n,k) = \bigcap_{i,j>n+1} \Big\{x \in X \,\Big|\, |f_i(x) - f_j(x)| < \frac{1}{k}\Big\} \cap \bigcap_{\substack{i=n+1 \\ j>n}} \Big\{x \in X \,\Big|\, |f_i(x) - f_j(x)| < \frac{1}{k}\Big\}$$
$$\cap \bigcap_{\substack{j=n+1 \\ i>n}} \Big\{x \in X \,\Big|\, |f_i(x) - f_j(x)| < \frac{1}{k}\Big\}$$

which implies that $S(n,k) \subseteq S(n+1,k)$.

Next, for each *fixed* $k \in \mathbb{N}$, we claim that

$$X = \bigcup_{n=1}^\infty S(n,k). \tag{3.104}$$

To this end, we see that the set inclusion

$$\bigcup_{n=1}^{\infty} S(n,k) \subseteq X$$

is evident. To prove the reverse direction, we note that for each $k \in \mathbb{N}$ and $x \in X$, we yield from our hypothesis $f_n(x) \to f(x)$ that there exists a positive integer N such that $i, j > N$ implies that

$$|f_i(x) - f_j(x)| < \frac{1}{k}.$$

In other words, we must have $x \in S(n,k)$, i.e.,

$$X \subseteq \bigcup_{n=1}^{\infty} S(n,k).$$

Thus these prove the validity of our claim (3.104) and Theorem 1.19(d) gives

$$\lim_{n \to \infty} \mu\big(S(n,k)\big) = \mu(X)$$

for every $k = 1, 2, \ldots$.

Given that $\epsilon > 0$. For each $k \in \mathbb{N}$, this fact allows us to choose n_k such that

$$\big|\mu\big(S(n_k, k)\big) - \mu(X)\big| < \frac{\epsilon}{2^k}.$$

Define

$$E = \bigcap_{k=1}^{\infty} S(n_k, k). \tag{3.105}$$

Then it is clear that

$$X \setminus E = \bigcup_{k=1}^{\infty} \big(X \setminus S(n_k, k)\big)$$

which implies that

$$\mu(X \setminus E) \le \sum_{k=1}^{\infty} \mu\big(X \setminus S(n_k, k)\big) \le \sum_{k=1}^{\infty} \big|\mu(X) - \mu\big(S(n_k, k)\big)\big| < \sum_{k=1}^{\infty} \frac{\epsilon}{2^k} = \epsilon.$$

Pick a k with $\frac{1}{k} < \epsilon$. If $x \in S(n_k, k)$, then we have

$$|f_i(x) - f_j(x)| < \epsilon \tag{3.106}$$

for all $i, j > n_k$. By the definition (3.105), we know that $E \subseteq S(n_k, k)$ for every $k \in \mathbb{N}$, so the inequality (3.106) holds on E. By the definition, $f_n \to f$ uniformly on E, proving Egoroff's Theorem.

- **A counterexample on a σ-finite space.** Recall from Definition 2.16 that X is said to be a σ-finite space if X is a countable union of sets E_i with $\mu(E_i) < \infty$. It is clear that

$$\mathbb{R} = \bigcup_{n=1}^{\infty} [n-1, n) \cup \bigcup_{n=1}^{\infty} (-n, 1-n],$$

so $\mathbb{R}$ is σ-finite. Let $f_n(x) = \frac{x}{n}$ on $\mathbb{R}$. It is clear that the sequence converges pointwise to 0 at every point of $\mathbb{R}$. Furthermore, we know that a measurable set $E \subseteq \mathbb{R}$ with $m(\mathbb{R} \setminus E) < 1$ must be unbounded. Otherwise, $E \subseteq [-M, M]$ for some $M > 0$ and this implies that

$$m(\mathbb{R} \setminus E) \ge m\big(\mathbb{R} \setminus [-M, M]\big) = \infty,$$

a contradiction. However, $\{f_n\}$ *cannot* converge uniformly to 0 on the unbounded set E. Thus this counterexample shows that Egoroff's Theorem *cannot* be extended to σ-finite spaces.

- **An extension of Egoroff's Theorem.** Suppose that $\{f_t\}$ is a family of complex measurable functions such that

  (i) $\lim\limits_{t\to\infty} f_t(x) = f(x)$ and

  (ii) Fix a $x \in X$. Then the function $F : (0, \infty) \to \mathbb{C}$ defined by

  $$F(t) = f_t(x)$$

  is continuous.

For each $n \in \mathbb{N}$, we consider the real function $g_n : X \to \mathbb{R}$ given by

$$g_n(x) = \sup_{t \geq n}\{|f_t(x) - f(x)|\}. \tag{3.107}$$

Then, for every $x \in X$, $g_n(x) \to 0$ as $n \to \infty$. Now it remains to show that each g_n is measurable. To this end, let $a \in \mathbb{R}$ and

$$\begin{aligned}
E_a(n) &= \{x \in X \mid g_n(x) < a\} \\
&= \{x \in X \mid |f_t(x) - f(x)| < a \text{ for all } t \geq n\} \\
&= \bigcap_{\substack{r \geq n \\ r \in \mathbb{Q}}} \{x \in X \mid |f_r(x) - f(x)| < a\}. \tag{3.108}
\end{aligned}$$

Since f_r is measurable, f is also measurable by Corollary (a) following Theorem 1.14. Thus $|f_r - f|$ is measurable by Proposition 1.9(b) and then the set

$$\{x \in X \mid |f_r(x) - f(x)| < a\}$$

is measurable for every $a \in \mathbb{R}$ by Definition 1.3(c). By the expression (3.108), since there are countable such measurable sets, Comment 1.6(c) implies that $E_a(n)$ is measurable for every real a. By Definition 1.3(c) again, each g_n is measurable and thus Egoroff's Theorem can be applied to conclude that for every $\epsilon > 0$, there exists a measurable set $E \subseteq X$ such that $\mu(X \setminus E) < \epsilon$ and $\{g_n\}$ converges uniformly to 0 on E. By the definition (3.107), $\{f_t\}$ converges to f uniformly on E.

This completes the proof of the problem. $\blacksquare$

Problem 3.17

Rudin Chapter 3 Exercise 17.

Proof.

(a) Since the inequality is clearly holds if either $\alpha = 0$ or $\beta = 0$, we assume, without loss of generality, that α and β cannot be both zero in the following discussion.

   Let $0 < p \leq 1$. We claim that the continuous function $\varphi : [0, 1] \to \mathbb{R}$ defined by

   $$\varphi(x) = x^p + (1 - x)^p$$

satisfies $\varphi(x) \geq 1$ on $[0, 1]$. To this end, we note that

$$\varphi'(x) = px^{p-1} - p(1-x)^{p-1} = 0$$

if and only if $x = \frac{1}{2}$. Since $p - 1 \leq 0$, we know that

$$\varphi'(x) \geq 0 \quad \text{and} \quad \varphi'(x) \leq 0$$

on $(0, \frac{1}{2})$ and $(\frac{1}{2}, 1)$ respectively. In other words, φ is increasing and decreasing on $(0, \frac{1}{2})$ and $(\frac{1}{2}, 1)$ respectively. By the continuity of φ, these are also valid on $[0, \frac{1}{2}]$ and $[\frac{1}{2}, 1]$ respectively. Thus we observe

$$\varphi(x) \geq \varphi(0) = 1 \text{ on } [0, \tfrac{1}{2}] \quad \text{and} \quad \varphi(x) \geq \varphi(1) = 1 \text{ on } [\tfrac{1}{2}, 1]$$

which mean that

$$x^p + (1-x)^p \geq 1 \tag{3.109}$$

on $[0, 1]$. For arbitrary nonzero complex numbers α and β, we put $x = \frac{|\alpha|}{|\alpha|+|\beta|}$ into the inequality (3.109) to get

$$\frac{|\alpha|^p}{(|\alpha| + |\beta|)^p} + \frac{|\beta|^p}{(|\alpha| + |\beta|)^p} \geq 1$$

which implies

$$(|\alpha| + |\beta|)^p \leq |\alpha|^p + |\beta|^p. \tag{3.110}$$

By the triangle inequality, it is true that $|\alpha - \beta| \leq |\alpha| + |\beta|$, so this and the inequality (3.110) give

$$|\alpha - \beta|^p \leq |\alpha|^p + |\beta|^p. \tag{3.111}$$

Next, let $1 < p < \infty$. The function $\psi : [0, 1] \to \mathbb{R}$ given by

$$\psi(x) = x^p$$

has second derivative $p(p - 1)x^{p-2} > 0$ on $(0, 1)$. By Definition 3.1, it is convex on $(0, 1)$, i.e.,

$$[(1 - \lambda)x + \lambda y]^p \leq (1 - \lambda)x^p + \lambda y^p, \tag{3.112}$$

where $x, y \in (0, 1)$ and $\lambda \in [0, 1]$. Put $\lambda = \frac{1}{2}$, $x = \frac{|\alpha|}{|\alpha|+|\beta|}$ and $y = \frac{|\beta|}{|\alpha|+|\beta|}$ into the inequality (3.112), we obtain

$$\left(\frac{1}{2} \cdot \frac{|\alpha| + |\beta|}{|\alpha| + |\beta|} \right)^p \leq \frac{1}{2} \cdot \frac{|\alpha|^p}{(|\alpha| + |\beta|)^p} + \frac{1}{2} \cdot \frac{|\beta|^p}{(|\alpha| + |\beta|)^p}$$

which reduces to

$$(|\alpha| + |\beta|)^p \leq 2^{p-1}(|\alpha|^p + |\beta|^p).$$

Again, the triangle inequality and this show that

$$|\alpha - \beta|^p \leq 2^{p-1}(|\alpha|^p + |\beta|^p). \tag{3.113}$$

Hence the desired inequality follows from combining the inequalities (3.111) and (3.113).

(b)   (i) If $\mu(X) = 0$, then Proposition 1.24(e) implies that $\|f\|_p = 0$ for every $f \in L^p(\mu)$ and there is nothing to prove. Thus, without loss of generality, we may assume that $\mu(X) > 0$. We divide the proof into several steps:

*** Step 1: Lemma 3.7 and its application.**

Lemma 3.7

For every $\epsilon > 0$, there exists a $\delta > 0$ such that for every $E \in \mathfrak{M}$ with $\mu(E) < \delta$, we have

$$\int_E |f|^p \, \mathrm{d}\mu \le \epsilon. \tag{3.114}$$

Proof of Lemma 3.7. Note that $|f|$ must be bounded a.e. on X. Otherwise, there exists an $E \in \mathfrak{M}$ such that $\mu(E) > 0$ and $|f| = \infty$ on E. Then Proposition 1.24(b) shows that

$$\int_X |f|^p \, \mathrm{d}\mu \ge \int_E |f|^p \, \mathrm{d}\mu = \infty,$$

a contradiction. Thus there exists a positive constant M such that $|f| \le M$ a.e. on X. Given $\epsilon > 0$. Now for every $E \in \mathfrak{M}$ with $\mu(E) < \frac{\epsilon}{M^p}$, we have

$$\int_E |f|^p \, \mathrm{d}\mu \le M^p \mu(E) \le \epsilon \quad \text{a.e. on } X$$

which is exactly what we want. $\qquad\blacksquare$

By Lemma 3.7, there exists a $\delta > 0$ such that the inequality (3.114) holds for every $E \in \mathfrak{M}$ with $\mu(E) < \delta$. For this $\delta > 0$, suppose that we can find a measurable set F such that $f_n \to f$ a.e. on F and $\mu(F) < \infty$. (The existence of such a F will be clear at the end of **Step 2** below.) Then Egoroff's Theorem guarantees that there exists a measurable set $B \subseteq F$ such that $\mu(F \setminus B) < \delta$ and $\{f_n\}$ converges uniformly to f on B. Since $F \setminus B \in \mathfrak{M}$ and $\mu(F \setminus B) < \delta$, we also have

$$\int_{F \setminus B} |f|^p \, \mathrm{d}\mu \le \frac{\epsilon}{2}. \tag{3.115}$$

Furthermore, since $f_n \to f$ uniformly on B, there is a positive integer N such that $n \ge N$ implies that

$$|f_n(x) - f(x)| \le \left(\frac{\epsilon}{\mu(B)} \right)^{\frac{1}{p}}$$

for all $x \in B$ and this means that

$$\int_B |f_n - f|^p \, \mathrm{d}\mu \le \epsilon \tag{3.116}$$

for $n \ge N$.

*** Step 2: Lemma 3.8 and its application.**

Lemma 3.8

If $f \in L^1(\mu)$, then for every $\epsilon > 0$, there exists an $E \in \mathfrak{M}$ such that $F = X \setminus E$, $\mu(F) < \infty$ and

$$\int_E |f| \, \mathrm{d}\mu \le \epsilon.$$

Proof of Lemma 3.8. Since $|f|$ is measurable, Theorem 1.17 (The Simple Function Approximation Theorem) ensures that there exists a sequence of simple measurable functions $\{s_k\}$ on X such that

$$0 \le s_1 \le s_2 \le \cdots \le |f| \tag{3.117}$$

and s_k converges to $|f|$ pointwisely. By Theorem 1.26 (Lebesgue's Monotone Convergence Theorem), we have

$$\lim_{k \to \infty} \int_X s_k \, d\mu = \int_X |f| \, d\mu.$$

Hence, for every $\epsilon > 0$, there exists a positive integer N such that

$$\int_X (|f| - s_k) \, d\mu = \left| \int_X (|f| - s_k) \, d\mu \right| \le \epsilon. \tag{3.118}$$

By Definition 1.16, we have

$$s_k = \sum_{i=1}^{n_k} \alpha_i \chi_{F_i},$$

where each α_i is positive, F_i is measurable and $F_i \cap F_j = \varnothing$ for all $i \ne j$. Recall that $|f| \in L^1(\mu)$, so the hypothesis (3.117) implies that $s_k \in L^1(\mu)$ for every $n \in \mathbb{N}$. Since each α_i is positive, μ is positive on X (i.e., $\mu(E) \ge 0$ for every $E \in \mathfrak{M}$) and

$$\sum_{i=1}^{n_k} \alpha_i \mu(F_i) = \int_X s_k \, d\mu < \infty,$$

they force that $\mu(F_i) < \infty$ for every $i = 1, 2, \ldots, n_k$.

We suppose that

$$E = \{x \in X \mid s_k(x) = 0\} \quad \text{and} \quad F = \bigcup_{i=1}^{n_k} F_i.$$

Then the previous analysis shows that $\mu(F) < \infty$. Furthermore, it is clear that $s_k(x) > 0$ if and only if $x \in F_i$ for some i, so this fact implies that

$$X \setminus E = F.$$

Thus we deduce from the estimate (3.118) and the definition of E that

$$\int_E |f| \, d\mu = \int_E (|f| - s_k) \, d\mu + \int_E s_k \, d\mu \le \int_X (|f| - s_k) \, d\mu + \int_E s_k \, d\mu \le \epsilon.$$

This proves Lemma 3.8. $\qquad\blacksquare$

Since $f \in L^p(\mu)$, we have $|f|^p \in L^1(\mu)$. Then Lemma 3.8 implies that there exists an $E \in \mathfrak{M}$ such that $F = X \setminus E$, $\mu(F) < \infty$ and

$$\int_E |f|^p \, d\mu \le \frac{\epsilon}{2}. \tag{3.119}$$

$\ast$ **Step 3: Constructions of A and B satisfying the hypotheses.** Let E and F be defined as in Lemma 3.8 so that the estimate (3.119) holds. We remark that

$$X = E \cup F = E \cup (F \setminus B) \cup B = A \cup B,$$

where B is the measurable set guaranteed by Egoroff's Theorem in **Step 1** and $A = E \cup (F \setminus B)$. Recall that $B \subseteq F$, so it is easy to see that

$$\mu(B) \le \mu(F) < \infty$$

and since $E \cap (F \setminus B) = \varnothing$, we obtain from Theorem 1.29 and the two estimates (3.119) and (3.115) that

$$\int_A |f|^p \, d\mu = \int_E |f|^p \, d\mu + \int_{F \setminus B} |f|^p \, d\mu \le \epsilon. \tag{3.120}$$

Since $f_n \to f$ a.e. on X and particularly on B, this and Theorem 1.28 (Fatou's Lemma) together show that

$$\int_B |f|^p \, d\mu = \int_B \liminf_{n \to \infty} |f_n|^p \, d\mu \le \liminf_{n \to \infty} \int_B |f_n|^p \, d\mu \tag{3.121}$$

Next, we know from the definition that $A = (E \cup F) \setminus B = X \setminus B$ so that $A \cap B = \varnothing$. Since $\|f_n\|_p \to \|f\|_p$ as $n \to \infty$, it deduces from this, Problem 1.4 and the inequalities (3.120) and (3.121) that

$$\begin{aligned}
\limsup_{n \to \infty} \int_A |f_n|^p \, d\mu &\le \limsup_{n \to \infty} \int_X |f_n|^p \, d\mu + \limsup_{n \to \infty} \left(- \int_B |f_n|^p \, d\mu \right) \\
&= \limsup_{n \to \infty} \int_X |f_n|^p \, d\mu - \liminf_{n \to \infty} \int_B |f_n|^p \, d\mu \\
&\le \int_X |f|^p \, d\mu - \int_B |f|^p \, d\mu \\
&= \int_A |f|^p \, d\mu \\
&\le \epsilon. \tag{3.122}
\end{aligned}$$

* **Step 4: The establishment of $\|f_n - f\|_p \to 0$ as $n \to \infty$.** By part (a), the estimates (3.116) and (3.122), we see that

$$\begin{aligned}
\limsup_{n \to \infty} \int_X |f_n - f|^p \, d\mu &\le \limsup_{n \to \infty} \int_A |f_n - f|^p \, d\mu + \limsup_{n \to \infty} \int_B |f_n - f|^p \, d\mu \\
&\le \gamma_p \limsup_{n \to \infty} \int_A (|f|^p + |f_n|^p) \, d\mu + \epsilon \\
&\le 2\gamma_p \epsilon + \epsilon.
\end{aligned}$$

Since ϵ is arbitrary, we conclude that

$$\lim_{n \to \infty} \|f_n - f\|_p = \lim_{n \to \infty} \left\{ \int_X |f_n - f|^p \, d\mu \right\}^{\frac{1}{p}} = 0.$$

(ii) Put $h_n = \gamma_p(|f|^p + |f_n|^p) - |f - f_n|^p$. By part (a), we have $h_n \ge 0$ for all $n \in \mathbb{N}$. Since f and f_n are measurable, each h_n is also measurable. By Theorem 1.28 (Fatou's Lemma), we get

$$\int_X \left(\liminf_{n \to \infty} h_n \right) d\mu \le \liminf_{n \to \infty} \int_X h_n \, d\mu. \tag{3.123}$$

Since $f_n \to f$ a.e. on X, $|f_n - f|^p \to 0$ and $|f_n|^p \to |f|^p$ a.e. on X. Thus we must have

$$\lim_{n \to \infty} h_n = 2\gamma_p |f|^p \quad \text{a.e. on } X. \tag{3.124}$$

Putting the limit (3.124) into the inequality (3.123), we obtain from Problem 1.4 that

$$
\begin{aligned}
\int_X 2\gamma_p |f|^p \, d\mu &= \int_X \left(\lim_{n\to\infty} h_n \right) d\mu \\
&= \int_X \left(\liminf_{n\to\infty} h_n \right) d\mu \\
&= \liminf_{n\to\infty} \int_X h_n \, d\mu \\
&\leq \liminf_{n\to\infty} \left[\int_X \gamma_p(|f|^p + |f_n|^p) \, d\mu - \int_X |f_n - f|^p \, d\mu \right] \\
&\leq \gamma_p \liminf_{n\to\infty} \left[\int_X (|f|^p + |f_n|^p) \, d\mu \right] - \limsup_{n\to\infty} \int_X |f_n - f|^p \, d\mu \\
&= \gamma_p \liminf_{n\to\infty} \left(\|f_n\|_p^p + \|f\|_p^p \right) - \limsup_{n\to\infty} \int_X |f_n - f|^p \, d\mu. \qquad (3.125)
\end{aligned}
$$

Since $\|f_n\|_p \to \|f\|_p$ as $n \to \infty$, we can further reduce the inequality (3.125) to

$$
0 \leq \limsup_{n\to\infty} \int_X |f_n - f|^p \, d\mu \leq \gamma_p \left(\lim_{n\to\infty} \|f_n\|_p^p + \|f\|_p^p \right) - 2\gamma_p \|f\|_p^p = 0.
$$

Hence this leads to the following

$$
\lim_{n\to\infty} \|f_n - f\|_p = \limsup_{n\to\infty} \|f_n - f\|_p = 0
$$

as desired.

(c) Consider $X = (0,1)$ and $\mu = m$ the Lebesgue measure. For each $n = 1, 2, \ldots$, we let $E_n = (0, \frac{1}{n^p})$ and $f_n = n\chi_{E_n} : (0,1) \to \mathbb{R}$. If $x \in (0,1)$, then there exists a positive integer N such that $x \notin E_n$ for all $n \geq N$. In this case, we have $f_n(x) = 0$ for all $n \geq N$. In other words, we have

$$
f_n(x) \to f(x) \equiv 0
$$

for every $x \in (0,1)$. It is clear that $f \in L^p\big((0,1)\big)$ and $\|f\|_p = 0$. Besides, we have

$$
\|f_n\|_p = \left\{ \int_0^1 |f_n(x)|^p \, dx \right\}^{\frac{1}{p}} = \left\{ \int_{E_n} n^p \, dx \right\}^{\frac{1}{p}} = 1
$$

for every $n \in \mathbb{N}$. Thus we know that

$$
\|f_n\|_p \nrightarrow \|f\|_p
$$

as $n \to \infty$. Finally, since $\|f_n - f\|_p = \|f_n\|_p = 1$, we see that $\|f_n - f\|_p \nrightarrow 0$ as $n \to \infty$, i.e., the conclusion of part (b) is false.

We have completed the proof of the problem. ∎

Remark 3.1

We remark that there is a short and elementary proof of Problem 3.17(b) in [45].

3.5   Convergence in Measure and the Essential Range of $f \in L^\infty(\mu)$

> **Problem 3.18**
>
> *Rudin Chapter 3 Exercise 18.*

Proof.

(a) Given small $\epsilon > 0$. For each positive integer k, we let

$$E_k = \{x \in X \mid |f_n(x) - f(x)| \leq \epsilon \text{ for all } n \geq k\} \quad \text{and} \quad E = \bigcup_{k=1}^\infty E_k.$$

The definitions of E_k and E guarantee that

$$X \setminus E = \bigcap_{k=1}^\infty (X \setminus E_k) = \bigcap_{k=1}^\infty \{x \in X \mid |f_n(x) - f(x)| > \epsilon \text{ for some } n \geq k\}. \tag{3.126}$$

Since $f_n(x) \to f(x)$ a.e. on X, we deduce from the expression (3.126) that

$$\mu(X \setminus E) = 0.$$

Furthermore, we have $E_1 \subseteq E_2 \subseteq \cdots$, so Theorem 1.19 implies that $\mu(E_k) \to \mu(E)$ as $k \to \infty$. Thus it follows from this fact and the result $\mu(X \setminus E) = 0$ that

$$\lim_{k \to \infty} \mu(E_k) = \mu(E) = \mu(X) < \infty.$$

This result ensures that there exists a $N \in \mathbb{N}$ such that $\mu(X \setminus E_N) < \epsilon$.

$$\{x \in X \mid |f_n(x) - f(x)| > \epsilon \text{ for all } n > N\} \subseteq X \setminus E_N.$$

Hence we must have

$$\mu(\{x \in X \mid |f_n(x) - f(x)| > \epsilon \text{ for all } n > N\}) < \epsilon,$$

i.e., $f_n \to f$ in measure.

(b) Suppose that $1 \leq p < \infty$. Since $\|f_n - f\|_p \to 0$ as $n \to \infty$, we must have $f_n - f \in L^p(\mu)$ for every $n \in \mathbb{N}$. Since $f_n \in L^p(\mu)$, we also have $f = (f - f_n) + f_n \in L^p(\mu)$ by Theorem 3.9. Given $\epsilon > 0$, there exists a positive integer N such that $n \geq N$ implies that

$$\|f_n - f\|_p < \epsilon^{p+1}. \tag{3.127}$$

Let $F_n = \{x \in X \mid |f_n(x) - f(x)| > \epsilon\}$ for each $n = 1, 2, \ldots$. Now it is easy to check that[i]

$$\int_{F_n} |f_n(x) - f(x)|^p \, d\mu \geq \int_{F_n} \epsilon^p \, d\mu = \epsilon^p \mu(F_n). \tag{3.128}$$

---

[i]The inequality (3.128) is a consequence of the so-called **Chebyshev's Inequality**: If f is a nonnegative, extended real-valued measurable function on X with measure μ, $0 < p < \infty$ and $\epsilon > 0$, then

$$\mu(\{x \in X \mid f(x) \geq \epsilon\}) \leq \frac{1}{\epsilon^p} \int_X f^p \, d\mu.$$

See [22, Theorem 6.17, p. 193].

Hence we establish from the inequalities (3.127) and (3.128) that

$$\mu(\{x \in X \mid |f_n(x) - f(x)| > \epsilon\}) = \mu(F_n) < \epsilon$$

for $n \geq N$. By the definition, $f_n \to f$ in measure.

Next, we suppose that $p = \infty$. Now $\|f_n - f\|_\infty \to 0$ means the existence of a positive integer N such that $n \geq N$ implies $|f_n(x) - f(x)| < \epsilon$ a.e. on X. By the definition, we have $\mu(F_n) = 0 < \epsilon$ for all $n \geq N$, i.e., $f_n \to f$ in measure in this case.

(c) Suppose that $f_n \to f$ in measure, i.e., for every $\epsilon > 0$, there exists a positive integer N such that $n \geq N$ implies that

$$\mu(\{x \in X \mid |f_n(x) - f(x)| > \epsilon\}) < \epsilon.$$

In particular, for each positive integer k, there exists a positive integer N_k such that

$$\mu(\{x \in X \mid |f_n(x) - f(x)| > 2^{-k}\}) < 2^{-k} \tag{3.129}$$

for all $n \geq N_k$. Now we may choose $n_k > N_k$ *freely* in the estimate (3.129) and consider the subsequence $\{f_{n_k}\}$. We define

$$E_k = \{x \in X \mid |f_{n_k}(x) - f(x)| > 2^{-k}\} \quad \text{and} \quad E = \bigcap_{m=1}^{\infty} \bigcup_{k \geq m}^{\infty} E_k. \tag{3.130}$$

Then it follows from the estimate (3.129) that

$$\mu(E) \leq \mu\left(\bigcup_{k \geq m}^{\infty} E_k \right) \leq \sum_{k=m}^{\infty} \mu(E_k) < \sum_{k=m}^{\infty} \frac{1}{2^k} = \frac{1}{2^{m-1}},$$

where $m = 1, 2, \ldots$. As a result, we must have $\mu(E) = 0$. Take any $p \in X \setminus E$, the definitions (3.130) say that

$$p \in \bigcap_{k \geq m}^{\infty} (X \setminus E_k) = \bigcap_{k \geq m}^{\infty} \{x \in X \mid |f_{n_k}(x) - f(x)| \leq 2^{-k}\}$$

for some $m \in \mathbb{N}$ and then

$$|f_{n_k}(p) - f(p)| \leq 2^{-k}$$

for all $k \geq m$. In other words, we have

$$\lim_{k \to \infty} f_{n_k}(p) = f(p)$$

which is our desired result.

Here we propose two examples which say that the converses of parts (a) and (b) fail:

- **Failure of the converse of part (a).** By Problem 2.9, there exists a sequence of continuous functions $f_n : [0, 1] \to \mathbb{R}$ such that $0 \leq f_n \leq 1$ and

$$\|f_n - 0\|_1 \to 0$$

as $n \to \infty$, but there is no $x \in [0, 1]$ such that $\{f_n(x)\}$ converges. In other words, $f_n(x)$ *does not* converge to 0 a.e. on $[0, 1]$. However, since each f_n is continuous $[0, 1]$, it is $\mathscr{R}$ on $[0, 1]$ so that $f_n \in L^1([0, 1])$ by [49, Theorem 11.33, p. 323]. Thus the sequence of functions $\{f_n\}$ satisfies the hypotheses of part (b) which shows that $f_n \to 0$ in measure. Hence this example implies the the converse of part (a) is false.

- **Failure of the converse of part (b).** Suppose that $X = [0, 1]$, μ is the Lebesgue measure and

$$f_n = e^n \chi_{[0, \frac{1}{n}]}$$

for each $n = 1, 2, \ldots$. Then $f_n \to 0$ a.e. on $[0, 1]$ and part (a) implies that $f_n \to 0$ in measure. However, we note that

$$\|f_n\|_p = \left\{ \int_X |f_n(x)|^p \, d\mu \right\}^{\frac{1}{p}} = \left\{ \int_0^{\frac{1}{n}} e^{np} \, dx \right\}^{\frac{1}{p}} = \frac{e^n}{n^{\frac{1}{p}}} \to \infty$$

as $n \to \infty$, so the sequence $\{f_n\}$ fails to converge in $L^p([0, 1])$.

Finally, by examining the proofs of parts (b) and (c), we see that they remain to be true in the case $\mu(X) = \infty$. However, part (a) *does not* hold in this case. For instance, let $X = [0, \infty)$ and $E_n = [0, n]$ for each $n \in \mathbb{N}$. Consider the functions $f_n = \chi_{E_n}$. Then for every $x \in [0, \infty)$, there exists a positive integer N such that $x \in E_n$ for all $n \geq N$, so it is true that $f_n(x) = 1$ for all $n \geq N$, i.e., $f_n(x) \to f(x) \equiv 1$ on $[0, \infty)$. However, for every $\epsilon \in (0, 1)$, since $f_n(x) = 0$ if and only if $x \in (n, \infty)$, we see that

$$\{x \in [0, \infty) \,|\, |f_n(x) - 1| > \epsilon\} = (n, \infty)$$

for every $n \in \mathbb{N}$. Hence it leads that

$$m(\{x \in [0, \infty) \,|\, |f_n(x) - 1| > \epsilon\}) = \infty$$

for every $n \in \mathbb{N}$, i.e., f_n does not converge in measure to f.

Hence we end the proof of the problem. $\blacksquare$

Problem 3.19

Rudin Chapter 3 Exercise 19.

Proof. For every $\epsilon > 0$, define $E_\epsilon(w) = \{x \in X \,|\, |f(x) - w| < \epsilon\}$ and

$$R_f = \{w \in \mathbb{C} \,|\, \mu(E_\epsilon(w)) > 0 \text{ for every } \epsilon > 0\} = \bigcap_{\epsilon > 0} \{w \in \mathbb{C} \,|\, \mu(E_\epsilon(w)) > 0\}. \tag{3.131}$$

We are going to prove the assertions one by one:

- **R_f is compact.** Given $\{w_n\} \subseteq R_f$ and $w \in \mathbb{C}$ such that $w_n \to w$ as $n \to \infty$. For every $\epsilon > 0$, there exists a positive integer N such that $|w_n - w| < \frac{\epsilon}{2}$. By the triangle inequality, we have

$$|f(x) - w| \leq |f(x) - w_n| + |w_n - w| < |f(x) - w_n| + \frac{\epsilon}{2}$$

which means that $E_{\frac{\epsilon}{2}}(w_n) \subseteq E_\epsilon(w)$ or equivalently

$$\mu(E_\epsilon(w)) > \mu(E_{\frac{\epsilon}{2}}(w_n)) > 0. \tag{3.132}$$

Since ϵ is arbitrary, we know from the estimate (3.132) that $w \in R_f$. In other words, R_f is closed in $\mathbb{C}$.

Recall that $\|f\|_\infty < \infty$, so we pick a complex number w_0 such that $|w_0| > \|f\|_\infty$ and then consider the positive number $\epsilon = |w_0| - \|f\|_\infty$. If $x \in E_\epsilon(w_0)$, then the triangle inequality shows that

$$|w_0| - |f(x)| < |f(x) - w_0| < \epsilon = |w_0| - \|f\|_\infty$$

which implies that $|f(x)| > \|f\|_\infty$. Thus we have $E_\epsilon(w_0) \subseteq \{x \in X \,|\, |f(x)| > \|f\|_\infty\}$. By Definition 3.7, $\mu(\{x \in X \,|\, |f(x)| > \|f\|_\infty\}) = 0$, so $\mu(E_\epsilon(w_0)) = 0$ too. By the definition (3.131), we have $w_0 \notin R_f$ and hence the set R_f is bounded by $\|f\|_\infty$. By the Heine–Borel Theorem, we conclude that R_f is compact.

- **A relation between R_f and $\|f\|_\infty$.** By the previous analysis, it is obvious that the relation

$$R_f \subseteq \{w \in \mathbb{C} \,|\, |w| \le \|f\|_\infty\} \tag{3.133}$$

holds, i.e., R_f lies in $\overline{B(0, \|f\|_\infty)}$. We claim that

$$\|f\|_\infty = \max\{|z| \,|\, z \in R_f\}. \tag{3.134}$$

Given $w \in \mathbb{C}$ and $\epsilon > 0$. Let $B(w, \epsilon) = \{z \in \mathbb{C} \,|\, |z - w| < \epsilon\}$. Then we have

$$f^{-1}(B(w, \epsilon)) = \{x \in X \,|\, f(x) \in B(w, \epsilon)\} = \{x \in X \,|\, |f(x) - w| < \epsilon\} = E_\epsilon(w).$$

By the definition, we see that

$$R_f = \bigcap_{\epsilon > 0} \{w \in \mathbb{C} \,|\, \mu(f^{-1}(B(w, \epsilon))) > 0\} = \mathbb{C} \setminus V, \tag{3.135}$$

where $V = \{w \in \mathbb{C} \,|\, \mu(f^{-1}(B(w, \epsilon))) = 0 \textit{ for some } \epsilon > 0\}$. Now we have the following two facts:

- **Fact 1:** It is easy to show that V is the *largest* open subset of $\mathbb{C}$ such that

$$\mu(f^{-1}(V)) = \mu(\{x \in X \,|\, f(x) \in V\}) = 0,$$

  i.e., V is the largest open subset of $\mathbb{C}$ such that $f(x) \notin V$ for almost all $x \in X$. By this fact and the relation (3.135), we conclude that R_f is the *smallest* closed subset of $\mathbb{C}$ such that $f(x) \in R_f$ for almost all $x \in X$.

- **Fact 2:** Recall from Definition 3.7 that the number $\|f\|_\infty$ is the minimum of the set $\{\alpha \ge 0\}$ such that $\mu(\{x \in X \,|\, |f(x)| > \alpha\}) = 0$, so we see from the relation (3.133) that it is the *minimum* radius of the closed disc centered at 0 containing the set R_f.

Hence we follow immediately from these two observations that the equality actually holds in the set relation (3.133). Since they are equal, our claim (3.134) is established.

- **Relations between A_f and R_f.** By the definition, we have

$$A_f = \left\{ \frac{1}{\mu(E)} \int_E f \, d\mu \,\middle|\, E \in \mathfrak{M} \text{ and } \mu(E) > 0 \right\}. \tag{3.136}$$

We claim that $R_f \subseteq \overline{A_f}$. Given $\epsilon > 0$ and $w \in R_f$. If $w \in A_f$, then there is nothing to prove. Therefore, we may assume that $w \notin A_f$. Consider $E_\epsilon = \{x \in X, \,|\, |f(x) - w| < \epsilon\}$. By the definition of R_f, we have $\mu(E_\epsilon) > 0$. If $\mu(E_\epsilon) = \infty$ for every ϵ, then $f(x) = w$ a.e. on X. In this case, we have

$$\frac{1}{\mu(E_\epsilon)} \int_{E_\epsilon} f \, d\mu = w \in A_f,$$

a contradiction. Therefore, we may assume that $\mu(E_\epsilon) < \infty$ *for infinitely many* $\epsilon > 0$. In fact, we may take $\epsilon = \frac{1}{n}$ and let $E(n) = E_{\frac{1}{n}}$. Then, since $f \in L^\infty(\mu)$, we have $f - w \in L^1(E(n))$, where

$$L^1(E(n)) = \left\{ f : X \to \mathbb{C} \,\middle|\, \int_{E(n)} |f| \, d\mu < \infty \right\}.$$

Consequently, it follows from Theorem 1.33 and the definition of E that

$$\left| \frac{1}{\mu\big(E(n)\big)} \int_{E(n)} f \, d\mu - w \right| \leq \frac{1}{\mu\big(E(n)\big)} \int_{E(n)} |f - w| \, d\mu < \frac{1}{n}.$$

In other words, w is a limit point of A_f and this means that $w \in \overline{A_f}$. Hence this proves the claim that $R_f \subseteq \overline{A_f}$.

- A_f **is not always closed.** For example, we consider $X = [0, 1]$, $f(x) = x$ with $\mu = m$ the Lebesgue measure. By Definition 3.7, $f \in L^\infty\big([0, 1]\big)$. In addition, for a measurable set E in $[0, 1]$ with $m(E) > 0$, we have

$$w(E) = \frac{1}{m(E)} \int_E f(x) \, dx. \tag{3.137}$$

We claim that $w(E)$ can take any value in $(0, 1)$. Indeed, if $a, b \in (0, 1)$ and $E = (a, b)$, then we have $m(E) = b - a > 0$ and it follows from the definition (3.137) that

$$w(E) = \frac{1}{m(E)} \int_E f \, dx = \frac{1}{b - a} \int_a^b x \, dx = \frac{1}{b - a} \times \frac{b^2 - a^2}{2} = \frac{a + b}{2}$$

which implies the claim.

Next, we claim that $w(E) \neq 0$. Assume that $w(E_0) = 0$ for a measurable set $E_0 \subseteq [0, 1]$ with $m(E_0) > 0$. Then we have

$$\int_{E_0} f(x) \, dx = 0,$$

but Theorem 1.39(a) implies that $f(x) = 0$ a.e. on E_0 which contradicts the fact that $f(x) = 0$ only at $x = 0$. Hence 0 is not a limit point of A_f and A_f is not closed in $\mathbb{R}$.

- **A measure μ such that A_f is convex for every $f \in L^\infty(\mu)$.** Consider $X = \{0\}$ and μ the counting measure (see Example 1.20(a)). Then we have $\mathfrak{M} = \{\varnothing, X\}$ and $\mu(X) = \mu(\{0\}) = 1 > 0$. By the definition, $\|f\|_\infty = |f(0)| < \infty$ for every $f \in L^\infty(\mu)$ so that $f(0)$ is either $\|f\|_\infty$ or $-\|f\|_\infty$. Now for every $f \in L^\infty(\mu)$, we have

$$w(X) = \frac{1}{\mu(X)} \int_X f \, d\mu = f(0).$$

Therefore, the definition (3.136) gives either $A_f = \{\|f\|_\infty\}$ or $A_f = \{-\|f\|_\infty\}$, but both cases are also convex sets.

- **A measure μ such that A_f is not convex for some $f \in L^\infty(\mu)$.** We consider the set $X = \{0, 1\}$, μ the counting measure and $f(x) = x$. Then we have $\mathfrak{M} = \{\varnothing, \{0\}, \{1\}, X\}$, $\mu(\{0\}) = \mu(\{1\}) = 1 > 0$ and $\mu(X) = 2 > 0$. Now we have

$$w(\{0\}) = \frac{1}{\mu(\{0\})} \int_{\{0\}} f \, d\mu = f(0) = 0, \quad w(\{1\}) = \frac{1}{\mu(\{1\})} \int_{\{1\}} f \, d\mu = f(1) = 1,$$

$$w(X) = \frac{1}{\mu(X)} \int_X f \, d\mu = \frac{1}{2}[f(0)\mu(\{0\}) + f(1)\mu(\{1\})] = \frac{1}{2}[f(0) + f(1)] = \frac{1}{2}.$$

Thus we have $A_f = \{0, 1, \frac{1}{2}\}$ is *not* convex.

- **The situations when $L^\infty(\mu)$ is replaced by $L^1(\mu)$.** We consider $f : (0, \infty) \to \mathbb{C}$ by

$$f(x) = \begin{cases} \dfrac{1}{\sqrt{x}}, & \text{if } x \in (0, 1]; \\[2mm] 0, & \text{if } x > 1 \end{cases}$$

and take $\mu = m$ the Lebesgue measure. By Lemma 3.1, we have $f \in L^1\big((0,\infty)\big)$. For any large positive integer N, if $x \in (0, \frac{1}{N^2})$, then we have $f(x) > N$ and

$$m(\{x \in (0,\infty) \,|\, f(x) > N\}) = m\Big(\Big(0, \frac{1}{N^2}\Big)\Big) = \frac{1}{N^2} > 0.$$

By Definition 3.7, we have $\|f\|_\infty = \infty$, i.e., $f \notin L^\infty\big((0,\infty)\big)$. Given that $\epsilon > 0$ and sufficiently large $N \in \mathbb{N}$.

- R_f **is unbounded and not compact.** If $x \in \big(\frac{1}{(N+\epsilon)^2}, \frac{1}{(N-\epsilon)^2}\big) \subseteq (0,1)$, then we have $|f(x) - N| < \epsilon$ so that

$$m(\{x \in (0,\infty) \,|\, |f(x) - N| < \epsilon\}) = m\Big(\Big(\frac{1}{(N+\epsilon)^2}, \frac{1}{(N-\epsilon)^2}\Big)\Big) > 0.$$

  Thus $N \in R_f$ and then $\mathbb{N} \subseteq R_f$, i.e., R_f is unbounded and not compact.

- A_f **is unbounded.** Next, if $E = (0, \frac{4}{N^2})$, then $m(E) = \frac{4}{N^2} > 0$ and

$$w(E) = \frac{1}{m(E)} \int_E f(x)\,\mathrm{d}x = \frac{N^2}{4} \int_0^{\frac{4}{N^2}} x^{-\frac{1}{2}}\,\mathrm{d}x = \frac{N^2}{4} \times 2\sqrt{x}\,\Big|_0^{\frac{4}{N^2}} = N.$$

  Therefore, $\mathbb{N} \subseteq A_f$, i.e., A_f is unbounded.

- A_f **is not always closed.** In fact, the previous example that $X = [0,1]$, $f(x) = x$ and $\mu = m$ satisfies both $f \in L^1\big([0,1]\big)$ and $f \in L^\infty\big([0,1]\big)$. Thus the same conclusion that 0 is *not* a limit point of A_f is achieved.

- **A measure** μ **such that** A_f **is convex for every** $f \in L^1(\mu)$**.** The example that $X = \{0\}$ and μ the counting measure also works for this case because if $f \in L^1(\mu)$, then $|f(0)| = \|f\|_1 < \infty$. Thus we have either $A_f = \{\|f\|_1\}$ or $A_f = \{-\|f\|_1\}$.

- **A measure** μ **such that** A_f **is not convex for some** $f \in L^1(\mu)$**.** The example we considered for the case $L^\infty(\mu)$ also works in this case.

We complete the proof of the problem. $\blacksquare$

3.6　A Converse of Jensen's Inequality

Problem 3.20

Rudin Chapter 3 Exercise 20.

Proof. We check Definition 3.1. Let $p, q \in \mathbb{R}$ and $\lambda \in [0,1]$. By the trick of Proposition 1.24(f), we see that

$$\lambda p + (1-\lambda)q = p \int_{[0,1]} \chi_{[0,\lambda]}(x)\,\mathrm{d}x + q \int_{[0,1]} \chi_{[\lambda,1]}(x)\,\mathrm{d}x$$

$$= \int_0^1 [p\chi_{[0,\lambda]}(x) + q\chi_{[\lambda,1]}(x)]\,\mathrm{d}x. \tag{3.138}$$

Suppose that

$$f(x) = p\chi_{[0,\lambda]}(x) + q\chi_{[\lambda,1]}(x) \tag{3.139}$$

which is clearly a real and bounded function. Besides, since $[0,\lambda]$ and $[\lambda,1]$ are Borel sets in $[0,1]$, $\chi_{[0,\lambda]}$ and $\chi_{[\lambda,1]}$ are measurable by Proposition 1.9(d). By the last paragraph in [51, §1.22,

p. 19], the function f, as the sum of two measurable functions, is also measurable. Hence, by substituting the expressions (3.138) and (3.139) into the inequality in question, we obtain

$$\varphi\big(\lambda p + (1-\lambda)q\big) \le \int_0^1 \varphi\big(p\chi_{[0,\lambda]}(x) + q\chi_{[\lambda,1]}(x)\big)\,\mathrm{d}x$$
$$= \int_0^\lambda \varphi(p)\,\mathrm{d}x + \int_\lambda^1 \varphi(q)\,\mathrm{d}x$$
$$= \lambda\varphi(p) + (1-\lambda)\varphi(q).$$

Hence φ is convex in $\mathbb{R}$, completing the proof of the problem.

Remark 3.2

We note that Problem 3.20 is a converse of Theorem 3.3 (Jensen's Inequality).

3.7   The Completeness/Completion of a Metric Space

Problem 3.21

Rudin Chapter 3 Exercise 21.

Proof. Let (X,d) and (X^*,ρ) be two metric spaces. An *isometry* of X into X^* is a mapping $\varphi : X \to Y$ such that

$$\rho\big(\varphi(p),\varphi(q)\big) = d(p,q) \tag{3.140}$$

for all $p,q \in X$. We notice immediately that an isometry φ is necessarily injective and continuous. If φ is surjective, then we call φ an *isomorphism*. Two metric spaces X and Y are called *isomorphic* if there is an isomorphism between them. Before stating the statement, we need:

Lemma 3.9

Let (X,d) be a metric space with metric d and $p \in X$. Then the function $f : X \to \mathbb{R}$ defined by $f(x) = d(x,p)$ is continuous.

Proof of Lemma 3.9. Given $\epsilon > 0$. Let $x \in X$. If $y \in X$ satisfies $d(x,y) < \epsilon$, then the triangle inequality implies that

$$|f(x) - f(y)| = |d(x,p) - d(y,p)| \le d(x,y) < \epsilon.$$

Hence f is continuous, completing the proof of Lemma 3.9.

Lemma 3.10

Suppose that (X,d) and (Y,ρ) are metric spaces and Y is complete. If E is dense in X and $f : E \to Y$ is an isometry, then there exists an isometry $g : X \to Y$ such that $g|_E = f$.

Proof of Lemma 3.10. If $x \in E$, we define $g(x) = f(x)$. Let $x \in X \setminus E$. Since E is dense in X, we can choose a sequence $\{x_n\} \subseteq E$ such that $x_n \to x$. By [49, Theorem 3.11(a)], $\{x_n\}$ is Cauchy in X. Thus, given $\epsilon > 0$, there exists a positive integer N such that $n, m \geq N$ imply that

$$d(x_m, x_n) < \epsilon.$$

Since f is an isometry, it follows from the expression (3.140) that

$$\rho\big(f(x_m), f(x_n)\big) < \epsilon$$

for all $n, m \geq N$. In other words, $\{f(x_n)\}$ is Cauchy in Y. Since Y is complete, $\{f(x_n)\}$ converges to a limit y in Y and we define $g(x) = y$ in this case.

Now this y is *uniquely determined* by x. Indeed, let $\{x'_n\} \subseteq E$ be another sequence converging to x and y' be the limit of $\{f(x'_n)\}$ in Y. For each $m \in \mathbb{N}$, it follows from Lemma 3.9 and the expression (3.140) that

$$\rho\big(y, f(x'_m)\big) = \lim_{n \to \infty} \rho\big(f(x_n), f(x'_m)\big) = \lim_{n \to \infty} d(x_n, x'_m) = d(x, x'_m).$$

Then this implies that

$$\rho(y, y') = \lim_{m \to \infty} \rho\big(y, f(x'_m)\big) = \lim_{m \to \infty} d(x, x'_m) = 0.$$

Since ρ is a metric, we must have $y = y'$ and so we obtain a mapping $g : X \to Y$ given by

$$g(x) = \begin{cases} f(x), & \text{if } x \in E; \\[2mm] \lim_{n \to \infty} f(x_n), & \text{if } x \in X \setminus E, \{x_n\} \subseteq X \text{ and } x_n \to x. \end{cases} \tag{3.141}$$

Next, we show that g is an isometry and we consider the following situations:

- **Case (i):** $x, y \in E$. In this case, we know from the definition (3.141) that $g(x) = f(x)$ and $g(y) = f(y)$ and then the use of the expression (3.140) implies that
$$\rho\big(g(x), g(y)\big) = \rho\big(f(x), f(y)\big) = d(x, y).$$

- **Case (ii):** $x \in X \setminus E$ **and** $y \in E$. In this case, we have
$$\rho\big(g(x), g(y)\big) = \rho\big(g(x), f(y)\big).$$

Choose a sequence $\{x_n\}$ in E such that $x_n \to x$. By Lemma 3.9 and the expression (3.140), we establish that

$$\rho\big(g(x), f(y)\big) = \lim_{n \to \infty} \rho\big(f(x_n), f(y)\big) = \lim_{n \to \infty} d(x_n, y) = d(x, y). \tag{3.142}$$

- **Case (iii):** $x, y \in X \setminus E$. Let $\{x_n\}, \{y_n\} \subseteq E$ be sequences converging to x and y respectively. Then we deduce from the definition (3.141) and the expression (3.142) that
$$\rho\big(g(x), f(y_n)\big) = d(x, y_n). \tag{3.143}$$

Apply Lemma 3.9 to (3.143), we see that

$$\rho\big(g(x), g(y)\big) = \lim_{n \to \infty} \rho\big(g(x), f(y_n)\big) = \lim_{n \to \infty} d(x, y_n) = d(x, y).$$

By the definition, g is an isometry and $g|_E = f$.   ■

> **Lemma 3.11**
>
> The mapping g in Lemma 3.9 is unique up to isometry.

Proof of Lemma 3.11. Let $g' : X \to Y$ be an isometry such that $g'|_E = f$. If $x \in E$, then

$$g(x) = f(x) = g'(x).$$

If $x \in X \setminus E$, then there is a sequence $\{x_n\} \subseteq E$ such that $x_n \to x$ as $n \to \infty$. The triangle inequality, the expressions (3.140) and (3.142) indicate that

$$
\begin{aligned}
\rho\big(g(x), g'(x)\big) &\leq \rho\big(g(x), g(x_n)\big) + \rho\big(g(x_n), g'(x_n)\big) + \rho\big(g'(x_n), g'(x)\big) \\
&= \rho\big(g(x), f(x_n)\big) + \rho\big(f(x_n), f'(x_n)\big) + \rho\big(f'(x_n), g'(x)\big) \\
&= \rho\big(g(x), f(x_n)\big) + d(x_n, x_n) + \rho\big(f'(x_n), g'(x)\big) \\
&= d(x, x_n) + d(x_n, x)
\end{aligned}
$$

Hence when $n \to \infty$, it is clear that $\rho(g(x), g'(x)) = 0$, so $g(x) = g'(x)$ on $X \setminus E$ and our result follows from this. $\blacksquare$

Let's return to the proof of the problem. Let (Y_1, ρ_1) and (Y_2, ρ_2) be complete metric spaces, $\varphi_1 : X \to Y_1$ and $\varphi_2 : X \to Y_2$ be isometries such that $\varphi_1(X)$ is dense in Y_1 and $\varphi_2(X)$ is dense in Y_2. We claim that (Y_1, ρ_1) and (Y_2, ρ_2) are isomorphic. Define $f : \varphi_1(X) \to Y_2$ by

$$f = \varphi_2 \circ \varphi_1^{-1}. \tag{3.144}$$

If $p, q \in X$, $x = \varphi_1(p)$ and $y = \varphi_1(q)$, then we have

$$
\begin{aligned}
\rho_2\big(f(x), f(y)\big) &= \rho_2\big(\varphi_2(\varphi_1^{-1}(x)), \varphi_2(\varphi_1^{-1}(y))\big) \\
&= \rho_2\big(\varphi_2(p), \varphi_2(q)\big) \\
&= d(p, q) \\
&= \rho_1\big(\varphi_1(p), \varphi_1(q)\big) \\
&= \rho_1(x, y).
\end{aligned}
$$

Consequently, f is an isometry. Since $\varphi_1(X)$ is dense in Y_1 and Y_2 is complete, Lemma 3.10 ensures that there exists an isometry $g : Y_1 \to Y_2$ such that

$$g|_{\varphi_1(X)} = f.$$

It remains to prove that g is surjective. To this end, let $y \in Y_2$. Since $\overline{\varphi_2(X)} = Y_2$, we can select a sequence $\{y_n\} \subseteq \varphi_2(X)$ such that $y_n \to y$ as $n \to \infty$. Furthermore, there exists a sequence $\{x_n\} \subseteq X$ such that

$$y_n = \varphi_2(x_n)$$

for all $n \in \mathbb{N}$. Consider the corresponding sequence

$$z_n = \varphi_1(x_n) \tag{3.145}$$

for $n \in \mathbb{N}$. Since $\{y_n\}$ is Cauchy in $\varphi_2(X)$ (and hence in Y_2) and φ_2 is an isometry, the expression (3.140) forces that $\{x_n\}$ is Cauchy in X. Then, since φ_1 is an isometry, the expression (3.140) again forces that $\{z_n\}$ is Cauchy in $\varphi_1(X)$ (and hence in Y_1). Since Y_1 is complete, we can find $z \in Y_1$ such that $z_n \to z$ as $n \to \infty$. We claim that

$$g(z) = y.$$

Thus we deduce from Lemma 3.10, the definition (3.145) and the fact $g|_{\varphi_1(X)} = f$ that

$$\rho_2\big(g(z), y\big) = \lim_{n\to\infty} \rho_2\big(g(z_n), y\big) = \lim_{n\to\infty} \rho_2\big(g(\varphi_1(x_n)), y\big) = \lim_{n\to\infty} \rho_2\big(f(\varphi_1(x_n)), y\big). \tag{3.146}$$

Finally, the definition (3.144) reduces the limit (3.146) to

$$\rho_2\big(g(z), y\big) = \lim_{n\to\infty} \rho_2\big(f(\varphi_1(x_n)), y\big) = \lim_{n\to\infty} \rho_2\big(\varphi_2(x_n), y\big) = \lim_{n\to\infty} \rho_2(y_n, y) = \rho_2(y, y) = 0.$$

Therefore, $g(z) = y$ and g is surjective. Hence g is an isomorphism and our claim follows, completing the proof of the problem. ∎

> ### Problem 3.22
>
> *Rudin Chapter 3 Exercise 22.*

Proof. This problem is proven in [63, Problem 3.20, p. 50]. ∎

3.8   Miscellaneous Problems

> ### Problem 3.23
>
> *Rudin Chapter 3 Exercise 23.*

Proof. If $\alpha_{n_0} = 0$ for some n_0, then we must have $f = 0$ a.e. on X by Theorem 1.39(a). However, this implies that the measure of

$$f^{-1}\big((\alpha, \infty]\big) = \{x \in X \mid f(x) > \alpha\}$$

is zero for every $\alpha > 0$. By Definition 3.7, it means that $\|f\|_\infty = 0$ which contradicts the hypothesis. Thus we have $\alpha_n \neq 0$ for all $n \in \mathbb{N}$ so that the limit in question is well-defined.

By Definition 3.7, we see that $|f(x)| \leq \|f\|_\infty$ holds for almost all $x \in X$. Therefore, we have

$$\alpha_{n+1} \leq \|f\|_\infty \cdot \alpha_n \quad \text{a.e. on } X \tag{3.147}$$

for all $n \in \mathbb{N}$. Since $\|f\|_\infty > 0$, we can find a $\epsilon > 0$ such that $\|f\|_\infty > \epsilon > 0$. By the definition, the measurable set

$$F = \{x \in X \mid |f(x)| \geq \|f\|_\infty - \epsilon > 0\}$$

satisfies $\mu(F) > 0$. Apply Theorem 3.5 (Hölder's Inequality) to the measurable functions f^n and 1 with $p = \frac{n+1}{n}$, we obtain

$$\alpha_n = \int_X |f|^n \, d\mu \leq \left\{ \int_X (|f|^n)^{\frac{n+1}{n}} \right\}^{\frac{n}{n+1}} \left\{ \int_X 1^{n+1} \, d\mu \right\}^{\frac{1}{n+1}} = (\alpha_{n+1})^{\frac{n}{n+1}} \times \mu(X)^{\frac{1}{n+1}}$$

so that

$$\frac{\alpha_{n+1}}{\alpha_n} \geq (\alpha_{n+1})^{\frac{1}{n+1}} \times \mu(X)^{\frac{-1}{n+1}}$$

$$\geq \left\{ \int_F |f|^{n+1} \, d\mu \right\}^{\frac{1}{n+1}} \times \mu(X)^{\frac{-1}{n+1}}$$

$$\geq \left[(\|f\|_\infty - \epsilon)^{n+1} \mu(F) \right]^{\frac{1}{n+1}} \times \mu(X)^{\frac{-1}{n+1}}$$

$$= \left(\frac{\mu(F)}{\mu(X)}\right)^{\frac{1}{n+1}} \times (\|f\|_\infty - \epsilon). \tag{3.148}$$

Thus we follow from the inequalities (3.147) and (3.148) that

$$\left(\frac{\mu(F)}{\mu(X)}\right)^{\frac{1}{n+1}} \times (\|f\|_\infty - \epsilon) \le \frac{\alpha_{n+1}}{\alpha_n} \le \|f\|_\infty \quad \text{a.e. on } X \tag{3.149}$$

for all $n \in \mathbb{N}$. Since $0 < \mu(F) \le \mu(X) < \infty$, the number $\left(\frac{\mu(F)}{\mu(X)}\right)^{\frac{1}{n+1}}$ tends to 1 as $n \to \infty$. By taking limit to both sides of the inequality (3.149), we see immediately that

$$\|f\|_\infty - \epsilon \le \liminf_{n\to\infty} \frac{\alpha_{n+1}}{\alpha_n} \le \limsup_{n\to\infty} \frac{\alpha_{n+1}}{\alpha_n} \le \|f\|_\infty \quad \text{a.e. on } X.$$

Remember that ϵ is arbitrary, so this implies that

$$\lim_{n\to\infty} \frac{\alpha_{n+1}}{\alpha_n} = \|f\|_\infty \quad \text{a.e. on } X$$

and we complete the proof of the problem. ∎

Problem 3.24

Rudin Chapter 3 Exercise 24.

Proof. Let $x \ge 0$ and $y \ge 0$. Put $\alpha - \beta = x$ and $\alpha = y$ into the equality (3.111), we gain

$$|x|^p - |y|^p \le |x - y|^p. \tag{3.150}$$

Next, we substitute $\alpha - \beta = y$ and $\alpha = x$ into the equality (3.111) to get

$$-|x - y|^p \le |x|^p - |y|^p. \tag{3.151}$$

Obviously, the inequalities (3.150) and (3.151) together imply that

$$|x^p - y^p| = \big||x|^p - |y|^p\big| \le |x - y|^p \tag{3.152}$$

if $0 < p < 1$.

If $x = y$, then the inequality

$$|x^p - y^p| \le p|x - y|(x^{p-1} + y^{p-1})$$

certainly holds. Without loss of generality, we may assume that $x < y$. For $p > 1$, we consider the function $\varphi(t) = t^p$ on $[x, y]$. Since φ is differentiable in (x, y), the Mean Value Theorem implies that

$$|y^p - x^p| \le |x - y||\varphi'(\xi)| \le p|x - y|\xi^{p-1} \tag{3.153}$$

for some $\xi \in (x, y)$. Clearly, we have $\xi^{p-1} \le x^{p-1} + y^{p-1}$, so it reduces from the inequality (3.153) that

$$|x^p - y^p| \le p|x - y|(x^{p-1} + y^{p-1}). \tag{3.154}$$

Hence we establish from the inequalities (3.152) and (3.154) that

$$|x^p - y^p| \le \begin{cases} |x - y|^p, & \text{if } 0 < p < 1; \\[2mm] p|x - y|(x^{p-1} + y^{p-1}), & \text{if } 1 \le p < \infty. \end{cases}$$

(a) We prove the assertions one by one:

- **The truth of the inequality.** Put $x = |f| \geq 0$ and $y = |g| \geq 0$ into the inequality (3.152) and then take integration, we get

$$\int \left| |f|^p - |g|^p \right| \, d\mu \leq \int \left| |f| - |g| \right|^p \, d\mu. \tag{3.155}$$

Since $\left| |a| - |b| \right| \leq |a - b|$ holds for every $a, b \in \mathbb{R}$, we follow immediately from the inequality (3.155) that

$$\int \left| |f|^p - |g|^p \right| \, d\mu \leq \int \left| |f| - |g| \right|^p \, d\mu \leq \int |f - g|^p \, d\mu, \tag{3.156}$$

where $0 < p < 1$.

- **Δ is a metric.** Define $\Delta : L^p(\mu) \times L^p(\mu) \to \mathbb{C}$ by

$$\Delta(f, g) = \int |f - g|^p \, d\mu.$$

By Remark 3.10, $L^p(\mu)$ is a complex vector space, so $f - g \in L^p(\mu)$ if $f, g \in L^p(\mu)$. Thus we have

$$0 \leq \Delta(f, g) = \int |f - g|^p \, d\mu < \infty$$

for all $f, g \in L^p(\mu)$. Next, Theorem 1.39(a) ensures that $\Delta(f, g) = 0$ if and only if $|f - g|^p = 0$ a.e. on X, i.e., $f = g$ a.e. on X. It is clear that

$$\Delta(f, g) = \int |f - g|^p \, d\mu = \int |g - f|^p \, d\mu = \Delta(g, f).$$

Finally, since $|f|^p - |g|^p \leq \left| |f|^p - |g|^p \right|$, it is clear from the inequality (3.156) that

$$\int (|f|^p - |g|^p) \, d\mu \leq \int \left| |f|^p - |g|^p \right| \, d\mu \leq \int |f - g|^p \, d\mu. \tag{3.157}$$

By replacing f and g by $f - g$ and $h - g$ in the inequality (3.157), we deduce that

$$\begin{aligned}
\Delta(f, g) - \Delta(h, g) &= \int (|f - g|^p - |h - g|^p) \, d\mu \\
&\leq \int |(f - g) - (h - g)|^p \, d\mu \\
&= \int |f - h|^p \, d\mu \\
&= \Delta(f, h).
\end{aligned}$$

After rearrangement, we have $\Delta(f, g) \leq \Delta(f, h) + \Delta(h, g)$. Hence, by the definition, Δ is in fact a metric.

- **$(L^p(\mu), \Delta)$ is a complete metric space.** Suppose that $\{f_n\}$ is a Cauchy sequence in $L^p(\mu)$ with respect to the metric Δ, i.e., for every $\epsilon > 0$, there exists a positive integer N such that $n, m \geq N$ implies that

$$\Delta(f_n, f_m) < \epsilon. \tag{3.158}$$

Although $\Delta(f, g) = \|f - g\|_p^p$ here, we cannot apply Theorem 3.11 directly to conclude that $L^p(\mu)$ is complete with respect to Δ because $p \geq 1$ in Theorem 3.11.

We imitate the proof of Theorem 3.11. To start with, there is a subsequence $\{f_{n_i}\}$ such that

$$\Delta(f_{n_{i+1}}, f_{n_i}) < 2^{-i} \tag{3.159}$$

for each $i = 1, 2, \ldots$. We define $g_k : X \to [0, \infty]$ and $g : X \to [0, \infty]$ by

$$g_k = \sum_{i=1}^{k} |f_{n_{i+1}} - f_{n_i}| \quad \text{and} \quad g = \sum_{i=1}^{\infty} |f_{n_{i+1}} - f_{n_i}|$$

respectively. It is clear that $g_k \geq 0$ and $g \geq 0$ and furthermore, it follows from Theorem 3.5 (Minkowski's Inequality) that

$$
\begin{aligned}
\Delta(g_k, 0) &= \Delta\Big(\sum_{i=1}^{k} |f_{n_{i+1}} - f_{n_i}|, 0 \Big) \\
&= \int \Big(\sum_{i=1}^{k} |f_{n_{i+1}} - f_{n_i}| \Big)^p \, d\mu \\
&\leq \Big[\sum_{i=1}^{k} \Big\{ \int |f_{n_{i+1}} - f_{n_i}|^p \, d\mu \Big\}^{\frac{1}{p}} \Big]^p \\
&= \Big[\sum_{i=1}^{k} \Delta(f_{n_{i+1}}, f_{n_i})^{\frac{1}{p}} \Big]^p
\end{aligned}
\tag{3.160}
$$

for each $k = 1, 2, \ldots$.

To proceed further, we need the following result:

> **Lemma 3.12**
>
> Let $p \in (0, 1)$. For nonnegative real numbers $a_1, a_2, \ldots, a_k$, we have
>
> $$(a_1 + a_2 + \cdots + a_k)^p \leq a_1^p + a_2^p + \cdots + a_k^p.$$

Proof of Lemma 3.12. Replace β by $-\beta$ in the inequality (3.111), we get

$$(\alpha + \beta)^p \leq |\alpha|^p + |\beta|^p$$

for $p \in (0, 1)$. In particular, if $\alpha \geq 0$ and $\beta \geq 0$, then we have

$$(\alpha + \beta)^p \leq \alpha^p + \beta^p \tag{3.161}$$

for $p \in (0, 1)$. By applying the inequality (3.161) repeatedly to $a_1, a_2, \ldots, a_k$, we derive the desired inequality. This completes the proof of Lemma 3.12. ∎

Apply Lemma 3.9 to the inequality (3.160) and then using the inequality (3.159) to get

$$\Delta(g_k, 0) \leq \sum_{i=1}^{k} \Delta(f_{n_{i+1}}, f_{n_i}) < \sum_{i=1}^{k} 2^{-i} \leq 1 \tag{3.162}$$

for each $k = 1, 2, \ldots$.

Since $g_k(x) \to g(x)$ as $k \to \infty$ for every $x \in X$, we have $g_k(x)^p \to g(x)^p$ as $k \to \infty$ on X. We note that since each f_{n_i} is measurable, each g_k^p is also measurable and we may apply Theorem 1.28 (Fatou's Lemma) to $\{g_k^p\}$ and the inequality (3.162) to get

$$\int g^p \, d\mu = \int \lim_{k\to\infty} g_k^p \, d\mu = \int \liminf_{k\to\infty} g_k^p \, d\mu \le \liminf_{k\to\infty} \int g_k^p \, d\mu = \liminf_{k\to\infty} \Delta(g_k, 0) \le 1,$$

implying that $g \in L^p(\mu)$. In particular, $g < \infty$ a.e. on X, so the series

$$f_{n_1}(x) + \sum_{i=1}^{\infty} [f_{n_{i+1}}(x) - f_{n_i}(x)] \tag{3.163}$$

converges absolutely for almost every $x \in X$. Denote the sum (3.163) by $f(x)$ for those x at which the sum (3.163) converges and put $f(x) = 0$ on the remaining set of measure zero.

Since we have

$$f_{n_1}(x) + \sum_{i=1}^{k-1} [f_{n_{i+1}}(x) - f_{n_i}(x)] = f_{n_k}(x),$$

it implies that $f_{n_i}(x) \to f(x)$ as $i \to \infty$ a.e. on X.

Now it remains to show that $f \in L^p(\mu)$ and $\Delta(f_n, f) \to 0$ as $n \to \infty$. Given $\epsilon > 0$. Since $\{f_n\}$ is Cauchy in $L^p(\mu)$ with respect to Δ, there exists a $N \in \mathbb{N}$ such that the estimate (3.158) holds for $n, m \ge N$. Take $m \ge N$, since $f_{n_i} \to f$ as $i \to \infty$ a.e. on X, Theorem 1.28 (Fatou's Lemma) again implies that

$$\int |f - f_m|^p \, d\mu = \int \left(\liminf_{i\to\infty} |f_{n_i} - f_m|^p \right) d\mu$$

$$\le \liminf_{i\to\infty} \left(\int |f_{n_i} - f_m|^p \, d\mu \right)$$

$$= \liminf_{i\to\infty} \Delta(f_{n_i}, f_m) < \epsilon. \tag{3.164}$$

so we conclude that $f - f_m \in L^p(\mu)$, hence that $f \in L^p(\mu)$.

Finally, according to the estimate (3.164), it follows that $\Delta(f, f_m) < \epsilon$ for all $m \ge N$. Since ϵ is arbitrary, we conclude that

$$\lim_{m\to\infty} \Delta(f, f_m) = 0.$$

This shows the completeness of the space $L^p(\mu)$ with respect to Δ.

(b) Similar to part (a), a direct substitution of $x = |f|$ and $y = |g|$ into the inequality (3.154) and the use of the fact $\big| |x| - |y| \big| \le |x - y|$ give

$$\int \big| |f|^p - |g|^p \big| \, d\mu \le p \int \big| |f| - |g| \big| \times (|f|^{p-1} + |g|^{p-1}) \, d\mu$$

$$\le p \int |f - g|(|f|^{p-1} + |g|^{p-1}) \, d\mu. \tag{3.165}$$

By Theorem 3.5 (Hölder's Inequality) and the facts that $\|f\|_p \le R$ and $\|g\|_p \le R$, we observe that

$$\int |f - g|(|f|^{p-1} + |g|^{p-1}) \, d\mu \le \int |f - g| \cdot |f|^{p-1} \, d\mu + \int |f - g| \cdot |g|^{p-1} \, d\mu$$

$$\le \left\{ \int |f - g|^p \, d\mu \right\}^{\frac{1}{p}} \left\{ \int (|f|^{p-1})^{\frac{p}{p-1}} \, d\mu \right\}^{\frac{p-1}{p}}$$

$$+ \left\{ \int |f - g|^p \, \mathrm{d}\mu \right\}^{\frac{1}{p}} \left\{ \int (|g|^{p-1})^{\frac{p}{p-1}} \, \mathrm{d}\mu \right\}^{\frac{p-1}{p}}$$

$$= \|f - g\|_p \cdot (\|f\|_p^{p-1} + \|g\|_p^{p-1})$$

$$\leq 2R^{p-1}\|f - g\|_p. \tag{3.166}$$

Combining the inequalities (3.165) and (3.166), we discover that

$$\int \left| |f|^p - |g|^p \right| \, \mathrm{d}\mu \leq 2pR^{p-1}\|f - g\|_p$$

which is our expected result.

We have completed the proof of the problem.

> **Problem 3.25**
>
> *Rudin Chapter 3 Exercise 25.*

Proof. Suppose that $E \subseteq X$ and $0 < \mu(E) < \infty$. Let $\mathfrak{M}_E$ be a σ-algebra in E. Then $(E, \mathfrak{M}_E, \mu)$ is a measure space.[j] Define

$$\varphi(F) = \int_F \frac{1}{\mu(E)} \, \mathrm{d}\mu \quad (F \in \mathfrak{M}_E).$$

Then we follow from Theorem 1.29 that φ is a measure on $\mathfrak{M}_E$ and

$$\int_E f \, \mathrm{d}\varphi = \frac{1}{\mu(E)} \int_E f \, \mathrm{d}\mu \tag{3.167}$$

for every measurable f on E with range in $[0, \infty]$. It is trivial that $\varphi(E) = 1$ and

$$0 \leq \int_E f \, \mathrm{d}\varphi \leq \frac{1}{\mu(E)} \int_X f \, \mathrm{d}\mu = \frac{1}{\mu(E)} < \infty$$

which means $f \in L^1(\varphi)$. Since $0 < f < \infty$ on E, $\log f$ is well-defined on E.

Define $\phi : (0, \infty) \to \mathbb{R}$ by $\phi(x) = -\log x$ which is convex on $(0, \infty)$. Apply Theorem 3.3 (Jensen's Inequality) with ϕ and measure φ to conclude that

$$-\log \left(\int_E f \, \mathrm{d}\varphi \right) \leq -\int_E \log f \, \mathrm{d}\varphi$$

or equivalently,

$$\int_E \log f \, \mathrm{d}\varphi \leq \log \left(\int_E f \, \mathrm{d}\varphi \right). \tag{3.168}$$

Now, by using the fact (3.167) to the inequality (3.168), we gain

$$\frac{1}{\mu(E)} \int_E \log f \, \mathrm{d}\mu \leq \log \left(\frac{1}{\mu(E)} \int_E f \, \mathrm{d}\mu \right) \leq \log \left(\frac{1}{\mu(E)} \int_X f \, \mathrm{d}\mu \right) = \log \frac{1}{\mu(E)}$$

which implies the first inequality.

For the second inequality, we define $\psi : (0, \infty) \to \mathbb{R}$ by

$$\psi(x) = -x^p,$$

---

[j]See the paragraphs following Proposition 1.24.

where $0 < p < 1$. Since $\psi''(x) = p(1-p)x^{p-2} > 0$ on $(0, \infty)$, we apply Theorem 3.3 (Jensen's Inequality) with ψ and measure φ to conclude that

$$-\Big\{ \int_E f \, \mathrm{d}\varphi \Big\}^p \le - \int_E f^p \, \mathrm{d}\varphi. \tag{3.169}$$

Next, we use the fact (3.167) to reduce the inequality (3.169) to

$$\frac{1}{\mu(E)} \int_E f^p \, \mathrm{d}\mu \le \Big\{ \frac{1}{\mu(E)} \int_E f \, \mathrm{d}\mu \Big\}^p \le \Big\{ \frac{1}{\mu(E)} \int_X f \, \mathrm{d}\mu \Big\}^p = \frac{1}{\mu(E)^p}$$

so that

$$\int_E f^p \, \mathrm{d}\mu \le \mu(E)^{1-p},$$

completing the proof of the problem. ◼

Problem 3.26

Rudin Chapter 3 Exercise 26.

Proof. Let $\mathfrak{M}$ be a σ-algebra in $[0,1]$. Suppose that there exists an $E \in \mathfrak{M}$ such that $f(x) = \infty$ and $m(E) > 0$, where m is the Lebesgue measure. Then

$$\int_0^1 f(x) \log f(x) \, \mathrm{d}x = \int_E f(x) \log f(x) \, \mathrm{d}x + \int_{[0,1]\setminus E} f(x) \log f(x) \, \mathrm{d}x. \tag{3.170}$$

On the one hand, since $f > 0$ on $[0,1] \setminus E$, we know from Proposition 1.24(a) that the second integral on the right-hand side of the expression (3.170) is nonnegative. Since $f(x) \log f(x) = \infty$ on E and $m(E) > 0$, we get from the expression (3.170) that

$$\int_0^1 f(x) \log f(x) \, \mathrm{d}x = \infty. \tag{3.171}$$

On the other hand, by using similar argument, we can show that

$$\int_0^1 f(s) \, \mathrm{d}s = \infty \quad \text{and} \quad \int_0^1 \log f(t) \, \mathrm{d}t = \infty. \tag{3.172}$$

Therefore, we conclude from the results (3.171) and (3.172) that

$$\int_0^1 f(x) \log f(x) \, \mathrm{d}x = \int_0^1 f(s) \, \mathrm{d}s \int_0^1 \log f(t) \, \mathrm{d}t$$

in this case.

Next, we suppose that $0 < f(x) < \infty$ a.e. on $[0,1]$. Define $\varphi : (0, \infty) \to \mathbb{R}$ and $\psi : (0, \infty) \to \mathbb{R}$ by

$$\varphi(x) = x \log x \quad \text{and} \quad \psi(x) = -\log x$$

respectively. Since $\varphi''(x) = \frac{1}{x} > 0$ and $\psi''(x) = x^{-2} > 0$ on $(0, \infty)$, they are convex on $(0, \infty)$. Since $0 < f(x) < \infty$ a.e. on $[0,1]$ and $m([0,1]) = 1$, we apply Theorem 3.3 (Jensen's Inequality) with φ to f to obtain

$$\Big\{ \int_0^1 f(x) \, \mathrm{d}x \Big\} \Big\{ \log \Big(\int_0^1 f(x) \, \mathrm{d}x \Big) \Big\} \le \int_0^1 f(x) \log f(x) \, \mathrm{d}x \tag{3.173}$$

Similarly, we apply Theorem 3.3 (Jensen's Inequality) with ψ to f to obtain

$$-\log\left(\int_0^1 f(x)\,dx\right) \le -\int_0^1 \log f(x)\,dx. \tag{3.174}$$

Combining the inequalities (3.173) and (3.174), we see that

$$\left\{\int_0^1 f(x)\,dx\right\}\left\{\int_0^1 \log f(x)\,dx\right\} \le \int_0^1 f(x)\log f(x)\,dx.$$

We have ended the proof of the problem. ∎

CHAPTER **4**

Elementary Hilbert Space Theory

In the following problems, we use $\langle x, y \rangle$ to denote the inner product of the complex vectors x and y and $\|\cdot\|$ the norm with respect to the Hilbert space H. Besides, we use span (S) to denote the span of a set S.

4.1 Basic Properties of Hilbert Spaces

> **Problem 4.1**
>
> *Rudin Chapter 4 Exercise 1.*

Proof. Let $x \in M$. Recall that $M^\perp = \{y \in H \mid y \perp x \text{ for all } x \in M\}$, so $x \perp y$ for every $y \in M^\perp$ which means that $x \in (M^\perp)^\perp$, i.e.,

$$M \subseteq (M^\perp)^\perp.$$

Conversely, suppose that $x \in (M^\perp)^\perp$. Since M is a closed subspace of H, it follows from Theorem 4.11(a) that $x = y + z$, where $y \in M$ and $z \in M^\perp$. On the one hand, since $x \in (M^\perp)^\perp$ and $z \in M^\perp$, we have $\langle x, z \rangle = 0$. On the other hand, we deduce from Definition 4.1 that

$$\langle x, z \rangle = \langle y + z, z \rangle = \langle y, z \rangle + \langle z, z \rangle = \langle z, z \rangle.$$

Since $\langle z, z \rangle = 0$, it must be $z = 0$ and then $x \in M$, i.e, $(M^\perp)^\perp \subseteq M$. In conclusion, we obtain $M = (M^\perp)^\perp$.

Suppose that M is a subspace of H which may *not* be closed. We claim that

$$\overline{M} = (M^\perp)^\perp.$$

To see this, we recall from §4.8 that $\overline{M}$ is a closed subspace of H, so the first assertion implies that

$$\overline{M} = (\overline{M}^\perp)^\perp. \tag{4.1}$$

Let $x \in \overline{M}^\perp$. By §4.9, we see that $x \perp \overline{M}$ so that $x \perp M$ particularly. In other words, we have

$$\overline{M}^\perp \subseteq M^\perp. \tag{4.2}$$

For the other direction, suppose that $x \in M^\perp$, we want to show that $x \in \overline{M}^\perp$, i.e., $x \perp \overline{M}$. To this end, given $y \in \overline{M}$, there exist a sequence $\{y_n\} \subseteq M$ such that $y_n \to y$ as $n \to \infty$. Since

123

$x \in M^{\perp}$ and $y_n \in M$, we must have $\langle y_n, x \rangle = 0$. By Theorem 4.6, the mapping $f : H \to \mathbb{C}$ defined by

$$f(y) = \langle y, x \rangle$$

is continuous on H so that

$$\langle y, x \rangle = f(y) = \lim_{n \to \infty} f(y_n) = \lim_{n \to \infty} \langle y_n, x \rangle = 0.$$

In other words, $x \perp \overline{M}$, i.e.,

$$M^{\perp} \subseteq \overline{M}^{\perp}. \tag{4.3}$$

Now the claim is derived by combining the set relations (4.2) and (4.3) and then using the result (4.1), completing the proof of the problem. ∎

> **Problem 4.2**
>
> *Rudin Chapter 4 Exercise 2.*

Proof. We note that this is actually the **Gram-Schmidt Process**. Suppose that $n = 1$. Then the set $\{u_1\}$ is clearly orthonormal and $\mathrm{span}\,(u_1) = \mathrm{span}\,(x_1)$. Thus the statement is true for $n = 1$.

Assume that the statement is true for $n = k$ for some $k \in \mathbb{N}$, i.e., $\{u_1, u_2, \ldots, u_k\}$ is an orthonormal set and

$$\mathrm{span}\,(u_1, u_2, \ldots, u_k) = \mathrm{span}\,(x_1, x_2, \ldots, x_k). \tag{4.4}$$

Let $n = k + 1$. Note that $x_{k+1} \notin \mathrm{span}\,(x_1, x_2, \ldots, x_k)$ because $\{x_1, x_2, \ldots\}$ is linearly independent. By the assumption (4.4), it implies that $x_{k+1} \notin \mathrm{span}\,(u_1, u_2, \ldots, u_k)$ so that $v_{k+1} \neq 0$. By this fact and Definition 4.1, for $j = 1, 2, \ldots, k$, we derive that

$$
\begin{aligned}
\langle u_{k+1}, u_j \rangle &= \left\langle \frac{v_{k+1}}{\|v_{k+1}\|}, u_j \right\rangle \\
&= \frac{1}{\|v_{k+1}\|} \left\langle x_{k+1} - \sum_{i=1}^{k} \langle x_{k+1}, u_i \rangle u_i, u_j \right\rangle \\
&= \frac{1}{\|v_{k+1}\|} \langle x_{k+1}, u_j \rangle - \frac{1}{\|v_{k+1}\|} \sum_{i=1}^{k} \langle x_{k+1}, u_i \rangle \langle u_i, u_j \rangle \\
&= \frac{1}{\|v_{k+1}\|} \langle x_{k+1}, u_j \rangle - \frac{1}{\|v_{k+1}\|} \langle x_{k+1}, u_j \rangle \langle u_j, u_j \rangle \\
&= 0.
\end{aligned}
$$

Therefore, $\{u_1, u_2, \ldots, u_{k+1}\}$ is orthonormal. Next, by the assumption (4.4), we know that $x \in \mathrm{span}\,(x_1, x_2, \ldots, x_k, x_{k+1})$ if and only if

$$x \in \mathrm{span}\,(u_1, u_2, \ldots, u_k, x_{k+1}). \tag{4.5}$$

Since

$$v_{k+1} = x_{k+1} - \sum_{i=1}^{k} \langle x_{k+1}, u_i \rangle u_i \quad \text{and} \quad u_{k+1} = \frac{v_{k+1}}{\|v_{k+1}\|},$$

the x_{k+1} in the result (4.5) can be replaced by v_{k+1} and ultimately by u_{k+1}. Thus we have

$$\mathrm{span}\,(u_1, u_2, \ldots, u_{k+1}) = \mathrm{span}\,(x_1, x_2, \ldots, x_{k+1}).$$

Hence the statement is true for $n = k + 1$ if it is true for $n = k$. By induction, the construction yields an orthonormal set $\{u_1, u_2, \ldots\}$ with

$$\text{span}\,(u_1, u_2, \ldots, u_n) = \text{span}\,(x_1, x_2, \ldots, x_n)$$

for all $n \in \mathbb{N}$. This completes the proof of the problem. $\blacksquare$

> **Problem 4.3**
>
> *Rudin Chapter 4 Exercise 3.*

Proof. Recall the definition that a space is separable if it contains a countable dense subset.

Suppose that $1 \le p < \infty$. Since T is compact, the second paragraph following Definition 3.16 says that $C_c(T) = C_0(T) = C(T)$. Thus we deduce from this and Theorem 3.14 that $C(T)$ is dense in $L^p(T)$ in the norm $\|\cdot\|_p$. Let $\mathcal{P}$ be the set of all trigonometric polynomials on T. By Theorem 4.25 (The Weierstrass Approximation Theorem) and Definitions 4.23 (or Remark 3.15), $\mathcal{P}$ is dense in $C(T)$ in the norm $\|\cdot\|_\infty$. By Definition 3.7, we know that $|f(x)| \le \|f\|_\infty$ a.e. on T and then

$$\|f\|_p = \left\{\frac{1}{2\pi}\int_{-\pi}^{\pi} |f(t)|^p\,dt\right\}^{\frac{1}{p}} \le \left\{\frac{1}{2\pi}\int_{-\pi}^{\pi} \|f\|_\infty^p\,dt\right\}^{\frac{1}{p}} = \|f\|_\infty. \tag{4.6}$$

In other words, the inequality (4.6) implies that $\mathcal{P}$ is also dense in $C(T)$ in the norm $\|\cdot\|_p$. By Lemma 2.10, we conclude that $\mathcal{P}$ is dense in $L^p(T)$ in the norm $\|\cdot\|_p$.

Next we suppose that $\mathcal{P}_\mathbb{Q}$ is the set of all trigonometric polynomials with rational coefficients. Note that[a]

$$\begin{aligned}
\mathcal{P}_\mathbb{Q} &= \left\{P(t) = \sum_{k=-n}^{n} (a_k + ib_k)e^{ikt}\,\bigg|\, a_{-n}, \ldots, a_n, b_{-n}, \ldots, b_n \in \mathbb{Q} \text{ and } n \in \mathbb{N}\right\} \\
&= \bigcup_{n=1}^{\infty} \left\{P(t) = \sum_{k=-n}^{n} (a_k + ib_k)e^{ikt}\,\bigg|\, a_{-n}, \ldots, a_n, b_{-n}, \ldots, b_n \in \mathbb{Q}\right\} \\
&\sim \bigcup_{n=1}^{\infty} \mathbb{Q}^{2(2n+1)}.
\end{aligned} \tag{4.7}$$

For each $n \in \mathbb{N}$, since each $\mathbb{Q}^{2(2n+1)}$ is countable, it follows from the set relation (4.7) that $\mathcal{P}_\mathbb{Q}$ is also countable. Given $f \in \mathcal{P}$. Since the set $\{p + iq \,|\, p, q \in \mathbb{Q}\}$ is dense in $\mathbb{C}$ and f must be in the form

$$f(t) = \sum_{k=-n}^{n} c_k e^{ikt}$$

for some positive integer n, there exists a sequence $\{f_{n_k}\} \subseteq \mathcal{P}_\mathbb{Q}$ such that $\|f_{n_k} - f\|_\infty \to 0$ as $n_k \to \infty$. By the inequality (4.6), it is true that

$$\|f_{n_k} - f\|_p \to 0$$

as $n_k \to \infty$. Consequently, $\mathcal{P}_\mathbb{Q}$ is dense in $\mathcal{P}$ in the norm $\|\cdot\|_p$ and an application of Lemma 2.10 shows that $\mathcal{P}_\mathbb{Q}$ is dense in $L^p(T)$ in the norm $\|\cdot\|_p$. Since $\mathcal{P}_\mathbb{Q}$ is countable, $L^p(T)$ is separable.

By Definition 4.23, every function defined on T can be identified with a 2π-periodic function on $\mathbb{R}$, so we may identity $L^\infty(T)$ with $L^\infty\big([0, 2\pi]\big)$. Consider the set

$$E = \{f_\theta = \chi_{[0,\theta]} \,|\, \theta \in [0, 2\pi]\}.$$

---

[a] We write $A \sim B$ if there is a bijection between the two sets A and B, see [49, Definition 2.3, p. 25].

It is clear that $\chi_{[0,\theta]} \in L^\infty([0, 2\pi])$ so that $E \subseteq L^\infty([0, 2\pi])$. Since $\chi_{[0,\theta_1]} \neq \chi_{[0,\theta_2]}$ if $\theta_1 \neq \theta_2$, $E \sim [0, 2\pi]$ and then it is *uncountable*. In fact, we have

$$\|\chi_{[0,\theta_1]} - \chi_{[0,\theta_2]}\|_\infty = 1 \tag{4.8}$$

if $\theta_1 < \theta_2$.

Assume that $L^\infty([0, 2\pi])$ was separable. Let $F = \{f_n\}$ be a countable dense subset of $L^\infty([0, 2\pi])$. Then we must have

$$E \subseteq L^\infty([0, 2\pi]) \subseteq \bigcup_{f_n \in F} B\left(f_n, \frac{1}{2}\right), \tag{4.9}$$

where $B(f_n, \frac{1}{2}) = \{f \in L^\infty([0, 2\pi]) \mid \|f - f_n\|_\infty < \frac{1}{2}\}$. If $\chi_{[0,\theta_1]}, \chi_{[0,\theta_2]} \in B(f_n, \frac{1}{2})$ with $\theta_1 \neq \theta_2$, then it follows from the result (4.8) that

$$1 = \|\chi_{[0,\theta_1]} - \chi_{[0,\theta_2]}\|_\infty \leq \|\chi_{[0,\theta_1]} - f_n\|_\infty + \|f_n - \chi_{[0,\theta_2]}\|_\infty < 1,$$

a contradiction. Thus each $B(f_n, \frac{1}{2})$ contains *at most* one element of E. Since F is countable but E is uncountable, we have

$$E \not\subseteq \bigcup_{f_n \in F} B\left(f_n, \frac{1}{2}\right)$$

which definitely contradicts the set relation (4.9). Hence $L^\infty([0, \pi])$ is not separable and we have completed the proof of the problem. ◼

Problem 4.4

Rudin Chapter 4 Exercise 4.

Proof. Suppose that H is separable. By the definition, it has a countable dense subset $\{u_n\}$. By Problem 4.2, we may assume that $\{u_n\}$ is also an orthonormal set. Let $\{u_{n_k}\}$ be a *maximal* linearly independent subset of $\{u_n\}$. Note that $\{u_{n_k}\}$ is at most countable. Since we have $\operatorname{span}(\{u_{n_k}\}) = \{u_n\}$, $\operatorname{span}(\{u_{n_k}\})$ is also dense in H. By Theorem 4.18, the set $\{u_{n_k}\}$ is in fact an at most countable maximal orthonormal set in H.

Conversely, we suppose that $E = \{u_n\}$ is an at most countable maximal orthonormal set in H. By Theorem 4.18 again, $\operatorname{span}(E)$ is dense in H. Let $\operatorname{span}_{\mathbb{Q}}(E)$ be the span of the set E with rational coefficients. It is easy to see that

$$\operatorname{span}_{\mathbb{Q}}(E) = \left\{ \sum_{k=1}^{n} (a_k + ib_k)u_k \,\bigg|\, a_1, \ldots, a_n, b_1, \ldots, b_n \in \mathbb{Q} \text{ and } n \in \mathbb{N} \right\}$$
$$= \bigcup_{n=1}^{\infty} \left\{ \sum_{k=1}^{n} (a_k + ib_k)u_k \,\bigg|\, a_1, \ldots, a_n, b_1, \ldots, b_n \in \mathbb{Q} \right\}$$
$$\sim \bigcup_{n=1}^{\infty} \mathbb{Q}^{2n}.$$

As a result, $\operatorname{span}_{\mathbb{Q}}(E)$ must be countable. Next, since $\mathbb{Q}$ is dense in $\mathbb{R}$, we conclude that $\operatorname{span}_{\mathbb{Q}}(E)$ is dense in $\operatorname{span}(E)$. By Lemma 2.10, $\operatorname{span}_{\mathbb{Q}}(E)$ is also dense in H and hence it is separable. This completes the proof of the problem. ◼

> **Problem 4.5**
>
> *Rudin Chapter 4 Exercise 5.*

Proof. Suppose that $M \neq H$. If $L = 0$, then $M = H$ so that $L \neq 0$. By Theorem 4.12 (The Riesz Representation Theorem for Hilbert Spaces), there is a unique $z \in H$ such that

$$L(x) = \langle x, z \rangle$$

for all $x \in H$. This z must be nonzero, otherwise we have $L = 0$ which contradicts our assumption. It is clear that

$$M = \{x \in H \mid \langle x, z \rangle = 0\} = L^{-1}(0), \tag{4.10}$$

so M is closed in H. Apart from this, if $x, y \in M$, then since L is a linear functional, Definition 2.1 shows that

$$L(x + y) = L(x) + L(y) = 0 \quad \text{and} \quad L(\alpha x) = \alpha L(x) = 0$$

for a scalar α. Thus M is a closed subspace of H.

Next, we note that $\operatorname{span}(z) = \{\alpha z \mid \alpha \in \mathbb{C}\}$ and so

$$\big(\operatorname{span}(z)\big)^{\perp} = \{x \in H \mid \langle x, \alpha z \rangle = 0 \text{ for every } \alpha \in \mathbb{C}\}. \tag{4.11}$$

Since $\langle x, \alpha z \rangle = \overline{\alpha} \langle x, z \rangle$, the two expressions (4.10) and (4.11) give

$$M = \big(\operatorname{span}(z)\big)^{\perp}$$

and then

$$M^{\perp} = \big(\big(\operatorname{span}(z)\big)^{\perp}\big)^{\perp}. \tag{4.12}$$

We claim that $\operatorname{span}(z)$ is a closed subspace of H. In fact, it is a subspace of H follows easily from the definition of $\operatorname{span}(z)$. Suppose that $\{x_n\}$ is a sequence in $\operatorname{span}(z)$. Then we know that each x_n has the form $x_n = \alpha_n z$, where $\alpha_n \in \mathbb{C}$. Let $\alpha = \lim\limits_{n \to \infty} \alpha_n$ and $x = \alpha z$. We want to show that $x_n \to x$ in H with respect to the norm $\| \cdot \|$. Indeed, we have

$$\|x_n - x\|^2 = \langle (\alpha_n - \alpha)z, (\alpha_n - \alpha)z \rangle = |\alpha_n - \alpha^2| \times \|z\|$$

so that $\|x_n - x\| \to 0$ as $n \to \infty$, i.e.,

$$\lim_{n \to \infty} x_n = x$$

as desired. In conclusion, $\operatorname{span}(z)$ contains all its limit points and therefore it is closed in H. By Problem 4.1, we have $\big(\big(\operatorname{span}(z)\big)^{\perp}\big)^{\perp} = \operatorname{span}(z)$. Combining this fact and the expression (4.12), we have

$$M^{\perp} = \operatorname{span}(z)$$

and since $\{z\}$ is a basis of $\operatorname{span}(z)$, we have $\dim\big(\operatorname{span}(z)\big) = 1$ and our required result follows. We end the proof of the problem. $\blacksquare$

> **Problem 4.6**
>
> *Rudin Chapter 4 Exercise 6.*

Proof. We prove the assertions one by one:

- $E = \{u_n\}$ **is closed and bounded, but not compact.** Since E is orthonormal, $\|u_n\| \le 1$ for every $n \in \mathbb{N}$. Thus E is bounded by 1. If $n \ne m$, then it follows from Definition 4.1 that

$$\|u_n - u_m\|^2 = \langle u_n - u_m, u_n - u_m \rangle = \langle u_n, u_n \rangle + \langle u_m, u_m \rangle = 2. \tag{4.13}$$

  Assume that u was a limit point of E. Then there exists a positive integer N such that $n \ge N$ implies that $\|u_n - u\| < \frac{\sqrt{2}}{2}$. If $n, m \ge N$, then we have

$$\|u_n - u_m\| \le \|u_n - u\| + \|u - x_m\| < \sqrt{2}$$

  which contradicts the result (4.13). In other words, E has *no* limit point and then it is closed in H.

  Suppose that $B(u_n, 1) = \{x \in H \mid \|u_n - x\| < 1\}$ which is clearly open in H. Furthermore, it is obvious that $\{B(u_n, 1)\}$ is an open cover of E. By the fact (4.13), we conclude that each $B(u_n, 1)$ contains *exactly* one element of E, so it is impossible to have

$$E \subseteq \bigcup_{i=1}^{k} B(u_{n_i}, 1)$$

  for any finite k, i.e., E is *not* compact.

- S **is compact if and only if** $\displaystyle\sum_{n=1}^{\infty} \delta_n^2 < \infty$. Suppose that $\displaystyle\sum_{n=1}^{\infty} \delta_n^2 < \infty$. It is well-known that a metric space X is compact if and only if X is *sequentially compact.*[b] As a subset of a metric space, S is itself a metric space. Let $\{x^k\} \subseteq S$ be a sequence of the form

$$x^k = \sum_{n=1}^{\infty} c_n^k u_n, \tag{4.14}$$

  where $k = 1, 2, \ldots$. Note that we have $|c_n^k| \le \delta_n$ for all $k \in \mathbb{N}$. Since $\{c_1^k\} \subseteq \overline{B(0, \delta_1)} \subseteq \mathbb{C}$ for all $k \in \mathbb{N}$ and $\overline{B(0, \delta_1)}$ is compact, $\overline{B(0, \delta_1)}$ is sequentially compact and then there exists a subsequence $k_1(m)$ such that

$$c_1^{k_1(m)} \to c_1 \in \overline{B(0, \delta_1)}$$

  as $m \to \infty$. Next, we consider the sequence $\{c_2^{k_1(m)}\} \subseteq \overline{B(0, \delta_2)}$. Then we can find a subsequence $\{k_2(m)\} \subseteq \{k_1(m)\}$ such that

$$c_2^{k_2(m)} \to c_2 \in \overline{B(0, \delta_2)}$$

  as $m \to \infty$. Furthermore, we also have

$$c_1^{k_2(m)} \to c_1 \in \overline{B(0, \delta_1)}$$

  as $m \to \infty$. By this observation, *for every* positive integer N, we can select a sequence $\{k_{N+1}(m)\} \subseteq \{k_N(m)\}$ such that for all $n \le N$,

$$c_n^{k_N(m)} \to c_n \in \overline{B(0, \delta_n)} \tag{4.15}$$

---

[b]We say that the space X is sequentially compact if every sequence of points of X has a convergent subsequence, see [42, Theorem 28.2, p. 179].

as $m \to \infty$. Now the limit (4.15) makes sure that

$$x = \sum_{n=1}^{\infty} c_n u_n \in S.$$

We claim that the sequence $\{x^{k_m(m)}\}$, which is clearly a subsequence of $\{x^k\}$ converges to the point x. To this end, given $\epsilon > 0$, there is a $N' \in \mathbb{N}$ such that

$$\sum_{n > N'} \delta_n^2 < \frac{\epsilon}{8}. \tag{4.16}$$

Then we obtain from the definition (4.14), the fact $|c_n^k| \le \delta_n$ for all k and the estimate (4.16) that

$$\begin{aligned}
\|x^{k_m(m)} - x\|^2 &= \left\| \sum_{n=1}^{\infty} [c_n^{k_m(m)} - c_n] u_n \right\|^2 \\
&= \sum_{n=1}^{\infty} |c_n^{k_m(m)} - c_n|^2 \\
&= \sum_{n=1}^{N'} |c_n^{k_m(m)} - c_n|^2 + \sum_{n=N'+1}^{\infty} |c_n^{k_m(m)} - c_n|^2 \\
&\le \sum_{n=1}^{N'} |c_n^{k_m(m)} - c_n|^2 + \sum_{n=N'+1}^{\infty} \left(|c_n^{k_m(m)}|^2 + 2|c_n^{k_m(m)}| \cdot |c_n| + |c_n|^2 \right) \\
&\le \sum_{n=1}^{N'} |c_n^{k_m(m)} - c_n|^2 + 4 \times \sum_{n=N'+1}^{\infty} \delta_n^2 \\
&< \sum_{n=1}^{N'} |c_n^{k_m(m)} - c_n|^2 + \frac{\epsilon}{2}. \tag{4.17}
\end{aligned}$$

To finish our proof, we have to find an upper bound of the summation in the inequality (4.17). Recall from the limit (4.15) that for each $n \le N'$

$$c_n^{k_{N'}(m)} \to c_n \in \overline{B(0, \delta_n)}$$

as $m \to \infty$. In other words, for each $n \le N'$, there is a $M_n \in \mathbb{N}$ such that $m \ge M_n$ implies

$$|c_n^{k_{N'}(m)} - c_n|^2 < \frac{\epsilon}{2N'}. \tag{4.18}$$

Since $\{c_n^{k_m(m)}\} \subseteq \{c_n^{k_{N'}(m)}\}$ for $m \ge N'$, this means that the inequality (4.18) also holds when $c_n^{k_{N'}(m)}$ is replaced by $c_n^{k_m(m)}$. Let $M = \max(M_1, M_n, \ldots, M_{N'})$. If $m \ge M$, then we conclude from the inequalities (4.17) and (4.18) that

$$\|x^{k_m(m)} - x\|^2 < \sum_{n=1}^{N'} \frac{\epsilon}{2N'} + \frac{\epsilon}{2} = \epsilon,$$

proving our claim. Hence S is sequentially compact and then compact.

Conversely, we suppose that S is compact. Consider the point

$$x_N = \sum_{n=1}^{N} \delta_n u_n,$$

where $N \in \mathbb{N}$. Then we have $\|x_N\|^2 = \sum_{n=1}^{N} \delta_n^2$. Since S is a compact metric space, it is bounded. Thus there exists a positive constant M such that

$$\sum_{n=1}^{N} \delta_n^2 \leq M$$

for all $N \in \mathbb{N}$ and hence $\sum_{n=1}^{\infty} \delta_n^2 < \infty$.

- Q **is compact.** Obviously, we have[c]

$$\sum_{n=1}^{\infty} |c_n|^2 \leq \sum_{n=1}^{\infty} \frac{1}{n^2} < \infty.$$

Therefore, the previous assertion implies that Q is compact.

- H **is not locally compact.** Recall that H is itself a vector space, so it contains the zero element 0. Assume that H was locally compact. Then the closure

$$\overline{B(0,r)} = \{x \in H \mid \|x\| \leq r\}$$

must be compact *for every* $r > 0$. Note that $\{ru_n\} \subseteq \overline{B(0,r)}$. Since $E_r = \{ru_n\}$ is closed in H, Theorem 2.4 implies that E_r is compact. Similar to part (a), $\{B(ru_n, r)\}$ is an open cover of E_r. A direct computation shows that

$$\|ru_n - ru_m\| = \sqrt{2}r. \tag{4.19}$$

If $ru_m \in B(ru_n, r)$ for $n \neq m$, then we have

$$\|ru_n - ru_m\| < r$$

which contradicts the fact (4.19). Therefore, each $B(ru_n, r)$ contains *exactly* one element of E_r so that $\{B(ru_n, r)\}$ *does not* have a finite subcover for E_r. In other words, E_r is not compact, a contradiction. Hence we establish the fact that H is not locally compact.

We have completed the proof of the problem. ∎

Problem 4.7

Rudin Chapter 4 Exercise 7.

Proof. We follow the suggestion given by Rudin. Assume that $\sum a_n^2 = \infty$. Therefore, for every $k \in \mathbb{N}$, there are disjoint sets $E_1, E_2, \ldots$ such that

$$\sum_{n \in E_k} a_n^2 > k > 0. \tag{4.20}$$

Now for $n \in E_k$, we define

$$b_n = \frac{a_n}{\left\{ \sum_{n \in E_k} a_n^2 \right\}^{\frac{1}{2}}} \times \frac{1}{k}.$$

---

[c]See [49, Theorem 3.28, p. 62].

On the one hand, we have

$$\sum_{n=1}^{\infty} b_n^2 = \sum_{k=1}^{\infty} \sum_{n \in E_k} b_n^2 = \sum_{k=1}^{\infty} \left(\frac{1}{\displaystyle\sum_{n \in E_k} a_n^2} \times \sum_{n \in E_k} a_n^2 \times \frac{1}{k^2} \right) = \sum_{k=1}^{\infty} \frac{1}{k^2} < \infty.$$

On the other hand, we obtain from the inequality (4.20) that

$$\sum_{n=1}^{\infty} a_n b_n = \sum_{k=1}^{\infty} \sum_{n \in E_k} a_n b_n = \sum_{k=1}^{\infty} \left(\sum_{n \in E_k} \frac{a_n^2}{\left\{ \displaystyle\sum_{n \in E_k} a_n^2 \right\}^{\frac{1}{2}}} \times \frac{1}{k} \right) = \sum_{k=1}^{\infty} \left\{ \sum_{n \in E_k} a_n^2 \right\}^{\frac{1}{2}} \frac{1}{k} > \sum_{k=1}^{\infty} \frac{1}{\sqrt{k}}$$

which is divergent, a contradiction. Hence we must have $\sum a_n^2 < \infty$ and it completes the proof of the problem. ∎

Problem 4.8

Rudin Chapter 4 Exercise 8.

Proof. By Problem 4.2, H contains an orthonormal set A and then Theorem 4.22 (The Hausdorff Maximality Theorem) implies that A is contained in a maximal orthonormal set in H. Suppose that $\{u_\alpha \mid \alpha \in A\}$ and $\{v_\beta \mid \beta \in B\}$ are maximal orthonormal sets in H_1 and H_2 respectively. Without loss of generality, we may assume that $|A| \le |B|$. Then we take a subset B' of B which is of the same cardinality as A. Now we consider

$$H = \overline{\text{span}\left(\{ v_\beta \mid \beta \in B' \} \right)} \subseteq H_2.$$

By Definition 4.13, $\text{span}\left(\{ v_\beta \mid \beta \in B' \} \right)$ is a subspace of H_2. Thus we follow from the last paragraph in §4.7 that H is a closed subspace of H_2. By the given hint, H is also Hilbert. Finally, we know from §4.19 that H_1 and H are *Hilbert space isomorphic* to $\ell^2(A)$ and $\ell^2(B')$ respectively. Since A and B have the same cardinality, $\ell^2(A)$ is Hilbert space isomorphic to $\ell^2(B')$. In other words, H_1 is Hilbert space isomorphic to $H \subseteq H_2$, completing the proof of the problem. ∎

Problem 4.9

Rudin Chapter 4 Exercise 9.

Proof. Since $A \subseteq [0, 2\pi]$ and A is measurable, we have $\chi_A \in L^2(T)$. By Theorem 3.11, $L^2(T)$ is a complete metric space, so it is a Hilbert space by Definition 4.4. Since $\{u_n \mid n \in \mathbb{Z}\}$ is a maximal orthonormal set in the Hilbert space $L^2(T)$, Theorem 4.18(iii) implies that

$$\sum_{n \in \mathbb{Z}} \left| \langle u_n, \chi_A \rangle \right|^2 = \|\chi_A\|^2 = \int_0^{2\pi} |\chi_A(x)|^2 \, \mathrm{d}x = m(A) < \infty.$$

By [49, Theorem 3.23, p. 60], we know that

$$\lim_{n \to \infty} \left| \langle u_n, \chi_A \rangle \right| = \lim_{n \to -\infty} \left| \langle u_n, \chi_A \rangle \right| = 0. \tag{4.21}$$

By Definition 4.1, we have

$$\langle \cos nx, \chi_A \rangle = \left\langle \frac{u_n + u_{-n}}{2}, \chi_A \right\rangle = \frac{1}{2}(\langle u_n, \chi_A \rangle + \langle u_{-n}, \chi_A \rangle),$$

$$\langle \sin nx, \chi_A \rangle = \left\langle \frac{u_n - u_{-n}}{2i}, \chi_A \right\rangle = \frac{1}{2i}(\langle u_n, \chi_A \rangle - \langle u_{-n}, \chi_A \rangle).$$

$$(4.22)$$

By applying the limits (4.21) to the inner products (4.22), we obtain that

$$\lim_{n \to \infty} \int_A \cos nt\, dt = \lim_{n \to \infty} \int_{-\pi}^{\pi} (\cos nt) \cdot \chi_A(t)\, dt = 2\pi \lim_{n \to \infty} \langle \cos nx, \chi_A \rangle = 0$$

and

$$\lim_{n \to \infty} \int_A \sin nt\, dt = \lim_{n \to \infty} \int_{-\pi}^{\pi} (\sin nt) \cdot \chi_A(t)\, dt = 2\pi \lim_{n \to \infty} \langle \sin nx, \chi_A \rangle = 0.$$

Thus we have completed the proof of the problem. ∎

> **Problem 4.10**
>
> *Rudin Chapter 4 Exercise 10.*

Proof. This problem is proven in [63, Problem 11.16, p. 352]. ∎

> **Problem 4.11**
>
> *Rudin Chapter 4 Exercise 11.*

Proof. For each $n = 1, 2, \ldots$, we define $x_n = \frac{n+1}{n} u_n$ and $E = \{x_n\}$ which is obviously a subset of $L^2(T)$ because, by Definition 4.23,

$$\|x_n\| = \|x_n\|_2 = \frac{n+1}{n} < \infty$$

for every $n \in \mathbb{N}$. It is clear that $E \neq \varnothing$. Now we are going to apply a similar argument as in the proof of the first assertion in Problem 4.6. In fact, we note that $\langle x_n, x_m \rangle = 0$ for $n \neq m$, so we follow from Definition 4.1 that

$$\begin{aligned}
\|x_m - x_n\|^2 &= \langle x_m - x_n, x_m - x_n \rangle \\
&= \langle x_m, x_m - x_n \rangle - \langle x_n, x_m - x_n \rangle \\
&= \langle x_m, x_m \rangle - \langle x_m, x_n \rangle - \langle x_n, x_m \rangle + \langle x_n, x_n \rangle \\
&= \left(\frac{m+1}{m}\right)^2 \langle u_m, u_m \rangle + \left(\frac{n+1}{n}\right)^2 \langle u_n, u_n \rangle \\
&= \left(\frac{m+1}{m}\right)^2 + \left(\frac{n+1}{n}\right)^2 \\
&\geq 2
\end{aligned}$$

which implies that

$$\|x_m - x_n\| \geq \sqrt{2}. \qquad (4.23)$$

Assume that x was a limit point of E. Then there exists a positive integer N such that $n \geq N$ implies $\|x_n - x\| < \frac{\sqrt{2}}{2}$. If $n, m \geq N$, then we have

$$\|x_n - x_m\| \leq \|x_n - x\| + \|x - x_m\| < \sqrt{2}$$

which contradicts the result (4.23). Thus E has *no* limit point and hence it is closed in $L^2(T)$.

Furthermore, assume that x_N was an element of E such that $\|x_N\| \le \|x_n\|$ for all $x_n \in E$. However, since we have

$$\|x_N\| = \frac{N+1}{N} > \frac{N+2}{N+1} = \|x_{N+1}\|,$$

this contradiction shows that E has no element of smallest norm. Hence we end the proof of the problem.

> **Remark 4.1**
>
> Problem 4.11 indicates that the condition "convexity" in Theorem 4.10 *cannot* be omitted.

> **Problem 4.12**
>
> *Rudin Chapter 4 Exercise 12.*

Proof. By the half-angle formula for cosine function, we have

$$1 = \frac{c_k}{\pi} \int_0^\pi \left(\frac{1+\cos t}{2}\right)^k \, dt = \frac{2c_k}{\pi} \int_0^{\frac{\pi}{2}} \cos^{2k} x \, dx. \tag{4.24}$$

A further substitution with $z = \sin x$ reduces the expression (4.24) to

$$1 = \frac{2c_k}{\pi} \int_0^1 (1 - z^2)^{k-\frac{1}{2}} \, dz. \tag{4.25}$$

If we substitute $y = z^2$ and use the integral form of the beta function (see [49, Theorem 8.20, p. 193]), then the expression (4.25) can be written as

$$1 = \frac{c_k}{\pi} \int_0^1 (1-y)^{k-\frac{1}{2}} y^{-\frac{1}{2}} \, dy = \frac{c_k}{\pi} \cdot \frac{\Gamma(k+\frac{1}{2})\Gamma(\frac{1}{2})}{\Gamma(k+1)} = \frac{c_k}{\sqrt{\pi}} \cdot \frac{\Gamma(k+\frac{1}{2})}{k\Gamma(k)}. \tag{4.26}$$

Finally, we recall from [49, Exercise 30, p. 203] that

$$\lim_{k \to \infty} \frac{\Gamma(k+c)}{k^c \Gamma(k)} = 1 \tag{4.27}$$

for every real constant c. Take $c = \frac{1}{2}$ in the limit (4.27) and then substitute the corresponding result into the expression (4.26), we get

$$\lim_{k \to \infty} k^{-\frac{1}{2}} c_k = \lim_{k \to \infty} \sqrt{\pi} \cdot \frac{k^{\frac{1}{2}}\Gamma(k)}{\Gamma(k+\frac{1}{2})} = \sqrt{\pi}.$$

This completes the proof of the problem.

> **Problem 4.13**
>
> *Rudin Chapter 4 Exercise 13.*

Proof. We follow the hint. Consider the function $f(t) = e^{2\pi i k t}$, where $k = 0, \pm 1, \pm 2, \ldots$. When $k = 0$, we have $f \equiv 1$ and then the formula holds trivially. Suppose that $k \neq 0$. Since α is irrational, $k\alpha \notin \mathbb{Z}$ and then $e^{2\pi i k \alpha} \neq 1$. By this, we have

$$\frac{1}{N} \sum_{n=1}^{N} e^{2\pi i k n \alpha} = \frac{1}{N} \sum_{n=1}^{N} (e^{2\pi i k \alpha})^n = \frac{e^{2\pi i k \alpha}}{N} \cdot \frac{e^{2\pi i k N \alpha} - 1}{e^{2\pi i k \alpha} - 1}$$

so that

$$\lim_{N \to \infty} \left| \frac{1}{N} \sum_{n=1}^{N} e^{2\pi i k n \alpha} \right| \leq \lim_{N \to \infty} \frac{2}{N|e^{2\pi i k \alpha} - 1|} = 0. \tag{4.28}$$

On the other hand, we have

$$\int_0^1 e^{2\pi i k t} \, dt = \frac{e^{2\pi i k t}}{2\pi i k} \Big|_0^1 = 0. \tag{4.29}$$

Hence we deduce from the inequality (4.28) and the integral (4.29) that

$$\lim_{N \to \infty} \frac{1}{N} \sum_{n=1}^{N} e^{2\pi i k n \alpha} = \int_0^1 e^{2\pi i k t} \, dt \tag{4.30}$$

for every $k = 0, \pm 1, \pm 2, \ldots$. Therefore the formula (4.30) holds for every trigonometric polynomials in the form

$$P(t) = \sum_{k=-N}^{N} c_k e^{2\pi i k t}, \tag{4.31}$$

where $c_{-N}, \ldots, c_N \in \mathbb{C}$.

Suppose that $f : \mathbb{R} \to \mathbb{R}$ is continuous and $f(x + 1) = f(x)$ for every $x \in \mathbb{R}$. By Theorem 4.25 (The Weierstrass Approximation Theorem), for every $\epsilon > 0$, there exists a trigonometric polynomial P in the form (4.31) such that

$$|f(t) - P(t)| < \frac{\epsilon}{3} \tag{4.32}$$

for every $t \in \mathbb{R}$. Next, the analysis in the previous paragraph shows that there is a positive integer N such that $n \geq N$ implies

$$\left| \frac{1}{N} \sum_{n=1}^{N} P(n\alpha) - \int_0^1 P(t) \, dt \right| < \frac{\epsilon}{3}. \tag{4.33}$$

Hence we follow from the estimates (4.32) and (4.33) that $n \geq N$ implies

$$\left| \frac{1}{N} \sum_{n=1}^{N} f(n\alpha) - \int_0^1 f(t) \, dt \right| = \left| \frac{1}{N} \sum_{n=1}^{N} f(n\alpha) - \frac{1}{N} \sum_{n=1}^{N} P(n\alpha) + \frac{1}{N} \sum_{n=1}^{N} P(n\alpha) \right.$$

$$\left. - \int_0^1 P(t) \, dt + \int_0^1 P(t) \, dt - \int_0^1 f(t) \, dt \right|$$

$$\leq \frac{1}{N} \sum_{n=1}^{N} \left| f(n\alpha) - P(n\alpha) \right| + \left| \frac{1}{N} \sum_{n=1}^{N} P(n\alpha) - \int_0^1 P(t) \, dt \right|$$

$$+ \int_0^1 |f(t) - P(t)| \, dt$$

$$< \epsilon.$$

Hence the formula (4.30) holds for every continuous function on $\mathbb{R}$ with period 1, completing the proof of the problem.

Remark 4.2

A question similar to Problem 4.13 is also proven in [63, Problem 8.19, pp. 194 – 196]. In fact, Problem 4.13 is a very important tool in the study of the Weyl's Equidistribution Theorem, see [57, §4.2, pp. 105 – 113].

4.2 Application of Theorem 4.14

Problem 4.14

Rudin Chapter 4 Exercise 14.

Proof. Let H be the Hilbert space $L^2\big([-1,1]\big)$. Define the inner product of $f, g \in H$ by

$$\langle f, g \rangle = \int_{-1}^{1} f(x)\overline{g(x)}\,\mathrm{d}x. \tag{4.34}$$

It is clear that $x^n \in H$ for every $n = 0, 1, 2, \ldots$ and the set $\{1, x, x^2, \ldots\}$ is a linearly independent set of vectors in H. By Problem 4.2, we can obtain an orthonormal set $\{u_1(x), u_2(x), \ldots\}$ such that $\{1, x, \ldots, x^{N-1}\}$ and $\{u_1(x), u_2(x), \ldots, u_N(x)\}$ have the same span for all positive integers N. Next, we let $F = \{0, 1, 2\}$ and we consider the closed subspace

$$M_F = \overline{\mathrm{span}\,(\{1, x, x^2\})} \tag{4.35}$$

of H. Then the orthonormal set $\{u_1(x), u_2(x), u_3(x)\}$ obtained by $\{1, x, x^2\}$ is given by

$$u_1(x) = \frac{1}{\sqrt{2}}, \quad u_2(x) = \sqrt{\frac{3}{2}}x \quad \text{and} \quad u_3(x) = \frac{3}{2}\sqrt{\frac{5}{2}}\Big(x^2 - \frac{1}{3}\Big).$$

Let $x^3 \in H$. Define

$$s_F(x^3) = \langle x^3, u_1(x) \rangle \cdot u_1(x) + \langle x^3, u_2(x) \rangle \cdot u_2(x) + \langle x^3, u_3(x) \rangle \cdot u_3(x).$$

By Theorem 4.14(b) and the inner product (4.34), we derive

$$\|x^3 - s_F\| = \min\{\|x^3 - s\| \mid s \in M_F\} = \Big[\min_{a,b,c} \int_{-1}^{1} |x^3 - a - bx - cx^2|^2\,\mathrm{d}x\Big]^{\frac{1}{2}}. \tag{4.36}$$

Since $\langle x^3, u_1(x) \rangle = 0$, $\langle x^3, u_2(x) \rangle = \frac{2}{5}\sqrt{\frac{3}{2}}$ and $\langle x^3, u_3(x) \rangle = 0$, we have

$$s_F(x^3) = \frac{3}{5}x$$

and then

$$\Big\|x^3 - \frac{3}{5}x\Big\| = \Big[\int_{-1}^{1} \Big(x^3 - \frac{3}{5}x\Big)^2\,\mathrm{d}x\Big]^{\frac{1}{2}} = \sqrt{\frac{8}{175}}. \tag{4.37}$$

Combining the results (4.36) and (4.37), it yields that

$$\min_{a,b,c} \int_{-1}^{1} |x^3 - a - bx - cx^2|^2\,\mathrm{d}x = \frac{8}{175}. \tag{4.38}$$

For the second assertion, we first notice that the definite integral must be a real number because we are considering its maximum. Secondly, the conditions on g actually mean that

$$\langle 1, \bar{g} \rangle = \langle x, \bar{g} \rangle = \langle x^2, \bar{g} \rangle = 0 \quad \text{and} \quad \|\bar{g}\| = 1.$$

Thus, by the discussion in §4.9, we have

$$\bar{g} \in M_F^{\perp} \quad \text{and} \quad \|\bar{g}\| = 1,$$

where M_F is defined by (4.35). Now we follow from these facts that

$$\max_{\substack{\bar{g} \in M_F^{\perp} \\ \|\bar{g}\|=1}} \int_{-1}^{1} x^3 g(x)\,dx \leq \max_{\substack{\bar{g} \in M_F^{\perp} \\ \|\bar{g}\|=1}} \left| \int_{-1}^{1} x^3 g(x)\,dx \right| = \max\{|\langle x^3, \bar{g} \rangle| \,|\, \bar{g} \in M_F^{\perp} \text{ and } \|\bar{g}\| = 1\}. \quad (4.39)$$

By Problem 4.16[d], we see that

$$\max\{|\langle x^3, \bar{g} \rangle| \,|\, \bar{g} \in M_F^{\perp} \text{ and } \|\bar{g}\| = 1\} = \min\{\|x^3 - s\| \,|\, s \in M_F\}. \quad (4.40)$$

Therefore, we combine the expressions (4.39) and (4.40) and then using the value (4.38) to get

$$\max_{\substack{\bar{g} \in M_F^{\perp} \\ \|\bar{g}\|=1}} \int_{-1}^{1} x^3 g(x)\,dx \leq \sqrt{\frac{8}{175}}. \quad (4.41)$$

If $g(x) = \sqrt{\frac{175}{8}}\left(x^3 - \frac{3}{5}x\right)$, then we have

$$\int_{-1}^{1} x^3 g(x)\,dx = \sqrt{\frac{8}{175}}$$

so that the equality in the inequality (4.41) is attainable. Hence we have completed the proof of the problem. ∎

> **Problem 4.15**
>
> *Rudin Chapter 4 Exercise 15.*

Proof. Now we work on the space $H = L^2([0, \infty])$ and it is easy to check that the formula given by

$$\langle f, g \rangle = \int_{0}^{\infty} f\bar{g}e^{-x}\,dx \quad (4.42)$$

satisfies Definition 4.1, so it is indeed an inner product for the space H.

Now we follow the idea of proof in Problem 4.14. Evidently, we have $x^n \in H$ for every $n = 0, 1, 2, \ldots$ and the set $\{1, x, x^2, \ldots\}$ is a linearly independent set of vectors in H. By Problem 4.2, we can obtain an orthonormal set $\{u_1(x), u_2(x), \ldots\}$ such that $\{1, x, \ldots, x^{N-1}\}$ and $\{u_1(x), u_2(x), \ldots, u_N(x)\}$ have the same span for all positive integers N. Next, we let $F = \{0, 1, 2\}$ and we consider the closed subspace

$$M_F = \overline{\text{span}\left(\{1, x, x^2\}\right)}$$

of H. Then the orthonormal set $\{u_1(x), u_2(x), u_3(x)\}$ obtained by $\{1, x, x^2\}$ is given by

$$u_1(x) = 1, \quad u_2(x) = x - 1 \quad \text{and} \quad u_3(x) = \frac{1}{2}(x^2 - 4x + 2).$$

---

[d]We assume its truth here.

Let $x^3 \in H$. Define

$$s_F(x^3) = \langle x^3, u_1(x) \rangle \cdot u_1(x) + \langle x^3, u_2(x) \rangle \cdot u_2(x) + \langle x^3, u_3(x) \rangle \cdot u_3(x).$$

By Theorem 4.14(b) and the inner product (4.42), we derive

$$\|x^3 - s_F\| = \min\{\|x^3 - s\| \mid s \in M_F\} = \left[\min_{a,b,c} \int_{-1}^{1} |x^3 - a - bx - cx^2|^2 e^{-x}\, dx \right]^{\frac{1}{2}}. \tag{4.43}$$

Since $\langle x^3, u_1(x) \rangle = 6$ and $\langle x^3, u_2(x) \rangle = \langle x^3, u_3(x) \rangle = 18$, we have

$$s_F(x^3) = 6 + 18(x - 1) + 9(x^2 - 4x + 2) = 9x^2 - 18x + 6$$

so that

$$\|x^3 - 9x^2 + 18x - 6\| = \left[\int_{-1}^{1} (x^3 - 9x^2 + 18x - 6)^2 e^{-x}\, dx \right]^{\frac{1}{2}} = 6. \tag{4.44}$$

Combining the results (4.43) and (4.44), it yields that

$$\min_{a,b,c} \int_{-1}^{1} |x^3 - a - bx - cx^2|^2 e^{-x}\, dx = 36. \tag{4.45}$$

For the second assertion, we observe again that the definite integral is a real number and the conditions on g imply that

$$g \in M_F^{\perp} \quad \text{and} \quad \|g\| = 1.$$

Therefore, it yields that

$$\max_{\substack{\overline{g} \in M_F^{\perp} \\ \|\overline{g}\|=1}} \int_{-1}^{1} x^3 g(x) e^{-x}\, dx \leq \max_{\substack{\overline{g} \in M_F^{\perp} \\ \|\overline{g}\|=1}} \left| \int_{-1}^{1} x^3 g(x) e^{-x}\, dx \right|$$

$$= \max\{ |\langle x^3, \overline{g} \rangle| \mid \overline{g} \in M_F^{\perp} \text{ and } \|\overline{g}\| = 1 \}. \tag{4.46}$$

By Problem 4.16 again, we see that

$$\max\{ |\langle x^3, \overline{g} \rangle| \mid \overline{g} \in M_F^{\perp} \text{ and } \|\overline{g}\| = 1 \} = \min\{\|x^3 - s\| \mid s \in M_F\}. \tag{4.47}$$

By combining the expressions (4.46) and (4.47) and then using the value (4.45), we gain

$$\max_{\substack{\overline{g} \in M_F^{\perp} \\ \|\overline{g}\|=1}} \int_{-1}^{1} x^3 g(x) e^{-x}\, dx \leq 6. \tag{4.48}$$

If $g(x) = \frac{1}{6}(x^3 - 9x^2 + 18x - 6)$, then we have

$$\int_{-1}^{1} x^3 g(x) e^{-x}\, dx = 6$$

so that the equality in the inequality (4.48) is attainable and we end the proof of the problem. ∎

4.3   Miscellaneous Problems

> **Problem 4.16**
>
> *Rudin Chapter 4 Exercise 16.*

Proof. Since M is a closed linear subspace of H, we follow from Theorem 4.11(a) that x_0 has the unique decomposition

$$x_0 = Px_0 + Qx_0,$$

where $Px_0 \in M$ and $Qx_0 \in M^\perp$. In addition, Theorem 4.11(b) ensures that

$$\|Px_0 - x_0\| = \min\{\|x - x_0\| \mid x \in M\}. \tag{4.49}$$

On the other hand, if $y \in M^\perp$ and $\|y\| = 1$, then we have

$$\langle x_0, y \rangle = \langle Px_0 + Qx_0, y \rangle = \langle Qx_0, y \rangle$$

and then Theorem 4.2 (The Schwarz Inequality) implies that

$$|\langle x_0, y \rangle| = |\langle Qx_0, y \rangle| \le \|Qx_0\| \times \|y\| = \|Qx_0\|. \tag{4.50}$$

In other words, the inequality (4.50) shows that

$$\max\{|\langle x_0, y \rangle| \mid y \in M^\perp \text{ and } \|y\| = 1\} \le \|Qx_0\|. \tag{4.51}$$

If $Qx_0 = 0$, then it means that $x_0 = Px_0 \in M$. In this case, we deduce from the expression (4.49) and the inequality (4.51) that

$$\min\{\|x - x_0\| \mid x \in M\} = \|Px_0 - x_0\| = \max\{|\langle x_0, y \rangle| \mid y \in M^\perp \text{ and } \|y\| = 1\} = 0.$$

If $Qx_0 \ne 0$, since $Qx_0 \in M^\perp$, we may consider $y_0 = \frac{Qx_0}{\|Qx_0\|}$ which satisfies the conditions $y_0 \in M^\perp$ and $\|y_0\| = 1$. A direct computation shows that

$$|\langle x_0, y_0 \rangle| = \frac{1}{\|Qx_0\|} \times |\langle x_0, Qx_0 \rangle| = \frac{1}{\|Qx_0\|} \times |\langle Qx_0, Qx_0 \rangle| = \|Qx_0\|.$$

Thus the inequality (4.51) actually becomes

$$\max\{|\langle x_0, y \rangle| \mid y \in M^\perp \text{ and } \|y\| = 1\} = \|Qx_0\|. \tag{4.52}$$

Hence our desired result follows from the comparison between the expressions (4.49) and (4.52) and the fact that $\|Px_0 - x_0\| = \|Qx_0\|$. We have ended the proof of the problem. $\blacksquare$

Problem 4.17

Rudin Chapter 4 Exercise 17.

Proof. Take $H = L^2([0,1])$ which is a Hilbert space with the inner product given by

$$\langle f, g \rangle = \int_0^1 f(t)\overline{g(t)}\, dt, \tag{4.53}$$

where $f, g \in H$, see Example 4.5(b). For $t \in [0,1]$, we note that $\chi_{[0,t]} \in H$ because

$$\|\chi_{[0,t]}\| = \left\{ \int_0^1 |\chi_{[0,t]}(x)|^2\, dx \right\}^{\frac{1}{2}} = \left\{ \int_0^t dx \right\}^{\frac{1}{2}} = \sqrt{t} \le 1.$$

Thus we can consider the mapping $\gamma : [0,1] \to H$ defined by

$$\gamma(t) = \chi_{[0,t]}$$

which is well-defined. It is trivial that γ is injective. Furthermore, for every $p \in [0,1]$, if $t > p$, then we have

$$\|\gamma(t) - \gamma(p)\| = \left\{ \int_0^1 |\chi_{(p,t]}(x)|^2 \, \mathrm{d}x \right\}^{\frac{1}{2}} = \left\{ \int_p^t \mathrm{d}x \right\}^{\frac{1}{2}} = (t-p)^{\frac{1}{2}}. \tag{4.54}$$

Similarly, if $t < p$, then we have

$$\|\gamma(t) - \gamma(p)\| = \left\{ \int_t^p \mathrm{d}x \right\}^{\frac{1}{2}} = (p-t)^{\frac{1}{2}}. \tag{4.55}$$

By combining the two expressions (4.54) and (4.55), we conclude that

$$\|\gamma(t) - \gamma(p)\| = |t-p|^{\frac{1}{2}}$$

and thus γ is continuous on $[0,1]$.

Let $0 \le a \le b \le c \le d \le 1$. If $a = b$, then $\gamma(a) - \gamma(b) = 0$ and

$$\langle \gamma(b) - \gamma(a), \gamma(d) - \gamma(c) \rangle = \langle 0, \gamma(d) - \gamma(c) \rangle = \int_0^1 0 \times \overline{\chi_{(c,d]}(x)} \, \mathrm{d}x = 0.$$

Similarly, if $c = d$, then $\langle \gamma(b) - \gamma(a), \gamma(d) - \gamma(c) \rangle = 0$. Without loss of generality, we may assume that $a < b$ and $c < d$. Then $(a,b] \cap (c,d] = \varnothing$ and so we have

$$\langle \gamma(b) - \gamma(a), \gamma(d) - \gamma(c) \rangle = \langle \chi_{(a,b]}, \chi_{(c,d]} \rangle = \int_0^1 \chi_{(a,b]}(x) \overline{\chi_{(c,d]}(x)} \, \mathrm{d}x = 0.$$

Hence the desired result holds in the special case that $H = L^2([0,1])$.

To treat the general case, we first consider the function $u_n(t) = \mathrm{e}^{2\pi i n t}$ for every $n \in \mathbb{Z}$. It is clear that $u_n \in H$ and

$$\langle u_n, u_m \rangle = \int_0^1 \mathrm{e}^{2\pi i(n-m)t} \, \mathrm{d}t = \begin{cases} 1, & \text{if } n = m; \\ 0, & \text{if } n \neq m. \end{cases}$$

Thus $\{u_n \,|\, n \in \mathbb{Z}\}$ is an orthonormal set in H. By Theorem 3.14, $C([0,1])$ is dense in H with respect to the norm induced by the inner product (4.53). By Theorem 4.25 (The Weierstrass Approximation Theorem) and Definition 4.23, we know that the set $\mathcal{P}$ of all trigonometric polynomials in the form

$$P(t) = \sum_{n=-N}^{N} c_n u_n(t) = \sum_{n=-N}^{N} c_n \mathrm{e}^{2\pi i n t}$$

is dense in $C([0,1])$ with respect to the norm $\|\cdot\|_\infty$. Note that $\mathcal{P} = \mathrm{span}\,(\{u_n \,|\, n \in \mathbb{Z}\})$. By Definition 3.7, it can be shown easily that $\mathcal{P}$ is also dense in $C([0,1])$ with respect to the norm induced by the inner product (4.53). Therefore, we derive from Lemma 2.10 that $\mathcal{P}$ is also dense in H in the norm induced by the inner product (4.53). Consequently, we conclude from Theorem 4.18(i) that $\{u_n \,|\, n \in \mathbb{Z}\}$ is in fact maximal and then[e]

$$H \cong \ell^2(\mathbb{N}). \tag{4.56}$$

Suppose that H' is a Hilbert space with an *infinite* maximal orthonormal set $\{v_\alpha \,|\, \alpha \in A\}$, where A is countable or uncountable infinite. Then we have

$$H' \cong \ell^2(A). \tag{4.57}$$

---

[e]We say $A \cong B$ if A is Hilbert space isomorphic to B.

Applying Problem 4.8 to the Hilbert space isomorphisms (4.56) and (4.57), we conclude that H must be Hilbert space isomorphic to a subspace of H' and then the special case is applicable to this general case. Indeed, let $\Lambda : H \to K$ be a Hilbert space isomorphism, where K is a Hilbert subspace of H'. If we define $\lambda : [0,1] \to H'$ by

$$\lambda = \Lambda \circ \gamma,$$

then λ is injective and continuous. Furthermore, since $\langle \Lambda(f), \Lambda(g) \rangle = \langle f, g \rangle$ for all $f, g \in H$, we deduce from the special case above that

$$
\begin{aligned}
\langle \lambda(b) - \lambda(a), \lambda(d) - \lambda(c) \rangle &= \big\langle \Lambda\big(\gamma(b)\big) - \Lambda\big(\gamma(a)\big), \Lambda\big(\gamma(d)\big) - \Lambda\big(\gamma(c)\big) \big\rangle \\
&= \big\langle \Lambda\big(\gamma(b)\big), \Lambda\big(\gamma(d)\big) \big\rangle - \big\langle \Lambda\big(\gamma(b)\big), \Lambda\big(\gamma(c)\big) \big\rangle \\
&\quad - \big\langle \Lambda\big(\gamma(a)\big), \Lambda\big(\gamma(d)\big) \big\rangle + \big\langle \Lambda\big(\gamma(a)\big), \Lambda\big(\gamma(c)\big) \big\rangle \\
&= \langle \gamma(b), \gamma(d) \rangle - \langle \gamma(b), \gamma(c) \rangle - \langle \gamma(a), \gamma(d) \rangle + \langle \gamma(a), \gamma(c) \rangle \\
&= \langle \gamma(b) - \gamma(a), \gamma(c) - \gamma(d) \rangle \\
&= 0.
\end{aligned}
$$

This completes the proof of the problem. $\blacksquare$

Problem 4.18

Rudin Chapter 4 Exercise 18.

Proof. If $r, s \in \mathbb{R}$ and $r \neq s$, then we have

$$\langle u_r, u_s \rangle = \lim_{A \to \infty} \frac{1}{2A} \int_{-A}^{A} e^{i(r-s)t} \, \mathrm{d}t = \lim_{A \to \infty} \frac{2i \sin[(r-s)A]}{2A(r-s)} = 0. \tag{4.58}$$

If $r = s$, then

$$\langle u_s, u_s \rangle = \lim_{A \to \infty} \frac{1}{2A} \int_{-A}^{A} e^{ist} e^{-ist} \, \mathrm{d}t = \lim_{A \to \infty} \frac{1}{2A} \int_{-A}^{A} \, \mathrm{d}t = 1. \tag{4.59}$$

Thus $\{u_s \,|\, s \in \mathbb{R}\}$ is an orthonormal set of H by Definition 4.13. Since f and g are finite combinations of u_s, we may suppose that

$$f(t) = \sum_{p=1}^{n} c_p e^{is_p t} \quad \text{and} \quad g(t) = \sum_{q=1}^{m} d_q e^{ir_q t}. \tag{4.60}$$

Therefore, we obtain

$$\lim_{A \to \infty} \frac{1}{2A} \int_{-A}^{A} \Big(\sum_{p=1}^{n} c_p e^{is_p t} \Big) \Big(\sum_{q=1}^{m} \overline{d_q} e^{-ir_q t} \Big) \, \mathrm{d}t = \lim_{A \to \infty} \frac{1}{2A} \int_{-A}^{A} \sum_{p=1}^{n} \sum_{q=1}^{m} c_p \overline{d_q} e^{i(s_p - r_q)t} \, \mathrm{d}t$$

which is undoubtedly finite by the values (4.58) and (4.59). As a consequence, we have shown that $\langle f, g \rangle$ is well-defined for all $f, g \in X$.

Next we are going to show that $\langle \cdot \rangle$ satisfies Definition 4.1. Firstly, it is clear that

$$\langle f, g \rangle = \lim_{A \to \infty} \frac{1}{2A} \int_{-A}^{A} f(t) \overline{g(t)} \, \mathrm{d}t = \overline{\lim_{A \to \infty} \frac{1}{2A} \int_{-A}^{A} g(t) \overline{f(t)} \, \mathrm{d}t} = \overline{\langle g, f \rangle}.$$

Secondly, for $f, g, h \in X$, we observe that

$$\langle f + g, h \rangle = \lim_{A \to \infty} \frac{1}{2A} \int_{-A}^{A} [f(t) + g(t)] \overline{h(t)} \, \mathrm{d}t$$

$$= \lim_{A\to\infty} \frac{1}{2A} \int_{-A}^{A} f(t)\overline{h(t)}\,\mathrm{d}t + \lim_{A\to\infty} \frac{1}{2A} \int_{-A}^{A} g(t)\overline{h(t)}\,\mathrm{d}t$$

$$= \langle f, h \rangle + \langle g, h \rangle.$$

Thirdly, for any scalar α, we have

$$\langle \alpha f, g \rangle = \lim_{A\to\infty} \frac{1}{2A} \int_{-A}^{A} \alpha f(t)\overline{h(t)}\,\mathrm{d}t = \alpha \lim_{A\to\infty} \frac{1}{2A} \int_{-A}^{A} f(t)\overline{h(t)}\,\mathrm{d}t = \alpha \langle f, g \rangle.$$

Fourthly, if f takes the representation (4.60), then we establish from the results (4.58) and (4.59) that

$$\langle f, f \rangle = \lim_{A\to\infty} \frac{1}{2A} \int_{-A}^{A} f(t)\overline{f(t)}\,\mathrm{d}t = \lim_{A\to\infty} \frac{1}{2A} \int_{-A}^{A} |f(t)|^2\,\mathrm{d}t = \sum_{p=1}^{n} |c_p|^2 \geq 0. \tag{4.61}$$

Finally, if $\langle f, f \rangle = 0$, then the result (4.61) definitely shows that $c_1 = c_2 = \cdots = c_p = 0$. In other words, we have $f \equiv 0$ in this case. Hence this inner product certainly makes X into a unitary space.

Let H be the completion of X, i.e., X is dense in H and $H = \overline{X}$. Since $\mathrm{span}\,(\{u_s \,|\, s \in \mathbb{R}\}) = X$, Theorem 4.18(i) says that the set $\{u_s \,|\, s \in \mathbb{R}\}$ is in fact a maximal orthonormal in H. By the discussion in §4.19, we have

$$H \cong \ell^2(\mathbb{R}). \tag{4.62}$$

Assume that H was separable. By Problem 4.4, H has an at most countable maximal orthonormal system $\{v_n \,|\, n \in \mathbb{N}\}$. By the discussion in §4.19 again, we know that

$$H \cong \ell^2(\mathbb{N}). \tag{4.63}$$

Now the two isomorphic relations (4.62) and (4.63) imply that $\ell^2(\mathbb{R}) \cong \ell^2(\mathbb{N})$, but then $\mathbb{R}$ and $\mathbb{N}$ have the same cardinal number, a contradiction. Hence we complete the proof of the problem. ∎

Problem 4.19

Rudin Chapter 4 Exercise 19.

Proof. If $N = 1$, then $\omega = e^{2\pi i} = 1$ and k can only take the value 0. Thus the orthogonality relation holds in this special case. Suppose that $N > 1$. If $k = 0$, then $\omega^0 = 1$ so that

$$\frac{1}{N} \sum_{n=1}^{N} \omega^0 = 1.$$

Let $k = 1, 2, \ldots, N - 1$. We remark that

$$z^N - 1 = (z - 1)(z^{N-1} + \cdots + z + 1). \tag{4.64}$$

Putting $z = \omega^k$ into the expression (4.64), we obtain

$$(\omega^k)^N - 1 = (\omega^k - 1)\big[(\omega^k)^{N-1} + \cdots + \omega^k + 1\big] = (\omega^k - 1)\big(\omega^{k(N-1)} + \cdots + \omega^k + 1\big). \tag{4.65}$$

It is clear that $(\omega^k)^N = (\omega^N)^k = 1$. Furthermore, since $1 \leq k \leq N - 1$, k is evidently not divisible by N which implies that $\omega^k \neq 1$. Therefore, it deduces from the expression (4.65) that

$$\omega^{k(N-1)} + \cdots + \omega^k + 1 = 0$$

or equivalently

$$\omega^{kN} + \omega^{k(N-1)} + \cdots + \omega^k = 0.$$

In conclusion, we have

$$\frac{1}{N} \sum_{n=1}^{N} \omega^{nk} = \begin{cases} 1, & \text{if } k = 0; \\ 0, & \text{if } 1 \le k \le N - 1. \end{cases}$$

Let $N \ge 3$. It is clear from the orthogonality relations and the properties in Definition 4.1 that

$$\frac{1}{N} \sum_{n=1}^{N} \|x + \omega^n y\|^2 \omega^n = \frac{1}{N} \sum_{n=1}^{N} \langle x + \omega^n y, x + \omega^n y \rangle \, \omega^n$$

$$= \frac{1}{N} \Big[\sum_{n=1}^{N} \langle x, x \rangle \, \omega^n + \sum_{n=1}^{N} \langle x, \omega^n y \rangle \, \omega^n + \sum_{n=1}^{N} \langle \omega^n y, x \rangle \, \omega^n$$

$$+ \sum_{n=1}^{N} \langle \omega^n y, \omega^n y \rangle \, \omega^n \Big]$$

$$= \frac{1}{N} \Big[\langle x, x \rangle \sum_{n=1}^{N} \omega^n + \sum_{n=1}^{N} \langle x, y \rangle \, \omega^{-n} \times \omega^n + \sum_{n=1}^{N} \langle y, x \rangle \, \omega^n \times \omega^n$$

$$+ \langle y, y \rangle \sum_{n=1}^{N} \omega^n \Big]$$

$$= \frac{1}{N} \Big[\langle x, x \rangle \sum_{n=1}^{N} \omega^n + N \langle x, y \rangle + \langle y, x \rangle \sum_{n=1}^{N} \omega^{2n} + \langle y, y \rangle \sum_{n=1}^{N} \omega^n \Big]$$

$$= \langle x, y \rangle \tag{4.66}$$

as desired.

Finally, Definition 4.1 gives

$$\|x + e^{i\theta} y\|^2 = \Big\langle x + e^{i\theta} y, x + e^{i\theta} y \Big\rangle = \langle x, x \rangle + e^{-i\theta} \langle x, y \rangle + e^{i\theta} \langle y, x \rangle + \langle y, y \rangle$$

so that

$$\frac{1}{2\pi} \int_{-\pi}^{\pi} \|x + e^{i\theta} y\|^2 e^{i\theta} \, d\theta = \frac{1}{2\pi} \int_{-\pi}^{\pi} \big[\|x\|^2 + \|y\|^2 + e^{-i\theta} \langle x, y \rangle + e^{i\theta} \langle y, x \rangle \big] e^{i\theta} \, d\theta$$

$$= \frac{1}{2\pi} \int_{-\pi}^{\pi} \langle x, y \rangle \, d\theta + \frac{1}{2\pi} \int_{-\pi}^{\pi} e^{2i\theta} \langle y, x \rangle \, d\theta$$

$$= \langle x, y \rangle . \tag{4.67}$$

This completes the proof of the problem. ◼

Remark 4.3

We note that the polarization identity (see [51, p. 86]) can be written in the form

$$\langle x, y \rangle = \frac{1}{4} \sum_{n=0}^{3} i^{-n} \|x + i^n y\|^2.$$

Thus the two identities (4.66) and (4.67) in Problem 4.19 are actually generalizations of this.

<h2>CHAPTER 5</h2>

Examples of Banach Space Techniques

5.1 The Unit Ball in a Normed Linear Space

> **Problem 5.1**
>
> *Rudin Chapter 5 Exercise 1.*

Proof. For $0 < p \leq \infty$, we suppose that

$$B = \{ f \in L^p(\mu) \mid \|f\|_p \leq 1 \}.$$

There are two cases:

- **Case (i):** $0 < p < \infty$. In this case, we have

$$
\begin{aligned}
\|f\|_p &= \left\{ \int_X |f|^p \, d\mu \right\}^{\frac{1}{p}} \\
&= \left(|f(a)|^p \times \mu(\{a\}) + |f(b)|^p \times \mu(\{b\}) \right)^{\frac{1}{p}} \\
&= \frac{1}{2^{\frac{1}{p}}} \{ |f(a)|^p + |f(b)|^p \}^{\frac{1}{p}}.
\end{aligned}
$$

Consequently, $\|f\|_p \leq 1$ if and only if

$$|f(a)|^p + |f(b)|^p \leq 2. \tag{5.1}$$

Therefore, it follows from the inequality (5.1) that the unit ball B is a circle if and only if $p = 2$. Besides, the unit ball B becomes a square if and only if $p = 1$.

- **Case (ii):** $p = \infty$. By Definition 3.7, we know that

$$\|f\|_\infty = \max(|f(a)|, |f(b)|) \tag{5.2}$$

so that $\|f\|_\infty \leq 1$ if and only if

$$\max(|f(a)|, |f(b)|) \leq 1 \tag{5.3}$$

which is a square.

See Figure 5.1 for an illustration.

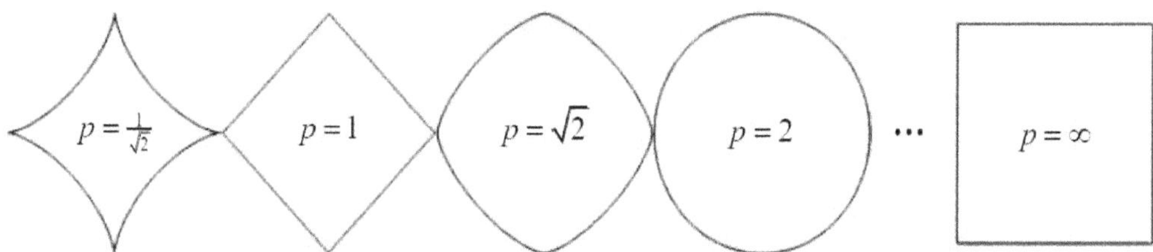

Figure 5.1: The unit circle in different p-norm.

If $\mu(\{a\}) \neq \mu(\{b\})$, then the expression (5.1) is replaced by

$$\mu(\{a\})|f(a)|^p + \mu(\{b\})|f(b)|^p \leq 1$$

which cannot be a circle or a square for any $p \in (0, \infty)$. However, we note that the expression (5.2) is still valid even in the case $\mu(\{a\}) \neq \mu(\{b\})$, so we have the inequality (5.3) and then B is a square when $p = \infty$. This ends the analysis of the problem.

Problem 5.2

Rudin Chapter 5 Exercise 2.

Proof. Let X be a normed linear space with norm $\|\cdot\|$ and $B(0,1) = \{x \in X \mid \|x\| < 1\}$ be the open unit ball. For every $x, y \in B(0,1)$ and $t \in [0,1]$, we follow from Definition 5.2 that

$$\|(1-t)x + ty\| \leq (1-t)\|x\| + t\|y\| < 1$$

so that $(1-t)x + ty \in B(0,1)$. By §4.8, $B(0,1)$ is convex. Similarly, the convexity of the closed unit ball $\overline{B(0,1)} = \{x \in X \mid \|x\| \leq 1\}$ can be proven similarly. Hence we complete the proof of the problem.

Problem 5.3

Rudin Chapter 5 Exercise 3.

Proof. Suppose that $\|f\|_p = \|g\|_p = 1$ and $f \neq g$. By Theorem 3.9, we always have

$$\|h\|_p = \left\|\frac{1}{2}(f+g)\right\|_p \leq \frac{1}{2}\|f\|_p + \frac{1}{2}\|g\|_p = 1.$$

Assume that $\|h\|_p = 1$. Then it follows from Problem 3.13 that $f = \lambda g$ a.e. on X for some $\lambda > 0$. Since $\lambda\|g\|_p = \|f\|_p = \|g\|_p$, we have $\lambda = 1$ implying $f = g$ a.e. on X, a contradiction.

However, strictly convexity *does not* hold for $L^1(\mu)$, $L^\infty(\mu)$ and $C(X)$. Ignore trivialities, we suppose that X contains more than one point. In fact, we can pick $E, F \in \mathfrak{M} \setminus \{\varnothing\}$ such that $E \cap F = \varnothing$ and $\mu(E), \mu(F) \in (0, \infty)$. Define $f : X \to \mathbb{C}$ and $g : X \to \mathbb{C}$ by

$$f(x) = \frac{1}{\mu(E)}\chi_E(x) \quad \text{and} \quad g(x) = \frac{1}{\mu(F)}\chi_F(x) \tag{5.4}$$

respectively. It is obvious that $f \neq g$. Furthermore, we have

$$\|f\|_1 = \int_X |f(x)|\,\mathrm{d}x = \frac{1}{\mu(E)}\int_E \chi_E(x)\,\mathrm{d}x = 1,$$

$$\|g\|_1 = \int_X |g(x)|\, dx = \frac{1}{\mu(F)} \int_F \chi_F(x)\, dx = 1,$$

$$\left\|\frac{f+g}{2}\right\|_1 = \int_X \left|\frac{f(x)+g(x)}{2}\right| dx = \frac{1}{2}\int_E f(x)\, dx + \frac{1}{2}\int_F g(x)\, dx = 1.$$

These show that the invalidity of strictly convexity in $L^1(\mu)$. For $L^\infty(\mu)$, we replace the functions f and g in the definition (5.4) by

$$f(x) = \chi_E(x) \quad \text{and} \quad g(x) = \chi_{E \cup F}(x)$$

respectively. Then it is also true that $f \neq g$. By Definition 3.7, we have

$$\|f\|_\infty = \|g\|_\infty = \left\|\frac{f+g}{2}\right\|_\infty = 1.$$

Therefore, the strictly convexity does not hold in $L^\infty(\mu)$.

Finally, we consider the space $C(X)$, where X is Hausdorff. By Definition 3.16, we see that X is assumed to be compact and so the norm $\|\cdot\|_\infty$ of $C(X)$ is given by

$$\|f\|_\infty = \sup_{x \in X} |f(x)|. \tag{5.5}$$

Take $f \in C(X)$ and suppose that f is nonconstant. Thus one can find a $x_0 \in X$ such that

$$|f(x_0)| < \|f\|_\infty. \tag{5.6}$$

Otherwise, we have $|f(x)| \geq \|f\|_\infty$ on X, but this and the definition (5.5) will imply that $|f(x)| = \|f\|_\infty$ on X, i.e., f is a constant which is a contradiction. Define

$$a = \frac{3}{4}\|f\|_\infty + \frac{1}{4}|f(x_0)| \quad \text{and} \quad b = \frac{1}{4}\|f\|_\infty + \frac{3}{4}|f(x_0)|.$$

Consider the sets

$$K = \left\{x \in X \,\middle|\, |f(x)| \geq a\right\} = |f|^{-1}([a,\infty)) \quad \text{and} \quad E = \left\{x \in X \,\middle|\, |f(x)| \leq b\right\} = |f|^{-1}((-\infty,b]).$$

It is clear that $x_0 \in E$, i.e., $E \neq \varnothing$. Since $|f| : X \to \mathbb{R}$ is continuous, the Extreme Value Theorem ([42, Theorem 27.4, p. 174]) implies that $\|f\|_\infty = \max_{x \in X} |f(x)| = |f(p)|$ for some $p \in X$. Consequently, we have

$$p \in K. \tag{5.7}$$

Furthermore, both K and E are closed in X by [42, Theorem 18.1, p. 104]. If $K \cap E \neq \varnothing$, then we have

$$\frac{3}{4}\|f\|_\infty + \frac{1}{4}|f(x_0)| \leq \frac{1}{4}\|f\|_\infty + \frac{3}{4}|f(x_0)|$$

but these imply that $\|f\|_\infty \leq |f(x_0)|$ which contradicts the hypothesis (5.6). Hence we have $K \cap E = \varnothing$.

What we have done in the previous paragraph is that $V = E^c$ is open in X and $K \subseteq V$. Since X is compact, Theorem 2.4 ensures that K is also compact. By Theorem 2.12 (Urysohn's Lemma), there exists a $h \in C_c(X)$ such that $K \prec h \prec V$, i.e., $0 \leq h(x) \leq 1$ on X, $h(x) = 1$ on K and $\operatorname{supp}(h) \subseteq V$ (or equivalently $h(x) = 0$ on E). If we define $g = h \times f : X \to \mathbb{C}$, then we must have $g \not\equiv f$ and furthermore, the fact (5.7) shows that

$$\|g\|_\infty = \sup_{x \in X} |h(x)f(x)| = |h(p)| \cdot |f(p)| = |f(p)| = \|f\|_\infty$$

and

$$\left\|\frac{f+g}{2}\right\|_\infty = \frac{1}{2}\|f + hf\|_\infty = \frac{1}{2}\sup_{x \in X}|[1 + h(x)]f(x)| = \frac{1}{2}|1 + h(p)| \cdot |f(p)| = |f(p)| = \|f\|_\infty.$$

Hence strictly convexity does not hold for $C(X)$, completing the proof of the problem.　∎

5.2 Failure of Theorem 4.10 and Norm-preserving Extensions

Problem 5.4

Rudin Chapter 5 Exercise 4.

Proof. Define $f : [0,1] \to \mathbb{C}$ by

$$
f(x) = \begin{cases} -8x + 4, & \text{if } x \in [0, \tfrac{1}{2}]; \\ \\ 0, & \text{if } x \in (\tfrac{1}{2}, 1]. \end{cases}
$$

Then it is easy to see that $f \in M$, i.e., M is nonempty. Suppose that $f, g \in M$, $t \in [0,1]$ and $h = (1-t)f + tg$. Then we have

$$
\int_0^{\frac{1}{2}} h(x)\,\mathrm{d}x - \int_{\frac{1}{2}}^1 h(x)\,\mathrm{d}x = \int_0^{\frac{1}{2}} [(1-t)f(x) + tg(x)]\,\mathrm{d}x - \int_{\frac{1}{2}}^1 [(1-t)f(x) + tg(x)]\,\mathrm{d}x
$$

$$
= (1-t)\Big[\int_0^1 f(x)\,\mathrm{d}x - \int_{\frac{1}{2}}^1 f(x)\,\mathrm{d}x \Big]
$$

$$
+ t\Big[\int_0^1 g(x)\,\mathrm{d}x - \int_{\frac{1}{2}}^1 g(x)\,\mathrm{d}x \Big]
$$

$$
= 1
$$

so that $h \in M$, i.e., M is convex.

Next, we note that C and M are metric spaces. By [49, Theorem 7.15, p. 151], we see that C is a complete metric space. Since $M \subseteq C$, M is also a complete metric space. Suppose that $\{f_n\} \subseteq M$ and

$$
\|f_n - f\| = \sup_{x \in [0,1]} |f_n(x) - f(x)| \to 0
$$

as $n \to \infty$. Then it follows from[a] [49, Theorem 7.9, p. 148] that $f_n \to f$ uniformly on $[0,1]$. Since each f_n is continuous on $[0,1]$, f is also continuous on $[0,1]$. Furthermore, we have

$$
\lim_{n \to \infty} \Big[\int_0^{\frac{1}{2}} f_n(t)\,\mathrm{d}t - \int_{\frac{1}{2}}^1 f_n(t)\,\mathrm{d}t \Big] = 1
$$

$$
\int_0^{\frac{1}{2}} \Big(\lim_{n \to \infty} f_n(t) \Big)\,\mathrm{d}t - \int_{\frac{1}{2}}^1 \Big(\lim_{n \to \infty} f_n(t) \Big)\,\mathrm{d}t = 1
$$

$$
\int_0^{\frac{1}{2}} f(t)\,\mathrm{d}t - \int_{\frac{1}{2}}^1 f(t)\,\mathrm{d}t = 1.
$$

In other words, $f \in M$ and M is closed in the space C.

Finally, we want to show that M contains *no* element of minimal norm. To this end, we notice that

$$
1 = \Big| \int_0^{\frac{1}{2}} f(t)\,\mathrm{d}t - \int_{\frac{1}{2}}^1 f(t)\,\mathrm{d}t \Big| \le \int_0^{\frac{1}{2}} |f(t)|\,\mathrm{d}t + \int_{\frac{1}{2}}^1 |f(t)|\,\mathrm{d}t = \|f\|_1 \le \|f\|_\infty \tag{5.8}
$$

---

[a] Or we can see the rephrased Theorem 7.9 on p. 151 in [49].

for every $f \in M$. Assume that there was a $f \in M$ such that $\|f\|_\infty = 1$. Then we deduce from the inequality (5.8) that

$$\int_0^1 |f(t)|\, dt = 1 \quad \text{or} \quad \int_0^1 (1 - |f(t)|)\, dt = 0. \tag{5.9}$$

Since $\|f\|_\infty = 1$, we have $|f(x)| \in [0,1]$ for all $x \in [0,1]$ so that the function $g(x) = 1 - |f(x)|$ is continuous and nonnegative on $[0,1]$. Applying Theorem 1.39(a) to the second integral in (5.9), we obtain

$$g(x) = 0 \quad \text{a.e. on } [0,1].$$

Now the continuity of g forces that $g(x) = 0$ on $[0,1]$, i.e., $|f(x)| = 1$ on $[0,1]$. In particular, we have

$$-1 \le \mathrm{Re}\left(f(x)\right) \le 1 \tag{5.10}$$

on $[0,1]$.

On the other hand, since $f \in M$, the definition gives

$$1 = \mathrm{Re}\left(\int_0^{\frac{1}{2}} f(t)\, dt - \int_{\frac{1}{2}}^1 f(t)\, dt \right) = \int_0^{\frac{1}{2}} \mathrm{Re}\left(f(t)\right) dt - \int_{\frac{1}{2}}^1 \mathrm{Re}\left(f(t)\right) dt$$

or equivalently,

$$\int_0^{\frac{1}{2}} \left[\mathrm{Re}\left(f(t)\right) - 1\right] dt + \int_{\frac{1}{2}}^1 \left[-1 - \mathrm{Re}\left(f(t)\right)\right] dt = 0. \tag{5.11}$$

Observe from the inequalities (5.10) that $\mathrm{Re}\left(f(x)\right) - 1 \le 0$ and $-1 - \mathrm{Re}\left(f(x)\right) \le 0$ on $[0,1]$. Therefore, the equation (5.11) and the continuities of $\mathrm{Re}\left(f(x)\right) - 1$ and $-1 - \mathrm{Re}\left(f(x)\right)$ imply that

$$\mathrm{Re}\left(f(x)\right) = \begin{cases} 1, & \text{if } [0, \frac{1}{2}]; \\ -1, & \text{if } [\frac{1}{2}, 1]. \end{cases}$$

However, this says that $\mathrm{Re}\left(f(x)\right)$ is discontinuous at $x = \frac{1}{2}$, a contradiction. Hence *no* such f exists and this completes the proof of the problem. ∎

Problem 5.5

Rudin Chapter 5 Exercise 5.

Proof. It is clear that $1 \in M$, so $M \ne \varnothing$. Suppose that $f, g \in M \subseteq L^1\left([0,1]\right)$, $t \in [0,1]$ and $h = (1 - t)f + tg$. Then we have

$$\int_0^1 |h(x)|\, dx \le (1 - t) \int_0^1 |f(x)|\, dx + t \int_0^1 |g(x)|\, dx < \infty$$

which means that $h \in L^1\left([0,1]\right)$. Besides, we also have

$$\int_0^1 h(x)\, dx = \int_0^1 [(1 - t)f(x) + tg(x)]\, dx = (1 - t) \int_0^1 f(x)\, dx + t \int_0^1 g(x)\, dx = 1$$

so that $h \in M$, i.e., M is convex. Next, suppose that $\{f_n\} \subseteq M$ and there exists a function f on $[0,1]$ such that $\|f_n - f\|_1 \to 0$ as $n \to \infty$. Thus $f_n - f \in L^1\left([0,1]\right)$ and it yields from Theorem 1.33 that

$$\left| \int_0^1 f_n(t)\, dt - \int_0^1 f(t)\, dt \right| \le \int_0^1 |f_n(t) - f(t)|\, dt = \|f_n - f\|_1$$

which implies that

$$\lim_{n \to \infty} \int_0^1 f_n(t)\,\mathrm{d}t = \int_0^1 f(t)\,\mathrm{d}t.$$

In other words, we have $f \in M$ and M is closed in $L^1\big([0,1]\big)$.

For every $f \in M \subseteq L^1\big([0,1]\big)$, we see from Theorem 1.33 that $\|f\|_1 \geq 1$. For each $n = 1, 2, \ldots$, we define $f_n : [0,1] \to \mathbb{C}$ by

$$f_n(x) = nx^{n-1}.$$

By direct checking, it is clear that

$$\|f_n\|_1 = \int_0^1 nt^{n-1}\,\mathrm{d}t = 1.$$

In addition, $f_n \neq f_m$ if $n \neq m$. Hence M contains infinitely many elements of minimal norm, completing the proof of the problem. ∎

Problem 5.6

Rudin Chapter 5 Exercise 6.

Proof. We note that M is *not* necessarily closed in H. Let $f : M \subseteq H \to \mathbb{C}$ be a bounded linear functional. If we check the proof of Theorem 5.4 carefully, we see that the bounded linear transformation $\Lambda : X \to Y$ is in fact uniformly continuous on X. Thus $f : M \to \mathbb{C}$ is *uniformly continuous* on M. Since H is a metric (in fact Hilbert) space, $\overline{M}$ is also a metric space. Since $\mathbb{C}$ is a complete metric space and M is dense in $\overline{M}$, we follow from [49, Exercises 4 & 13, pp. 98, 99] that f can be *uniquely* extended to a continuous function $\widetilde{f} : \overline{M} \to \mathbb{C}$.[b] Without loss of generality, we may assume that M is closed in H.

As a closed subspace of a Hilbert space H, M is itself a Hilbert space (see the note in Problem 4.8). By Theorem 4.12 (The Riesz Representation Theorem for Hilbert Spaces), there corresponds a *unique* $x_0 \in M$ such that

$$f(x) = \langle x, x_0 \rangle \tag{5.12}$$

for all $x \in M$. Recall from Definition 5.3 that $\|f\|$ is the *smallest* number such that

$$|f(x)| \leq \|f\| \cdot \|x\|$$

for every $x \in M$. In particular, we must have $\|f(x_0)\| \leq \|f\| \cdot \|x_0\|$. On the other hand, the representation (5.12) gives

$$|f(x_0)| = f(x_0) = \langle x_0, x_0 \rangle = \|x_0\|^2 = \|x_0\| \cdot \|x_0\|,$$

so we must have

$$\|x_0\| = \|f\|. \tag{5.13}$$

By Theorem 4.11, every $x \in H$ can be expressed uniquely as $x = Px + Qx$, where $Px \in M$ and $Qx \in M^{\perp}$. This means that $H = M \oplus M^{\perp}$.[c] Now if we define $F : H \to \mathbb{C}$ by

$$F(x) = \langle x, x_0 \rangle$$

---

[b] In fact, such $\widetilde{f}$ is uniformly continuous on $\overline{M}$. See also [42, Exercise 2, p. 270].

[c] Let U and W be two vector subspaces of the vector space V. Then V is said to be the *direct sum* of U and W, denoted by $V = U \oplus W$, if $V = U + W$ and $U \cap W$. See [4, pp. 95, 96].

for every $x \in H$, then we obtain

$$F(x) = \langle x, x_0 \rangle = \begin{cases} f(x), & \text{if } x \in M; \\ 0, & \text{if } x \in M^{\perp}. \end{cases}$$

Furthermore, by the same analysis in obtaining the representation (5.13), we get $\|F\| = \|x_0\|$ and then using the representation (5.13) again, we conclude that

$$\|F\| = \|f\| = \|x_0\|, \tag{5.14}$$

i.e., F is a norm-preserving extension on H of f which vanishes on $M^{\perp}$.

Suppose that $F' : H \to \mathbb{C}$ is *another* norm-preserving extension of f that vanishes on $M^{\perp}$. Then F' is bounded and thus continuous by Theorem 5.4. By Theorem 4.12 (The Riesz Representation Theorem for Hilbert Spaces), there corresponds a unique $x_1 \in H$ such that

$$F'(x) = \langle x, x_1 \rangle$$

for all $x \in H$ and also

$$\|F'\| = \|f\| = \|x_1\|. \tag{5.15}$$

Since $F'(x) = f(x)$ on M, we have

$$\langle x, x_1 \rangle = \langle x, x_0 \rangle$$

on M. By Definition 4.1, $\langle x, x_1 - x_0 \rangle = 0$ and then $x_1 - x_0 \in M^{\perp}$. Since $x_1 = x_0 + (x_1 - x_0)$, where $x_0 \in M$ and $x_1 - x_0 \in M^{\perp}$, Theorem 4.11 shows that x_1 and $x_1 - x_0$ are unique and

$$\|x_0\|^2 = \|x_1\|^2 + \|x_0 - x_1\|^2. \tag{5.16}$$

By putting the numbers (5.14) and (5.15) into the formula (5.16), we see that

$$\|f\|^2 = \|f\|^2 + \|x_0 - x_1\|^2$$

and this reduces to $\|x_0 - x_1\|^2 = 0$. By Definition 4.1, we must have $x_0 = x_1$, i.e., $F' = F$. This proves the uniqueness part and so we have completed the proof of the problem. $\blacksquare$

Problem 5.7

Rudin Chapter 5 Exercise 7.

Proof. Consider $X = [-1, 1]$ with $\mu = m$, the Lebesgue measure. Then the vector space $L^1([-1, 1])$ consists of all complex measurable function on $[-1, 1]$ such that

$$\|f\|_1 = \int_{-1}^{1} |f(x)| \, \mathrm{d}x < \infty.$$

We define

$$M = \{f \in L^1([-1, 1]) \mid f \text{ is real-valued and } f(x) = 0 \text{ on } [-1, 0]\}.$$

Now the function $\chi_{[0,1]}$ is obviously an element of M, so $M \neq \varnothing$ and also $\|\chi_{[0,1]}\|_1 = 1$. It is also clear that M is a subspace of $L^1([-1, 1])$. Next, we define the functional $\Lambda : M \to \mathbb{C}$ by

$$\Lambda(f) = \int_{-1}^{1} f(x) \, \mathrm{d}x = \int_{0}^{1} f(x) \, \mathrm{d}x \tag{5.17}$$

which is linear. By Definition 5.3, the definition (5.17) and the fact that $\chi_{[0,1]} \in M$, we have

$$\|\Lambda\| = \sup\{|\Lambda(f)| \,|\, f \in M \text{ and } \|f\|_1 = 1\}$$
$$= \sup\left\{ \int_0^1 f(x)\,\mathrm{d}x \,\Big|\, f \in M \text{ and } \int_{-1}^1 |f(x)|\,\mathrm{d}x = \int_0^1 f(x)\,\mathrm{d}x = 1 \right\}$$
$$= 1.$$

Thus Λ is bounded.

Let $\delta \in (0,1)$. If we define $\Lambda_\delta : L^1\big([-1,1]\big) \to \mathbb{C}$ by

$$\Lambda_\delta(f) = \int_{-1}^1 \mathrm{Re}\,[f(x)]\chi_{(-\delta,1)}(x)\,\mathrm{d}x = \int_{-\delta}^1 \mathrm{Re}\,[f(x)]\,\mathrm{d}x,$$

then it satisfies Definition 2.1 so that Λ_δ is linear on $L^1\big([-1,1]\big)$. It is also clear that $\Lambda_\delta|_M = \Lambda$. A direct computation also shows that

$$|\Lambda_\delta(f)| = \left| \int_{-\delta}^1 \mathrm{Re}\,[f(x)]\,\mathrm{d}x \right| \le \int_{-\delta}^1 |\mathrm{Re}\,[f(x)]|\,\mathrm{d}x \le \int_{-1}^1 |f(x)|\,\mathrm{d}x$$

which implies that

$$\|\Lambda_\delta\| = \sup\{|\Lambda_\delta(f)| \,|\, f \in L^1\big([-1,1]\big) \text{ and } \|f\|_1 = 1\} = 1 = \|\Lambda\|.$$

Thus Λ_δ is a norm-preserving extension on $L^1\big([-1,1]\big)$ of Λ. Hence if $\alpha \ne \beta$, then we have $\Lambda_\alpha \ne \Lambda_\beta$. This completes the proof of the problem.   ▨

5.3   The Dual Space of X

> **Problem 5.8**
>
> *Rudin Chapter 5 Exercise 8.*

Proof.

(a) We first show that X^* is a normed linear space by checking Definition 5.2. For all $f, g \in X^*$, since $|f(x) + g(x)| \le |f(x)| + |g(x)|$ for all $x \in X$, it must be true that

$$\|f + g\| \le \|f\| + \|g\|.$$

Next, for a scalar α and $f \in X^*$, we have $|\alpha f(x)| = |\alpha| \cdot |f(x)|$ for all $x \in X$ which implies that

$$\|\alpha f\| = |\alpha| \cdot \|f\|.$$

Finally, suppose that $f \in X^*$ is such that $\|f\| = 0$. By [51, Eqn. (3), p. 96], we have

$$|f(x)| \le \|f\| \cdot \|x\| = 0$$

holds for every $x \in X$. Thus $f \equiv 0$ on X. By Definition 5.2, X^* is a normed linear space.

It remains to show that X^* is complete. To this end, let $\{\Lambda_n\} \subseteq X^*$ be Cauchy. Then given $\epsilon > 0$, there corresponds a positive integer N such that $n, m \ge N$ imply that

$$\|\Lambda_n - \Lambda_m\| < \epsilon$$

and for *arbitrary but fixed* $x \in X$, this and [51, Eqn. (3), p. 96] together imply that

$$|\Lambda_n(x) - \Lambda_m(x)| = |(\Lambda_n - \Lambda_m)(x)| \le \|\Lambda_n - \Lambda_m\| \cdot \|x\| < \epsilon\|x\|. \tag{5.18}$$

Thus $\{\Lambda_n(x)\} \subseteq \mathbb{C}$ is Cauchy. Since $\mathbb{C}$ is complete, $\{\Lambda_n(x)\}$ converges to a point in $\mathbb{C}$ and it is reasonable to define the pointwise limit $\Lambda : X \to \mathbb{C}$ by

$$\Lambda(x) = \lim_{n \to \infty} \Lambda_n(x).$$

Now our target is to show that $\Lambda \in X^*$ and $\{\Lambda_n\}$ converges to Λ in X^*, i.e., $\|\Lambda_n - \Lambda\| \to 0$ as $n \to \infty$. Let $\alpha, \beta \in \mathbb{C}$ and $x, y \in X$. Since X is a normed linear space, $\alpha x + \beta y \in X$ and thus

$$\begin{aligned}
\Lambda(\alpha x + \beta y) &= \lim_{n \to \infty} \Lambda_n(\alpha x + \beta y) \\
&= \lim_{n \to \infty} [\alpha \Lambda_n(x) + \beta \Lambda_n(y)] \\
&= \alpha \lim_{n \to \infty} \Lambda_n(x) + \beta \lim_{n \to \infty} \Lambda_n(y) \\
&= \alpha \Lambda(x) + \beta \Lambda(y).
\end{aligned}$$

By Definition 2.1, Λ is linear. By the hypothesis, $\{\Lambda_n\}$ is a Cauchy sequence in X^* so that it is bounded, i.e., there exists a positive constant M such that

$$\|\Lambda_n\| \le M \tag{5.19}$$

for every $n = 1, 2, \ldots$. Therefore, it follows from the triangle inequality, then [51, Eqn. (3), p. 96] and finally the inequalities (5.19) that

$$\begin{aligned}
|\Lambda(x)| &\le |\Lambda_n(x)| + |\Lambda(x) - \Lambda_n(x)| \\
&\le \|\Lambda_n\| \cdot \|x\| + |\Lambda(x) - \Lambda_n(x)| \\
&\le M\|x\| + |\Lambda(x) - \Lambda_n(x)| \tag{5.20}
\end{aligned}$$

for all $n > N$ and for all $x \in X$. Recall that if we take $m \to \infty$ in the inequality (5.18), then we have

$$|\Lambda_n(x) - \Lambda(x)| \le \epsilon\|x\| \tag{5.21}$$

for all $n > N$. Thus, by putting the inequality (5.19) into the inequality (5.20), we get

$$|\Lambda(x)| \le (M + \epsilon)\|x\|$$

for every $x \in X$. Hence we obtain $\|\Lambda\| < \infty$, i.e., $\Lambda \in X^*$. Since the inequality (5.21) is true for all $x \in X$, it must be true for those x with $\|x\| \le 1$ and so $n > N$ implies that

$$\|\Lambda_n - \Lambda\| = \sup\{|\Lambda_n(x) - \Lambda(x)| \,|\, x \in X \text{ and } \|x\| \le 1\} \le \epsilon.$$

Since ϵ is arbitrary, we have $\{\Lambda_n\}$ converges to Λ in X^*. In other words, X^* is complete and we conclude that X^* is a Banach space.

(b) Fix the $x \in X$, define $\Lambda_x^* : X^* \to \mathbb{C}$ by

$$\Lambda_x^*(f) = f(x).$$

If $x = 0$, then it is clear that $\Lambda_0^*(f) = f(0) = 0$ on X^*. Therefore, Λ_0^* must be linear and of norm 0. Without loss of generality, we may assume that $x \neq 0$ in the following discussion. For any $f, g \in X^*$ and $\alpha, \beta \in \mathbb{C}$, we have

$$\Lambda_x^*(\alpha f + \beta g) = (\alpha f + \beta g)(x) = \alpha f(x) + \beta g(x) = \alpha \Lambda_x^*(f) + \beta \Lambda_x^*(g).$$

By Definition 2.1, Λ_x^* is a linear functional on X^*. On the one hand, by [51, Eqn. (3), p. 96], we know that

$$|\Lambda_x^*(f)| = |f(x)| \le \|f\| \cdot \|x\| \le \|x\| \tag{5.22}$$

for all $f \in X^*$ with $\|f\| = 1$. Thus it means that $\|\Lambda_x^*\| \le \|x\|$. On the other hand, since X is a normed linear space, and $x \neq 0$, Theorem 5.20 ensures the existence of a $g \in X^*$ such that $\|g\| = 1$ and $g(x) = \|x\|$, so we have

$$\|\Lambda_x^*\| \ge |\Lambda_x^*(g)| = |g(x)| \ge \|x\|, \tag{5.23}$$

where $\|g\| = 1$. Combining the inequalities (5.22) and (5.23), we get the desired result that

$$\|\Lambda_x^*\| = \|x\|.$$

(c) Suppose that $\{x_n\} \subseteq X$ and $\{f(x_n)\}$ is bounded for every $f \in X^*$. We consider the mapping $\Lambda_{x_n}^* : X^* \to \mathbb{C}$ given by

$$\Lambda_{x_n}^*(f) = f(x_n).$$

By part (a), X^* is a Banach space. By part (b), $\|\Lambda_{x_n}^*\| = \|x_n\|$ for every $n \in \mathbb{N}$. By the hypothesis, *for each* $f \in X^*$, the set $\{\Lambda_{x_n}^*(f)\} = \{f(x_n)\}$ is bounded by a positive constant M_n, so you can't find a $f \in X^*$ such that

$$\sup_{n \in \mathbb{N}} |\Lambda_{x_n}^*(f)| = \infty.$$

Therefore, we deduce from Theorem 5.8 (The Banach-Steinhaus Theorem) that

$$\|\Lambda_{x_n}^*\| \le M$$

for all $n \in \mathbb{N}$, where M is a positive constant. Hence the sequence $\{\|x_n\|\}$ is bounded. This completes the proof of the problem.

> **Problem 5.9**
>
> *Rudin Chapter 5 Exercise 9.*

Proof.

(a) We are going to prove the assertions one by one.

    – $\|\Lambda\| = \|y\|_1$. Fix $y = \{\eta_i\} \in \ell^1$. Define $\Lambda : c_0 \to \mathbb{C}$ by

$$\Lambda(x) = \sum_{i=1}^{\infty} \xi_i \eta_i, \tag{5.24}$$

for every $x = \{\xi_i\} \in c_0 \subseteq \ell^\infty$. Since $\xi_i \to 0$ as $i \to \infty$, the sequence $\{\xi_i\}$ is bounded by a positive constant M (see [49, Theorem 3.2(c), p. 48]). Since $y \in \ell^1$, we have

$$\left| \sum_{i=1}^{\infty} \xi_i \eta_i \right| \le \sum_{i=1}^{\infty} |\xi_i \eta_i| \le M \sum_{i=1}^{\infty} |\eta_i| < \infty$$

so that the series (5.24) converges (absolutely) and thus the map Λ is well-defined.

Let $z = \{\omega_i\} \in c_0$ and $\alpha, \beta \in \mathbb{C}$. Since c_0 is a vector space, we have $\alpha x + \beta z \in c_0$. By the definition, we have

$$\Lambda(x + z) = \sum_{i=1}^{\infty} (\alpha \xi_i + \beta \omega_i) \eta_i \tag{5.25}$$

By the analysis in the previous paragraph, the series (5.25) is allowed to split into two series so that

$$\Lambda(x + z) = \alpha \sum_{i=1}^{\infty} \xi_i \eta_i + \beta \sum_{i=1}^{\infty} \omega_i \eta_i = \alpha \Lambda(x) + \beta \Lambda(z),$$

i.e., Λ is linear. By the definition (5.24), it is easy to see that

$$|\Lambda(x)| \le \|x\|_\infty \cdot \|y\|_1$$

for every $x \in c_0$. Thus it is true that $\|\Lambda\| \le \|y\|_1$.

For the reverse direction, since $y \in \ell^1$, we know that $|\eta_i| \to 0$ as $i \to \infty$. Next, for each $n \in \mathbb{N}$, we consider the sequence $x_n = \{\xi_{n,i}\}$ defined by

$$\xi_{n,i} = \begin{cases} \dfrac{\overline{\eta_i}}{|\eta_i|}, & \text{if } i = 1, 2, \ldots, n \text{ and } \eta_i \neq 0; \\[2ex] 0, & \text{if } i > n \text{ or } \eta_i = 0. \end{cases} \tag{5.26}$$

Since $\left|\frac{\overline{\eta_i}}{|\eta_i|}\right| \le 1$ for all $i \in \mathbb{N}$, we have $x_n \in \ell^\infty$ for each $n \in \mathbb{N}$. In fact, we have $\|x_n\|_\infty \le 1$. Furthermore, *for every fixed* $n \in \mathbb{N}$, one can find $i \in \mathbb{N}$ such that $i > n$, thus we deduce easily from the definition (5.26) that

$$\xi_{n,i} \to 0$$

as $i \to \infty$, i.e., $x_n \in c_0$. Now we derive from Definition 5.2 that

$$\|\Lambda\| = \sup\{|\Lambda(x)| \mid x \in c_0 \text{ and } \|x\|_\infty \le 1\} \ge |\Lambda(x_n)| = \sum_{i=1}^{n} \xi_{n,i} \eta_i = \sum_{i=1}^{n} |\eta_i|$$

for every $n \in \mathbb{N}$, i.e., $\|\Lambda\| \ge \|y\|_1$. In conclusion, we have $\|\Lambda\| = \|y\|_1 < \infty$ and thus

$$\Lambda \in (c_0)^*.$$

– $(c_0)^*$ **is isometric isomorphic to** ℓ^1. In the first assertion, we have shown that given a $y \in \ell^1$, if Λ is defined in the form (5.24), then $\Lambda \in (c_0)^*$. More precisely, the mapping $\Phi : \ell^1 \to (c_0)^*$ given by

$$\Phi(y) = \Lambda_y$$

is well-defined, where Λ_y is in the form (5.24).

    ∗ **Step 1:** Φ **is linear.** For every $y = \{\eta_i\} \in \ell^1, z = \{\theta_i\} \in \ell^1$ and $\alpha, \beta \in \mathbb{C}$, we have $\alpha y + \beta z = \{\alpha \eta_i + \beta \theta_i\} \in \ell^1$. If $x = \{\xi_i\} \in c_0$, then we deduce from the definition (5.24) that

$$\Lambda_{\alpha y + \beta z}(x) = \sum_{i=1}^{\infty} \xi_i(\alpha \eta_i + \beta \theta_i) = \alpha \sum_{i=1}^{\infty} \xi_i \eta_i + \beta \sum_{i=1}^{\infty} \xi_i \theta_i = \alpha \Lambda_y(x) + \beta \Lambda_z(x).$$

Consequently, $\Phi(\alpha y + \beta z) = \alpha \Phi(y) + \beta \Phi(z)$.

* **Step 2: Φ is surjective.** Given $\Lambda \in (c_0)^*$. We want to show that *there exists* a $y \in \ell^1$ such that $\Phi(y) = \Lambda$.

It is well-known that the sequence $\{e_i\}_{i=1}^{\infty}$ of standard unit vectors of c_0 is a basis for c_0 and that $\{\xi_i\} = \sum \xi_i e_i$ whenever $\{\xi_i\}_{i=1}^{\infty} \in c_0$.[d] In fact, we have $e_i = \{\delta_{i,k}\}_{k=1}^{\infty}$, where $\delta_{i,k}$ is the Kronecker delta function. Then it is routine to check that $e_i \in c_0$ and $\|e_i\|_{\infty} = 1$.

Next, we define the sequence $y = \{\eta_i\}_{i=1}^{\infty}$ by

$$\eta_i = \Lambda(e_i) \tag{5.27}$$

for every $i = 1, 2, \ldots$. Our aim is to show that $y \in \ell^1$. If the sequence $x_n = \{\xi_{n,i}\}_{i=1}^{n}$ is given by the form (5.26), then we immediately know that $x_n \in c_0$ and $\|x_n\|_{\infty} = 1$. Since

$$x_n = \sum_{i=1}^{\infty} \xi_{n,i} e_i = \sum_{i=1}^{n} \xi_{n,i} e_i,$$

the linearity of Λ shows that

$$\Lambda(x_n) = \sum_{i=1}^{n} \xi_{n,i} \Lambda(e_i) = \sum_{i=1}^{n} \xi_{n,i} \eta_i = \sum_{i=1}^{n} |\eta_i|. \tag{5.28}$$

Therefore, we apply [51, Eqn. (3), p. 96] to the expression (5.28) to obtain

$$\sum_{i=1}^{n} |\eta_i| = |\Lambda(x_n)| \le \|\Lambda\| \cdot \|x_n\|_{\infty} = \|\Lambda\|$$

holds *for every* $n = 1, 2, \ldots$. In particular, we have

$$\|y\|_1 = \sum_{i=1}^{\infty} |\eta_i| \le \|\Lambda\| < \infty,$$

i.e, $y \in \ell^1$.

Finally, for every $x = \{\xi_i\} \in c_0$, we have

$$x = \sum_{i=1}^{\infty} \xi_i e_i$$

and the application of the expressions (5.27) and the linearity of Λ indicate that

$$\Lambda(x) = \Lambda\left(\sum_{i=1}^{\infty} \xi_i e_i \right) = \sum_{i=1}^{\infty} \xi_i \Lambda(e_i) = \sum_{i=1}^{\infty} \xi_i \eta_i$$

which is exactly in the form (5.24). Hence the map Φ is surjective.

* **Step 3: Φ is injective.** Let $y = \{\eta_i\} \in \ell^1, z = \{\theta_i\} \in \ell^1$. Suppose further that $\Phi(y) = \Phi(z)$. Since Φ is linear, we have $\Lambda_{y-z} = \Phi(y - z) = 0$. By this, for every $j = 1, 2, \ldots$, we get

$$\Lambda_{y-z}(e_j) = \sum_{i=1}^{\infty} \delta_{j,i}(\eta_i - \theta_i) = \eta_j - \theta_j = 0$$

so that $\eta_j = \theta_j$. In other words, we conclude that $y = z$, as desired.

---

[d]The sequence $\{e_i\}$ is a *Schauder basis* for c_0 (and also for ℓ^1). In fact, a sequence $\{x_i\}_{i=1}^{\infty}$ in a Banach space X is called a Schauder basis if for each $x \in X$, there is a unique sequence $\{\alpha_i\}$ of scalars such that $x = \sum_{i=1}^{\infty} \alpha_i x_i = \lim_{n \to \infty} \sum_{i=1}^{n} \alpha_i x_i$. See, for instance, [40, Example 4.1.3, p. 351].

* **Step 4: Φ is an isometry.** This follows from the first assertion directly because $\|\Phi(y)\| = \|\Lambda_y\| = \|y\|_1$.

Hence we have $(c_0)^* = \ell^1$.

(b) For each $y = \{\eta_i\} \in \ell^\infty$, we define $\Theta : \ell^1 \to \mathbb{C}$ by

$$\Theta(x) = \sum_{i=1}^\infty \xi_i \eta_i \tag{5.29}$$

for every $x = \{\xi_i\} \in \ell^1$. Similar to the proof of the first assertion in part (a), we can show that the series (5.29) converges (absolutely) and Θ is linear. It is clear from the definition that $|\Theta(x)| \le \|y\|_\infty \cdot \|x\|_1$, so we have

$$\|\Theta\| \le \|y\|_\infty < \infty. \tag{5.30}$$

This means that $\Theta \in (\ell^1)^*$.

On the other hand, suppose that $\Theta \in (\ell_1)^*$, i.e., $\|\Theta\| < \infty$. If we consider the Schader basis $\{e_i\}$ for ℓ^1 and the relationships (5.27) with Λ replaced by Θ, then we have

$$|\eta_i| = |\Theta(e_i)| \le \|\Theta\| \cdot \|e_i\|_1 = \|\Theta\| < \infty \tag{5.31}$$

for every $i = 1, 2, \ldots$. Put $z = \{\eta_i\}$. Then the inequality (5.31) immediately implies that $z \in \ell^\infty$. In fact, it shows further that $\|z\|_\infty \le \|\Theta\|$. Now for every $x = \{\xi_i\} \in \ell^1$, we have

$$x = \sum_{i=1}^\infty \xi_i e_i$$

and then

$$\Theta(x) = \sum_{i=1}^\infty \xi_i \eta_i$$

which is exactly in the form (5.29), so the inequality (5.30) holds for z. Therefore, we have $\|\Theta\| = \|z\|_\infty$, i.e., $\Theta \in \ell^\infty$.

If we define the mapping $\Psi : \ell^\infty \to (\ell^1)^*$ by

$$\Psi(y) = \Theta_y,$$

where Θ_y is given by the form (5.29), then the previous paragraph actually shows that Ψ is surjective. By using similar argument as in the proofs of **Step 1**, **Step 3** and **Step 4** in part (a), it is easy to conclude that Ψ is an isometric vector space isomorphism.

(c) Let $y = \{\eta_i\} \in \ell^1$. Define the functional $\Lambda : \ell^\infty \to \mathbb{C}$ by

$$\Lambda(x) = \sum_{i=1}^\infty \xi_i \eta_i,$$

where $x = \{\xi_i\} \in \ell^\infty$. Since this series converges (absolutely), it is a linear functional. Since $|\Lambda(x)| \le \|x\|_\infty \cdot \|y\|_1$, we know from [51, Eqn. (2), p. 96] that $\|\Lambda\| \le \|y\|_1$, i.e., $\Lambda \in (\ell^\infty)^*$.

Assume that ℓ^1 was isometric isomorphic to $(\ell^\infty)^*$. It is clear that $x = \{1, 1, \ldots\} \in \ell^\infty \setminus c_0$. By [40, Example 1.2.13, p. 14], we know that c_0 is a *proper* closed subspace of

ℓ^∞ so that $\ell^\infty \setminus c_0 \neq \varnothing$. Take $x_0 \in \ell^\infty \setminus c_0$. Then Theorem 5.19 implies that there is a $f \in (\ell^\infty)^*$ such that $\|f\| = 1$, $f(x) = 0$ on c_0 and

$$f(x_0) \neq 0. \tag{5.32}$$

If $f \in \ell^1$, then we have $f \in (c_0)^*$ by part (a). Since $f(x) = 0$ on c_0, f is in fact the trivial bounded functional on c_0. By the assumption and part (a), $(c_0)^*$ is isometric isomorphic to $(\ell^\infty)^*$ so that the trivial bounded functional on c_0 corresponds to the trivial bounded functional on ℓ^∞. In other words, $f(x) = 0$ on ℓ^∞ which contradicts the result (5.32). Hence we conclude that ℓ^1 is *not* isometric isomorphic to $(\ell^\infty)^*$.

(d) We prove the assertions one by one.

    – c_0 **is separable.** For each $k \in \mathbb{N}$, we define

$$\mathcal{S}_k = \{\{\xi_1, \ldots, \xi_k, 0, 0, \ldots\} \mid \xi_1, \ldots, \xi_k \in \mathbb{Q}\} \quad \text{and} \quad \mathcal{S} = \bigcup_{k=1}^{\infty} \mathcal{S}_k.$$

Since each $\mathcal{S}_k$ is countable, $\mathcal{S}$ is also countable. We claim that $\mathcal{S}$ is dense in c_0. Pick $y = \{\eta_i\} \in c_0$. Given $\epsilon > 0$. Since $\eta_i \to 0$ as $i \to \infty$, there exists a positive integer N such that $i \geq N$ implies

$$|\eta_i| < \epsilon. \tag{5.33}$$

By the density of $\mathbb{Q}$ in $\mathbb{R}$, we can take $x = \{\xi_1, \ldots, \xi_N, 0, 0, \ldots\} \in \mathcal{S}_N$ such that

$$|\xi_i - \eta_i| < \epsilon \tag{5.34}$$

for all $i = 1, 2, \ldots, N$. Thus we obtain from the definition of $\| \cdot \|_\infty$, the inequalities (5.33) and (5.34) that

$$\|x - y\|_\infty = \sup_{i \in \mathbb{N}} |\xi_i - \eta_i| \leq \epsilon.$$

Hence this proves the claim and then c_0 is separable.

    – ℓ^1 **is separable.** Take $y = \{\eta_i\} \in \ell^1$. Given that $\epsilon > 0$. Since $\displaystyle\sum_{i=1}^{\infty} |\eta_i| < \infty$, the Cauchy Criterion (see [49, Theorem 3.22, p. 59]) shows that one can find a positive integer N' such that $i \geq N'$ implies

$$\sum_{i=N'}^{\infty} |\eta_i| < \frac{\epsilon}{2}. \tag{5.35}$$

By the density of $\mathbb{Q}$ in $\mathbb{R}$ again, we can take $x = \{\xi_1, \ldots, \xi_{N'}, 0, 0, \ldots\} \in \mathcal{S}_{N'}$ such that

$$\sum_{i=1}^{N'} |\xi_i - \eta_i| < \frac{\epsilon}{2}. \tag{5.36}$$

Combining the inequalities (5.35) and (5.36), we get

$$\|x - y\|_1 = \sum_{i=1}^{\infty} |\xi_i - \eta_i| \leq \sum_{i=1}^{N'} |\xi_i - \eta_i| + \sum_{i=N'}^{\infty} |0 - \eta_i| < \frac{\epsilon}{2} + \frac{\epsilon}{2} = \epsilon.$$

Hence $\mathcal{S}$ is also dense in ℓ^1 and thus ℓ^1 is separable.

- ℓ^∞ **is not separable.** Let $\mathcal{A}$ be the set of all sequences $x = \{\xi_i\}$ such that $\xi_i \in \{0, 1\}$. Clearly, $\mathcal{A}$ is uncountable (see [49, Theorem 2.14]) and $\mathcal{A} \subseteq \ell^\infty$. Let $x, y \in \mathcal{A}$ be such that $x = \{\xi_i\}$ and $y = \{\eta_i\}$. If $x \neq y$, then there is *at least one* i such that $\xi_i \neq \eta_i$ which implies that

$$\|x - y\|_\infty = 1.$$

By this and the fact that $B(x, \frac{1}{2}) \cap B(y, \frac{1}{2}) = \varnothing$ if $x \neq y$, we conclude that the collection of open balls $\{B(x, \frac{1}{2}) \mid x \in \mathcal{A}\}$ is uncountable. If S was a dense subset of ℓ^∞, then we have

$$S \cap B\left(x, \frac{1}{2}\right) \neq \varnothing$$

for every $x \in \mathcal{A}$ so that S must be uncountable. Hence ℓ^∞ is *not* separable.

Remark 5.1

We say two normed linear spaces X and Y *isometric isomorphism* if there is a bijective linear map $T : X \to Y$ such that

$$\|T(x)\| = \|x\|$$

for every $x \in X$.

5.4   Applications of Baire's and other Theorems

Problem 5.10

Rudin Chapter 5 Exercise 10.

Proof. For each positive integer n, we define $\Lambda_n : c_0 \to \mathbb{C}$ by

$$\Lambda_n(x) = \sum_{i=1}^{n} \alpha_i \xi_i, \tag{5.37}$$

where $x = \{\xi_i\} \in c_0$. By the hypotheses of Problem 5.9, we know that c_0 is a Banach space. On the one hand, we note from the definition (5.37) that Λ_n is linear and the analysis in Problem 5.9(a) first assertion shows that

$$|\Lambda_n(x)| \leq \|x\|_\infty \cdot \sum_{i=1}^{n} |\alpha_i|$$

for every $x \in c_0 \subseteq \ell^\infty$ and therefore,

$$\|\Lambda_n\| \leq \sum_{i=1}^{n} |\alpha_i|. \tag{5.38}$$

On the other hand, if we define the sequence $x_n = \{\xi_{n,i}\}$ by (5.26) with η_i replaced by α_i, then, using the argument as in the proof of Problem 5.9(a), we conclude that $x_n \in c_0$. Thus we obtain from Definition 5.2 that

$$\|\Lambda_n\| = \sup\{|\Lambda_n(x)| \mid x \in c_0 \text{ and } \|x\|_\infty \leq 1\} \geq |\Lambda_n(x_n)| = \sum_{i=1}^{n} |\alpha_i| \tag{5.39}$$

for every $n \in \mathbb{N}$. Combining the inequalities (5.38) and (5.39), we conclude that

$$\|\Lambda_n\| = \sum_{i=1}^{n} |\alpha_i| \tag{5.40}$$

which is bounded for each $n \in \mathbb{N}$. By the hypothesis and the definition (5.37), there is *no* $x \in c_0$ such that $\sup_{n \in \mathbb{N}} |\Lambda_n(x)| = \infty$. Thus Theorem 5.8 (The Banach-Steinhaus Theorem) implies that there exists a positive constant M such that

$$\|\Lambda_n\| \leq M \tag{5.41}$$

for every $n \in \mathbb{N}$. Consequently, we deduce from the two results (5.40) and (5.41) that

$$\sum_{i=1}^{\infty} |\alpha_i| < \infty,$$

as required. We have completed the proof of the problem. ▨

> **Problem 5.11**
>
> *Rudin Chapter 5 Exercise 11.*

Proof. The proof will be divided into several steps:

- **Step 1: Lip α is a vector space.** For every $f, g \in \mathrm{Lip}\,\alpha$ and $\mu, \nu \in \mathbb{C}$, we have

$$
\begin{aligned}
M_{\mu f + \nu g} &= \sup \frac{|\mu f(s) + \nu g(s) - \mu f(t) - \nu g(t)|}{|s - t|^\alpha} \\
&\leq |\mu| \cdot \sup \frac{|f(s) - f(t)|}{|s - t|^\alpha} + |\nu| \cdot \sup \frac{|g(s) - g(t)|}{|s - t|^\alpha} \\
&= |\mu| \cdot M_f + |\nu| \cdot M_g \\
&< \infty
\end{aligned}
\tag{5.42}
$$

which implies that $\mu f + \nu g \in \mathrm{Lip}\,\alpha$, i.e., $\mathrm{Lip}\,\alpha$ is a vector space.

- **Step 2: Lip α is a normed linear space with the mentioned norms.** If we define $\|f\|_1 = |f(a)| + M_f$, then for $f, g \in \mathrm{Lip}\,\alpha$, it is easy to see from the inequality (5.42) that

$$\|f + g\|_1 = |f(a) + g(a)| + M_{f+g} \leq |f(a)| + |g(a)| + M_f + M_g = \|f\|_1 + \|g\|_1.$$

Next, if $\mu \in \mathbb{C}$, then we have

$$\|\mu f\|_1 = |\mu f(a)| + M_{\mu f} = |\mu| \cdot |f(a)| + |\mu| \cdot M_f = |\mu| \cdot \|f\|_1.$$

Finally, if $\|f\|_1 = 0$, then $|f(a)| = M_f = 0$ so that $|f(s) - f(a)| = 0$ for all $s \in (a, b]$. Thus we establish that $f(x) = 0$ on $[a, b]$ and we can say that $\mathrm{Lip}\,\alpha$ is a normed linear space with respect to the first norm.

For the second norm, if $f, g \in \mathrm{Lip}\,\alpha$, then we see that

$$\|f+g\|_2 = M_{f+g} + \sup_{x \in [a,b]} |f(x)+g(x)| \leq M_f + M_g + \sup_{x \in [a,b]} |f(x)| + \sup_{x \in [a,b]} |g(x)| = \|f\|_2 + \|g\|_2.$$

Next, if $\mu \in \mathbb{C}$, then we have

$$\|\mu f\|_2 = M_{\mu f} + \sup_{x\in[a,b]} |\mu f(x)| = |\mu| \cdot M_f + |\mu| \cdot \sup_{x\in[a,b]} |f(x)| = |\mu| \cdot \|f\|_2.$$

Finally, if $\|f\|_2 = 0$, then we have $M_f = \sup_{x\in[a,b]} |f(x)| = 0$ implying that $f(x) = 0$ on $[a,b]$. Hence $\operatorname{Lip}\alpha$ is also a normed linear space with respect to the second norm.

- **Step 3: $\operatorname{Lip}\alpha$ is complete.** Suppose that $\{f_n\} \subseteq \operatorname{Lip}\alpha$ is Cauchy with respect to the norm $\|\cdot\|_2$. Given $\epsilon > 0$. There exists a positive integer N such that $n, m \geq N$ imply that

$$\sup_{x\in[a,b]} |f_n(x) - f_m(x)| + M_{f_n - f_m} = \|f_n - f_m\|_2 < \frac{\epsilon}{2}. \tag{5.43}$$

Particularly, we have

$$|f_n(x) - f_m(x)| < \frac{\epsilon}{2} \tag{5.44}$$

for all $x \in [a,b]$. Thus $\{f_n(x)\}$ is a Cacuhy sequence in $\mathbb{C}$. Since $\mathbb{C}$ is complete, there exists a function $f : [a,b] \to \mathbb{C}$ such that

$$f(x) = \lim_{n\to\infty} f_n(x) \tag{5.45}$$

for each $x \in [a,b]$.

We claim that $\|f_n - f\|_2 \to 0$ as $n \to \infty$ and $f \in \operatorname{Lip}\alpha$. On the one hand, we take $m \to \infty$ in the inequality (5.44) to get

$$|f_n(x) - f(x)| \leq \frac{\epsilon}{2} \tag{5.46}$$

for all $x \in [a,b]$. Thus we have

$$\sup_{x\in[a,b]} |f_n(x) - f(x)| \leq \frac{\epsilon}{2}. \tag{5.47}$$

On the other hand, we know from the inequality (5.43) that $M_{f_n - f_m} < \frac{\epsilon}{2}$, or explicitly

$$\frac{|f_n(s) - f_m(s) - f_n(t) + f_m(t)|}{|s - t|^\alpha} < \frac{\epsilon}{2} \tag{5.48}$$

for all $s, t \in [a,b]$ with $s \neq t$. Taking $m \to \infty$ in the inequality (5.48), we observe that

$$\frac{|f_n(s) - f(s) - f_n(t) + f(t)|}{|s - t|^\alpha} \leq \frac{\epsilon}{2}$$

for all $s, t \in [a,b]$ with $s \neq t$. In other words, we have

$$M_{f_n - f} \leq \frac{\epsilon}{2}. \tag{5.49}$$

Hence we combine the inequalities (5.47) and (5.49) to establish that

$$\|f_n - f\|_2 = \sup_{x\in[a,b]} |f_n(x) - f(x)| + M_{f_n - f} \leq \epsilon,$$

i.e., $\|f_n - f\|_2 \to 0$ as $n \to \infty$. Furthermore, for $s, t \in [a,b]$ with $s \neq t$, since we have

$$\frac{|f(s) - f(t)|}{|s - t|^\alpha} = \frac{|f(s) - f_N(s) + f_N(t) - f(t) + f_N(s) - f_N(t)|}{|s - t|^\alpha}$$

$$\leq \frac{|f(s) - f_N(s) + f_N(t) - f(t)|}{|s-t|^\alpha} + \frac{|f_N(s) - f_N(t)|}{|s-t|^\alpha}$$

$$\leq M_{f_N - f} + M_{f_N},$$

where N is the positive integer which makes the inequality (5.43) holds. Therefore, we deduce from the definition and the inequality (5.49) that

$$M_f \leq M_{f_N - f} + M_{f_N} \leq \frac{\epsilon}{2} + M_{f_N} < \infty,$$

i.e., $f \in \mathrm{Lip}\,\alpha$. In other words, $\mathrm{Lip}\,\alpha$ is complete with respect to the norm $\|\cdot\|_2$.

For the first norm $\|\cdot\|_1$, we note that

$$\|f\|_1 = |f(a)| + M_f \leq \sup_{x\in[a,b]} |f(x)| + M_f = \|f\|_2. \tag{5.50}$$

If $\{f_n\} \subseteq \mathrm{Lip}\,\alpha$ is Cauchy with respect to the norm $\|\cdot\|_1$, then we deduce from the inequality (5.50) that

$$\|f_n - f\|_1 \leq \|f_n - f\|_2 < \epsilon$$

for large enough n, where the f is defined by the limit (5.45). Since $f \in \mathrm{Lip}\,\alpha$, we have shown that $\mathrm{Lip}\,\alpha$ is complete with respect to the norm $\|\cdot\|_1$.

Hence $\mathrm{Lip}\,\alpha$ is also a Banach space with respect to the two norms and this completes the proof of the problem. ∎

Problem 5.12

Rudin Chapter 5 Exercise 12.

Proof. We have $A = \{f : K \to \mathbb{R}\,|\,f(x,y) = \alpha x + \beta y + \gamma,\ \text{where}\ \alpha, \beta, \gamma \in \mathbb{R}\}$.

- **Case (i): K is a triangle.** Suppose that $P = (x_1, y_1), Q = (x_2, y_2), R = (x_3, y_3)$ and $H = \{P, Q, R\}$. Then it is well-known that

$$\begin{aligned}
K &= \mathrm{Conv}\,(P, Q, R) \\
&= \{\lambda P + \tau Q + \nu R\,|\,\lambda, \tau, \nu \geq 0\ \text{and}\ \lambda + \tau + \nu = 1\} \\
&= \{(\lambda x_1 + \tau x_2 + \nu x_3, \lambda y_1 + \tau y_2 + \nu y_3)\,|\,\lambda, \tau, \nu \geq 0\ \text{and}\ \lambda + \tau + \nu = 1\}.
\end{aligned}$$

Therefore, for each $(x_0, y_0) \in K$, we have the representation

$$(x_0, y_0) = \lambda P + \tau Q + \nu R = (\lambda x_1 + \tau x_2 + \nu x_3, \lambda y_1 + \tau y_2 + \nu y_3) \tag{5.51}$$

and thus

$$\begin{aligned}
f(x_0, y_0) &= f(\lambda x_1 + \tau x_2 + \nu x_3, \lambda y_1 + \tau y_2 + \nu y_3) \\
&= \alpha(\lambda x_1 + \tau x_2 + \nu x_3) + \beta(\lambda y_1 + \tau y_2 + \nu y_3) + (\lambda + \tau + \nu)\gamma \\
&= \lambda(\alpha x_1 + \beta y_1 + \gamma) + \tau(\alpha x_2 + \beta y_2 + \gamma) + \nu(\alpha x_3 + \beta y_3 + \gamma) \\
&= \lambda f(x_1, y_1) + \tau f(x_2, y_2) + \nu f(x_3, y_3). \tag{5.52}
\end{aligned}$$

If we define $\mu(P) = \mu(x_1, y_1) = \lambda$, $\mu(Q) = \mu(x_2, y_2) = \tau$ and $\mu(R) = \mu(x_3, y_3) = \nu$, then the expression (5.52) becomes

$$f(x_0, y_0) = \int_H f\,d\mu. \tag{5.53}$$

To prove the uniqueness of μ, we suppose that the integral representation (5.53) holds for the measure μ'. Then the integral (5.53) implies that

$$f(x_0, y_0) = \mu'(P)f(x_1, y_1) + \mu'(Q)f(x_2, y_2) + \mu'(R)f(x_3, y_3)$$
$$= f\big(\mu'(P)x_1 + \mu'(Q)x_2 + \mu'(R)x_3, \mu'(P)y_1 + \mu'(Q)y_2 + \mu'(R)y_3\big). \qquad (5.54)$$

By the definition of f, the expression (5.54) means that

$$(x_0, y_0) = \mu'(P) \cdot P + \mu'(Q) \cdot Q + \mu'(R) \cdot R. \qquad (5.55)$$

Comparing the expressions (5.51) and (5.55), we derive that

$$[\mu'(P) - \lambda]P + [\mu'(Q) - \tau]Q + [\mu'(R) - \nu]R = 0. \qquad (5.56)$$

Since the set $\{P, Q, R\}$ is linearly independent, it follows from the equation (5.56) that

$$\mu'(P) = \lambda, \quad \mu'(Q) = \tau \quad \text{and} \quad \mu'(R) = \nu.$$

In other words, we establish the fact that $\mu' = \mu$.

- **Case (ii):** K **is a square.** Without loss of generality, we may assume that the vertices of K are $P(0,0), Q(1,0), R(1,1)$ and $S(0,1)$. Let $A(a,b)$ be a point of K, see Figure 5.2 below.

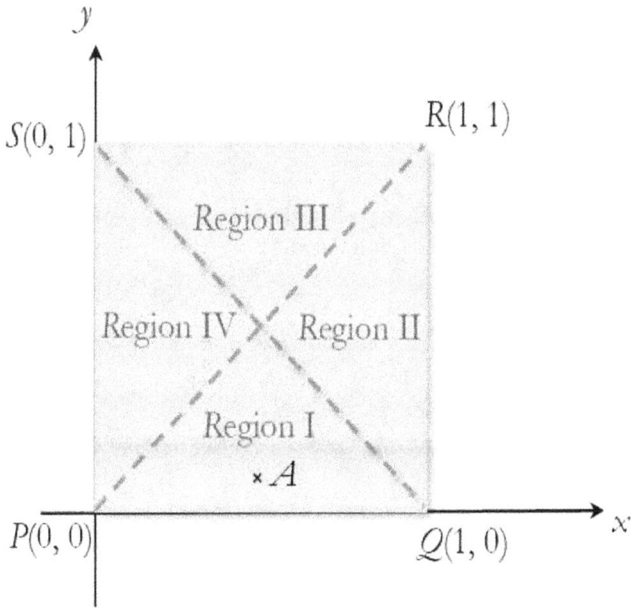

Figure 5.2: The square K.

Suppose that $(a, b) = \lambda P + \tau Q + \nu R + \theta S$, where $\lambda, \tau, \nu, \theta \in \mathbb{R}$. Then we have

$$(a, b) = \tau(1, 0) + \nu(1, 1) + \theta(0, 1) = (\tau + \nu, \nu + \theta).$$

Thus we have $\nu + \theta = b$ and $\tau + \nu = a$. Solving the equations, we have

$$\nu = b - \theta \quad \text{and} \quad \tau = a - b + \theta. \qquad (5.57)$$

Now we have five subcases below:

– **Subcase (i):** *A* **belongs to Region I.** Here we have $0 < b \le a < 1$ and $a+b-1 \le 0$. If $\theta = 0$, then we obtain from the equations (5.57) that $\nu = b$ and $\tau = a - b \ge 0$. If $\lambda = 1 - a \ge 0$, then it is easy to see that

$$(a, b) = (1 - a)P + (a - b)Q + bR + 0 \cdot S.$$

Similarly, $\theta = b$, $\nu = 0$, $\tau = a$ and $\lambda = 1 - a - b \ge 0$ imply that

$$(a, b) = (1 - a - b)P + aQ + 0 \cdot R + bS.$$

Hence, every point in Region I has two different convex combinations of P, Q, R and S.

– **Subcase (ii):** *A* **belongs to Region II.** In this case, we have $0 < b \le a < 1$ and $a + b - 1 \ge 0$. Since we have

$$\begin{aligned}
(a, b) &= (1 - a)P + (a - b)Q + bR + 0 \cdot S \\
&= 0 \cdot P + (1 - b)Q + (a + b - 1)R + (1 - a)S,
\end{aligned}$$

every point in Region II has two distinct convex combinations of P, Q, R and S.

– **Subcase (iii):** *A* **belongs to Region III.** In this case, we have $0 < a \le b < 1$ and $a + b - 1 \ge 0$. It is easy to see that

$$\begin{aligned}
(a, b) &= 0 \cdot P + (1 - b)Q + (a + b - 1)R + (1 - a)S \\
&= (1 - b)P + 0 \cdot Q + aR + (b - a)S,
\end{aligned}$$

so every point in Region III also has two different convex combinations of P, Q, R and S.

– **Subcase (iv):** *A* **belongs to Region IV.** Now we have $0 < a \le b < 1$ and $a + b - 1 \le 0$. Then we certainly have

$$\begin{aligned}
(a, b) &= (1 - b)P + 0 \cdot Q + aR + (b - a)S \\
&= (1 - a - b)P + aQ + 0 \cdot R + bS
\end{aligned}$$

which means that it has two different convex combinations of P, Q, R and S in Region IV.

– **Subcase (v):** *A* **belongs to the boundary of** K**.** Suppose that $A = (a, b)$. We note that

$$(a, 0) = (1 - a) \cdot P + aQ + 0 \cdot R + 0 \cdot S = 1 \cdot P + aQ + 0 \cdot R + 0 \cdot S,$$

thus every point on PQ can be expressed as two distinct linear (but *not* convex) combinations of P, Q, R and S. Similarly, all points lie on QR, RS and SP also have two different linear combinations of P, Q, R and S.

In conclusion, every point (x_0, y_0) in the square K has *at least two* distinct linear combinations of P, Q, R and S. Now, instead of the expression (5.52), we have

$$f(x_0, y_0) = \lambda f(P) + \tau f(Q) + \nu f(R) + \theta f(S) = \lambda' f(P) + \tau' f(Q) + \nu' f(R) + \theta' f(S),$$

where $\{\lambda, \tau, \nu, \theta\}$ and $\{\lambda', \tau', \nu', \theta'\}$ are two sets of *distinct* numbers corresponding to two *distinct* linear combinations of P, Q, R and S. Thus $\{\lambda, \tau, \nu, \theta\}$ and $\{\lambda', \tau', \nu', \theta'\}$ give *distinct* measures μ and μ' such that the representation (5.53) holds.

- **Case (iii): The general situation.** Suppose that $H = \{v_0, v_1, \ldots, v_n\}$ and

$$K = \{\lambda_0 v_0 + \lambda_1 v_1 + \cdots + \lambda_n v_n \subseteq \mathbb{R}^{n+1} \mid \lambda_0, \lambda_1, \ldots, \lambda_n \geq 0 \text{ and } \lambda_0 + \cdots + \lambda_n = 1\}.$$

Let $f : K \to \mathbb{R}$ be defined by

$$f(v_0, v_1, \ldots, v_n) = \alpha_0 v_0 + \alpha_1 v_1 + \cdots + \alpha_n v_n + \gamma,$$

where $\alpha_0, \alpha_1, \ldots, \alpha_n, \gamma$ are real. For each $(v_0, v_1, \ldots, v_n) \in K$, there corresponds a measure μ on H such that

$$f(v_0, v_1, \ldots, v_n) = \int_H f \, d\mu.$$

Furthermore, the measure μ is unique if K is a n-simplex, i.e., $\{v_1 - v_0, v_2 - v_0, \ldots, v_n - v_0\}$ is linearly independent, see [32, pp. 102, 103].

We have completed the proof of the problem. ▪

Problem 5.13

Rudin Chapter 5 Exercise 13.

Proof.

(a) For each $N = 1, 2, \ldots$, we define

$$E_N = \{x \in X \mid |f_n(x)| \leq N \text{ for all } n \in \mathbb{N}\}.$$

Since each f_n is continuous on X, each E_N is closed in X. On the one hand, we have

$$E_N \subseteq X. \tag{5.58}$$

On the other hand, for each *fixed* $x \in X$, since $f(x)$ is well-defined as a complex number, it must be true that

$$|f(x)| \leq m$$

for some $m > 0$. By the convergence of $\{f_n(x)\}$, there is a positive integer N_0 such that $n \geq N_0$ implies that

$$|f(x) - f_n(x)| < 1$$

Thus, for $n \geq N_0$, the triangle inequality shows that

$$|f_n(x)| \leq |f(x)| + |f(x) - f_n(x)| < m + 1. \tag{5.59}$$

Take $m' = \max\big(f_1(x), f_2(x), \ldots, f_{N_0-1}(x)\big)$. Combining this and the inequality (5.59), we obtain that

$$|f_n(x)| < \max(m', m + 1)$$

for all $n \in \mathbb{N}$. Hence we have $x \in E_{N_1}$ for some $N_1 \geq \max(m', m + 1)$ and then we follow from the set relation (5.58) that

$$X = \bigcup_{N=1}^{\infty} E_N.$$

By Theorem 5.6 (Baire's Theorem), X is *not* of the first category, i.e., X is not a countable union of nowhere dense sets. Therefore, *some* $E_{N'}$ contains a nonempty open subset V of X, i.e., for all $x \in V$ and $n = 1, 2, \ldots$, we have

$$|f_n(x)| \leq N',$$

where $N' < \infty$.

(b) We follow the hint. For each $N = 1, 2, 3, \ldots$, we put

$$A_N = \{x \in X \mid |f_m(x) - f_n(x)| \leq \epsilon \text{ for all } n, m \geq N\}.$$

Since f_m and f_n are continuous on X, $f_m - f_n$ is also continuous on X. Thus for each pair $n, m \in \mathbb{N}$, the set

$$S(m, n) = (f_m - f_n)^{-1}\big([-\epsilon, \epsilon]\big) = \{x \in X \mid |f_m(x) - f_n(x)| \leq \epsilon\}$$

is closed in X. Since

$$A_N = \bigcap_{n, m \geq N} S(m, n),$$

A_N is definitely closed in X.

It is clear that

$$A_N \subseteq X \tag{5.60}$$

for each $N = 1, 2, \ldots$. Now for every $x \in X$, by the convergence of $\{f_n(x)\}$, it is Cauchy in $\mathbb{C}$, i.e., there exists a $N_0 \in \mathbb{N}$ such that $n, m \geq N_0$ imply that

$$|f_m(x) - f_n(x)| \leq \epsilon$$

or equivalently, $x \in A_{N_0}$. In other words, we have

$$X \subseteq \bigcup_{N=1}^{\infty} A_N. \tag{5.61}$$

Combining the set relations (5.60) and (5.61), we conclude that

$$X = \bigcup_{N=1}^{\infty} A_N.$$

By Theorem 5.6 (Baire's Theorem), X is *not* of the first category, i.e., X is not a countable union of nowhere dense sets. Therefore, *some* $A_{N'}$ contains a nonempty open subset V of X. Particularly, if $x \in V$ and $n, m \geq N'$, then we have

$$|f_m(x) - f_n(x)| \leq \epsilon. \tag{5.62}$$

Taking $m \to \infty$ in the inequality (5.62), we have established that

$$|f(x) - f_n(x)| \leq \epsilon$$

for all $x \in V$ and $n \geq N'$.

This has completed the proof of the problem. ◾

Problem 5.14

Rudin Chapter 5 Exercise 14.

Proof. Let $C = \{f : [0, 1] \to \mathbb{R} \mid f \text{ is continuous on } [0, 1]\}$ and $\|f\| = \sup\limits_{x \in [0,1]} |f(x)|$. Furthermore, for a positive integer n, let

$$X_n = \{f \in C \mid \text{there exists a } t \in I \text{ such that } |f(s) - f(t)| \leq n|s - t| \text{ for all } s \in I\}.$$

To prove the first assertion, we fix $n \in \mathbb{N}$. We follow the hint given by Rudin and the construction of g will be shown in several steps below.

Let V be an open set in C and $f \in V$. Then there exists a $\epsilon > 0$ such that $B(f, 2\epsilon) \subseteq V$. Since f is continuous on the compact $[0, 1]$, it is uniformly continuous on $[0, 1]$, i.e., there exists a $k \in \mathbb{N}$ such that

$$|f(x) - f(y)| < \epsilon \tag{5.63}$$

for all $x, y \in [0, 1]$ with $|x - y| \leq \frac{1}{k}$. In particular, we consider the partition $P = \{0 = x_0, x_1, \ldots, x_k = 1\}$ of $[0, 1]$, where $x_i = \frac{i}{k}$ for $i = 0, 1, \ldots, k$. Then we know from the inequality (5.63) that

$$|f(x) - f(x_i)| < \epsilon \tag{5.64}$$

for all $x \in [x_{i-1}, x_{i+1}]$. To prove the first assertion, we are going to divide the proof into several steps:

- **Step 1: Construction of a $g \in C$ with $\|g - f\| < \epsilon$.** Suppose that

$$g_i(x) = g_i\big(\lambda_x x_i + (1 - \lambda_x) x_{i+1}\big) = \lambda_x f(x_i) + (1 - \lambda_x) f(x_{i+1}) \tag{5.65}$$

  on $[x_i, x_{i+1}]$, where $\lambda_x \in [0, 1]$. Since $x = \lambda_x x_i + (1 - \lambda_x) x_{i+1}$, we have

$$\lambda_x = \frac{x - x_{i+1}}{x_i - x_{i+1}} = -k(x - x_{i+1}) = -kx + (i + 1). \tag{5.66}$$

  Thus λ_x is a continuous function of x. Therefore, each zigzag function g_i is continuous on $[x_i, x_{i+1}]$ and furthermore, we have

$$g_i(x_{i+1}) = f(x_{i+1}) = g_{i+1}(x_{i+1}).$$

  Thus if we "glue" all the zigzag functions $g_0, g_1, \ldots, g_k$ together and let it be g, then we have $g \in C$. In addition, if $x \in [x_i, x_{i+1}]$, then we see from the definition (5.65) and the inequality (5.64) that

$$|g(x) - f(x)| \leq \lambda_x \cdot |f(x_i) - f(x)| + (1 - \lambda_x) \cdot |f(x) - f(x_{i+1})| < \epsilon.$$

  In other words, $\|g - f\| < \epsilon$ as required.

- **Step 2: Construction of a continuous function with large one-sided derivatives.** To complete the construction, we have to consider the function $\varphi : \mathbb{R} \to [0, 1]$ defined by $\varphi(x + 1) = \varphi(x)$, where

$$\varphi(x) = \begin{cases} 2x, & \text{if } x \in [0, \frac{1}{2}]; \\[2mm] 2 - 2x, & \text{if } x \in [\frac{1}{2}, 1]. \end{cases}$$

  Clearly, φ is a zigzag continuous function on $\mathbb{R}$.[e] Now our desired function is constructed in the following lemma:

> **Lemma 5.1**
>
> For each $n \in \mathbb{N}$, if we define
>
> $$\varphi_n(x) = 2^{-n} \varphi(4^n x),$$
>
> then φ_n is a zigzag continuous function on $\mathbb{R}$ such that $|\varphi_n'(x+)| \geq 2^n$ and $|\varphi_n'(x-)| \geq 2^n$ for all $x \in \mathbb{R}$.

---

[e] Such zigzag function φ is something looks like the graph in Figure 2.1.

Proof of Lemma 5.1. Obviously, φ_n is a zigzag continuous function on $\mathbb{R}$. Suppose that $x \in [N, N+1]$ for some $N \in \mathbb{N}$. Then we have $4^n x \in [4^n N, 4^n(N+1)]$. Since φ is a function of period 1, we have

$$\varphi(4^n x) = \varphi(4^n x - 4^n N), \qquad (5.67)$$

where $4^n x - 4^n N \in [0, 4^n]$. It is clear that there exists $N' \in \mathbb{N} \cup \{0\}$ such that $4^n x - 4^n N - N' \in [0, 1]$, then the expression (5.67) and the periodicity of φ again shows that

$$\varphi(4^n x) = \varphi(4^n x - 4^n N) = \varphi(4^n x - 4^n N - N').$$

In other words, we may assume that $4^n x \in [0, 1)$. Let $h > 0$ be so small such that $4^n(x+h) \in [0, 1]$. Then we obtain

$$
\begin{aligned}
&\frac{\varphi_n(x+h) - \varphi_n(x)}{h} \\
&= \frac{1}{2^n} \cdot \frac{\varphi\big(4^n(x+h)\big) - \varphi(4^n x)}{h} \\
&= \begin{cases} \dfrac{1}{2^n} \cdot \dfrac{2 \cdot 4^n(x+h) - 2 \cdot 4^n x}{h}, & \text{if } 4^n x, 4^n(x+h) \in [0, \tfrac{1}{2}]; \\[2ex] \dfrac{1}{2^n} \cdot \dfrac{1 - 2 \cdot 4^n h - 1}{h}, & \text{if } 4^n x = \tfrac{1}{2} \text{ and } 4^n(x+h) = \tfrac{1}{2} + 4^n h; \\[2ex] \dfrac{1}{2^n} \cdot \dfrac{-2 \cdot 4^n(x+h) + 2 \cdot 4^n x}{h}, & 4^n x, 4^n(x+h) \in [\tfrac{1}{2}, 1], \end{cases} \\[2ex]
&= \begin{cases} 2 \cdot 2^n, & \text{if } 4^n x, 4^n(x+h) \in [0, \tfrac{1}{2}]; \\[2ex] -2 \cdot 2^n, & \text{if } 4^n x = \tfrac{1}{2} \text{ and } 4^n(x+h) = \tfrac{1}{2} + 4^n h; \\[2ex] -2 \cdot 2^n, & 4^n x, 4^n(x+h) \in [\tfrac{1}{2}, 1] \end{cases}
\end{aligned}
$$

which implies that $|\varphi_n'(x+)| \geq 2^n$. Similarly, $|\varphi_n'(x-)| \geq 2^n$, completing the proof of Lemma 5.1. $\blacksquare$

- **Step 3: Construction of a $\widetilde{g} \in C$ with large slopes and $\|\widetilde{g} - f\| < 2\epsilon$.** We define

$$\widetilde{g} = g + \varphi_k,$$

where the k will be determined later. We claim that $\widehat{g}$ satisfies the required properties. To this end, let

$$M_k = \max_{0 \leq i \leq k-1} k|f(x_{i+1}) - f(x_i)|.$$

Now for every $x \in [0, 1]$, if we take $h > 0$ small enough such that $x + h \in [x_i, x_{i+1}]$, then we see from the definition (5.65) and the expression (5.66) that

$$\left| \frac{g(x+h) - g(x)}{h} \right| = \frac{|\lambda_{x+h} - \lambda_x| \cdot |f(x_{i+1}) - f(x_i)|}{h} = k|f(x_{i+1}) - f(x_i)| \leq M_k \qquad (5.68)$$

which implies that $|g'(x+)| \leq M_k$ on $[0, 1]$. Similarly, we have $|g'(x-)| \leq M_k$ for every $x \in [0, 1]$.[f]

---

[f] If $x = 1$, then $g'(1+)$ does not exist. Similarly, if $x = 0$, then $g'(0-)$ does not exist.

Suppose that $N' \in \mathbb{N}$. Since f is bounded on $[0,1]$, we must have $M_k \leq kM$ for some constant $M > 0$ and then k can be chosen so large that

$$2^k \geq kM \geq M_k + N' \quad \text{and} \quad \frac{1}{2^k} < \frac{\epsilon}{2}. \tag{5.69}$$

Since g and φ_k are continuous on $[0,1]$, $\widetilde{g}$ is also continuous on $[0,1]$. Next, for $x \in [0,1)$ and $h > 0$ so small such that $x + h \in [0,1]$, we derive from Lemma 5.1 and the inequality (5.68) that

$$\begin{aligned}
\left| \frac{\widetilde{g}(x+h) - \widetilde{g}(x)}{h} \right| &= \left| \frac{g(x+h) - g(x)}{h} + \frac{\varphi(x+h) - \varphi(x)}{h} \right| \\
&\geq \left| \frac{\varphi(x+h) - \varphi(x)}{h} \right| - \left| \frac{g(x+h) - g(x)}{h} \right| \\
&\geq 2^k - M_k \\
&> N' \tag{5.70}
\end{aligned}$$

which yields $|\widetilde{g}'(x+)| > N'$ for all $x \in [0,1)$. Similarly, we have $|\widetilde{g}'(x-)| > N'$ for all $x(0,1]$. Since N' can be chosen arbitrary large, our $\widetilde{g}$ is continuous function on $[0,1]$ with large slopes.

Finally, it remains to show that $\|\widetilde{g} - f\| < \epsilon$. On $[0,1]$, we know from **Steps 1** and **2** that

$$|\widetilde{g}(x) - f(x)| \leq |g(x) - f(x)| + |\varphi_k(x)| < \epsilon + 2^{-k}|\varphi(4^k x)| < \epsilon + \frac{\epsilon}{2} < 2\epsilon.$$

Thus we conclude that $\|\widetilde{g} - f\| < 2\epsilon$. By the definition of X_n and the lower bound (5.70), we note that $\widetilde{g} \notin X_n$ if we pick $N' \geq n$.

- **Step 4: X_n is closed in C.** Let $\{f_k\} \subseteq X_n$ and $f_k \to f$ in C. Then one can find a sequence $\{t_k\} \subseteq [0,1]$ such that

$$|f_k(s) - f_k(t_k)| \leq n|s - t_k| \tag{5.71}$$

for every $s \in [0,1]$ and $k \in \mathbb{N}$. Since $[0,1]$ is compact, we may assume that $t_k \to t \in [0,1]$. By Theorem 3.17, C is a metric space and then we recall from the rephrased Theorem 7.9 [49, p. 151] that $f_k \to f$ uniformly on $[0,1]$. Hence it follows from [49, Exercise 9, p. 166] that

$$\lim_{k \to \infty} f_k(t_k) = f(t)$$

and we deduce from the inequality (5.71) that

$$|f(s) - f(t)| \leq n|s - t|,$$

for every $s \in [0,1]$. In other words, we conclude that $f \in X_n$ which means X_n is closed in C.

- **Step 5: Construction of $B(\widetilde{g}, \epsilon') \subseteq C$ such that $B(\widetilde{g}, \epsilon') \cap X_n = \varnothing$.** By **Step 3** and our choice of $\epsilon > 0$, we know that $\widetilde{g} \in B(f, 2\epsilon) \subseteq V$ and $\widetilde{g} \notin X_n$. Since $\widetilde{g} \in X_n^c$ and X_n^c is open in C by **Step 4**. There exists a $\delta > 0$ such that $B(\widetilde{g}, \delta) \subseteq X_n^c$, i.e., $B(\widetilde{g}, \delta) \cap X_n = \varnothing$. If we take $\epsilon' = \frac{1}{2} \min(\|\widetilde{g} - f\|, 2\epsilon - \|\widetilde{g} - f\|)$ and $h \in B(\widetilde{g}, \epsilon')$, then we have

$$\|h - f\| \leq \|h - \widetilde{g}\| + \|\widetilde{g} - f\| < \epsilon' + \|\widetilde{g} - f\| \leq \frac{1}{2}(2\epsilon - \|\widetilde{g} - f\|) + \|\widetilde{g} - f\| < 2\epsilon$$

which means $B(\widetilde{g}, \epsilon') \subseteq B(f, 2\epsilon) \subseteq V$.

Hence we have shown the first assertion.

To prove the second assertion, we notice that $X_n^\circ = \varnothing$; otherwise, since X_n° is open in C, the first assertion shows the existence of a non-empty open set $G \subseteq X_n^\circ \subseteq X_n$ such that $G \cap X_n = \varnothing$, a contradiction. Hence each X_n^c is open and dense in C. Since C is a complete metric space, we establish from Theorem 5.6 (Baire's Theorem) that the set

$$X = \bigcap_{n=1}^{\infty} X_n^c$$

is a dense G_δ in C. Suppose that $f \in C$ and f is differentiable at $p \in [0,1)$. Then there exists a $\delta > 0$ such that for all $0 < |h| < \delta$, we have[g]

$$\left| \frac{f(p+h) - f(p)}{h} - f'(p) \right| \le 1$$

which implies that

$$\left| \frac{f(p+h) - f(p)}{h} \right| \le 1 + |f'(p)|. \tag{5.72}$$

Obviously, if $|h| \ge \delta$, then

$$\left| \frac{f(p+h) - f(p)}{h} \right| \le \frac{|f(p+h)| + |f(p)|}{h} \le \frac{2\|f\|}{\delta}. \tag{5.73}$$

Combining the inequalities (5.72) and (5.73), we obtain that

$$|f(x) - f(p)| \le N|x - p|$$

if $N = \max(1 + |f'(p)|, 2\delta^{-1}\|f\|)$. In the case that $p = 1$, we will replace $f(p+h)$ by $f(p-h)$ in the inequalities (5.72) and (5.73).

Consequently, $f \in X_n$ for every $n \ge N$. By this, we must have $f \notin X$. Otherwise, $f \in X_N^c$ or equivalently, $f \notin X_N$ which is a contradiction. Hence we have completed the proof of the problem. ∎

Problem 5.15

Rudin Chapter 5 Exercise 15.

Proof. Recall from Problem 5.9 that c_0 is the subspace of ℓ^∞ consisting of all $x = \{\xi_i\} \in \ell^\infty$ for which $\xi_i \to 0$ as $i \to \infty$. By the definition, we have

$$\{\sigma_i\} = \mathbf{A}(x)$$

whenever the series converges (but ξ_i *not* necessarily converging to 0). We are going to prove the results one by one:

- **Necessity part:** Suppose that $\mathbf{A}$ transforms every convergent sequence $\{s_j\}$ to a sequence $\{\sigma_i\}$ which converges to the same limit.

    To prove **Condition (a)**, it suffices to prove the case when both $\{s_j\}$ and $\{\sigma_i\}$ converge to 0 because we may replace s_j and σ_i by $s_j - L$ and $\sigma_i - L$ respectively, where L is the

---

[g]Of course, only one-sided derivative $f'(0+)$ will be considered if $p = 0$.

common limit of $\{s_j\}$ and $\{\sigma_i\}$. In other words, we work on the space c_0. In fact, for each *fixed* $i \in \mathbb{N}$, we define $\Lambda_i : c_0 \to \mathbb{C}$ by

$$\Lambda_i(x) = \sigma_i = \sum_{j=0}^{\infty} a_{ij} s_j$$

for all $x \in c_0$ so that $\mathbf{A}(x) = \{\Lambda_i(x)\}$. Since $\{\sigma_i\}$ is convergent, it is bounded so that

$$|\Lambda_i(x)| \le M \tag{5.74}$$

for all $x \in c_0$ and $i = 1, 2, \ldots$ for some positive constant M. For every $k = 0, 1, 2, \ldots$, we take $s_j = \delta_{jk}$ (the Kronecker delta function). Then we have $s_j \to 0$ as $j \to \infty$ and we obtain

$$0 = \lim_{i \to \infty} \sigma_i = \lim_{i \to \infty} \sum_{j=0}^{\infty} a_{ij} s_j = \lim_{i \to \infty} a_{ik}$$

which is **Condition (a)**.

If $s_j = 1$ for all $j \in \mathbb{N}$, then $s_j \to 1$ as $j \to \infty$ and thus

$$1 = \lim_{i \to \infty} \sigma_i = \lim_{i \to \infty} \sum_{j=0}^{\infty} a_{ij}$$

which is **Condition (c)**.

To prove **Condition (b)**, we first show that $\displaystyle\sum_{j=0}^{\infty} |a_{ij}| < \infty$ for every $i = 1, 2, \ldots$. Assume that it was not the case. Then one is able to find a strictly increasing sequence of integers $\{n_k\}$ such that

$$\sum_{j=n_k+1}^{n_{k+1}} |a_{ij}| > k.$$

Let $k \in \mathbb{N}$ and define

$$s_j = \begin{cases} 0, & \text{if } j = 1, 2, \ldots, n_k; \\[2mm] \dfrac{\operatorname{sgn}(a_{ij})}{k}, & \text{if } j = n_k + 1, n_k + 2, \ldots, n_{k+1}. \end{cases} \tag{5.75}$$

Obviously, $j \to \infty$ if and only if $k \to \infty$. Then we have $s_j \to 0$ as $j \to \infty$. Thus $x = \{s_j\} \in c_0$ and so $\displaystyle\lim_{i \to \infty} \sigma_i = 0$. However, we notice from the construction (5.75) that

$$|\Lambda_i(x)| = |\sigma_i| = \left| \sum_{j=0}^{\infty} a_{ij} s_j \right| = \sum_{k=1}^{\infty} \sum_{j=n_k+1}^{n_{k+1}} \frac{|a_{ij}|}{k} = \infty$$

which contradicts the inequality (5.74). Thus we must have

$$\sum_{j=0}^{\infty} |a_{ij}| < \infty.$$

By the inequality (5.74) again, we have $\Lambda_i \in c_0^*$ and since c_0 is Banach, we may apply Theorem 5.8 (The Banach-Steinhaus Theorem) to conclude that

$$\|\Lambda_i\| = \sup\{|\Lambda_i(x)| \,|\, x \in c_0 \subseteq \ell^{\infty} \text{ and } \|x\|_{\infty} = 1\} \le M \tag{5.76}$$

for all $i \in \mathbb{N}$. Recall that $\sum_{j=0}^{\infty} |a_{ij}| < \infty$ for every $i = 1, 2, \ldots$, so $y_i = \{a_{ij}\} \in \ell^1$ for every $i = 1, 2, \ldots$ and Problem 5.9(a) implies that

$$\|\Lambda_i\| = \|y_i\|_1 = \sum_{j=0}^{\infty} |a_{ij}| \tag{5.77}$$

for every $i = 1, 2, \ldots$. Hence **Condition (b)** follows from the inequality (5.76) and the equality (5.77).

- **Sufficiency part:** Let $x = \{s_j\}$ and $s_j \to L$ as $j \to \infty$. By **Condition (b)**, the series $\sigma_i = \sum_{j=0}^{\infty} a_{ij} s_j$ converges absolutely for every $i \in \mathbb{N}$. We write

$$\sigma_i = \sum_{j=0}^{\infty} a_{ij}(s_j - L) + L \cdot \sum_{j=0}^{\infty} a_{ij}. \tag{5.78}$$

By **Condition (c)**, we have

$$\lim_{i \to \infty} L \cdot \sum_{j=0}^{\infty} a_{ij} = L. \tag{5.79}$$

Let the supremum in **Condition (b)** be M. Given $\epsilon > 0$, we choose a $N \in \mathbb{N}$ such that $|s_j - L| < \frac{\epsilon}{M}$ for all $n \geq N$. Thus we see that

$$\left| \sum_{j=0}^{\infty} a_{ij}(s_j - L) \right| \leq \sum_{j=0}^{N} |a_{ij}| \cdot |s_j - L| + \sum_{n=N+1}^{\infty} |a_{ij}| \cdot |s_j - L| \leq \sum_{j=0}^{N} |a_{ij}| \cdot |s_j - L| + \epsilon.$$

By **Condition (a)**, we know that

$$\lim_{i \to \infty} \sum_{j=0}^{N} |a_{ij}| \cdot |s_j - L| = 0$$

so that

$$\lim_{i \to \infty} \left| \sum_{j=0}^{\infty} a_{ij}(s_j - L) \right| \leq \epsilon.$$

Since ϵ is arbitrary, we conclude from the representation (5.78) and the limit (5.79) that

$$\lim_{i \to \infty} \sigma_i = L.$$

- **Two examples.** It is easy to check that the examples satisfy the conditions of the problem.

   To show the last assertion, we consider the sequence $\{s_j\}$ defined by

$$s_j = \begin{cases} \sqrt{k}, & \text{if } j = 2k; \\[2ex] -\sqrt{k}, & \text{if } j = 2k - 1. \end{cases}$$

Then it is clear that $\{s_j\}$ is unbounded and direct computation shows that, for each $i = 1, 2, \ldots$,

$$\sigma_i = \sum_{j=0}^{\infty} a_{ij} s_j$$

$$= \sum_{j=0}^{i} \frac{1}{i+1} s_j$$

$$= \begin{cases} \dfrac{1}{2k+1} \displaystyle\sum_{j=0}^{2k} s_j, & \text{if } i = 2k; \\[2ex] \dfrac{1}{2k} \displaystyle\sum_{j=0}^{2k-1} s_j, & \text{if } i = 2k - 1; \end{cases}$$

$$= \begin{cases} 0, & \text{if } i = 2k; \\[2ex] -\dfrac{1}{2\sqrt{k}}, & \text{if } i = 2k - 1. \end{cases}$$

Since $i \to \infty$ if and only if $k \to \infty$, we have $\sigma_i \to 0$ as $i \to \infty$, i.e., $\{\sigma_i\}$ is convergent.

For the other example, pick δ to be a number such that $r_i < \delta < 1$ for every $i \in \mathbb{N}$. If $s_j = (-\delta)^{-j}$, then $\{s_j\}$ is divergent and we have

$$\sigma_i = \sum_{j=0}^{\infty} a_{ij} s_j = (1 - r_i) \sum_{j=0}^{\infty} \left(-\frac{r_i}{\delta} \right)^{j} = \frac{1 - r_i}{1 + \delta^{-1} r_i} \to 0$$

as $i \to \infty$.

We complete the proof of the problem. ∎

Remark 5.2

Classically, Problem 5.15 is called the Silverman-Toepliotz Theorem. For more information or other proofs about this theorem, please refer to [30, §3.2], [44, Chap. 4] and [68, §1.2].

5.5   Miscellaneous Problems

Problem 5.16

Rudin Chapter 5 Exercise 16.

Proof. We follow Rudin's hint. Let $X \oplus Y = \{(x, y) \mid x \in X \text{ and } y \in Y\}$ with addition and scalar multiplication defined componentwise. We define the norm $\| \cdot \|$ on $X \oplus Y$ by[h]

$$\|(x, y)\| = \|x\|_X + \|y\|_Y. \tag{5.80}$$

We check Definition 5.2. For $(x_1, y_1), (x_2, y_2) \in X \oplus Y$, we have

$$\begin{aligned} \|(x_1, y_1) + (x_2, y_2)\| &= \|(x_1 + x_2, y_1 + y_2)\| \\ &= \|x_1 + x_2\|_X + \|y_1 + y_2\|_Y \\ &\leq \|x_1\|_X + \|x_2\|_X + \|y_1\|_Y + \|y_2\|_Y \\ &= \|(x_1, y_1)\| + \|(x_2, y_2)\|. \end{aligned}$$

---

[h] Of course, $\| \cdot \|_X$ and $\| \cdot \|_Y$ denote the norms in X and Y respectively.

Next, if $(x, y) \in X \oplus Y$ and α is a scalar, then we have

$$\|\alpha(x, y)\| = \|(\alpha x, \alpha y)\| = \|\alpha x\|_X + \|\alpha y\|_Y = |\alpha| \cdot \|x\|_X + |\alpha| \cdot \|y\|_Y = |\alpha| \cdot \|(x, y)\|.$$

Finally, if $\|(x, y)\| = 0$, then $\|x\|_X + \|y\|_Y = 0$. Since $\|x\|_X$ and $\|y\|_Y$ are nonnegative for every $x \in X$ and $y \in Y$, $\|x\|_X + \|y\|_Y = 0$ implies that $\|x\|_X = \|y\|_Y = 0$ and thus $(x, y) = (0, 0)$. Hence $X \oplus Y$ is a normed linear space.

Suppose that $\{(x_n, y_n)\} \subseteq X \oplus Y$ is Cauchy. Given $\epsilon > 0$, Then there exists a positive integer N such that $n, m \geq N$ imply that

$$\|(x_n, y_n) - (x_m, y_m)\| < \epsilon.$$

By the definition (5.80), we must have $\|x_n - x_m\|_X < \epsilon$ and $\|y_n - y_m\|_Y < \epsilon$ for all $n, m \geq N$. In other words, $\{x_n\} \subseteq X$ and $\{y_n\} \subseteq Y$ are also Cauchy. Since X and Y are Banach, there exist $x \in X$ and $y \in Y$ such that

$$\|x_n - x\|_X \to 0 \quad \text{and} \quad \|y_n - y\|_Y \to 0$$

as $n \to \infty$. Now these limits and the definition (5.80) show that

$$\|(x_n, y_n) - (x, y)\| = \|(x_n - x, y_n - y)\| = \|x_n - x\|_X + \|y_n - y\|_Y \to 0$$

as $n \to \infty$. By Definition 5.2, $X \oplus Y$ ia Banach.

Suppose that $G = \{(x, \Lambda(x)) \,|\, x \in X\} \subseteq X \oplus Y$. Furthermore, we let $\alpha, \beta \in \mathbb{C}$ and $(x, \Lambda(x)), (y, \Lambda(y)) \in G$. Then the linearity of Λ says that

$$\alpha(x, \Lambda(x)) + \beta(y, \Lambda(y)) = (\alpha x + \beta y, \alpha \Lambda(x) + \beta \Lambda(y)) = (\alpha x + \beta y, \Lambda(\alpha x + \beta y)) \in G.$$

Thus G is a linear subspace of $X \oplus Y$ and it is also a metric space. We claim that G is closed in $X \oplus Y$. To see this, let $\{(x_n, \Lambda(x_n))\} \subseteq G$ and $\|(x_n, \Lambda(x_n)) - (x, y)\| \to 0$ as $n \to \infty$. By the definition (5.80), we have $\|x_n - x\|_X \to 0$ and $\|\Lambda(x_n) - y\|_Y \to 0$ as $n \to \infty$, i.e.,

$$x = \lim_{n \to \infty} x_n \quad \text{and} \quad y = \lim_{n \to \infty} \Lambda(x_n).$$

By the hypothesis, we gain that $y = \Lambda(x)$ and so $(x, y) = (x, \Lambda(x)) \in G$ by the definition. Therefore, we have shown the claim that G is closed in $X \oplus Y$. Since $X \oplus Y$ is complete, we deduce from [49, Theorem 3.11, p. 53] that G is also complete and thus Banach.

Consider the mapping $\Phi : G \to X$ defined by

$$\Phi(x, \Lambda(x)) = x.$$

Then it is clear that Φ is linear and bijective. Furthermore, for every $x \in X$, we see that

$$\|\Phi(x, \Lambda(x))\|_X = \|x\|_X \leq \|x\|_X + \|\Lambda(x)\|_Y = \|(x, \Lambda(x))\|$$

which implies that

$$\|\Phi\| = \sup\{\|\Phi(x, \Lambda(x))\|_X \,|\, (x, \Lambda(x)) \in G \text{ and } \|(x, \Lambda(x))\| = 1\} \leq 1,$$

i.e., Φ is bounded. By Theorem 5.10, Φ^{-1} is also a bounded linear transformation of X onto G and it follows from Theorem 5.10's proof that there exists a positive constant δ such that

$$\|\Phi^{-1}\| \leq \frac{1}{\delta}.$$

Besides, Theorem 5.4 shows that Φ^{-1} is continuous. Combining these two facts and [51, Eqn. (3), p. 96], we obtain

$$\|x\|_X + \|\Lambda(x)\|_Y = \|(x, \Lambda(x))\| = \|\Phi^{-1}(x)\| \le \|\Phi^{-1}\| \cdot \|x\|_X \le \frac{1}{\delta}\|x\|_X < \infty \qquad (5.81)$$

for every $x \in X$. Since $\|x\|_X$ and $\|\Lambda(x)\|_Y$ are nonnegative, we get from the inequality (5.81) that $\delta < 1$ and

$$0 \le \|\Lambda(x)\|_Y \le \left(\frac{1}{\delta} - 1\right) \cdot \|x\|_X \qquad (5.82)$$

on X. Hence we conclude from the inequality (5.82) that

$$\|\Lambda\| = \sup\{\|\Lambda(x)\|_Y \mid x \in X \text{ and } \|x\|_X = 1\} \le \frac{1}{\delta} < \infty - 1.$$

Consequently, Λ is bounded and it is continuous on X by Theorem 5.4, completing the proof of the problem. ∎

Problem 5.17

Rudin Chapter 5 Exercise 17.

Proof. We prove the assertions one by one.

- $\|M_f\| \le \|f\|_\infty$. Given $f \in L^\infty(\mu)$. Define $M_f : L^2(\mu) \to L^2(\mu)$ by

$$M_f(g) = fg.$$

  By Definition 3.7, we have $|fg| \le \|f\|_\infty \cdot |g|$ for all $f \in L^2(\mu)$ and then Remark 3.10 shows that

$$\|M_f(g)\|_2 = \|fg\|_2 = \left\{\int_X |fg|^2 \, d\mu\right\}^{\frac{1}{2}} \le \left\{\int_X \|f\|_\infty^2 \cdot |g|^2 \, d\mu\right\}^{\frac{1}{2}} = \|f\|_\infty \cdot \|g\|_2. \qquad (5.83)$$

  By the inequality (5.83) and Definition 5.3, we obtain immediately that

$$\|M_f\| = \sup\{\|M_f(g)\|_2 \mid g \in L^2(\mu) \text{ and } \|g\|_2 = 1\} \le \|f\|_\infty. \qquad (5.84)$$

- **Measures μ with $\|M_f\| = \|f\|_\infty$ for all $f \in L^\infty(\mu)$.** We call the measure μ **semifinite** if for each $E \in \mathfrak{M}$ with $\mu(E) = \infty$ one can find a $F \in \mathfrak{M}$ with $F \subset E$ such that $0 < \mu(F) < \infty$, see [22, p. 25].

  Now we claim that $\|M_f\| = \|f\|_\infty$ for all $f \in L^\infty(\mu)$ if and only if the measure μ is semifinite.

  - Suppose that μ is semifinite. Since $f \in L^\infty(\mu)$, we have $\|f\|_\infty < \infty$. Let $\alpha = \|f\|_\infty$. If $\alpha = 0$, then the inequality (5.84) forces that

$$\|M_f\| = \|f\|_\infty = 0$$

    and we are done. If $\alpha > 0$, then we see from Problem 3.19 that $\alpha = \max\{|z| \mid z \in R_f\}$, where R_f denotes the essential range of f. Thus there exists a $z_0 \in \mathbb{C}$ such that $\alpha = |z_0|$. Without loss of generality, we may assume that $z_0 = \alpha$ and so

$$\mu\{x \in X \mid |f(x) - \alpha| < \epsilon\} > 0 \qquad (5.85)$$

    for every $\epsilon > 0$. Combining Definition 3.7 and the result (5.85), we establish that the measure of the set

$$E = \{x \in \mathbb{R} \mid |f(x)| > \alpha - \epsilon\}$$

    is nonzero.

$*$ **Case (i):** $\mu(E) < \infty$. Now it is trivial that the function

$$g = \frac{1}{\sqrt{\mu(E)}}\chi_E$$

satisfies the conditions that $g \in L^2(\mu)$ and $\|g\|_2 = 1$. Furthermore, we derive from the definition that

$$\|M_f(g)\|_2 = \left\{\frac{1}{\mu(E)}\int_E |f\chi_E|^2 \, \mathrm{d}\mu\right\}^{\frac{1}{2}} > \left\{\frac{(\alpha-\epsilon)^2}{\mu(E)}\int_E \mathrm{d}\mu\right\}^{\frac{1}{2}} = \alpha - \epsilon. \qquad (5.86)$$

Since ϵ is arbitrary, the estimate (5.86) implies that

$$\|M_f\| = \sup\{\|M_f(g)\|_2 \mid g \in L^2(\mu) \text{ and } \|g\|_2 = 1\} \geq \alpha$$

and hence $\|M_f\| = \|f\|_\infty$ by the inequality (5.84).

$*$ **Case (ii):** $\mu(E) = \infty$. Then the hypothesis ensures that there exists a $F \in \mathfrak{M}$ with $F \subset E$ and $0 < \mu(F) < \infty$. Now the function

$$g = \frac{1}{\sqrt{F}}\chi_F$$

also satisfies the estimate (5.86) and thus $\|M_f\| = \|f\|_\infty$ holds in this case.

– Suppose that μ is *not* semifinite. Then there is a $E \in \mathfrak{M}$ such that $\mu(E) = \infty$ and every $F \subset E$ satisfies either $\mu(F) = 0$ or $\mu(F) = \infty$. Take $g \in L^2(\mu)$ so that

$$\int_E |g(x)|^2 \, \mathrm{d}\mu \leq \int_X |g(x)|^2 \, \mathrm{d}\mu < \infty.$$

If there is a $F \subset E$ such that $|g(x)| > 0$ on F and $g(x) = 0$ on $E \setminus F$, then $\mu(F) \neq \infty$ by Proposition 1.24(a). Recall that μ is not semifinite, we have $\mu(F) = 0$. In other words, it must be true that $g = 0$ a.e. on E. Let $f = \chi_E$. Then we have

$$fg = 0 \text{ a.e. on } E$$

for every $g \in L^2(\mu)$. By the definition, we obtain

$$\|M_f\| = 0.$$

On the other hand, $\|f\|_\infty = 1$ so that $\|M_f\| < \|f\|_\infty$.

Hence we have proven the claim.

- **Functions $f \in L^\infty(\mu)$ such that M_f is onto.** Suppose that μ satisfies the condition that every measurable set E of positive measure contains a measurable subset F with $0 < \mu(F) < \infty$.[i] We claim that the map M_f is onto if and only if $\frac{1}{f} \in L^\infty(\mu)$.

  – If M_f is onto, then we let

$$E = \{x \in X \mid f(x) = 0\}. \qquad (5.87)$$

We claim that $\mu(E) = 0$. Assume that $\mu(E) > 0$. By the hypothesis, there exists a $F \subseteq E$ with $0 < \mu(F) < \infty$. Thus $\chi_F \in L^2(\mu)$. Since $fg = 0$ on F for every $g \in L^2(\mu)$, it implies that

$$\chi_F \notin M_f\big(L^2(\mu)\big)$$

---

[i]Some books take this as the definition of a semifinite measure μ. See, for example, [7, Exercise 25.9]

which contradicts the surjective property of M_f. Thus we conclude that $\mu(E) = 0$.

If $g \in L^2(\mu)$ is such that $M_f(g) = fg = 0$, then it follows from the definition (5.87) that $g = 0$ a.e. on E^c. Since $\mu(E) = 0$, we obtain fact that

$$g = 0 \text{ a.e. on } X$$

and this means that M_f is one-to-one. Since M_f is assumed to be onto, it is bijective.

On the one hand, recall that $L^2(\mu)$ is Banach and $\|M_f\| \le \|f\|_\infty < \infty$ by the first assertion, Theorem 5.10 ensures that there corresponds a $\delta > 0$ such that

$$\|M_f(g)\|_2 \ge \delta \|g\|_2 \tag{5.88}$$

for every $g \in L^2(\mu)$. On the other hand, we consider $F = \{x \in X \,|\, |f(x)| < \frac{\delta}{2}\}$. If $\mu(F) \ne 0$, then our hypothesis tells us that there is a $G \subset F$ such that $0 < \mu(G) < \infty$ and thus

$$\|\chi_G\|_2 = [\mu(G)]^{\frac{1}{2}}$$

which verifies that

$$\|M_f(\chi_G)\|_2 = \|f\chi_G\|_2 = \left\{ \int_G |f|^2 \, d\mu \right\}^{\frac{1}{2}} < \left\{ \int_G \frac{\delta^2}{4} \, d\mu \right\}^{\frac{1}{2}} = \frac{\delta}{2}[\mu(G)]^{\frac{1}{2}} = \frac{\delta}{2}\|\chi_G\|_2,$$

but it contradicts the inequality (5.88). Hence we have $\mu(F) = 0$ if and only if $|f(x)| \ge \frac{\delta}{2} > 0$ a.e. on X if and only if $\frac{1}{|f(x)|} \le \frac{2}{\delta}$ a.e. on X if and only if $\frac{1}{f} \in L^\infty(\mu)$.

– Suppose that $\frac{1}{f} \in L^\infty(\mu)$. Then it is clear that $M_{\frac{1}{f}}$ is the inverse operator of M_f so that M_f is bijective. In particular, M_f is surjective.

Hence we have completed the proof of the problem. $\blacksquare$

> **Problem 5.18**
>
> *Rudin Chapter 5 Exercise 18.*

Proof. Let $x \in X$ and $\epsilon > 0$. Since E is dense in the normed linear space X, we can find $y \in E$ such that

$$\|x - y\| < \frac{\epsilon}{2M}. \tag{5.89}$$

Since $\{\Lambda_n(y)\}$ converges in the Banach space Y, there exists a positive integer N such that $n, m \ge N$ imply that

$$\|\Lambda_n(y) - \Lambda_m(y)\| < \frac{\epsilon}{2}. \tag{5.90}$$

Therefore, if $n, m \ge N$, then we deduce immediately from the inequalities (5.89) and (5.90) that

$$\begin{aligned}
\|\Lambda_n(x) - \Lambda_m(x)\| &\le \|\Lambda_n(x) - \Lambda_n(y)\| + \|\Lambda_n(y) - \Lambda_m(y)\| + \|\Lambda_m(y) - \Lambda_m(x)\| \\
&< \|\Lambda_n(x - y)\| + \frac{\epsilon}{2} + \|\Lambda_m(y - x)\| \\
&\le \|\Lambda_n\| \cdot \|x - y\| + \|\Lambda_m\| \cdot \|x - y\| + \frac{\epsilon}{2} \\
&< \frac{\epsilon}{2} + \frac{\epsilon}{2} \\
&= \epsilon.
\end{aligned}$$

This means that $\{\Lambda_n(x)\}$ is Cauchy in Y. Since Y is Banach, $\{\Lambda_n(x)\}$ converges in Y and we complete the analysis of the problem. $\blacksquare$

Problem 5.19

Rudin Chapter 5 Exercise 19.

Proof. Given $f \in C(T)$, i.e., f is a continuous complex function on T. Recall that

$$s_n(f;x) = \frac{1}{2\pi} \int_{-\pi}^{\pi} f(t)D_n(x-t)\,\mathrm{d}t \quad \text{and} \quad D_n(t) = \sum_{k=-n}^{n} e^{ikt} = \frac{\sin(n+\tfrac{1}{2})t}{\sin\frac{t}{2}}. \tag{5.91}$$

By §5.11, $C(T)$ is Banach relative to the supremum norm $\|f\|_\infty$.[j] Next, since $f \in C(T)$, we follow from the equations (5.91) that $s_n(f;x) \in C(T)$ for each $n \in \mathbb{N}$.

Let $X = Y = C(T)$ and for each $n = 2, 3, \ldots$, we define $\Lambda_n : C(T) \to C(T)$ by

$$\Lambda_n(f) = \frac{s_n(f;x)}{\log n}.$$

Since $f \in C(T)$, there exists a $M > 0$ such that $|f(x)| \le M$ for all $x \in T$ so that

$$|s_n(f;x)| \le \frac{1}{2\pi} \int_{-\pi}^{\pi} |f(t)| \cdot |D_n(x-t)|\,\mathrm{d}t \le \frac{M}{2\pi} \int_{-\pi}^{\pi} |D_n(x-t)|\,\mathrm{d}t. \tag{5.92}$$

Since $D_n(x-t) = e^{ikx}D_n(t)$ and $D_n(t)$ is an even function, the estimate (5.92) can be replaced by

$$|s_n(f;x)| \le \frac{M}{2\pi} \int_{-\pi}^{\pi} |D_n(t)|\,\mathrm{d}t = \frac{M}{\pi} \int_0^{\pi} |D_n(t)|\,\mathrm{d}t = \frac{M}{\pi} \int_0^{\pi} \left| \frac{\sin(n+\tfrac{1}{2})t}{\sin\frac{t}{2}} \right|\,\mathrm{d}t. \tag{5.93}$$

Next, we consider $f(x) = \sin\frac{x}{2} - \frac{x}{4}$ on $[0,\pi]$. Using differentiation, we can show that $\sin\frac{x}{2} \ge \frac{x}{4}$ on $[0,\pi]$. Thus the estimate (5.93) can be further written as

$$\begin{aligned}
|s_n(f;x)| &\le \frac{4M}{\pi} \int_0^{\pi} \frac{|\sin(n+\tfrac{1}{2})t|}{t}\,\mathrm{d}t \\
&= \frac{4M}{\pi} \int_0^{(n+\frac{1}{2})\pi} \frac{|\sin t|}{t}\,\mathrm{d}t \\
&\le \frac{4M}{\pi} \left(\int_0^{\pi} \frac{\sin t}{t}\,\mathrm{d}t + \sum_{k=1}^{n} \int_{k\pi}^{(k+1)\pi} \frac{|\sin t|}{t}\,\mathrm{d}t \right).
\end{aligned} \tag{5.94}$$

On the interval $[k\pi, (k+1)\pi]$, it is easy to see that $\frac{1}{t} \le \frac{1}{k\pi}$, where $k = 1, 2, \ldots, n$, so we have

$$\sum_{k=1}^{n} \int_{k\pi}^{(k+1)\pi} \frac{|\sin t|}{t}\,\mathrm{d}t \le \sum_{k=1}^{n} \frac{1}{k\pi} \left(\int_{k\pi}^{(k+1)\pi} |\sin t|\,\mathrm{d}t \right) = \frac{2}{\pi} \sum_{k=1}^{n} \frac{1}{k}. \tag{5.95}$$

Besides, we get from [49, Problem 8.6, p. 197] that $\frac{2}{\pi} < \frac{\sin t}{t} < 1$ for $0 < t < \frac{\pi}{2}$, so the integral

$$\int_0^{\frac{\pi}{2}} \frac{\sin t}{t}\,\mathrm{d}t$$

is bounded by $\frac{\pi}{2}$ and thus

$$\int_0^{\pi} \frac{\sin t}{t}\,\mathrm{d}t = \int_0^{\frac{\pi}{2}} \frac{\sin t}{t}\,\mathrm{d}t + \int_{\frac{\pi}{2}}^{\pi} \frac{\sin t}{t}\,\mathrm{d}t \le \frac{\pi}{2} + \int_{\frac{\pi}{2}}^{\pi} \frac{\mathrm{d}t}{t} = \frac{\pi}{2} + \ln\pi - \ln\frac{\pi}{2} = \frac{\pi}{2} + \ln 2. \tag{5.96}$$

---

[j]To see this, since T is compact, we follow from Definition 3.16 that $C(T) = C_0(T)$ and Theorem 3.17 implies that $C(T)$ is complete. By Definition 5.2, we see that $C(T)$ is Banach.

Now, by putting the estimates (5.95) and (5.96) into the estimate (5.94), we derive that

$$|\Lambda_n(f)| = \left|\frac{s_n(f;x)}{\log n}\right| \le \frac{4M}{\pi \log n} \times \left(\frac{\pi}{2} + \ln 2\right) + \frac{8M}{\pi^2 \log n} \cdot \sum_{k=1}^{n} \frac{1}{k}. \tag{5.97}$$

When $\|f\|_\infty = 1$, we may take $M = 1$ in the estimate (5.97). It is well-known (see [63, Eqn. (6.27), p. 124]) that $\log n$ and $\sum_{k=1}^{n} \frac{1}{k}$ are of the same growth as $n \to \infty$, so we follow from the estimate (5.97) that there exists a $M' > 0$ such that

$$\|\Lambda_n\| \le M'$$

for all $n = 2, 3, \ldots$. That is, $\{\Lambda_n\}$ satisfies the first hypothesis of Problem 5.18.

By Theorem 4.25 (The Weierstrass Approximation Theorem), the set of all trigonometric polynomials, namely $\mathcal{P}$, is dense in $C(T)$. Let $P_m(t) = e^{imt}$ for some $m \in \mathbb{Z}$. If $n \ge m$, then we follow from the result [51, Eqn. (8), p. 89] that

$$s_n(P_m; x) = \frac{1}{2\pi} \int_{-\pi}^{\pi} e^{-imt} \sum_{k=-n}^{n} e^{ikt}\, dt = \sum_{k=-n}^{n} \frac{1}{2\pi} \int_{-\pi}^{\pi} e^{i(k-m)t}\, dt = 1. \tag{5.98}$$

Therefore, if $P(t) = \sum_{m=-N}^{N} c_m P_m(t)$, then we get immediately from the result (5.98) that

$$s_n(P; x) = \sum_{m=-N}^{N} c_m s_n(P_m; x) = \sum_{m=-N}^{N} c_m$$

for every $n \ge N$ and this implies that

$$\Lambda_n(P) = \frac{s_n(P;x)}{\log n} = \frac{1}{\log n} \cdot \sum_{m=-N}^{N} c_m \to 0$$

as $n \to \infty$. Thus $\{\Lambda_n\}$ satisfies the second hypothesis of Problem 5.18. Hence we establish from Problem 5.18 that $\{\Lambda_n(f)\}$ converges to 0 for each $f \in C(T)$ and as a consequence, we have

$$\lim_{n \to \infty} \frac{\|s_n\|_\infty}{\log n} = \lim_{n \to \infty} \|\Lambda_n\| = 0.$$

For the second assertion, we note that if we define

$$\Lambda_n f = \frac{s_n(f;0)}{\lambda_n},$$

then we may apply an argument similar to that used in §5.11 to show that

$$\|\Lambda_n\| = \frac{\|D_n\|_1}{|\lambda_n|}$$

holds. Furthermore, we follow from the hypothesis $\frac{\lambda_n}{\log n} \to 0$ as $n \to \infty$ that[k]

$$\frac{\|D_n\|_1}{|\lambda_n|} \ge \frac{4}{\pi^2 |\lambda_n|} \sum_{k=1}^{n} \frac{1}{k} \ge \frac{4}{\pi^2 |\lambda_n|}(\log n + \gamma) = \frac{4}{\pi^2}\left(\frac{\log n}{|\lambda_n|} + \frac{\gamma}{|\lambda_n|}\right) \to \infty \tag{5.99}$$

---

[k]We have applied the estimate of $\|D_n\|_1$ used in [51, p. 102] in the first inequality in (5.99).

as $n \to \infty$, where γ is the famous Euler constant. Hence Theorem 5.8 (the Banach-Steinhaus Theorem) ensures that the sequence

$$\left\{ \frac{s_n(f;0)}{\lambda_n} \right\}$$

is unbounded for every f in some dense G_δ set in $C(T)$, completing the proof of the problem. ■

Problem 5.20

Rudin Chapter 5 Exercise 20.

Proof.

(a) Assume that such a sequence of continuous positive functions $\{f_n\}$ existed. Let $x \in \mathbb{R}$ and $n, k \in \mathbb{N}$. Furthermore, let

$$U = \{x \in \mathbb{R} \mid \{f_n(x)\} \text{ is unbounded}\} \quad \text{and} \quad U_k = \{x \in \mathbb{R} \mid f_n(x) > k \text{ for some } n \in \mathbb{N}\}.$$

We claim that

$$U = \bigcap_{k=1}^{\infty} U_k. \tag{5.100}$$

On the one hand, if $x \in U_k$ for all $k \in \mathbb{N}$, then for every $k \in \mathbb{N}$, there exists a $n \in \mathbb{N}$ such that $f_n(x) > k$. In other words, we have $x \in U$. On the other hand, if $x \in U$, then since $\{f_n(x)\}$ is unbounded, for every $k \in \mathbb{N}$, there exists a $n \in \mathbb{N}$ such that $f_n(x) > k$ which is equivalent to saying that $x \in U_k$ for all $k \in \mathbb{N}$. Thus this proves the claim. Next, it is easy to see that

$$U_k = \bigcup_{n=1}^{\infty} \{x \in \mathbb{R} \mid f_n(x) > k\}.$$

Since each f_n is continuous, the set $\{x \in \mathbb{R} \mid f_n(x) > k\}$ is open in $\mathbb{R}$ and therefore, each U_k is open in $\mathbb{R}$. By the definition (5.100), U is a G_δ set. Thus we follow from the assumption that

$$\mathbb{Q} = U$$

is also a G_δ set. We deduce from the definition (5.100) that

$$\mathbb{R} = \mathbb{Q} \cup \mathbb{Q}^c = \mathbb{Q} \cup \bigcup_{k=1}^{\infty} U_k^c = \bigcup_{k=1}^{\infty} \{q_k\} \cup \bigcup_{k=1}^{\infty} U_k^c \tag{5.101}$$

Clearly, each q_k is a nowhere dense subset of $\mathbb{R}$. Furthermore, it is trivial that each U_k^c is closed in $\mathbb{R}$. Assume that $U_{k_0}^c$ was not a nowhere dense subset of $\mathbb{R}$ for some k_0. Let V_{k_0} be a nonempty open subset of $U_{k_0}^c$. Then there exists a $p \in V_{k_0}$ and a $\delta > 0$ such that

$$(p - \delta, p + \delta) \subseteq V_{k_0} \subseteq U_{k_0}^c \subseteq \mathbb{Q}^c,$$

but this means that $\mathbb{Q} \cap \mathbb{Q}^c \neq \varnothing$, a contradiction. Thus every U_k^c is also a nowhere dense subset of $\mathbb{R}$ and the representation (5.101) shows that $\mathbb{R}$ is a set of the first category which contradicts Theorem 5.6 (Baire's Theorem) that *no complete metric space is of the first category.*

(b) Suppose that $\mathbb{Q} = \{q_1, q_2, \ldots\}$ and for each $n \in \mathbb{N}$, we define $f_n : \mathbb{R} \to \mathbb{R}$ by

$$f_n(x) = \min(n|x - q_1| + 1, n|x - q_2| + 2, \ldots, n|x - q_n| + n). \tag{5.102}$$

It is clear that f_n is continuous on $\mathbb{R}$ and $f_n(x) \geq 1$ for all $x \in \mathbb{R}$, i.e., $\{f_n\}$ is a sequence of continuous positive functions on $\mathbb{R}$.

Suppose that θ is irrational and $N \in \mathbb{N}$. Consider the number

$$\alpha = \min(|\theta - q_1|, |\theta - q_2|, \ldots, |\theta - q_N|) > 0.$$

Then the Archimedean Property ([49, Theorem 1.20(a)]) implies that there is a positive integer n such that $n\alpha > N$ and thus

$$f_n(\theta) = \min(n|\theta - q_1| + 1, n|\theta - q_2| + 2, \ldots, n|\theta - q_n| + n) > N + 1.$$

Since N is arbitrary, we have $f_n(\theta) \to \infty$ as $n \to \infty$ so that $\{f_n(\theta)\}$ is unbounded.

To prove the other direction (i.e., $\{f_n(x)\}$ is unbounded implies x is irrational), we prove its contrapositive. Let q_k be a rational number. If $n \geq k$, then we have

$$f_n(q_k) = \min(n|q_k - q_1| + 1, \ldots, n|q_k - q_k| + k, \ldots, n|q_k - q_n| + n) \leq k$$

so that $f_n(q_k) \leq \min\big(f_1(q_k), \ldots, f_{k-1}(q_k), k\big)$. In other words, the set $\{f_n(q_k)\}$ is bounded.

(c) The sequence of the functions given by (5.102) shows that the assertion is true for irrationals. For the rational numbers, we first prove the case on $[0, 1]$.[1] To begin with, suppose that $Q_n = \{q_1, q_2, \ldots, q_n\} \subseteq [0, 1] \cap \mathbb{Q}$, where $q_1 = 0$. Furthermore, we let

$$\delta_n = \frac{1}{2^{n+1}} \min\{|q_i - q_j| \,|\, 1 \leq i < j \leq n\} > 0.$$

Clearly, we have

$$\lim_{n \to \infty} \delta_n = 0 \quad \text{and} \quad (q_i - \delta_n, q_i + \delta_n) \cap (q_j - \delta_n, q_j + \delta_n) = \varnothing \qquad (5.103)$$

for all $1 \leq i < j \leq n$. (If $q_i = 0$ or $q_i = 1$, then $(q_i - \delta_n, q_i + \delta_n)$ are replaced by $[0, \delta_n)$ or $(1 - \delta_n, 1]$ respectively.)

Suppose that $E_n = \displaystyle\bigcup_{i=1}^{n} (q_i - \delta_n, q_i + \delta_n)$ and $f_n : [0, 1] \to \mathbb{R}$ is defined by

$$f_n(x) = \begin{cases} \dfrac{n}{\delta_n}(x - q_i + \delta_n), & \text{if } x \in (q_i - \delta_n, q_i] \text{ and } 1 \leq i \leq n; \\[2mm] -\dfrac{n}{\delta_n}(x - q_i - \delta_n), & \text{if } x \in [q_i, q_i + \delta_n) \text{ and } 1 \leq i \leq n; \\[2mm] 0, & \text{if } x \in [0, 1] \setminus E_n. \end{cases}$$

Thus f_n is a continuous function on $[0, 1]$, zig-zag on $(q_i - \delta_n, q_i + \delta_n)$ and $f_n(q_i) = n$ for each $i = 1, 2, \ldots, n$.[m] Therefore, if $x \in [0, 1] \cap \mathbb{Q}$, then $x = q_k$ for some $k \in \mathbb{N}$ and thus we obtain

$$\lim_{n \to \infty} f_n(x) = \infty.$$

Next, suppose that x is irrational. Assume that there was an $N \in \mathbb{N}$ such that $x \in E_n$ for all $n \geq N$. By the definition, it means that

$$x \in (q_k - \delta_N, q_k + \delta_N) \qquad (5.104)$$

---

[1]The following argument is stimulated by the papers of Fabrykowski [19] and Myerson [43].
[m]The graph of the f_n looks like the graph shown in Figure 2.1.

for some $k \in \{1, 2, \ldots, N\}$. Since $x \in E_n$ for all $n \geq N$, the set relation (5.104) shows that

$$x \in (q_k - \delta_n, q_k + \delta_n)$$

for all $n \geq N$. By the limit (5.103), we know that $x = q_k \in \mathbb{Q}$, a contradiction. Hence we must have $x \notin E_n$ for infinitely many n, i.e., $f_n(x) = 0$ for *infinitely many* n so that $\lim_{n \to \infty} f_n(x) = \infty$ is impossible.

In conclusion, we have constructed the sequence of continuous functions $f_n : [0, 1] \to \mathbb{R}$ which satisfies $f_n(x) \to \infty$ as $n \to \infty$ if and only if $x \in \mathbb{Q}$. If we suppose that f_n is a function of period 1, then the domain of f_n can be extended to $\mathbb{R}$ and we obtain the desired result.

Hence we have completed the proof of the problem.

Problem 5.21

Rudin Chapter 5 Exercise 21.

Proof. Since $\mathbb{Q}$ is a countable union of closed sets of $\mathbb{R}$, we have $\mathbb{Q} \in \mathscr{B}$. It is well-known that $m(\mathbb{Q}) = 0$. Obviously, the translate $\mathbb{Q} + \sqrt{2}$ does not intersect $\mathbb{Q}$. This gives an affirmative answer to the first assertion.

Assume that there was a homeomorphism[n] $h : \mathbb{R} \to \mathbb{R}$ such that

$$\widetilde{E} = h(E) \cap E \neq \varnothing$$

for every measurable $E \subseteq \mathbb{R}$ with $m(E) = 0$. Given $\mathbb{Q} = \{q_k\}$ and $\epsilon > 0$. We first construct a particular measurable set E with $m(E) = 0$. To this end, for all $q_k \in \mathbb{Q}$, we consider the neighborhoods $(q_k - 2^{-k}\epsilon, q_k + 2^{-k}\epsilon)$ and their union

$$E(\epsilon) = \bigcup_{k=1}^{\infty} (q_k - 2^{-k}\epsilon, q_k + 2^{-k}\epsilon).$$

It is easy to see that every $E(\epsilon)$ is nonempty open in $\mathbb{R}$. In addition, since $\mathbb{Q} \subseteq E(\epsilon)$, each $E(\epsilon)$ is dense in $\mathbb{R}$ with $m\big(E(\epsilon)\big) \leq 2\epsilon$. Hence it follows from Theorem 5.6 (Baire's Theorem) that the set

$$E = \bigcap_{n=1}^{\infty} E\Big(\frac{1}{n}\Big) \tag{5.105}$$

is a dense G_δ set in $\mathbb{R}$ and $m(E) = 0$.

Suppose that E is the set (5.105). Since h is a homeomorphism, it is an open map which implies that

$$h(E) = \bigcap_{n=1}^{\infty} h\Big(E\Big(\frac{1}{n}\Big)\Big)$$

is also a dense G_δ set in $\mathbb{R}$ with $m\big(h(E)\big) = 0$. Thus the intersection $\widetilde{E}$ is also a dense G_δ set of measure zero. Since $E\big(\frac{1}{n}\big)$ is countable for all $n \in \mathbb{N}$, E is also countable and thus $\widetilde{E} \subseteq E$ is a countable dense G_δ in $\mathbb{R}$ of measure zero. However, $\mathbb{R}$ is a complete metric space which has *no* isolated points, the existence of $\widetilde{E}$ certainly contradicts Theorem 5.13. Hence no such homeomorphism h exists and this completes the proof of the problem.

---

[n] By the definition, h is a continuous bijection and h^{-1} is continuous.

Problem 5.22

Rudin Chapter 5 Exercise 22.

Proof. Suppose that $f \in C(T)$, $f \in \mathrm{Lip}\,\alpha$ and $f(0) = 0$. We need to show that

$$\lim_{n \to \infty} s_n(f; x) = f(x). \tag{5.106}$$

To achieve the goal, we first quote the following stronger form of the Riemann-Lebesgue Lemma whose proof can be found in [3, Theorem 11.16, p. 313].

Lemma 5.2 (Riemann-Lebesgue Lemma)

For every $f \in L^1(T)$, we define

$$\widehat{f}(n) = \frac{1}{2\pi} \int_{-\pi}^{\pi} f(t) \mathrm{e}^{-i(n+\beta)t}\, \mathrm{d}t,$$

where $\beta \in \mathbb{R}$. Then we have $\widehat{f}(n) \to 0$ as $|n| \to \infty$.

To begin with, since $f \in C(T)$, we have $f \in L^1(T)$. Furthermore, the hypothesis $f \in \mathrm{Lip}\,\alpha$ implies that $|f(s) - f(t)| \le M_f |s - t|^\alpha$ for all $s, t \in [-\pi, \pi]$, where M_f is finite and $\alpha \in (0, 1]$. Particularly, take $s = 0$ so that $|f(t)| \le M_f |t|^\alpha$ for all $t \in [-\pi, \pi]$ which implies

$$\int_{-\pi}^{\pi} \left| \frac{f(t)}{t} \right| \mathrm{d}t \le M_f \int_{-\pi}^{\pi} |t|^{\alpha-1}\, \mathrm{d}t = M_f \left[\int_{-\pi}^{0} (-1)^{\alpha-1} t^{\alpha-1}\, \mathrm{d}t + \int_{0}^{\pi} t^{\alpha-1}\, \mathrm{d}t \right] = \frac{2M_f \pi^\alpha}{\alpha} < \infty.$$

In other words, we have $\frac{f(t)}{t} \in L^1(T)$. Since we have

$$\sin\left(n + \frac{1}{2}\right)t = \frac{1}{2i}\left[\mathrm{e}^{i(n+\frac{1}{2})t} - \mathrm{e}^{-i(n+\frac{1}{2})t} \right],$$

it yields

$$\left| \frac{1}{\pi} \int_{-\pi}^{\pi} f(t) \cdot \frac{\sin(n+\frac{1}{2})t}{t}\, \mathrm{d}t \right| = \left| \frac{1}{2\pi i} \int_{-\pi}^{\pi} \frac{f(t)}{t} \mathrm{e}^{i(n+\frac{1}{2})t}\, \mathrm{d}t - \frac{1}{2\pi i} \int_{-\pi}^{\pi} \frac{f(t)}{t} \mathrm{e}^{-i(n+\frac{1}{2})t}\, \mathrm{d}t \right|$$
$$\le \left| \frac{1}{2\pi} \int_{-\pi}^{\pi} \frac{f(t)}{t} \mathrm{e}^{i(n+\frac{1}{2})t}\, \mathrm{d}t \right| + \left| \frac{1}{2\pi} \int_{-\pi}^{\pi} \frac{f(t)}{t} \mathrm{e}^{-i(n+\frac{1}{2})t}\, \mathrm{d}t \right|. \tag{5.107}$$

By Lemma 5.2, each of the integrals on the right-hand side of the inequality (5.107) tends to 0 as $|n| \to \infty$. Thus it is true that

$$\frac{1}{\pi} \int_{-\pi}^{\pi} f(t) \cdot \frac{\sin(n+\frac{1}{2})t}{t}\, \mathrm{d}t \to 0 = f(0)$$

as $|n| \to \infty$.[o]

Next, we claim that

$$s_n(f; 0) \to f(0) \tag{5.108}$$

---

[o]Integrals of the form $\displaystyle\int_{0}^{b} g(t) \frac{\sin \alpha t}{t}\, \mathrm{d}t$ are called **Dirichlet integrals**, where $\alpha > 0$ and g is defined on $[0, b]$.

as $n \to \infty$. To this end, we consider

$$
\left| s_n(f;0) - \frac{1}{\pi} \int_{-\pi}^{\pi} f(t) \cdot \frac{\sin(n+\frac{1}{2})t}{t} \, \mathrm{d}t \right| = \left| \frac{1}{2\pi} \int_{-\pi}^{\pi} f(t) D_n(t) \, \mathrm{d}t - \frac{1}{\pi} \int_{-\pi}^{\pi} f(t) \frac{\sin(n+\frac{1}{2})t}{t} \, \mathrm{d}t \right|
$$

$$
= \left| \frac{1}{2\pi} \int_{-\pi}^{\pi} \left(\frac{1}{\sin \frac{t}{2}} - \frac{1}{\frac{t}{2}} \right) f(t) \sin\left(n+\frac{1}{2}\right) t \, \mathrm{d}t \right|
$$

$$
= \left| \frac{1}{2\pi} \int_{-\pi}^{\pi} F(t) f(t) \sin\left(n+\frac{1}{2}\right) t \, \mathrm{d}t \right|, \tag{5.109}
$$

where $F : [0,\pi] \to \mathbb{R}$ is defined by

$$
F(t) = \begin{cases} \frac{1}{\sin \frac{t}{2}} - \frac{1}{\frac{t}{2}}, & \text{if } t \in [-\pi,\pi] \setminus \{0\}; \\[2ex] 0, & \text{if } t = 0. \end{cases}
$$

Clearly, F is continuous on $[-\pi,\pi]$ and so $Ff \in L^1(T)$. Using Lemma 5.2 to the right-hand side of the equation (5.109), we see that

$$
\lim_{n \to \infty} \frac{1}{2\pi} \int_{-\pi}^{\pi} F(t) f(t) \sin\left(n+\frac{1}{2}\right) t \, \mathrm{d}t = 0
$$

and this guarantees the validity of the claim (5.108).

For the general case, we consider the function $g : [-\pi,\pi] \to \mathbb{R}$ defined by

$$
g(t) = f(x+t) - f(x) \tag{5.110}
$$

for every $x \in \mathbb{R}$. Clearly, the real function g satisfies the conditions $g \in C(T)$, $g \in \mathrm{Lip}\,\alpha$ and $g(0) = 0$. By the above argument, we obtain

$$
\lim_{n \to \infty} s_n(g;0) = 0. \tag{5.111}
$$

By the definition (5.110), we gain

$$
s_n(g;0) = s_n\big(f(x+t) - f(x);0\big)
$$

$$
= \frac{1}{2\pi} \int_{-\pi}^{\pi} f(x+t) D_n(-t) \, \mathrm{d}t - \frac{1}{2\pi} \int_{-\pi}^{\pi} f(x) D_n(-t) \, \mathrm{d}t
$$

$$
= \frac{1}{2\pi} \int_{-\pi}^{\pi} f(x+t) D_n(-t) \, \mathrm{d}t - f(x)
$$

$$
= \frac{1}{2\pi} \int_{-\pi}^{\pi} f(x-t) D_n(t) \, \mathrm{d}t - f(x)
$$

$$
= s_n(f;x) - f(x). \tag{5.112}
$$

By applying the limit (5.111) to the equation (5.112), we obtain our desired result (5.106), completing the proof of the problem. ∎

CHAPTER **6**

Complex Measures

6.1 Properties of Complex Measures

> **Problem 6.1**
>
> *Rudin Chapter 6 Exercise 1.*

Proof. By the definition, we have

$$\lambda(E) = \sup\Big\{ \sum_{k=1}^{n} |\mu(E_i)| \,\Big|\, E_1, E_2, \ldots, E_k \text{ are mutually disjoint and } E = \bigcup_{j=1}^{k} E_j \Big\}.$$

By Definition 6.1, we have

$$|\mu|(E) = \sup\Big\{ \sum_{k=1}^{\infty} |\mu(E_i)| \,\Big|\, E_1, E_2, \ldots \text{ are mutually disjoint and } E = \bigcup_{j=1}^{\infty} E_j \Big\}.$$

Obviously, we have $\lambda(E) \le |\mu|(E)$ for every $E \in \mathfrak{M}$.

For the other direction, we suppose that $\{E_i\}$ is a partition of $E \in \mathfrak{M}$. Given $\epsilon > 0$. Since μ is a complex measure, we get from Definition 6.1 that the series

$$\sum_{i=1}^{\infty} \mu(E_i)$$

converges absolutely. Thus there exists a $N \in \mathbb{N}$ such that

$$\sum_{n=N}^{\infty} |\mu(E_n)| < \epsilon. \tag{6.1}$$

Define $E'_1 = E_1, E'_2 = E_2, \ldots, E'_{N-1} = E_{N-1}$ and $E'_N = E_N \cup E_{N+1} \cup \cdots$. By Definition 1.3(a), we have $E'_N \in \mathfrak{M}$. Then we have

$$\bigcup_{i=1}^{\infty} E_i = E'_1 \cup E'_2 \cup \cdots \cup E'_{N-1} \cup E'_N$$

which implies, with the aid of the estimate (6.1), that

$$\sum_{i=1}^{\infty} |\mu(E_i)| = \sum_{i=1}^{N-1} |\mu(E'_i)| + \sum_{i=N}^{\infty} |\mu(E_i)| < \sum_{i=1}^{N-1} |\mu(E'_i)| + |\mu(E'_N)| + \epsilon \le \lambda(E) + \epsilon. \tag{6.2}$$

Since the inequality (6.2) holds for every partition of E, we have $|\mu|(E) \leq \lambda(E) + \epsilon$. Since ϵ is arbitrary, we obtain $|\mu|(E) \leq \lambda(E)$. Hence we conclude that $\lambda = |\mu|$ which completes the proof of the problem. ∎

> **Problem 6.2**
>
> *Rudin Chapter 6 Exercise 2.*

Proof. Let μ be Lebesgue measure on $(0,1)$ and λ be the counting measure on the σ-algebra $\mathfrak{M}$ of all Lebesgue measurable sets in $(0,1)$. Assume that λ was σ-finite. Then we have

$$(0,1) = \bigcup_{n=1}^{\infty} E_n, \tag{6.3}$$

where $\lambda(E_n) < \infty$ for each $n \in \mathbb{N}$. Clearly, we have $\lambda(E) < \infty$ if and only if E is finite. Therefore, the representation (6.3) implies that $(0,1)$ is countable, a contradiction. Hence λ is not σ-finite.

Next, we check $\mu \ll \lambda$. Suppose that $\lambda(E) = 0$, where $E \in \mathfrak{M}$. By the definition of λ, we know that $E = \varnothing$ which implies definitely that $\mu(E) = \mu(\varnothing) = 0$. In addition, the definition of μ ensures that $\mu\big((0,1)\big) = 1$, i.e., μ is a bounded measure.

Assume that there was a $h \in L^1(\lambda)$ such that $\mathrm{d}\mu = h \,\mathrm{d}\lambda$, i.e.,

$$\mu(E) = \int_E h \,\mathrm{d}\lambda \tag{6.4}$$

for every $E \in \mathfrak{M}$. Note that $\mu(E) \geq 0$, so $h(x) \geq 0$ a.e. $[\lambda]$ on E.[a] Particulary, if $E = (0,1)$, then $h(x) \geq 0$ a.e. $[\lambda]$ on $(0,1)$. Suppose that $E_n = \{x \in (0,1) \,|\, h(x) \geq \frac{1}{n}\}$ for every $n \in \mathbb{N}$. Since $h \in L^1(\lambda)$, Proposition 1.24(a) implies that

$$\infty > \int_{E_n} h \,\mathrm{d}\lambda \geq \frac{1}{n}\lambda(E_n) \geq 0.$$

Thus we must have $\lambda(E_n) \in [0, \infty)$ or equivalently, E_n is finite or $E_n = \varnothing$. Let

$$E = \bigcup_{n=1}^{\infty} E_n = \{x \in (0,1) \,|\, h(x) > 0\}. \tag{6.5}$$

Therefore, the set E is countable. There are two cases for consideration:

- **Case (i):** $E \neq \varnothing$. Let $E = \{x_1, x_2, \ldots\}$. On the one hand, we know from [49, Remark 11.11(f), p. 309] that $\mu(F) = 0$ *for every* countable subset $F \subset (0,1)$, so we deduce from the representation (6.4) and the definition (6.5) that

$$0 = \mu(E) = \int_E h \,\mathrm{d}\lambda = \sum_{n=1}^{\infty} h(x_n) \neq 0,$$

  a contradiction.

- **Case (ii):** $E = \varnothing$. In this case, we have $h(x) = 0$ a.e. $[\lambda]$ on $(0,1)$ and the representation (6.4) again shows that

$$1 = \mu\big((0,1)\big) = \int_{(0,1)} h \,\mathrm{d}\lambda = 0,$$

  a contradiction.

---

[a] See Definition 1.35 for the meaning of the notation a.e. $[\lambda]$.

Hence *no* such h exists and this ends the proof of the problem.   ∎

> **Problem 6.3**
>
> *Rudin Chapter 6 Exercise 3.*

Proof. We first show that if μ and λ are complex regular Borel measures, then both $\mu + \lambda$ and $\alpha\mu$ are complex regular Borel measures too. By §6.18, it is equivalent to show that both $|\mu + \nu|$ and $|\alpha\mu|$ are regular Borel measures. To this end, we follow from §6.18 again that both $|\mu|$ and $|\lambda|$ are regular Borel measures. Let $E \in \mathscr{B}$ and $\epsilon > 0$. By Definition 2.15, there exist open sets $V_1, V_2 \supseteq E$ such that

$$|\mu|(V_1) < |\mu|(E) + \frac{\epsilon}{2} \quad \text{and} \quad |\lambda|(V_2) < |\lambda|(E) + \frac{\epsilon}{2}. \tag{6.6}$$

Similarly, there are compact sets $K_1, K_2 \subseteq E$ such that

$$|\mu|(E) < |\mu|(K_1) + \frac{\epsilon}{2} \quad \text{and} \quad |\lambda|(E) < |\lambda|(K_2) + \frac{\epsilon}{2}. \tag{6.7}$$

The triangle inequality certainly implies $|\mu(E) + \lambda(E)| \leq |\mu(E)| + |\lambda(E)|$ and then Definition 6.1 gives

$$|\mu + \lambda|(E) = \sup \sum_{i=1}^{\infty} |\mu(E) + \lambda(E)| \leq \sup \sum_{i=1}^{\infty} \big[|\mu(E)| + |\lambda(E)|\big], \tag{6.8}$$

where the supremum being taken over all partitions $\{E_i\}$ of E. Since the series $\sum_{i=1}^{\infty} |\mu(E)|$ and $\sum_{i=1}^{\infty} |\lambda(E)|$ converge, we deduce from the expression (6.8) that

$$|\mu + \lambda|(E) \leq \sup \sum_{i=1}^{\infty} |\mu(E)| + \sup \sum_{i=1}^{\infty} |\lambda(E)| = |\mu|(E) + |\lambda|(E).$$

Thus we have

$$|\mu + \lambda| \leq |\mu| + |\lambda|. \tag{6.9}$$

Let $V = V_1 \cap V_2$ and $K = K_1 \cup K_2$. Then K is compact and V is open in X. Now we follow from the estimates (6.6) and (6.7) and the inequality (6.9) that

$$\begin{aligned}
|\mu + \lambda|(V) &= |\mu + \lambda|(E) + |\mu + \lambda|(V \setminus E) \\
&\leq |\mu + \lambda|(E) + |\mu|(V \setminus E) + |\lambda|(V \setminus E) \\
&\leq |\mu + \lambda|(E) + |\mu|(V_1 \setminus E) + |\lambda|(V_2 \setminus E) \\
&< |\mu + \lambda|(E) + \epsilon.
\end{aligned}$$

Since ϵ and E are arbitrary, the measure $|\mu + \lambda|(E)$ is in fact outer regular. Similarly, we have

$$\begin{aligned}
|\mu + \lambda|(K) &= |\mu + \lambda|(E) - |\mu + \lambda|(E \setminus K) \\
&\geq |\mu + \lambda|(E) - |\mu + \lambda|(E \setminus K_1) - |\mu + \lambda|(E \setminus K_2) \\
&> |\mu + \lambda|(E) - \epsilon.
\end{aligned}$$

In other words, $|\mu + \lambda|$ is also inner regular. By Definition 2.15, $\mu + \lambda$ is a regular complex Borel measure, i.e., $\mu + \lambda \in M(X)$. Now the regularity of $|\alpha\mu|$ is easy to prove, so we omit the details here.

Now it is time to prove the assertion in the question. Since X is a locally compact Hausdorff space, Theorem 6.19 (The Riesz Representation Theorem) ensures every $\Phi \in C_0(X)^*$ is represented by a unique $\mu_\Phi \in M(X)$ in the sense that

$$\Phi(f) = \int_X f \, \mathrm{d}\mu_\Phi \tag{6.10}$$

for every $f \in C_0(X)$. By [51, Eqn. (3), p. 130], we see that

$$\int_X f \, \mathrm{d}\mu_{\Phi+\Psi} = (\Phi + \Psi)(f) = \Phi(f) + \Psi(f) = \int_X f \, \mathrm{d}\mu_\Phi + \int_X f \, \mathrm{d}\mu_\Psi = \int_X f \, \mathrm{d}(\mu_\Phi + \mu_\Psi)$$

and

$$\int_X f \, \mathrm{d}\mu_{\alpha\Phi} = \alpha\Phi(f) = \int_X f \, \mathrm{d}(\alpha\mu_\Phi).$$

Therefore, they imply that

$$\mu_{\Phi+\Psi} = \mu_\Phi + \mu_\Psi \quad \text{and} \quad \mu_{\alpha\Phi} = \alpha\mu_\Phi. \tag{6.11}$$

By the previous analysis, we see that $\mu_{\Phi+\Psi}, \mu_{\alpha\mu} \in M(X)$, so we may define the mapping $F : C_0(X)^* \to M(X)$ by

$$F(\Phi) = \mu_\Phi$$

and it is easy to see from the results (6.11) that

$$F(\Phi + \Psi) = F(\Phi) + F(\Psi) \quad \text{and} \quad F(\alpha\Phi) = \alpha F(\Phi).$$

Furthermore, Theorem 6.19 (The Riesz Representation Theorem) also implies that F is a bijection and $\|\Phi\| = |\mu_\Phi|(X) = \|\mu_\Phi\| = \|F(\Phi)\|$. Consequently, F is actually an isometric vector space isomorphism, i.e.,

$$C_0(X)^* \cong M(X).$$

Since $C_0(X)$ is a Banach space with the supremum norm, Problem 5.8 guarantees that $C_0(X)^*$ is Banach. Hence $M(X)$ is Banach and we have completed the proof of the problem. $\blacksquare$

6.2 Dual Spaces of $L^p(\mu)$

> **Problem 6.4**
>
> *Rudin Chapter 6 Exercise 4.*

Proof. Since μ is positive and σ-finite, we can write

$$X = \bigcup_{n=1}^{\infty} X_n, \tag{6.12}$$

where $\{X_n\}$ is an increasing sequence of measurable sets and $\mu(X_n) < \infty$ for all $n \in \mathbb{N}$, see [54, Definition 4.22, p. 22]. Let $A = \{x \in X \mid |g(x)| = \infty\}$ and $A_n = \{x \in X_n \mid |g(x)| = \infty\}$, where $n \in \mathbb{N}$. Since g is measurable, every A_n and A are measurable by Problem 1.5. Besides, it is evident that $A_1 \subseteq A_2 \subseteq \cdots$ and if $x \in A$, then $x \in X_{N_0}$ for some $N_0 \in \mathbb{N}$ so that $x \in A_{N_0}$. As a result, we get

$$A = \bigcup_{n=1}^{\infty} A_n.$$

Now we are going to divide the proof into several steps:

- **Step 1:** $\mu(A) = 0$. Otherwise, it follows from the construction and Theorem 1.19(c) that

$$\lim_{n\to\infty} \mu(A_n) = \mu(A) > 0.$$

This means that one can find a $N_1 \in \mathbb{N}$ such that $0 < \mu(A_{N_1}) \le \mu(X_{N_1}) < \infty$. We know that $\chi_{A_{N_1}} \in L^p(\mu)$, so our hypothesis implies that $\chi_{A_{N_1}} g \in L^1(\mu)$, but

$$\|\chi_{A_{N_1}} g\|_1 = \int_X |\chi_{A_{N_1}} g|\,\mathrm{d}\mu = \int_{A_{N_1}} |g|\,\mathrm{d}\mu = \mu(A_{N_1}) \cdot \infty = \infty$$

which is a contradiction. Hence $\mu(A) = 0$, i.e., g is finite a.e. on X.

- **Step 2: $g \in L^q(\mu)$ when $1 < p < \infty$.** In this case, we have $q \in (1, \infty)$. Here we define

$$E_n = \{x \in X_n \,|\, |g(x)| \le n\}.$$

Obviously, $\{E_n\}$ is also an increasing sequence of measurable sets. Furthermore, if $x_0 \in X \setminus A$, then $x_0 \in X_{N_2}$ for some $N_2 \in \mathbb{N}$ by the definition (6.12) so that $x_0 \in X_n$ for all $n \ge N_2$. Since g is finite a.e. on X, there exists a positive integer N_3 such that $|g(x)| \le N_3$ for almost all $x \in X$. Take $N_4 = \max(N_2, N_3)$, then $x_0 \in X_{N_4}$ and $|g(x_0)| \le N_4$ which imply that $x_0 \in E_{N_4}$. Consequently, we have shown that

$$X = \bigcup_{n=1}^{\infty} E_n.$$

Let $g_n = \chi_{E_n} g : X \to \mathbb{C}$, where $n = 1, 2, \ldots$. Then each g_n is measurable on X and

$$\|g_n\|_q^q = \int_X |g_n|^q\,\mathrm{d}\mu = \int_{E_n} |g|^q\,\mathrm{d}\mu \le n^q \mu(E_n) \le n^q \mu(X_n) < \infty \tag{6.13}$$

for each $n = 1, 2, \ldots$. Thus we have $g_n \in L^q(\mu)$ for all $n \in \mathbb{N}$. Next, we define $\Lambda_n : L^p(\mu) \to \mathbb{C}$ by

$$\Lambda_n(f) = \int_X f g_n\,\mathrm{d}\mu$$

which is linear for $n \in \mathbb{N}$. Since $f g_n = (f \chi_{E_n}) g$ and $f \chi_{E_n} \in L^p(\mu)$, we have $f g_n \in L^1(\mu)$. By Theorem 1.33 and Theorem 3.8, we obtain

$$\left| \int_X f g_n\,\mathrm{d}\mu \right| \le \int_X |f g_n|\,\mathrm{d}\mu = \|f g_n\|_1 \le \|f\|_p \times \|g_n\|_q, \tag{6.14}$$

where $n = 1, 2, \ldots$. By Definition 5.3 and the inequality (6.14), we know that

$$\|\Lambda_n\| = \sup\{|\Lambda_n(f)| \,|\, f \in L^p(\mu) \text{ and } \|f\|_p = 1\} \le \|g_n\|_q, \tag{6.15}$$

where $n = 1, 2, \ldots$. Recall that $1 < q < \infty$, so we may consider

$$f_0 = \|g_n\|_q^{-\frac{q}{p}} |g_n|^{q-2} \overline{g_n}.$$

Then we have $|f_0|^p = \|g_n\|_q^{-q} |g_n|^{(q-1)p}$ so that

$$\|f_0\|_p = \int_X |f_0|^p\,\mathrm{d}\mu = \|g_n\|_q^{-q} \int_X |g_n|^{(q-1)p}\,\mathrm{d}\mu = \|g_n\|_q^{-q} \int_X |g_n|^q\,\mathrm{d}\mu = 1$$

and

$$\Lambda_n(f_0) = \int_X \|g_n\|_q^{-\frac{q}{p}} |g_n|^{q-2} \overline{g_n} \cdot g_n\,\mathrm{d}\mu$$

$$= \|g_n\|_q^{-\frac{q}{p}} \int_X |g_n|^q \, d\mu$$

$$= \|g_n\|_q^{-\frac{q}{p}} \times \|g_n\|_q^q$$

$$= \|g_n\|_q \tag{6.16}$$

for all $n \in \mathbb{N}$. Combining the results (6.13), (6.15) and (6.16), we have established the fact that

$$\|\Lambda_n\| = \|g_n\|_q < \infty, \tag{6.17}$$

where $n \in \mathbb{N}$. Hence $\{\Lambda_n\}$ is a family of bounded linear transformations of $L^p(\mu)$ into $\mathbb{C}$.

Since $L^p(\mu)$ is Banach (see Definition 5.2) and

$$|\Lambda_n(f)| = \left| \int_X f g_n \, d\mu \right| \leq \int_X |f g_n| \, d\mu = \int_{E_n} |fg| \, d\mu \leq \int_X |fg| \, d\mu = \|fg\|_1 < \infty$$

for every $f \in L^p(\mu)$ and $n \in \mathbb{N}$, Theorem 5.8 (The Banach-Steinhaus Theorem) implies that there is a $M > 0$ such that

$$\|\Lambda_n\| \leq M \tag{6.18}$$

for all $n \in \mathbb{N}$. Using the results (6.17) and (6.18), we conclude that

$$\|g_n\|_q \leq M \tag{6.19}$$

for all $n \in \mathbb{N}$.

By **Step 1** and the definition of g_n, we have $|g_n(x)| \leq |g(x)| < \infty$ a.e. on X for all $n \in \mathbb{N}$ and

$$|g_1(x)| \leq |g_2(x)| \leq \cdots$$

for every $x \in X$. Now the Monotone Convergence Theorem [49, Theorem 3.14, p. 55] implies that $|g_n(x)| \to |g(x)| < \infty$ a.e. on X as $n \to \infty$. Hence we gain from Theorem 1.26 (Lebesgue's Monotone Convergence Theorem) that

$$\lim_{n \to \infty} \|g_n\|_q^q = \lim_{n \to \infty} \int_X |g_n|^q \, d\mu = \int_X |g|^q \, d\mu = \|g\|_q^q. \tag{6.20}$$

Hence it follows from the inequality (6.19) and the limit (6.20) that

$$\|g\|_q \leq M < \infty,$$

i.e. $g \in L^q(\mu)$.

- **Step 3: $g \in L^\infty(\mu)$ when $p = 1$.** In this case, $q = \infty$. We recall from **Step 1** that g is finite a.e. on X, so there exists a $M > 0$ such that $|g(x)| \leq M$ a.e. on X. By Definition 3.7, we assert that $\|g\|_\infty \leq M$, i.e., $g \in L^\infty(\mu)$.

- **Step 4: $g \in L^1(\mu)$ when $p = \infty$.** Define $f_1 : X \to \mathbb{C}$ by

$$f_1(x) = \begin{cases} \dfrac{\overline{g(x)}}{|g(x)|}, & \text{if } g(x) \neq 0; \\[2mm] 0, & \text{otherwise.} \end{cases}$$

Since $|f_1(x)| \leq 1$ for all $x \in X$, we get from Definition 3.7 that $\|f_1\|_\infty \leq 1$, i.e., $f_1 \in L^\infty(\mu)$. By the hypothesis, we have $f_1 g \in L^1(\mu)$. Since $f_1 g = |g|$, we conclude that $g \in L^1(\mu)$.

Now we have completed the proof of the problem.

> **Problem 6.5**
>
> *Rudin Chapter 6 Exercise 5.*

Proof. The answer is negative. On the one hand, if $f_1, f_2 : X \to \mathbb{C}$ are defined by

$$f_1(a) = 0, \quad f_1(b) = 1 \quad \text{and} \quad f_2(a) = 1, \quad f_2(b) = 0, \tag{6.21}$$

then we have $f_1, f_2 \in L^\infty(\mu)$ and $f_1 \not\equiv f_2$. For every $f \in L^\infty(\mu)$, we have $f(a) = A + Bi$ and $f(b) = C + Di$ for some $A, B, C, D \in \mathbb{R}$. Obviously, we get from the definition (6.21) the representation

$$f = (C + Di)f_1 + (A + Bi)f_2.$$

As a result, $L^\infty(\mu)$ is a two-dimensional space spanned by f_1 and f_2, i.e., $\dim L^\infty(\mu) = 2$.

On the other hand, if $f \in L^1(\mu)$, then $\|f\|_1 < \infty$ and we follow from Definition 3.6 that $f(b) = 0$ and $\|f\|_1 = |f(a)| = |f(a)|f_2(a)$. Therefore, $L^1(\mu)$ is an one-dimensional space spanned by f_2. As a vector space (see Remark 3.10), we have

$$\dim L^1(\mu)^* = \dim L^1(\mu) = 1.$$

Hence, $L^\infty(\mu) \neq L^1(\mu)^*$, completing the proof of the problem. $\blacksquare$

> **Problem 6.6**
>
> *Rudin Chapter 6 Exercise 6.*

Proof. We want to show that $L^p(\mu)^* \cong L^q(\mu)$ for $1 < p < \infty$. Equivalently, we have to show that for each $\Lambda \in L^p(\mu)^*$, there exists a unique $g \in L^q(\mu)$ such that

$$\Lambda f = \int_X fg \, d\mu \tag{6.22}$$

for all $f \in L^p(\mu)$. The following proof follows mainly Follan's argument ([22, p. 190]):

- **Step 1: μ is finite.** Let s be a simple function on X. Since μ is finite, we have $\mu(\{x \in X \mid s(x) \neq 0\}) \leq \mu(X) < \infty$ and then it follows from Theorem 3.13 that $s \in L^p(\mu)$. Let $\Lambda \in L^p(\mu)^*$, E be measurable and $\lambda(E) = \Lambda(\chi_E)$. For every partition $\{E_i\}$ of E, if we let $F_n = E_1 \cup E_2 \cup \cdots \cup E_n$, then we obtain

$$\int_X |\chi_E - \chi_{F_n}|^p \, d\mu = \int_X |\chi_{E \setminus F_n}|^p \, d\mu = \int_X \chi_{E \setminus F_n} \, d\mu = \mu(E \setminus F_n) \tag{6.23}$$

for all $n \in \mathbb{N}$. In fact, the expression (6.23) can be rewritten as

$$\|\chi_E - \chi_{F_n}\|_p = \mu(E \setminus F_n)^{\frac{1}{p}}. \tag{6.24}$$

Since $E \setminus F_1 \supseteq E \setminus F_2 \supseteq \cdots$ and $\mu(E \setminus F_1)$ is finite, it establishes from Theorem 1.19(e) that

$$\lim_{n \to \infty} \mu(E \setminus F_n) = \mu\Big(\bigcap_{n=1}^{\infty} (E \setminus F_n) \Big) = \mu(\varnothing) = 0. \tag{6.25}$$

Combining the results (6.24) and (6.25) and using the fact $p \in (1, \infty)$, we derive that

$$\lim_{n \to \infty} \|\chi_{F_n} - \chi_E\|_p = 0.$$

Since $\chi_E \in L^p(\mu)$ and each $\chi_{E_i} g$ is measurable, we deduce from the representation (6.22) and then Theorem 1.27 that

$$\lambda(E) = \Lambda(\chi_E) = \int_X \chi_E g \, \mathrm{d}\mu = \int_X \sum_{i=1}^{\infty} (\chi_{E_i} g) \, \mathrm{d}\mu = \sum_{i=1}^{\infty} \int_X \chi_{E_i} g \, \mathrm{d}\mu = \sum_{i=1}^{\infty} \lambda(\chi_{E_i}).$$

By Definition 6.1, λ is a complex measure.

If $E \in \mathfrak{M}$ satisfies $\mu(E) = 0$, then $\chi_E = 0$ in $L^p(\mu)$ and so $\lambda(E) = \Lambda(0) = 0$. By Definition 6.7, we have $\lambda \ll \mu$. Thus we know from Theorem 6.10 (The Lebesgue-Radon-Nikodym Theorem) that there exists a unique $h \in L^1(\mu)$ such that

$$\Lambda(\chi_E) = \lambda(E) = \int_E g \, \mathrm{d}\mu = \int_X \chi_E g \, \mathrm{d}\mu$$

for every $E \in \mathfrak{M}$. Recall that $\Lambda \in L^p(\mu)^*$, Λ is linear and bounded so that

$$\Lambda(s) = \int_X s g \, \mathrm{d}\mu$$

for every simple function $s \in L^p(\mu)$. By [51, Eqn. (3), p. 96], we know that

$$\left| \int_X s g \, \mathrm{d}\mu \right| = \|\Lambda(s)\| \leq \|\Lambda\| \cdot \|s\|_p < \infty.$$

By [22, Theorem 6.14, p. 189], we conclude that $g \in L^q(\mu)$. Given that $f \in L^p(\mu)$. By Theorem 3.8, we have $fg \in L^1(\mu)$. By Theorem 3.13, there exists a sequence of simple functions $\{s_n\}$ such that $|s_n| \leq |f|$ for every $n \in \mathbb{N}$ and $s_n \to f$ in $L^p(\mu)$ as $n \to \infty$. Since $|s_n g| \leq |fg|$ and $fg \in L^1(\mu)$, we obtain from Theorem 1.34 (Lebesgue's Dominated Convergence Theorem) and Theorem 5.4 that

$$\Lambda(f) = \lim_{n \to \infty} \Lambda(s_n) = \lim_{n \to \infty} \int_X s_n g \, \mathrm{d}\mu = \int_X fg \, \mathrm{d}\mu.$$

- **Step 2:** μ **is** σ**-finite.** This is exactly the hypothesis of Theorem 6.16, so we have $L^p(\mu)^* \cong L^q(\mu)$ in this case.

- **Step 3:** μ **is any measure and** $p > 1$**.** In this case, $q \in (1, \infty)$. As in **Step 2**, for each σ-finite subset E of X, there corresponds a *unique* $g_E \in L^q(E)$ such that

$$\Lambda(f) = \int_X f g_E \, \mathrm{d}\mu$$

for all $f \in L^p(E)$, where $L^p(E) = \{ f \in L^p(\mu) \,|\, f(x) = 0 \text{ for all } x \notin E \}$ and $L^q(E)$ is defined similarly. By Theorem 6.16, we know that

$$\|g_E\|_q = \big\| \Lambda|_{L^p(E)} \big\| \leq \|\Lambda\|. \tag{6.26}$$

If F is a σ-finite subset of X containing E, then the uniqueness of g_E implies that $g_F = g_E$ a.e. on E so that

$$\|g_E\|_q \leq \|g_F\|_q. \tag{6.27}$$

Now we define

$$M = \sup\{ \|g_E\|_q \,|\, E \text{ a } \sigma\text{-finite subset of } X \}.$$

Then it follows from the inequality (6.26) that

$$M \leq \|\Lambda\|$$

holds. By the definition, we may select a sequence $\{E_n\}$ of σ-finite subsets of X such that $\|g_{E_n}\|_q \to M$ as $n \to \infty$. Define $F = \bigcup_{n=1}^{\infty} E_n$ which is obviously σ-finite subset of X so that $\|g_F\|_q \leq M$. Furthermore, we deduce from the inequality (6.27) that

$$\|g_{E_n}\|_q \leq \|g_F\|_q$$

for all $n \in \mathbb{N}$. Therefore, the definition of M implies that $M \leq \|g_F\|_q$ and then

$$M = \|g_F\|_q.$$

Finally, if A is a σ-finite subset of X containing F, we get

$$\int_X |g_F|^q \, d\mu + \int_X |g_{A \setminus F}|^q \, d\mu = \int_X |g_A|^q \, d\mu = \|g_A\|_q^q \leq M^q = \int_X |g_F|^q \, d\mu$$

which means that $g_{A \setminus F} = 0$ a.e. on X or equivalently,

$$g_A = g_F \text{ a.e. on } X. \tag{6.28}$$

If $f \in L^p(\mu)$, then the set $A = F \cup \{x \in X \mid f(x) \neq 0\}$ is clearly σ-finite. Therefore, we observe from the result (6.28) that

$$\Lambda(f) = \int_X f g_A \, d\mu = \int_X f g_F \, d\mu.$$

Hence we may pick $g = g_F$ and the above argument makes sure this g is *unique*.

We complete the proof of the problem. $\blacksquare$

6.3   Fourier Coefficients of Complex Borel Measures

> **Problem 6.7**
>
> *Rudin Chapter 6 Exercise 7.*

Proof. If μ is a real measure, then we have

$$\widehat{\mu}(-n) = \int e^{int} \, d\mu(t) = \overline{\int e^{-int} \, d\mu(t)} = \overline{\widehat{\mu}(n)}$$

for every $n \in \mathbb{Z}$. Therefore, our result follows in this case.

Since μ is a complex Borel measure, Theorem 6.12 implies the existence of a Borel measurable function h such that $|h(t)| = 1$ on $[0, 2\pi)$ and $d\mu = h \, d|\mu|$. Thus we have

$$d|\mu| = 1 \cdot d|\mu| = \overline{h} \cdot h \, d|\mu| = \overline{h} \, d\mu. \tag{6.29}$$

Since $|\mu|$ is real, the previous paragraph and the expression (6.29) imply that

$$\overline{\widehat{\mu}(-n)} = \overline{\int e^{int} h \, d|\mu|} = \int e^{-int} \overline{h(t)} \, d|\mu|(t) = \int e^{-int} [\overline{h(t)}]^2 \, d\mu(t)$$

for every $n \in \mathbb{Z}$. By Theorem 3.14, $C(T)$ is dense in $L^1(|\mu|)$. Using Theorem 4.25 (The Weierstrass Approximation Theorem) and then Lemma 2.10, we deduce that the set of all trigonometric polynomials $\mathcal{P}$ is dense in $L^1(|\mu|)$.

We want to apply Problem 5.18. For any $f \in L^1(|\mu|)$, we define $\Lambda_n : L^1(|\mu|) \to \mathbb{C}$ by

$$\Lambda_n(f) = \int f(t) e^{-int} \, d\mu(t)$$

for all $n \in \mathbb{N}$. Then we have

$$|\Lambda_n(f)| \le \int |f| \cdot |\, d\mu(t)| = \int |f| \cdot |h| \, d|\mu(t)| = \int |f| \, d|\mu(t)| = \|f\|_1. \tag{6.30}$$

By Definition 5.3, we see from the inequality (6.30) that

$$\|\Lambda_n\| = \sup\{|\Lambda_n(f)| \mid f \in L^1(|\mu|) \text{ and } \|f\|_1 = 1\} \le 1$$

for every $n \in \mathbb{N}$. Next, if $g(t) = e^{ikt}$, then we have

$$\Lambda_n(g) = \int g(t) e^{-int} \, d\mu(t) = \int e^{-i(n-k)t} \, d\mu(t).$$

Now our hypothesis ensures that $\Lambda_n(g) \to 0$ as $n \to \infty$ so that

$$\Lambda_n(f) \to 0 \tag{6.31}$$

as $n \to \infty$ for each $f \in \mathcal{P}$. Therefore, Problem 5.18 says that the limit (6.31) also holds for every $f \in L^1(|\mu|)$. Obviously, we have $\overline{h}^2 \in L^1(|\mu|)$ and so

$$\overline{\widehat{\mu}(-n)} = \int e^{-int} [\overline{h}(t)]^2 \, d\mu(t) = \Lambda_n(\overline{h}^2).$$

Hence we conclude from the result (6.31) that

$$\lim_{n \to \infty} \overline{\widehat{\mu}(-n)} = \lim_{n \to \infty} \Lambda_n(\overline{h}^2) = 0.$$

This completes the proof of the problem. $\blacksquare$

Problem 6.8

Rudin Chapter 6 Exercise 8.

Proof. Suppose that $X = [0, 2\pi)$, $\mathscr{B}$ is the collection of all Borel sets in X and $\widehat{\mu}$ is periodic with period $k \in \mathbb{Z}$. Denote $d\lambda(t) = (e^{-ikt} - 1) \, d\mu(t)$. Clearly, we have

$$\widehat{\mu}(n+k) - \widehat{\mu}(n) = \int e^{-int}(e^{-ikt} - 1) \, d\mu(t) = \int e^{-int} \, d\lambda(t) = \widehat{\lambda}(n) \tag{6.32}$$

for all $n \in \mathbb{Z}$. Particularly, take $n = 0$ in the result (6.32) and use the periodicity of $\widehat{\mu}$ and Theorem 6.12 to get

$$\int h \, d|\lambda| = \int d\lambda = 0, \tag{6.33}$$

where h is a measurable function such that $|h(x)| = 1$ on X. If $|\lambda|(X) \ne 0$, then Theorem 1.39(a) and the form (6.33) together imply that $h = 0$ a.e. on X, a contradiction. This means that $|\lambda|(X) = 0$ and we assert from the fact $|\lambda(E)| \le |\lambda|(X) = 0$ that

$$\lambda(E) = 0 \tag{6.34}$$

for every $E \in \mathscr{B}$.

Using Theorem 6.12 to write $\mathrm{d}\mu = H\,\mathrm{d}|\mu|$, where H is a measurable function with $|H(x)| = 1$ on X. Next, we recall the meaning of the notation $\mathrm{d}\lambda(t) = (\mathrm{e}^{-ikt} - 1)\,\mathrm{d}\mu(t)$ and we observe from the result (6.34) that

$$0 = \lambda(E) = \int_E (\mathrm{e}^{-ikt} - 1)\,\mathrm{d}\mu = \int_E (\mathrm{e}^{-ikt} - 1)H\,\mathrm{d}|\mu| \tag{6.35}$$

for every $E \in \mathscr{B}$. Let $A_1 = \{t \in X \mid \cos kt - 1 = 0\}$, $A_2 = \{t \in X \mid \sin kt = 0\}$ and

$$A = \{t \in X \mid \mathrm{e}^{-ikt} - 1 = 0\} = A_1 \cap A_2.$$

Since $f_1(t) = \cos kt - 1$, $f_2(t) = \sin kt$ and $g(t) = 0$ are continuous on X, they are Borel measurable.[b] By Problem 1.5(a), we have $A_1, A_2 \in \mathscr{B}$ and also $A \in \mathscr{B}$.

We claim that μ is concentrated on A. Assume that it was not the case, i.e., there is $E_0 \in \mathscr{B}$ such that a $E_0 \cap A = \varnothing$ but $\mu(E_0) \neq 0$. We recall from the result (6.35) that $(\mathrm{e}^{-ikt} - 1)H = 0$ a.e. on *any* E with $|\mu|(E) > 0$. Since $|H| = 1$ on X, we have $\mathrm{e}^{-ikt} = 1$ a.e. on any E with $|\mu|(E) > 0$. Particularly, since $\mu(E_0) \neq 0$, we have $|\mu|(E_0) > 0$ and then

$$\mathrm{e}^{-ikt} = 1 \tag{6.36}$$

a.e. on E_0. However, $E_0 \cap A = 0 = \varnothing$ means that $\mathrm{e}^{-ikt} \neq 1$ on E_0 which contradicts the expression (6.36). Hence we have proven our claim that μ is concentrated on A.

On the other hand, if μ is concentrated on A, then for every $E \in \mathscr{B}$, we write $E = (E \setminus A) \cup A$ and then

$$\int_E (\mathrm{e}^{-ikt} - 1)\,\mathrm{d}\mu = \int_{E \setminus A} (\mathrm{e}^{-ikt} - 1)\,\mathrm{d}\mu + \int_A (\mathrm{e}^{-ikt} - 1)\,\mathrm{d}\mu. \tag{6.37}$$

Notice that $\mathrm{e}^{-ikt} - 1 = 0$ on A and $(E \setminus A) \cap A = \varnothing$ so that $\mu(E \setminus A) = 0$. Thus we deduce from Proposition 1.24(d) and (e) that all the integrals in the expression (6.37) are zero. This certainly implies that $\lambda(E) = 0$ for every $E \in \mathscr{B}$ and then

$$\widehat{\lambda}(n) = \int \mathrm{e}^{-int}\,\mathrm{d}\lambda = 0 \tag{6.38}$$

for every $n \in \mathbb{Z}$. Combining the expression (6.32) and the result (6.38), we may conclude that

$$\widehat{\mu}(n + k) = \widehat{\mu}(n)$$

for all $n \in \mathbb{Z}$.

Hence we have shown that μ is a complex Borel measure with periodic Fourier coefficient $\widehat{\mu}$ with period k if and only if μ is concentrated on

$$A = \{t \in X \mid \mathrm{e}^{-ikt} - 1 = 0\} = \left\{ t = \frac{2n\pi}{k} \,\middle|\, n \in \mathbb{N} \cup \{0\} \right\},$$

completing the proof of the problem.

> **Problem 6.9**
>
> *Rudin Chapter 6 Exercise 9.*

---

[b] See §1.11.

Proof. The assertion is false. Take $\mu = m$ the Lebesgue measure on I. Since $m \ll m$, if $m \perp m$, then Proposition 6.8(g) implies that $m = 0$, a contradiction. Now it remains to construct a sequence of $\{g_n\}$ satisfying the required properties.

Suppose that $n, k \in \mathbb{N}$ and $k = 1, 2, \ldots, n$. Let $\delta_n = \frac{2}{n(2n+3)}$. Define $g_{n,k} : [0,1] \to \mathbb{R}$ and $g_n : [0,1] \to \mathbb{R}$ by

$$
g_{n,k}(x) = \begin{cases}
n+1, & \text{if } x \in \left[\dfrac{k}{n+1} - \dfrac{\delta_n}{2}, \dfrac{k}{n+1} + \dfrac{\delta_n}{2}\right]; \\[2ex]
\text{linear}, & \text{if } x \in \left(\dfrac{k}{n+1} - \dfrac{\delta_n}{2} - \dfrac{\delta_n}{2n+2}, \dfrac{k}{n+1} - \dfrac{\delta_n}{2}\right); \\[2ex]
\text{linear}, & \text{if } x \in \left(\dfrac{k}{n+1} + \dfrac{\delta_n}{2}, \dfrac{k}{n+1} + \dfrac{\delta_n}{2} + \dfrac{\delta_n}{2n+2}\right); \\[2ex]
0, & \text{otherwise}
\end{cases}
$$

and

$$
g_n(x) = \sum_{k=1}^{n} g_{n,k}(x)
$$

respectively. By direct computation, it is easy to check that

$$
0 < \frac{1}{n+1} - \frac{\delta_n}{2} - \frac{\delta_n}{2n+2}, \quad \frac{n}{n+1} + \frac{\delta_n}{2} + \frac{\delta_n}{2n+2} < 1
$$

and

$$
\frac{k}{n+1} + \frac{\delta_n}{2} + \frac{\delta_n}{2n+2} < \frac{k+1}{n+1} - \frac{\delta_n}{2} - \frac{\delta_n}{2n+2}
$$

for $k = 1, 2, \ldots, n-1$. See Figure 6.1 for $g_{n,1}(x)$ and $g_{n,2}(x)$ below:

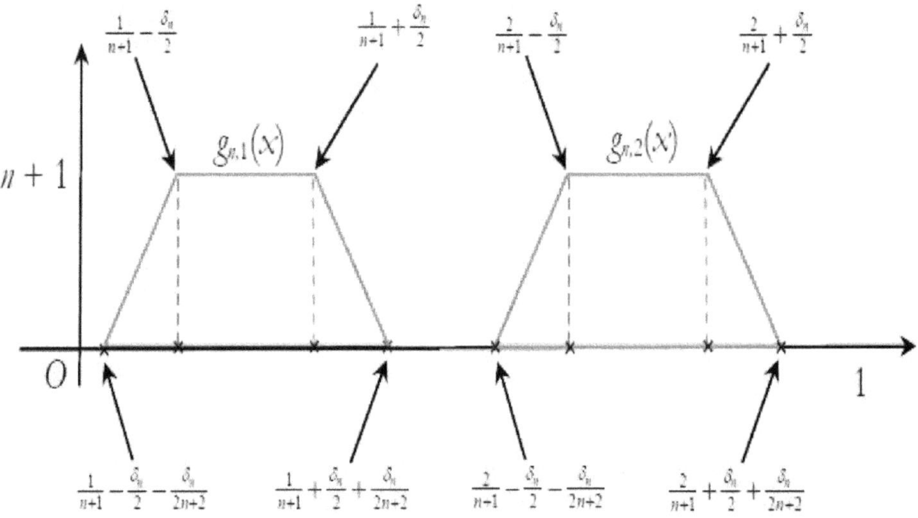

Figure 6.1: The graphs of $g_{n,1}(x)$ and $g_{n,2}(x)$.

Now the Lebesgue measure of the support of each $g_{n,k}$ is given by

$$
\frac{k}{n+1} + \frac{\delta_n}{2} + \frac{\delta_n}{2n+2} - \frac{k}{n+1} + \frac{\delta_n}{2} + \frac{\delta_n}{2n+2} = \left(1 + \frac{1}{n+1}\right)\delta_n = \frac{2(n+2)}{n(n+1)(2n+3)}
$$

so that the Lebesgue measure of the support of g_n is given by

$$\sum_{k=1}^{n} \frac{2(n+2)}{n(n+1)(2n+3)} = \frac{2(n+2)}{(n+1)(2n+3)} \to 0$$

as $n \to \infty$. This shows that condition (i) holds for this g_n. Since each $g_{n,k}$ is in fact a trapezium (see Figure 6.1 again) with area

$$\int_0^1 g_{n,k}(x)\,\mathrm{d}x = \frac{(n+1)}{2}\left(\frac{n+2}{n+1}\delta_n + \delta_n\right) = \frac{2n+3}{2}\delta_n = \frac{1}{n}$$

which implies that

$$\int_0^1 g_n(x)\,\mathrm{d}x = \sum_{k=1}^{n}\int_0^1 g_{n,k}(x)\,\mathrm{d}x = 1.$$

Thus this g_n satisfies condition (ii). Finally, note that for large n, we have

$$\frac{1}{(n+1)^2\delta_n} = \frac{n(2n+3)}{2(n+1)^2} \approx 1.$$

Thus for every $f \in C(I)$ and sufficiently large n, we have

$$\left|\int_0^1 f g_n\,\mathrm{d}x - \frac{1}{n+1}\sum_{k=1}^{n+1} f\left(\frac{k}{n+1}\right)\right| = \left|\sum_{k=1}^{n}\int_0^1 f g_{n,k}\,\mathrm{d}x - \frac{1}{n+1}\sum_{k=1}^{n} f\left(\frac{k}{n+1}\right) + \frac{f(1)}{n+1}\right|$$

$$\leq \left|\sum_{k=1}^{n}\left[\int_0^1 f g_{n,k}\,\mathrm{d}x - \frac{1}{n+1}f\left(\frac{k}{n+1}\right)\right]\right| + \frac{|f(1)|}{n+1}. \tag{6.39}$$

Since $f \in C(I)$, there is a $M > 0$ such that $|f(x)| \leq M$ on I. On the linear part, we have $g_{n,k} = a_k x + b_k$ for some $a_k, b_k \in \mathbb{R}$. Then it asserts that, for example,

$$\left|\int_{\frac{k}{n+1}-\frac{\delta_n}{2}-\frac{\delta_n}{2n+2}}^{\frac{k}{n+1}-\frac{\delta_n}{2}} (a_k x + b_k)f(x)\,\mathrm{d}x\right| \leq M|a_k| \cdot \left|\int_{\frac{k}{n+1}-\frac{\delta_n}{2}-\frac{\delta_n}{2n+2}}^{\frac{k}{n+1}-\frac{\delta_n}{2}} x\,\mathrm{d}x\right| + M|b_k| \cdot \frac{\delta_n}{2n+2} \tag{6.40}$$

for every $k = 1, 2, \ldots, n$. By routine computation, it can be shown easily that the integral on the right-hand side of the inequality (6.40) is of the growth $\frac{M'}{n^4}$, so the integral on the left-hand side of the inequality (6.40) is bounded by $\frac{M'}{n^3}$ for some positive constant M'. Using this fact, the inequality (6.39) can be further reduced to

$$\left|\int_0^1 f g_n\,\mathrm{d}x - \frac{1}{n+1}\sum_{k=1}^{n+1} f\left(\frac{k}{n+1}\right)\right|$$

$$\leq \left|\sum_{k=1}^{n}\left[\int_{\frac{k}{n+1}-\frac{\delta_n}{2}}^{\frac{k}{n+1}+\frac{\delta_n}{2}} (n+1)f(x)\,\mathrm{d}x - \frac{1}{n+1}f\left(\frac{k}{n+1}\right)\right]\right| + \frac{\widetilde{M}}{n^2} + \frac{\|f\|_\infty}{n+1}$$

$$= \left|\sum_{k=1}^{n}\left[\int_{\frac{k}{n+1}-\frac{\delta_n}{2}}^{\frac{k}{n+1}+\frac{\delta_n}{2}} (n+1)f(x)\,\mathrm{d}x - \frac{1}{(n+1)\delta_n}\int_{\frac{k}{n+1}-\frac{\delta_n}{2}}^{\frac{k}{n+1}+\frac{\delta_n}{2}} f\left(\frac{k}{n+1}\right)\mathrm{d}x\right]\right| + \frac{\widetilde{M}}{n^2} + \frac{\|f\|_\infty}{n+1}$$

$$\leq \sum_{k=1}^{n}\left[\int_{\frac{k}{n+1}-\frac{\delta_n}{2}}^{\frac{k}{n+1}+\frac{\delta_n}{2}} (n+1)\left|f(x) - \frac{1}{(n+1)^2\delta_n}f\left(\frac{k}{n+1}\right)\right|\mathrm{d}x\right] + \frac{\widetilde{M}}{n^2} + \frac{\|f\|_\infty}{n+1}$$

$$\leq \sum_{k=1}^{n}\left[(n+1)\int_{\frac{k}{n+1}-\frac{\delta_n}{2}}^{\frac{k}{n+1}+\frac{\delta_n}{2}} \left|f(x) - f\left(\frac{k}{n+1}\right)\right|\mathrm{d}x\right] + \frac{\widetilde{M}}{n^2} + \frac{\|f\|_\infty}{n+1}, \tag{6.41}$$

where $\widetilde{M}$ is a positive constant.

Applying the Mean-Value Theorem for Integrals (see [3, Theorem 7.30, pp. 160, 161]) to the integral in the inequality (6.39), we see that

$$\int_{\frac{k}{n+1}-\frac{\delta_n}{2}}^{\frac{k}{n+1}+\frac{\delta_n}{2}} \left| f(x) - f\left(\frac{k}{n+1}\right) \right| \, dx = \left| f(\xi) - f\left(\frac{k}{n+1}\right) \right| \delta_n, \tag{6.42}$$

where $\xi_k \in [\frac{k}{n+1} - \frac{\delta_n}{2}, \frac{k}{n+1} + \frac{\delta_n}{2}]$. Substituting the expression (6.42) into the inequality (6.41), we get

$$\left| \int_0^1 f g_n \, dx - \frac{1}{n+1} \sum_{k=1}^{n+1} f\left(\frac{k}{n+1}\right) \right| \leq \sum_{k=1}^{n} (n+1) \left| f(\xi_k) - f\left(\frac{k}{n+1}\right) \right| \delta_n + \frac{\|f\|_\infty}{n+1}$$

$$\leq n(n+1)\delta_n M_n + \frac{\widetilde{M}}{n^2} + \frac{\|f\|_\infty}{n+1}, \tag{6.43}$$

where

$$M_n = \max \left(\left| f(\xi_1) - f\left(\frac{1}{n+1}\right) \right|, \left| f(\xi_2) - f\left(\frac{2}{n+1}\right) \right|, \ldots, \left| f(\xi_n) - f\left(\frac{n}{n+1}\right) \right| \right).$$

When $n \to \infty$, $\xi_k \to \frac{k}{n+1}$ for each $k = 1, 2, \ldots, n$ so that $M_n \to 0$ as $n \to \infty$. By this observation and the fact $\lim\limits_{n\to\infty} n(n+1)\delta_n = 1$, we derive from the inequality (6.43) that

$$\lim_{n\to\infty} \int_0^1 f g_n \, dx = \int_0^1 f \, dx$$

which is exactly condition (iii).

We have completed the proof of the problem. $\blacksquare$

6.4   Problems on Uniformly Integrable Sets

> **Problem 6.10**
>
> *Rudin Chapter 6 Exercise 10.*

Proof.

(a) Let $\Phi = \{f_1, f_2, \ldots, f_N\} \subseteq L^1(\mu)$ for some $N \in \mathbb{N}$ and $\epsilon > 0$. By Problem 1.12, there exists a $\delta_k > 0$ such that

$$\int_{E_k} |f_k| \, d\mu < \epsilon \tag{6.44}$$

whenever $\mu(E_k) < \delta_k$, where $k = 1, 2, \ldots, N$. Suppose that $\delta = \min(\delta_1, \delta_2, \ldots, \delta_k)$. Then it yields from Theorem 1.33 and the estimates (6.44) that if $\mu(E) < \delta \leq \delta_k$, then

$$\left| \int_E f_k \, d\mu \right| \leq \int_E |f_k| \, d\mu < \epsilon$$

for each $k = 1, 2, \ldots, N$. By the definition, Φ is uniformly integrable.

(b) We first show that

> **Lemma 6.1**
>
> Suppose that $\Phi \subseteq L^1(\mu)$ is uniformly integrable. Given $\epsilon > 0$, there exists a $\delta > 0$ such that
>
> $$\int_E |f|\,\mathrm{d}\mu < \epsilon$$
>
> whenever $\mu(E) < \delta$ and $f \in \Phi$.

Proof of Lemma 6.1. Suppose that Φ is a collection of *real* measurable functions in $L^1(\mu)$ and $E \in \mathfrak{M}$. Given $\epsilon > 0$. Since Φ is uniformly integrable, there corresponds a $\delta > 0$ such that

$$\left| \int_E f\,\mathrm{d}\mu \right| < \frac{\epsilon}{4} \tag{6.45}$$

whenever $f \in \Phi$ and $\mu(E) < \delta$.

Define $E_+ = \{x \in E \mid f(x) \geq 0\}$ and $E_- = \{x \in E \mid f(x) < 0\}$. It is clear that $E = E_+ \cup E_-$, $E_+ \cap E_- = \varnothing$ and $\mu(E_\pm) \leq \mu(E) < \delta$. Thus we gain from Theorem 1.29 that

$$\int_E |f|\,\mathrm{d}\mu = \int_{E_+} |f|\,\mathrm{d}\mu + \int_{E_-} |f|\,\mathrm{d}\mu = \int_{E_+} f\,\mathrm{d}\mu - \int_{E_-} f\,\mathrm{d}\mu$$

and then the estimate (6.45) yields that

$$\int_E |f|\,\mathrm{d}\mu = \left| \int_E |f|\,\mathrm{d}\mu \right| \leq \left| \int_{E_+} f\,\mathrm{d}\mu \right| + \left| \int_{E_-} f\,\mathrm{d}\mu \right| < \frac{\epsilon}{2} \tag{6.46}$$

for all $f \in \Phi$ and $\mu(E) < \delta$. In other words, Lemma 6.1 holds for subsets of *real* measurable functions in $L^1(\mu)$.

Next, suppose that Φ is a collection of *complex* measurable functions in $L^1(\mu)$. If $f \in \Phi$, then $f = u + iv$, where u and v are real measurable functions in $L^1(\mu)$. Let

$$\Psi = \{\mathrm{Re}\,(f) \mid f \in \Phi\} \quad \text{and} \quad \Omega = \{\mathrm{Im}\,(f) \mid f \in \Phi\}.$$

Since $0 < |u| \leq |f|$ and $0 < |v| \leq |f|$, we see that $u, v \in L^1(\mu)$ and then $\Psi, \Omega \subseteq L^1(\mu)$. Besides, by the estimate (6.46), we have

$$\int_E |u|\,\mathrm{d}\mu < \frac{\epsilon}{2} \quad \text{and} \quad \int_E |v|\,\mathrm{d}\mu < \frac{\epsilon}{2} \tag{6.47}$$

when $\mu(E) < \delta$ and for all $u \in \Psi$ and $v \in \Omega$. By the definition, the sets Ψ and Ω are uniformly integrable. Since $|f| \leq |u| + |v|$, we deduce from the estimates (6.47) that

$$\int_E |f|\,\mathrm{d}\mu \leq \int_E |u|\,\mathrm{d}\mu + \int_E |v|\,\mathrm{d}\mu < \epsilon$$

whenever $\mu(E) < \delta$ and for all $f \in \Phi$. This completes the proof of Lemma 6.1. $\blacksquare$

Now it is time to return to the proof of the problem.

- **Proof of** $\|f\|_1 < \infty$. Since $\{f_n\}$ is uniformly integrable, it follows from Lemma 6.1 that given $\epsilon > 0$, there is a $\delta > 0$ such that

$$\int_E |f_n|\,\mathrm{d}\mu < \epsilon \tag{6.48}$$

whenever $\mu(E) < \delta$ and for all $n \in \mathbb{N}$. Employing Theorem 1.28 (Fatou's Lemma) to the estimate (6.48), we know that

$$\int_E |f|\,\mathrm{d}\mu \le \liminf_{n\to\infty} \int_E |f_n|\,\mathrm{d}\mu \le \epsilon \tag{6.49}$$

whenever $\mu(E) < \delta$.

By hypothesis (iii), there exists a $E_1 \in \mathfrak{M}$ such that $\mu(X\backslash E_1) = 0$ and $f_n(x) \to f(x)$ pointwisely a.e. on E_1. Since $\mu(X) < \infty$ and hypothesis (iii) again, Egoroff's Theorem asserts the existence of a measurable set $E_2 \subseteq X$ with $\mu(X \setminus E_2) < \delta$ such that

$$f_n(x) \to f(x) \tag{6.50}$$

uniformly a.e. on E_2 as $n \to \infty$. Since uniform convergence implies pointwise convergence, we may assume that $E_2 \subseteq E_1$. Otherwise, we can replace E_2 by the set $E_2 \cap E_1$. In this case, De Morgan's Laws give

$$\mu\big(X \setminus (E_1 \cap E_2)\big) = \mu\big((X \setminus E_1) \cup (X \setminus E_2)\big) \le \mu(X \setminus E_1) + \mu(X \setminus E_2) < \delta$$

so that the limit (6.50) also holds on $E_2 \cap E_1$.

By hypothesis (iv), there exists a $E_3 \in \mathfrak{M}$ such that $\mu(X \setminus E_3) = 0$ and $|f(x)| < \infty$ on E_3. Take $E = E_2 \cap E_3 \subseteq E_2$ so that the uniform convergence (6.50) also holds on E. This means that there exists a positive integer N such that $n \ge N$ implies

$$|f_n(x) - f(x)| < \epsilon$$

a.e. on E and then hypothesis (i) implies

$$\int_E |f_n - f|\,\mathrm{d}\mu < \epsilon\mu(E) \le \epsilon\mu(X) < \infty \tag{6.51}$$

for all $n \ge N$. By De Morgan's Laws again, we obtain

$$\mu(X \setminus E) \le \mu(X \setminus E_2) + \mu(X \setminus E_3) < \delta,$$

so we deduce from the estimates (6.49) that

$$\int_{X\backslash E} |f|\,\mathrm{d}\mu \le \epsilon. \tag{6.52}$$

Now using the estimates (6.51), (6.52) and the fact $f_N \in L^1(\mu)$ that

$$\int_X |f|\,\mathrm{d}\mu = \int_E |f|\,\mathrm{d}\mu + \int_{X\backslash E} |f|\,\mathrm{d}\mu \le \int_E |f_N - f|\,\mathrm{d}\mu + \int_E |f_N|\,\mathrm{d}\mu + \int_{X\backslash E} |f|\,\mathrm{d}\mu < \infty.$$

In other words, we have $f \in L^1(\mu)$.

– **Proof of $\|f_n - f\|_1 \to 0$ as $n \to \infty$.** Again the estimates (6.48), (6.51) and (6.52) together show that

$$\int_X |f_n - f|\,\mathrm{d}\mu = \int_E |f_n - f|\,\mathrm{d}\mu + \int_{X\backslash E} |f_n - f|\,\mathrm{d}\mu$$

$$< \epsilon\mu(X) + \int_{X\backslash E} |f_n|\,\mathrm{d}\mu + \int_{X\backslash E} |f|\,\mathrm{d}\mu$$

$$< [2 + \mu(X)]\cdot\epsilon$$

for all $n \ge N$. Hence we conclude that $\|f_n - f\|_1 \to 0$ as $n \to \infty$.

(c) For each $n \in \mathbb{N}$, define $f_n : \mathbb{R} \to \mathbb{R}$ by

$$f_n(x) = \begin{cases} \dfrac{1}{n}, & \text{if } x \in [0, n]; \\[2mm] 0, & \text{otherwise.} \end{cases}$$

Then it is easy to see that

$$\|f_n\|_1 = \int_{\mathbb{R}} |f_n| \, dm = \int_0^n \frac{1}{n} \, dm = 1, \tag{6.53}$$

i.e., $\{\|f_n\|_1\}$ is bounded. Next, for every $x \in \mathbb{R}$, we have $f_n(x) \to 0$ as $n \to \infty$. So we let $f(x) = 0$. Since $|f_n(x)| \le 1$ for all $n \in \mathbb{N}$ and $x \in \mathbb{R}$, for each $\epsilon > 0$, if $\delta = \epsilon$, then for every $E \in \mathscr{B}$ with $m(E) < \delta$, we obtain

$$\left| \int_E f_n \, dm \right| \le \int_E |f_n| \, dm \le m(E) < \epsilon.$$

Thus $\{f_n\}$ is uniformly integrable. However, the expression (6.53) indicates that

$$\lim_{n \to \infty} \int_{\mathbb{R}} |f_n - 0| \, dm \neq 0.$$

Hence hypothesis (i) cannot be omitted in part (b).

(d) We construct two examples showing that hypothesis (iv) is redundant in some cases, but not in other cases.

  – **Example 1.** Let $\mathscr{B}$ be the collection of Borel sets of $[0, 1]$. We employ the concept of **atomicless measures** . A set $E \in \mathfrak{M}$ is called an **atom** of the measure μ if $\mu(E) > 0$ and every measurable subset $F \subset E$ has measure either 0 or $\mu(E)$. If μ has no atoms, then it is called atomless. Lebesgue measure m on $[0, 1]$ is atomless. In other words, for every $\epsilon \in (0, 1)$, if $E \in \mathscr{B}$ satisfies $m(E) = \epsilon > 0$, then there exists a $F \in \mathscr{B}$ and $F \subset E$ such that $0 < m(F) < \epsilon$. See [9, Definition 1.12.7, Example 1.12.8, p. 55].

    We claim that Vitali's Convergence Theorem holds for m without the hypothesis (iv). To this end, it suffices to show that hypotheses (i) to (iii) imply hypothesis (iv). Suppose that

$$E = \{x \in [0, 1] \,|\, |f(x)| = \infty\}.$$

If $m(E) = 0$, then there is nothing to prove. So we assume that $m(E) > 0$. Since $\{f_n\}$ is uniformly integrable, Lemma 6.1 implies the existence of a $\delta > 0$ such that

$$\int_E |f_n| \, dx < 1$$

whenever $m(E) < \delta$ and all $n \in \mathbb{N}$. Since m is atomless and $m(E) > 0$, there exists a $F \in \mathscr{B}$ such that $0 < m(F) < \delta$ and thus

$$\int_F |f_n| \, dx < 1 \tag{6.54}$$

for all $n \in \mathbb{N}$. By Theorem 1.28 (Fatou's Lemma) and the fact that $f_n(x) \to f(x)$ a.e. on $[0, 1]$ as $n \to \infty$, we deduce from the inequality (6.54) that

$$\int_F |f| \, dx \le \liminf_{n \to \infty} \int_F |f_n| \, dx \le 1. \tag{6.55}$$

However, since $F \subset E$ and $m(F) > 0$, the left-most integral in the inequality (6.55) is actually ∞, a contradiction. Hence $m(E) = 0$ or equivalently, $|f(x)| < \infty$ a.e. on $[0, 1]$.

– **Example 2.** Let $\mathfrak{M}$ be the collection of all sets $E \subseteq \mathbb{R}$ such that either E or E^c is at most countable and define $\mu(E) = 0$ in the first case, $\mu(E) = 1$ in the second. By Problem 1.6, we see that $\mathfrak{M}$ is a σ-algebra in $\mathbb{R}$ and μ is a measure on $\mathfrak{M}$. Clearly, μ is a finite measure. We define $f(x) \equiv \infty$ on $\mathbb{R}$ and $f_n : \mathbb{R} \to \mathbb{R}$ by $f_n(x) = n$ for every $n \in \mathbb{N}$ so that $f_n(x) \to f(x)$ as $n \to \infty$ on $\mathbb{R}$, but hypothesis (iv) *does not* hold.

To each $\epsilon > 0$, we pick $\delta = 1$, so if $E \in \mathfrak{M}$ satisfies $\mu(E) < 1$, then it must be $\mu(E) = 0$ and thus Proposition 1.24(e) implies that

$$\left| \int_E f_n \, d\mu \right| = 0 < \epsilon$$

for all $n \in \mathbb{N}$. By the definition, $\{f_n\}$ is uniformly integrable. However, it is clear that $f \notin L^1(\mu)$ because $\mathbb{R}^c = \varnothing$ is at most countable. Consequently, conclusion of part (b) is false.

(e) Suppose that the hypotheses of Theorem 1.34 (Lebesgue's Dominated Convergence Theorem) hold. Then hypotheses (i) and (iii) are true definitely. On the one hand, since $g \in L^1(\mu)$, Problem 1.12 says that each $\epsilon > 0$ there exists a $\delta > 0$ such that

$$\int_E |g| \, d\mu < \epsilon \tag{6.56}$$

whenever $\mu(E) < \delta$. On the other hand, $|f_n(x)| \leq g(x)$ for all $n \in \mathbb{N}$ will imply that

$$\int_X |f_n| \, d\mu \leq \int_X |g| \, d\mu < \infty. \tag{6.57}$$

Therefore, $\{f_n\} \subseteq L^1(\mu)$ and then we gain from this, Theorem 1.33 and the estimate (6.56) that

$$\left| \int_E f_n \, d\mu \right| \leq \int_E |f_n| \, d\mu < \epsilon$$

whenever $\mu(E) < \delta$ and all $n \in \mathbb{N}$. Hence $\{f_n\}$ is uniformly integrable, i.e., hypothesis (ii) is true. Since $|f_n|$ are measurable by Proposition 1.9(b), it follows from Theorem 1.28 (Fatou's Lemma) and the estimate (6.57) that

$$\int_X |f| \, d\mu \leq \liminf_{n \to \infty} \int_X |f_n| \, d\mu < \infty.$$

Thus $|f(x)| < \infty$ a.e. on X, i.e., hypothesis (iv) is valid. Hence part (b) verifies that Vitali's Convergence Theorem implies Theorem 1.34 (Lebesgue's Dominated Convergence Theorem) if $\mu(X) < \infty$.

Now we are going to construct an example in which Vitali's Theorem applies, but *not* Theorem 1.34 (Lebesgue's Dominated Convergence Theorem). For every $n \in \mathbb{N}$, we consider $f_n : (0, 1) \to \mathbb{R}$ defined by

$$f_n(x) = \frac{1}{x} \chi_{(\frac{1}{n+1}, \frac{1}{n})} \geq 0.$$

Evidently, we have $m\big((0, 1)\big) = 1$ and $|f(x)| \leq$ on $(0, 1)$ which are hypotheses (i) and (iv) respectively. For each $x \in (0, 1)$, we have $x \notin \chi_{(\frac{1}{n+1}, \frac{1}{n})}$ for all sufficiently large n so that $f_n(x) \to 0 = f(x)$ as $n \to \infty$. Thus we have hypothesis (iii). It is trivial to check that

$$\int_0^1 |f_n| \, dx = \int_{\frac{1}{n+1}}^{\frac{1}{n}} \frac{1}{x} \, dx = \ln\left(1 + \frac{1}{n}\right) < \infty \tag{6.58}$$

so that $\{f_n\} \in L^1((0,1))$. Given $\epsilon > 0$. One can find a positive integer N such that $\ln(1 + \frac{1}{n}) < \epsilon$ for all $n \geq N$. By this result, we observe from the representation (6.58) that

$$\left| \int_E f_n \, dx \right| \leq \left| \int_0^1 f_n \, dx \right| \leq \int_0^1 |f_n| \, dx < \epsilon$$

for every $n \geq N$ and every measurable subset E of $(0,1)$. By part (a), the finite set $\{f_1, f_2, \ldots, f_{N-1}\}$ is uniformly integrable. Therefore, these two facts confirm that $\{f_n\}$ is uniformly integrable, i.e., hypothesis (ii). Hence the set $\{f_n\}$ satisfies all the hypotheses of Vitali's Convergence Theorem.

However, for every $x \in (0,1)$, we note from Definition 1.13 that

$$\sup_{n \in \mathbb{N}}(f_n(x)) = \frac{1}{x},$$

so if $|f_n(x)| \leq g(x)$ for all $n \in \mathbb{N}$, then $\frac{1}{x} \leq g(x)$. Since $\frac{1}{x} \notin L^1$, $g \notin L^1$ too and this means that $\{f_n\}$ *does not* satisfy the hypotheses of Theorem 1.34 (Lebesgue's Dominated Convergence Theorem).

(f) For each $n \in \mathbb{N}$, we consider $f_n : [0,1] \to \mathbb{R}$ defined by

$$f_n(x) = n\chi_{(0,\frac{1}{n})}(x) - n\chi_{(1-\frac{1}{n},1)}(x).$$

Notice that each f_n is measurable and satisfies $f_n(0) = f_n(1) = 0$. If $x \in (0,1)$, then there exists a positive integer N such that $x \notin (0,\frac{1}{n})$ and $x \notin (1-\frac{1}{n},1)$ for all $n \geq N$. Thus we obtain $f_n(x) = 0$ for all $n \geq N$. Next,

$$\int_0^1 f_n(x) \, dx = \int_0^{\frac{1}{n}} n \, dx - \int_{1-\frac{1}{n}}^1 n \, dx = n \times \frac{1}{n} - n \times \frac{1}{n} = 0$$

for each $n \in \mathbb{N}$.

However, $\{f_n\}$ is *not* uniformly integrable. To see this, assume that there corresponds a $\delta > 0$ such that

$$\left| \int_E f_n \, dx \right| < 1 \tag{6.59}$$

whenever $m(E) < \delta$ and for all $n \in \mathbb{N}$. In particular, fix $n = N > \frac{1}{\delta}$ and take $E_N = (0, \frac{1}{N})$ so that $m(E_n) = \frac{1}{N} < \delta$ and

$$\left| \int_{E_N} f_N \, dx \right| = \left| \int_0^{\frac{1}{N}} f_N \, dx \right| = \left| \int_0^{\frac{1}{N}} N \, dx \right| = 1$$

which contradicts to the inequality (6.59).

(g) If $\mu(X) = 0$, then there is nothing to prove. Without loss of generality, we may assume that $\mu(X) > 0$. We follow the hint and prove it into several steps.

    – **Step 1: ρ is a metric (modulo sets of measure 0)^{c}.** Define

$$\rho(A, B) = \int_X |\chi_A - \chi_B| \, d\mu \geq 0. \tag{6.60}$$

---

^{c}Remember that sets A and B with $A = B$ a.e. on X are treated to be identical.

Since $\mu(X) > 0$, $\rho(A, B) = 0$ if and only if $\chi_A = \chi_B$ a.e. on X if and only if $A = B$ a.e. on X. It is trivial that $\rho(A, B) = \rho(B, A)$. Finally, since

$$|\chi_A - \chi_B| \leq |\chi_A - \chi_C| + |\chi_C - \chi_B|$$

for every $A, B, C \in \mathfrak{M}$, we must have

$$\rho(A, B) \leq \rho(A, C) + \rho(C, B).$$

By the definition, ρ is a metric (modulo sets of measure 0).

– **Step 2: $(\mathfrak{M}, \rho)$ is a complete metric space (modulo sets of measure 0).** Let $\{E_n\} \subseteq \mathfrak{M}$ be a Cauchy sequence, i.e., given $\epsilon > 0$, there exists a positive integer N such that $n, m \geq N$ imply that

$$\rho(E_n, E_m) < \epsilon.$$

We note from the definition (6.60) that $\rho(A, B) = \|\chi_A - \chi_B\|_1$, so we obtain

$$\|\chi_{E_n} - \chi_{E_m}\|_1 < \epsilon$$

for all $n, m \geq N$, i.e., $\{\chi_{E_n}\}$ is Cauchy in $L^1(\mu)$. By Theorem 3.11, there exists $f \in L^1(\mu)$ such that $\{\chi_{E_n}\}$ converges to f in $L^1(\mu)$. By Theorem 3.12, $\{\chi_{E_n}\}$ has a subsequence $\{\chi_{E_{n_k}}\}$ such that

$$\chi_{E_{n_k}}(x) \to f(x)$$

pointwise a.e. on X as $n \to \infty$. By the definition of $\chi_{E_{n_k}}$, we get immediately that either $f(x) = 0$ or 1 for almost every $x \in X$. Now if we define $E = \{x \in X \mid f(x) = 1\}$, then $E \in \mathfrak{M}$, $f = \chi_E$ a.e. on X and therefore

$$\rho(E_n, E) = \|\chi_{E_n} - \chi_E\|_1 \to 0$$

as $n \to \infty$. Hence $(\mathfrak{M}, \rho)$ is a complete metric space (modulo sets of measure 0).

– **Step 3: The map $E \to \int_E f_n \, d\mu$ is continuous for each n.** Denote this map by φ_n. Recall that $f_n \in L^1(\mu)$, so Problem 1.12 asserts that for every $\epsilon > 0$, there corresponds a $\delta > 0$ such that

$$\int_E |f_n| \, d\mu < \epsilon \tag{6.61}$$

whenever $\mu(E) < \delta$. Suppose that $E, F \in \mathfrak{M}$ such that $\rho(E, F) < \delta$. We deduce from the fact $E \cup F = (E \setminus F) \cup (E \cap F) \cup (F \setminus E)$ that

$$\begin{aligned}
\rho(E, F) &= \int_X |\chi_E - \chi_F| \, d\mu \\
&= \int_{E \cup F} |\chi_E - \chi_F| \, d\mu \\
&= \int_{E \setminus F} |\chi_E - \chi_F| \, d\mu + \int_{F \setminus E} |\chi_E - \chi_F| \, d\mu \\
&= \mu(E \setminus F) + \mu(F \setminus E) \tag{6.62}
\end{aligned}$$

so that $\mu(E \setminus F) < \delta$ and $\mu(F \setminus E) < \delta$. Since $(\chi_E - \chi_F)f_n \in L^1(\mu)$, Theorem 1.33 and the estimate (6.61) imply that

$$|\varphi_n(E) - \varphi_n(F)| = \left| \int_E f_n \, d\mu - \int_F f_n \, d\mu \right|$$

$$\begin{aligned}
&= \left| \int_X \chi_E f_n \, \mathrm{d}\mu - \int_X \chi_F f_n \, \mathrm{d}\mu \right| \\
&= \left| \int_X (\chi_E - \chi_F) f_n \, \mathrm{d}\mu \right| \\
&\le \int_X |\chi_E - \chi_F| \cdot |f_n| \, \mathrm{d}\mu \\
&= \int_{E \setminus F} |f_n| \, \mathrm{d}\mu + \int_{F \setminus E} |f_n| \, \mathrm{d}\mu \\
&< 2\epsilon.
\end{aligned}$$

Hence φ_n is continuous on $\mathfrak{M}$ for every $n \in \mathbb{N}$.

- **Step 4: Completion of the proof.** Now $\{\varphi_n\}$ is a sequence of continuous complex functions on the complete metric space $(\mathfrak{M}, \rho)$. Denote

$$\varphi(E) = \lim_{n \to \infty} \varphi_n(E)$$

as a complex number for every $E \in \mathfrak{M}$. If $\epsilon > 0$, it follows from Problem 5.13 (particularly the inequality (5.62)) that there exist $E_0 \in \mathfrak{M}$, $\delta > 0$ and $N \in \mathbb{N}$ such that

$$\left| \int_E (f_n - f_N) \, \mathrm{d}\mu \right| = |\varphi_n(E) - \varphi_N(E)| < \epsilon \tag{6.63}$$

for all $E \in \mathfrak{M}$ with $\rho(E, E_0) < \delta$ and $n \ge N$.

If $\mu(A) < \delta$ and $B = E_0 \setminus A$, then the expression (6.62) indicates that

$$\rho(B, E_0) = \mu\big((E_0 \setminus A) \setminus E_0\big) + \mu(E_0 \setminus (E_0 \setminus A)) = \mu(E_0 \cap A) \le \mu(A) < \delta.$$

Similarly, if $C = E_0 \cup A$, then we know that

$$\rho(C, E_0) = \mu\big((E_0 \cup A) \setminus E_0\big) + \mu(E_0 \setminus (E_0 \cup A)) = \mu(A) < \delta.$$

Thus the estimate (6.63) holds with B and C in place of E.

Since $E_0 \cup A = (E_0 \setminus A) \cup A$ and $(E_0 \setminus A) \cap A = \varnothing$, we get

$$\int_{E_0 \cup A} (f_n - f_N) \, \mathrm{d}\mu = \int_A (f_n - f_N) \, \mathrm{d}\mu + \int_{E_0 \setminus A} (f_n - f_N) \, \mathrm{d}\mu$$

which implies that

$$\begin{aligned}
\left| \int_A (f_n - f_N) \, \mathrm{d}\mu \right| &= \left| \int_{E_0 \cup A} (f_n - f_N) \, \mathrm{d}\mu - \int_{E_0 \setminus A} (f_n - f_N) \, \mathrm{d}\mu \right| \\
&\le \left| \int_{E_0 \cup A} (f_n - f_N) \, \mathrm{d}\mu \right| + \left| \int_{E_0 \setminus A} (f_n - f_N) \, \mathrm{d}\mu \right| \\
&< 2\epsilon
\end{aligned} \tag{6.64}$$

for all $n > N$. By part (a), the set $\{f_1, f_2, \ldots, f_N\}$ is uniformly integrable, i.e., there corresponds a $\delta' > 0$ such that

$$\left| \int_A f_n \, \mathrm{d}\mu \right| < \epsilon \tag{6.65}$$

whenever $\mu(A) < \delta'$ and $n = 1, 2, \ldots, N$. Define $\delta'' = \min(\delta, \delta')$. If $\mu(A) < \delta''$, then since $\mu(A) < \delta'$, the inequality (6.65) still holds for $n = 1, 2, \ldots, N$. Next, for

$n = N+1, N+2, \ldots$, since $\mu(A) < \delta$, it yields from the estimates (6.64) and (6.65) that

$$\left| \int_A f_n \, d\mu \right| \leq \left| \int_A (f_n - f_N) \, d\mu \right| + \left| \int_A f_N \, d\mu \right| < 3\epsilon. \tag{6.66}$$

By combining the results (6.65) and (6.66), we conclude that

$$\left| \int_A f_n \, d\mu \right| < 3\epsilon$$

if $\mu(A) < \delta''$ and for all $n \in \mathbb{N}$, i.e., $\{f_n\}$ is uniformly integrable.

We have completed the proof of the problem. ∎

> **Problem 6.11**
>
> *Rudin Chapter 6 Exercise 11.*

Proof. We note that if $\mu(X) = 0$, then Proposition 1.24(e) shows that the result holds trivially. Therefore, without loss of generality, we may assume that $\mu(X) \neq 0$. Now we want to apply Problem 6.10(b). By our hypotheses, it suffices to show that $\Phi = \{f_n\}$ is uniformly integrable and $|f(x)| < \infty$ a.e. on X.

Since $f_n \in L^1(\mu)$ for every $n \in \mathbb{N}$, we apply Theorem 1.33 and then Theorem 3.5 (Hölder's Inequality) to obtain

$$\left| \int_E f_n \, d\mu \right| \leq \int_E |f_n| \, d\mu \leq \left\{ \int_E |f_n|^p \, d\mu \right\}^{\frac{1}{p}} \left\{ \int_E d\mu \right\}^{\frac{1}{q}} < C^{\frac{1}{p}} \times \mu(E)^{\frac{1}{q}} \tag{6.67}$$

for every $E \in \mathfrak{M}$. Since $p > 1$, $q < \infty$. Take $\delta = \epsilon^q C^{-\frac{q}{p}}$. If $\mu(E) < \delta$, then it follows from the inequality (6.67) that

$$\left| \int_E f_n \, d\mu \right| < \epsilon.$$

Thus the set $\Phi = \{f_n\}$ is uniformly integrable.

It remains to show that $|f(x)| < \infty$ a.e. on X. Since $f_n(x) \to f(x)$ a.e. on X, $|f_n(x)| \to |f(x)|$ a.e. on X. By Theorem 1.28 (Fatou's Lemma) and the inequality (6.67), we know that

$$\int_X |f| \, d\mu \leq \liminf_{n \to \infty} \int_X |f_n| \, d\mu < C^{\frac{1}{p}} \times \mu(X)^{\frac{1}{q}} < \infty. \tag{6.68}$$

Since $\mu(X) \neq 0$, the inequality (6.68) ensures that $|f(x)| < \infty$ a.e. on X. Hence we deduce from Problem 6.10(b) that

$$\lim_{n \to \infty} \int_X |f - f_n| \, d\mu = 0$$

holds, completing the proof of the problem. ∎

6.5 Dual Spaces of $L^p(\mu)$ Revisit

> **Problem 6.12**
>
> *Rudin Chapter 6 Exercise 12.*

Proof. We note that Problem 1.6 establishes that $\mathfrak{M}$ is a σ-algebra. Let $E = (\frac{1}{4}, \frac{1}{2})$ which is open in $[0, 1]$. It is clear that $g^{-1}(E) = E$. Since E and E^c are uncountable, we have $g^{-1}(E) \notin \mathfrak{M}$ and thus g is *not* $\mathfrak{M}$-measurable by Definition 1.3(c). We have completed the proof of the problem. ◼

> **Problem 6.13**
>
> *Rudin Chapter 6 Exercise 13.*

Proof. Let $C(I)$ be the space of all continuous functions on I. It is clear that $\chi_{[0,\delta]} \in L^\infty$ but $\chi_{[0,\delta]} \notin C(I)$ for some $\delta > 0$. We need the following result:

> **Lemma 6.2**
>
> The space $C(I)$ is closed in the metric space L^∞.

Proof of Lemma 6.2. Let $\{f_n\} \subseteq C(I)$ and $\{f_n\}$ converges to f in L^∞. Given $\epsilon > 0$. It means that there exists a positive integer N such that $n \geq N$ implies

$$|f_n(x) - f(x)| \leq \sup_{x \in I} |f_n(x) - f(x)| = \|f_n - f\|_\infty < \epsilon$$

for all $x \in I$. In other words, $\{f_n\}$ converges uniformly to f on I and thus f is continuous on I (see [49, Theorem 7.12, p. 150]), i.e., $f \in C(I)$. Hence $C(I)$ is closed in L^∞, completing the proof of the lemma. ◼

We return to the proof of the problem. Since L^∞ is a Banach space, we get from Lemma 6.2 and the equivalent form of Theorem 5.19 that *there exists* a bounded linear functional Λ on L^∞ such that

$$\Lambda(f) = 0 \tag{6.69}$$

for every $f \in C(I)$, but $\Lambda(\chi_{[0,\delta]}) \neq 0$.

Assume that there was a $g \in L^1(\mu)$ such that

$$\Lambda(f) = \int_I fg \, dm \tag{6.70}$$

for every $f \in L^\infty$. In particular, the result (6.69) and the representation (6.70) imply that

$$\int_I fg \, dm = 0 \tag{6.71}$$

on $C(I)$. Since $f \in C(I)$ and $g \in L^1(\mu)$, we have $fg \in L^1(\mu)$ and then we apply Theorem 1.39(b) to the integral (6.71) to conclude that $fg = 0$ a.e. on I. If we take $f(x) = e^x$ on I which belongs to $C(I)$, then we obtain the fact that

$$g = 0$$

a.e. on I. Hence we follow from the representation (6.70) that the result (6.69) actually holds on the whole L^∞, a contradiction to the fact that $\Lambda(\chi_{[0,\delta]}) \neq 0$. This completes the proof of the problem. ◼

Differentiation

7.1 Lebesgue Points and Metric Densities

Problem 7.1

Rudin Chapter 7 Exercise 1.

Proof. Let x be a Lebesgue point of f. Since $f \in L^1(\mathbb{R}^k)$, we deduce from the triangle inequality and Theorem 1.33 that

$$
\begin{aligned}
|f(x)| &\leq \left| f(x) - \frac{1}{m(B_r)} \int_{B_r} f(y) \, \mathrm{d}y \right| + \left| \frac{1}{m(B_r)} \int_{B_r} f(y) \, \mathrm{d}y \right| \\
&\leq \left| \frac{1}{m(B_r)} \int_{B_r} [f(x) - f(y)] \, \mathrm{d}y \right| + \frac{1}{m(B_r)} \int_{B_r} |f(y)| \, \mathrm{d}y \\
&\leq \frac{1}{m(B_r)} \int_{B_r} |f(y) - f(x)| \, \mathrm{d}y + (Mf)(x).
\end{aligned}
\tag{7.1}
$$

Since x is a Lebesgue point of f, if we let $r \to 0+$ in the inequality (7.1), then we obtain

$$
|f(x)| \leq (Mf)(x)
$$

as required, completing the proof of the problem. $\blacksquare$

Problem 7.2

Rudin Chapter 7 Exercise 2.

Proof. We have $I = (-\delta, \delta)$. Now we are going to prove the existence of a measurable set $E \subseteq \mathbb{R}$ such that

$$
\liminf_{\delta \to 0} \frac{m\big(E \cap I(\delta)\big)}{2\delta} = \alpha \quad \text{and} \quad \limsup_{\delta \to 0} \frac{m\big(E \cap I(\delta)\big)}{2\delta} = \beta.
\tag{7.2}
$$

In fact, it suffices to prove that

$$
\liminf_{\delta \to 0} \frac{m\big(E \cap [0, \delta)\big)}{\delta} = \alpha \quad \text{and} \quad \limsup_{\delta \to 0} \frac{m\big(E \cap [0, \delta)\big)}{\delta} = \beta,
\tag{7.3}
$$

where $E \subseteq [0, \infty)$ because we can get the desired result (7.2) by reflecting about the origin. To start our proof, we consider five cases:

- **Case (i):** $0 < \alpha < \beta < 1$. In this case, we consider the numbers

$$\theta = \frac{\alpha(1-\beta)}{\beta(1-\alpha)} < \frac{\alpha}{\beta} < 1 \quad \text{and} \quad d_n = \frac{\beta-\alpha}{1-\alpha}\cdot\theta^n > 0$$

for every $n \in \mathbb{N}$. Then it is easy to see that

$$\theta^n - \theta^{n+1} = \theta^n\left[1 - \frac{\alpha(1-\beta)}{\beta(1-\alpha)}\right] = \frac{\beta-\alpha}{\beta(1-\alpha)}\theta^n = \frac{d_n}{\beta} > d_n. \tag{7.4}$$

Next, for each $n = 1, 2, \ldots$, we define the closed interval

$$E_n = [\theta^n - d_n, \theta^n].$$

We claim that the measurable set

$$E = \bigcup_{n=1}^{\infty} E_n$$

satisfies the requirement (7.3). To see this, we first note from the inequality (7.4) that $d_n \to 0$ as $n \to \infty$ and $E_n \cap E_{n+1} = \varnothing$ for every $n \in \mathbb{N}$, see Figure 7.1 below:

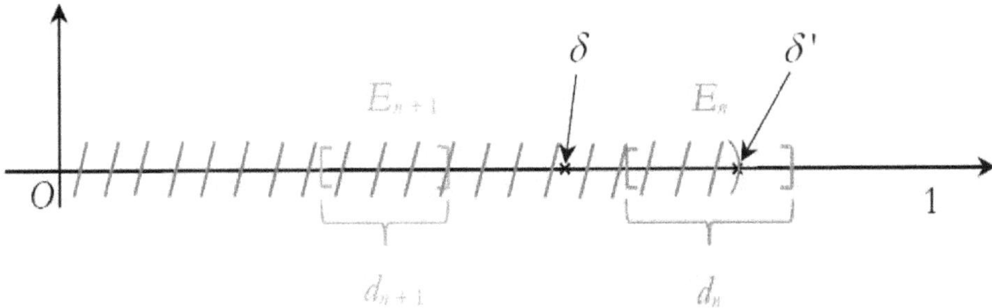

Figure 7.1: The closed intervals E_n and E_{n+1}.

Next, choose $\delta' > \delta > 0$ as shown in Figure 7.1, then it is easy to know that

$$[0, \theta^n - d_n) \cap E = \bigcup_{k=n+1}^{\infty} E_k, \quad [0, \theta^n) \cap E = \bigcup_{k=n}^{\infty} E_k, \quad [0, \delta) \cap E = \bigcup_{k=n+1}^{\infty} E_k$$

and

$$[0, \delta') \cap E = [\theta^n - d_n, \delta') \cup \bigcup_{k=n+1}^{\infty} E_k$$

so that

$$\frac{m\big([0, \theta^n - d_n) \cap E\big)}{\theta^n - d_n} = \frac{1-\alpha}{(1-\beta)\theta^n}\sum_{k=n+1}^{\infty} d_k = \frac{\beta-\alpha}{1-\beta}\cdot\frac{\theta}{1-\theta} = \alpha,$$

$$\frac{m\big([0, \theta^n) \cap E\big)}{\theta^n} = \frac{1}{\theta^n}\sum_{k=n}^{\infty} d_k = \frac{\beta-\alpha}{1-\alpha}\cdot\frac{1}{1-\theta} = \beta,$$

$$\frac{m\big([0, \delta) \cap E\big)}{\delta} = \frac{m\big([0, \theta^{n+1}) \cap E\big)}{\delta} = \frac{1}{\delta}\sum_{k=n+1}^{\infty} d_k = \frac{\beta\theta^{n+1}}{\delta}$$

and

$$\frac{m\big([0,\delta')\cap E\big)}{\delta'} = \frac{m\big([0,\theta^n)\cap E\big) - m\big([\delta',\theta^n]\big)}{\delta'} = \frac{\beta\theta^n + \delta' - \theta^n}{\delta'} = 1 - (1-\beta)\frac{\theta^n}{\delta'}.$$

Therefore, the above analysis shows trivially that the function $f : (0,\infty) \to \mathbb{R}$ defined by

$$f(\delta) = \frac{m([0,\delta)\cap E)}{\delta} \tag{7.5}$$

attains the values α and β at the end-points of each E_n, decreases from β to α in the interval $[\theta^{n+1},\theta^n - d_n]$ and increases from α to β in E_n. Hence, we have

$$\alpha = \min_{x\in(0,\delta)} f(x) \quad \text{and} \quad \beta = \max_{x\in(0,\delta)} f(x)$$

which mean that the limits (7.3) hold.

- **Case (ii): $\alpha = 0$ and $\beta = 1$.** In this case, we defined the measurable set

$$E = \bigcup_{n=1}^{\infty} \left[\frac{1}{(2n)!}, \frac{1}{(2n-1)!}\right].$$

By similar analysis as in **Case (i)**, we see that

$$0 \le \frac{m\big(E\cap[0,\frac{1}{(2n)!})\big)}{\frac{1}{(2n)!}} \le (2n)! \times m\left(\left[0,\frac{1}{(2n+1)!}\right)\right) = (2n)! \times \frac{1}{(2n+1)!} = \frac{1}{2n+1}$$

and

$$\frac{2n-1}{2n} = (2n-1)! \times m\left(\left[\frac{1}{(2n)!},\frac{1}{(2n-1)!}\right)\right) \le \frac{m\big(E\cap[0,\frac{1}{(2n-1)!})\big)}{\frac{1}{(2n-1)!}} \le 1.$$

Hence we have proven the limits (7.3) in this case.

- **Case (iii): $0 \le \alpha = \beta \le 1$.** For each $n \in \mathbb{N}$, we take

$$E_n = \left[\frac{\alpha}{n+1} + \frac{1-\alpha}{n}, \frac{1}{n}\right].$$

Thus it is easy to see that $E_n \cap E_{n+1} = \varnothing$ for all $n \in \mathbb{Z}$. We define $E = \bigcup_{n=1}^{\infty} E_n$. If we take $\delta = \frac{1}{n}$, then we follow similar argument as in **Case (i)** that

$$m\left(E\cap\left[0,\frac{1}{n}\right)\right) = \sum_{k=n}^{\infty} m(E_k) = \alpha \sum_{k=n}^{\infty} \frac{1}{k(k+1)}. \tag{7.6}$$

Since $\displaystyle\sum_{k=n}^{N} \frac{1}{k(k+1)} = \frac{1}{n} - \frac{1}{N}$, we have

$$\sum_{k=n}^{\infty} \frac{1}{k(k+1)} = \frac{1}{n}. \tag{7.7}$$

Now we put the infinite series (7.7) into the expression (7.6) to get

$$\frac{m\big(E\cap[0,\frac{1}{n})\big)}{\frac{1}{n}} = \alpha$$

which induces exactly the limits (7.3).

- **Case (iv):** $0 = \alpha < \beta < 1$. Similar to the construction in **Case (i)**, we suppose that

$$E_n = \left[\frac{1-\beta}{(2n)!} + \frac{\beta}{(2n+2)!}, \frac{1}{(2n)!}\right] \tag{7.8}$$

where $n \in \mathbb{N}$. Similarly, we have $E_n \cap E_{n+1} = \varnothing$ for all $n \in \mathbb{N}$. Let $E = \bigcup_{n=1}^{\infty} E_n$. On the one hand, since

$$m\left(E \cap \left[0, \frac{1-\beta}{(2n)!} + \frac{\beta}{(2n+2)!}\right)\right) = \beta \sum_{k=n+1}^{\infty} \left[\frac{1}{(2k)!} - \frac{1}{(2k+2)!}\right] = \frac{\beta}{(2n+2)!},$$

we have

$$\frac{m\left(E \cap \left[0, \frac{1-\beta}{(2n)!} + \frac{\beta}{(2n+2)!}\right)\right)}{\frac{1-\beta}{(2n)!} + \frac{\beta}{(2n+2)!}} = \frac{\frac{\beta}{(2n+2)!}}{\frac{1-\beta}{(2n)!} + \frac{\beta}{(2n+2)!}} = \frac{\beta}{(1-\beta)(2n+2)(2n+1)+\beta} \to 0$$

as $n \to \infty$. On the other hand, we have

$$m\left(E \cap \left[0, \frac{1}{(2n)!}\right)\right) = \beta \sum_{k=n}^{\infty} \left[\frac{1}{(2k)!} - \frac{1}{(2k+2)!}\right] = \frac{\beta}{(2n)!},$$

we have

$$\frac{m\left(E \cap \left[0, \frac{1}{(2n)!}\right)\right)}{\frac{1}{(2n)!}} = \frac{\frac{\beta}{(2n)!}}{\frac{1}{(2n)!}} = \beta.$$

Besides, the function (7.5) decreases from β to 0 in the interval $\left[\frac{1}{(2n+2)!}, \frac{1-\beta}{(2n)!} + \frac{\beta}{(2n+2)!}\right]$ and increases from 0 to β in E_n. Hence these imply the results (7.3).

- **Case (v):** $0 < \alpha < \beta \le 1$. Instead of the intervals (7.8), we consider the intervals

$$E_n = \left[\frac{1}{(2n)!}, \frac{\alpha}{(2n)!} + \frac{1-\alpha}{(2n+2)!}\right],$$

where $n \in \mathbb{N}$. Thus it can be shown similar to **Case (iv)** that the limits (7.3) hold for the measurable set $E = \bigcup_{n=1}^{\infty} E_n$ and we omit the details here.

This completes the proof of the problem. ▪

> **Remark 7.1**
>
> The limits defined in (7.2) are called the upper and lower densities of E at 0 respectively.

7.2   Periods of Functions and Lebesgue Measurable Groups

> **Problem 7.3**
>
> *Rudin Chapter 7 Exercise 3.*

Proof. We follow Rudin's hint. Pick $\alpha \in \mathbb{R}$ and put

$$F(x) = m(E \cap [\alpha, x])$$

for $x > \alpha$. If $y > x > \alpha + p_i$, then we deduce from Theorem 2.20(c) and the fact

$$[\alpha - p_i, \alpha + p_i] = [\alpha - p_i, x] \setminus (\alpha + p_i, x]$$

that

$$
\begin{aligned}
F(x + p_i) - F(x - p_i) &= m(E \cap [\alpha, x + p_i]) - m(E \cap [\alpha, x - p_i]) \\
&= m(E \cap [\alpha, x + p_i] - p_i) - m(E \cap [\alpha, x - p_i] + p_i) \\
&= m(E \cap [\alpha - p_i, x]) - m(E \cap (\alpha + p_i, x]) \\
&= m(E \cap [\alpha - p_i, \alpha + p_i]) \\
&= F(y + p_i) - F(y - p_i).
\end{aligned}
\tag{7.9}
$$

If F is differentiable at x, then we have

$$
\lim_{i \to \infty} \frac{F(x + p_i) - F(x - p_i)}{2p_i} = \lim_{i \to \infty} \frac{1}{2}\left[\frac{F(x + p_i) - F(x)}{p_i} + \frac{F(x - p_i) - F(x)}{-p_i}\right] = F'(x).
\tag{7.10}
$$

Combining the results (7.9) and (7.10), we see that $F'(x) = F'(y)$ for all $x, y > \alpha$ and thus

$$F'(x) = c_\alpha \tag{7.11}$$

for all $x > \alpha$, where c_α is a constant. Next, we recall from the definition of F that

$$F(x) = m(E \cap [\alpha, x]) = \int_\alpha^x \chi_E(t)\,dt.$$

Since $\chi_E \in L^1(\mathbb{R})$, it establishes from Theorem 7.11 that F is differentiable and $F'(x) = \chi_E(x)$ a.e. for $x > \alpha$. Using this and the fact (7.11), we get

$$\chi_E(x) = c_\alpha \tag{7.12}$$

a.e. for $x > \alpha$. Since χ_E takes only 0 or 1, we see that $c_\alpha \in \{0, 1\}$. If $x > \beta > \alpha$, then

$$c_\beta = F'(x) = \chi_E(x) = F'(x) = c_\alpha,$$

so c_α *does not* depend on α and we can simply replace c_α by c in the results (7.11) and (7.12). Finally, if $c = 0$, then we conclude from the result (7.12) that

$$m(E) = \int_\mathbb{R} \chi_E\,dm = 0.$$

Otherwise, $c = 1$ and the result (7.12) implies that $\chi_{E^c}(x) = 0$ and hence $m(E^c) = 0$. We have completed the proof of the problem. ∎

> **Problem 7.4**
>
> *Rudin Chapter 7 Exercise 4.*

Proof. Our proof does not follow Rudin's hint. In fact, we mainly follow the argument of Cignoli and Hounie [15] and now we first quote one of their results that will be used very soon:

> **Lemma 7.1**
>
> Let D be a dense subset of $\mathbb{R}$ and μ a Borel measure on $\mathbb{R}$ such that
>
> $$\mu\big(d + [a,b)\big) = \mu\big([a,b)\big)$$
>
> for every $d \in D$ and $a, b \in \mathbb{R}$. Then we have $\mu = km$, where $k = \mu\big((0,1]\big)$.

Next, it is clear that the set $G(f) = \{ns + mt \,|\, n, m \in \mathbb{Z}\}$ is an additive group of $\mathbb{R}$. Without loss of generality, we assume that $t > 0$. Since $\alpha = s/t$ is irrational, it follows from the Kronecker's Approximation Theorem[a] that for every $\theta \in \mathbb{R}$ and every $\epsilon t^{-1} > 0$, there exists $n, m \in \mathbb{Z}$ such that

$$\left| n\alpha + m - \frac{\theta}{t} \right| < \frac{\epsilon}{t}$$

which is equivalent to saying that

$$|ns + mt - \theta| < \epsilon.$$

Hence $G(f)$ is dense in $\mathbb{R}$.

We first consider the special case that f is nonnegative and bounded a.e. on $\mathbb{R}$. Note that if $F(x) = f(xt)$, then $F(x+1) = f(xt + t) = f(xt) = F(x)$, so F is a function of period 1 and we may assume that f has period 1. Consequently, it suffices to prove the problem when f is a nonnegative, bounded a.e. function on $[0,1]$ having period 1. Suppose that $\mu : \mathscr{B} \to [0, \infty)$ is defined by

$$\mu(E) = \int_E f(x)\,\mathrm{d}x \tag{7.13}$$

for all $E \in \mathscr{B}$. By Theorem 1.29, μ is a Borel measure. Let p be a period of f. Then we deduce from the representation (7.13) that

$$\mu(E + p) = \int_{E+p} f(x)\,\mathrm{d}x = \int_E f(x+p)\,\mathrm{d}x = \int_E f(x)\,\mathrm{d}x = \mu(E).$$

In other words, μ has the same periods as f. Furthermore, if $p \in G(f)$, then we obtain from Theorem 7.26 (The Change-of-variables Theorem)[b] that

$$\mu\big(p + [a,b)\big) = \mu\big([a+p, b+p)\big) = \int_{a+p}^{b+p} f(x)\,\mathrm{d}x = \int_a^b f(x+p)\,\mathrm{d}x = \int_a^b f(x)\,\mathrm{d}x = \mu\big([a,b)\big).$$

By Lemma 7.1, we have

$$\mu = km, \tag{7.14}$$

where $0 < k = \mu\big((0,1]\big) < \infty$. By combining the results (7.13) and (7.14), we establish that

$$\int_E [f(x) - k]\,\mathrm{d}x = 0$$

for every measurable subset E of $[0,1]$. Since f is bounded a.e. on $[0,1]$, it is clear that $f - k \in L^1\big([0,1]\big)$. Now Theorem 1.39(b) implies that $f = k$ a.e. on $[0,1]$ and the periodicity of f shows that

$$f = k$$

---

[a] See [63, Lemma 4.6, p. 82]

[b] It is applied to the third equality. Recall that we are discussing Lebesgue integral, *not* Riemann integral.

a.e. on $\mathbb{R}$ in this special case.

For the general case, the composite function $g(x) = \frac{\pi}{2} + \tan^{-1}\big(f(x)\big)$ is clearly a measurable function on $\mathbb{R}$. Furthermore, g is also nonnegative and bounded a.e. on $\mathbb{R}$. If $f(x+1) = f(x)$, then $g(x+1) = \frac{\pi}{2} + \tan^{-1}\big(f(x+1)\big) = \frac{\pi}{2} + \tan^{-1}\big(f(x)\big) = g(x)$ so that g is also a function of period 1. Applying the previous part to g to conclude that $g = k$ a.e. on $\mathbb{R}$ for a constant k and this implies that

$$f(x) = \tan\left(k - \frac{\pi}{2}\right)$$

a.e. on $\mathbb{R}$.

Finally, if we define $G = \{ns + mt \mid m, n \in \mathbb{Z}\}$ and $f : \mathbb{R} \to \mathbb{R}$ by

$$f(x) = \begin{cases} 0, & \text{if } x \in G; \\ \\ 1, & \text{otherwise,} \end{cases}$$

then it is trivial that f has periods s and t as well as $f = 1$ a.e. on $\mathbb{R}$.[c] This completes the proof of the problem. ■

> ## Problem 7.5
>
> *Rudin Chapter 7 Exercise 5.*

Proof. We prove the assertions one by one.

- $A + B$ **contains a segment.** Since $m(A) > 0$ and $m(B) > 0$, A and B are Lebesgue measurable. By the comment in §7.12, the metric densities of A and B are 1 at *almost every* point of A and B. Thus we pick $a_0 \in A$ and $b_0 \in B$ such that

$$\lim_{r \to 0+} \frac{m\big(A \cap B(a_0, r)\big)}{m\big(B(a_0, r)\big)} = \lim_{r \to 0+} \frac{m\big(B \cap B(b_0, r)\big)}{m\big(B(b_0, r)\big)} = 1. \tag{7.15}$$

Since $\chi_A, \chi_B \in L^1(\mathbb{R})$, we may assume with the aid of Theorem 7.7 that a_0 and b_0 are Lebesgue points of χ_A and χ_B respectively.

Set $c_0 = a_0 + b_0$. For each $\epsilon > 0$, define

$$B_\epsilon = \{c_0 + \epsilon - b \mid b \in B \text{ and } |b - b_0| < \delta\}, \tag{7.16}$$

where $\delta > 0$ is small. For every $c_0 + \epsilon - b$, we have

$$|c_0 + \epsilon - b - (a_0 + \epsilon)| = |c_0 + \epsilon - b - (c_0 - b_0 + \epsilon)| = |b - b_0| < \delta$$

which implies that $B_\epsilon \subseteq (a_0 + \epsilon - \delta, a_0 + \epsilon + \delta)$. If we take $\delta = 2\epsilon$ and ϵ is sufficiently small, then we have

$$B_\epsilon \subseteq (a_0 + \epsilon - \delta, a_0 + \epsilon + \delta) = (a_0 - \epsilon, a_0 + 3\epsilon) \subseteq (a_0 - 4\epsilon, a_0 + 4\epsilon) = B(a_0, 4\epsilon). \tag{7.17}$$

By the definition (7.16), we note that

$$B_\epsilon - (c_0 + \epsilon) = \{-b \mid b \in B \text{ and } |b - b_0| < 2\epsilon\} = B \cap B(b_0, 2\epsilon).$$

Since $B_\epsilon - (c_0 + \epsilon)$ is just a translate of B_ϵ, Theorem 2.20(c) says that

$$m(B_\epsilon) = m\big(B_\epsilon - (c_0 + \epsilon)\big) = m\big(B \cap B(b_0, 2\epsilon)\big). \tag{7.18}$$

---

[c] By Problem 7.6, we have $m(G) = 0$.

By the metric density of B at b_0 in (7.15), we know that

$$m\big(B \cap B(b_0, 2\epsilon)\big) > \frac{m\big(B(b_0, 2\epsilon)\big)}{2} = 2\epsilon. \tag{7.19}$$

Combining the expression (7.18) and the inequality (7.19), we arrive at

$$m(B_\epsilon) > 2\epsilon > 0. \tag{7.20}$$

We claim that $A \cap B_\epsilon \neq \varnothing$ *for all* small enough $\epsilon > 0$. Otherwise, no matter how small $\epsilon > 0$ is, we can find a $\eta \in (0, \epsilon)$ such that $A \cap B_\eta = \varnothing$. In fact, we may pick η small enough so that the metric density of A at a_0 in (7.15) implies

$$m\big(A \cap B(a_0, 4\eta)\big) > \frac{7}{8} m\big(B(a_0, 4\eta)\big) = 7\eta. \tag{7.21}$$

Since $A \cap B_\eta = \varnothing$, we have $A \cap [B(a_0, 4\eta) \setminus B_\eta] = [A \cap B(a_0, 4\eta)] \setminus [A \cap B_\eta] = A \cap B(a_0, 4\eta)$ so that

$$m\big(A \cap B(a_0, 4\eta)\big) = m\big(A \cap [B(a_0, 4\eta) \setminus B_\eta]\big) \leq m\big(B(a_0, 4\eta) \setminus B_\eta\big). \tag{7.22}$$

Recall from the set relation (7.17), we have the fact $B(a_0, 4\eta) = [B(a_0, 4\eta) \setminus B_\eta] \cup B_\eta$. By this fact and the inequality (7.20), we may further reduce the inequality (7.22) to

$$m\big(A \cap B(a_0, 4\eta)\big) \leq m\big(B(a_0, 4\eta)\big) - m(B_\eta) < 8\eta - 2\eta = 6\eta. \tag{7.23}$$

However, the inequalities (7.21) and (7.23) are contradictory and this proves our claim that $A \cap B_\epsilon \neq \varnothing$ for all small $\epsilon > 0$.

Pick $x \in (c_0 - \epsilon', c_0 + \epsilon')$ for sufficiently small ϵ' and without loss of generality, suppose that $\epsilon = x - c_0 > 0$. Then we have $0 < \epsilon < \epsilon'$ and the above claim implies that

$$A \cap B_\epsilon \neq \varnothing.$$

Choose $a \in A \cap B_\epsilon$. By the definition (7.16) and then using the fact $\epsilon = x - c_0$, we have

$$a = c_0 + \epsilon - b = x - b$$

for some $b \in B$ with $|b - b_0| < 2\epsilon$. In other words, we have the representation

$$x = a + b \in A + B,$$

i.e., $(c_0 - \epsilon', c_0 + \epsilon') \subseteq A + B$.

- $\frac{C}{2} + \frac{C}{2} = [0, 1]$. Recall from [49, Exercise 19, p. 81] that if $c \in C$, then

$$c = \sum_{n=1}^{\infty} \frac{\alpha_n}{3^n},$$

where $\alpha_n = 0$ or 2. On the one hand, let $x \in [0, 1]$ and we have

$$x = \sum_{n=1}^{\infty} \frac{a_n}{3^n}, \tag{7.24}$$

where $a_n \in \{0, 1, 2\}$ for all $n = 1, 2, \ldots$. Define

$$\beta_n = \begin{cases} 0, & \text{if } a_n \in \{0, 1\}; \\ 1, & \text{otherwise,} \end{cases} \quad \text{and} \quad \gamma_n = \begin{cases} 0, & \text{if } a_n \in \{0\}; \\ 1, & \text{otherwise.} \end{cases}$$

Then it is always true that $a_n = \beta_n + \gamma_n$ for all $n = 1, 2, \ldots$. Thus the x given in (7.24) can be represented as

$$x = y + z.$$

Here y and z are given by

$$y = \sum_{n=1}^{\infty} \frac{\beta_n}{3^n} \quad \text{and} \quad z = \sum_{n=1}^{\infty} \frac{\gamma_n}{3^n}, \tag{7.25}$$

where $\beta_n, \gamma_n \in \{0, 1\}$ for all $n = 1, 2, \ldots$.[d] In other words, the analysis implies that $[0, 1] \subseteq \frac{1}{2}C + \frac{1}{2}C$.

On the other hand, let $y, z \in \frac{1}{2}C$ be expressed by the series (7.25), where $\beta_n, \gamma_n \in \{0, 1\}$. Then we have

$$y + z = \sum_{n=1}^{\infty} \frac{\beta_n + \gamma_n}{3^n},$$

where $\beta_n + \gamma_n \in \{0, 1, 2\}$ for all $n = 1, 2, \ldots$. If $\beta_n + \gamma_n = 0$ for all $n = 1, 2, \ldots$, then $y + z = 0$. If $\beta_n + \gamma_n = 2$ for all $n = 1, 2, \ldots$, then

$$y + z = \sum_{n=1}^{\infty} \frac{2}{3^n} = 1.$$

Otherwise, $y + z \in (0, 1)$. Therefore, we have $\frac{1}{2}C + \frac{1}{2}C \subseteq [0, 1]$. Hence we have obtained the required result that $\frac{1}{2}C + \frac{1}{2}C = [0, 1]$.

We have completed the proof of the problem. ◼

Remark 7.2

The result in Problem 7.5 is called the **Steinhaus Theorem**. It has a generalization in $\mathbb{R}^k$ which says that if $A \in \mathscr{B}^k$ and $m_k(A) > 0$, then $\mathbf{0}$ is an interior point of $A - A$, see [5, Theorem 26.6, p. 163] for details.

Problem 7.6

Rudin Chapter 7 Exercise 6.

Proof. Assume that $m(G) > 0$. By Problem 7.5, $G + G$ contains a segment (a, b). Since G is a group with addition, we have $G + G \subseteq G$ so that $(a, b) \subseteq G$. Let $m = \frac{a+b}{2} \in (a, b)$ so that $m \in G$. Since G is a group, $-m \in G$ so that $(a - m, b - m) \subseteq G$. Since m is the mid-point of a and b, the segment $(a - m, b - m)$ is in the form $(-c, c)$ for some *fixed* $c \in \mathbb{R} \setminus \{0\}$. Recall that $G + G \subseteq G$, so we have $(-2c, 2c) \subseteq G$. Repeat this kind of argument, we may conclude that

$$(-nc, nc) \subseteq G$$

for all $n \in \mathbb{N}$, but this implies that $G = \mathbb{R}$, a contradiction. Hence $m(G) = 0$ and this completes the proof of the problem. ◼

---

[d]Since the series (7.25) converge, we can add them together.

7.3   The Cantor Function and the Non-measurability of $f \circ T$

Problem 7.7

Rudin Chapter 7 Exercise 7.

Proof. The construction has been done in [26, Example 15, pp. 96 – 98]. We won't repeat it here. See also [26, Example 30, pp. 105, 195]. ∎

Problem 7.8

Rudin Chapter 7 Exercise 8.

Proof. We have $T : V = (a, b) \to \mathbb{R}$. Since V is a bounded segment, we define $\varphi(a) = \varphi(b) = 0$ so that the domain of T can be extended to $[a, b]$ easily.

(a) Obviously, V is open in $\mathbb{R}$ and Lebesgue measurable. Since $0 \le \varphi_n(x) < 2^{-n}$ in V and $\sum_{n=1}^{\infty} 2^{-n} < \infty$, the Weierstrass M-test [49, Theorem 7.10, p. 148] indicates that

$$\varphi = \sum_{n=1}^{\infty} \varphi_n \tag{7.26}$$

converges uniformly on V. Since each φ_n is continuous on V, φ is continuous on V. Furthermore, $0 \le \varphi(x) \le 1$ on V so that φ is also bounded on $[a, b]$. By [49, Theorem 11.33, p. 323], we see that $\varphi \in \mathscr{R}$ on $[a, x]$ and

$$T(x) = \int_a^x \varphi(t)\, \mathrm{d}t = \mathscr{R} \int_a^x \varphi(t)\, \mathrm{d}t$$

for every $x \in [a, b]$. Therefore, it follows from the First Fundamental Theorem of Calculus that T is differentiable at every point of the continuity of φ in $[a, x]$. Consequently, T is differentiable (and hence continuous) at every point of V.

Next, if $T(x) = T(y)$ for $x < y$, then we have

$$T(y) = \int_a^y \varphi(t)\, \mathrm{d}t = T(x) + \int_x^y \varphi(t)\, \mathrm{d}t$$

which implies that

$$\int_x^y \varphi(t)\, \mathrm{d}t = 0. \tag{7.27}$$

Since W is dense in V, $(x, y) \cap W \neq \varnothing$ so that $(x, y) \cap W_N \neq \varnothing$ *for some* positive integer N. Let $p \in (x, y) \cap W_N$. By the definition, we have $\varphi(p) > 0$ and the sign-preserving property of continuous functions (see [64, Problem 7.15, p. 112]) assures that there exists a $\delta > 0$ such that $(p - \delta, p + \delta) \subseteq (x, y) \cap W_n$ so that $\varphi(q) > 0$ for all $q \in (p - \delta, p + \delta)$, but this implies that

$$\int_x^y \varphi(t)\, \mathrm{d}t \ge \int_{p-\delta}^{p+\delta} \varphi(t)\, \mathrm{d}t > 0,$$

a contradiction to the result (7.27). Thus T is one-to-one and condition (ii) is satisfied. Since T is differentiable at every point of V, it is continuous on V and condition (i) is also satisfied. Condition (iii) is obvious because $X \setminus X = \varnothing$.

(b) By the First Fundamental Theorem of Calculus, we know that $T' = \varphi$ which is continuous on V. If $p \in K$, then $p \notin W_n$ for all $n \in \mathbb{N}$ which implies that $\varphi_n(p) = 0$ for all $n \in \mathbb{N}$. Therefore, we have

$$T'(p) = \varphi(p) = \sum_{n=1}^{\infty} \varphi_n(p) = 0. \tag{7.28}$$

For the third assertion, we first show that $T(K)$ is a Borel set. To this end, since T is injective, we note from [42, Exercise 2(h), p. 21] that

$$T(K) = T(V \setminus W) = T(V) \setminus T(W) = T(V) \setminus \bigcup_{n=1}^{\infty} T(W_n).$$

Since T is continuous and injective on V, the property of **Invariance of Domain** [32, Theorem 2B.3, p. 172] shows that T is an open map. Thus $T(V)$ and each $T(W_n)$ is open in $\mathbb{R}$ so that $T(K) \in \mathscr{B}$, as desired.

By the previous result, the function $f = \chi_{T(K)}$ is Lebesgue measurable. By part (a), T satisfies the hypotheses of Theorem 7.26 (The Change-of-variables Theorem), so we have

$$m\big(T(K)\big) = \int_{T(V)} \chi_{T(K)} \, \mathrm{d}m = \int_{T(V)} f \, \mathrm{d}m = \int_V (f \circ T)|J_T| \, \mathrm{d}m, \tag{7.29}$$

where $|J_T(x)|$ denotes the Jacobian of T at x. Recall that T is injective, so $T(x) \in T(K)$ if and only if $x \in K$ and then we reduce from the expression (7.29) that

$$m\big(T(K)\big) = \int_V \chi_{T(K)}\big(T(x)\big)|J_T| \, \mathrm{d}m = \int_K \chi_K(x)|J_T| \, \mathrm{d}m. \tag{7.30}$$

Notice from the result (7.28) that $|J_T(x)| = |\det T'(x)| = 0$ on K, so we obtain from the integral (7.30) that $m\big(T(K)\big) = 0$.

(c) Since $m(K) > 0$, it follows from Corollary to Theorem 2.22 that K contains a nonmeasurable subset E. Let $A = T(E)$. Since $E \subset K$, we have $A \subset T(K)$. Since $m\big(T(K)\big) = 0$, we recall from Theorem 1.36 that all subsets of sets of measure 0 are measurable, thus A is a measurable set and χ_A is a measurable function. By the definition, we know that

$$\chi_A\big(T(x)\big) = \chi_{T(E)}\big(T(x)\big) = \begin{cases} 1, & \text{if } x \in E; \\ \\ 0, & \text{otherwise.} \end{cases}$$

In other words, we have $\chi_A \circ T = \chi_E$. Since E is nonmeasurable, χ_E is not measurable.

(d) By part (b), we know that $T'(x) = \varphi(x)$ on V. Thus if we can construct a φ which is *infinitely differentiable*, then we are done. To see this, we are going to construct an infinitely differentiable function φ_n on W_n for each positive integer n. Let's recall the **bump function** $\psi : \mathbb{R} \to \mathbb{R}$ defined by

$$\psi(x) = \begin{cases} \exp\left(-\dfrac{1}{1-x^2}\right), & \text{if } x \in (-1, 1); \\ \\ 0, & \text{otherwise.} \end{cases}$$

The ψ has derivatives of all orders in $\mathbb{R}$, i.e., ψ is an element of $\mathscr{C}^\infty(\mathbb{R})$ the space of all infinitely differentiable functions on $\mathbb{R}$. Since each W_n is a segment, we let $W_n = (a_n, b_n)$ for some $a_n, b_n \in (a, b)$ with $a_n < b_n$. Therefore the bump function $\varphi_n : V \to \mathbb{R}$ given by

$$\varphi_n(x) = 2^{-n}\psi\left(\frac{2}{b_n - a_n}(x - b_n) + 1\right) = 2^{-n}\psi\left(\frac{x - \frac{a_n+b_n}{2}}{\frac{b_n-a_n}{2}}\right)$$

is what we want. Put $U_1 = W_1$ and $U_n = W_n \setminus (W_1 \cup W_2 \cup \cdots \cup W_{n-1})$ for $n = 2, 3, \ldots$. Then we have $U_i \cap U_j = \varnothing$ for all $i \neq j$ and their union is also W. Since each W_n is a segment, each U_n is a union of at most countable *disjoint* segments. In other words, we may assume that $W_i \cap W_j = \varnothing$ for all $i \neq j$. Therefore, the function $\varphi : V \to \mathbb{R}$ defined by (7.26) satisfies

$$\varphi(x) = \begin{cases} \varphi_n(x), & \text{if } x \in W_n \subseteq W; \\[2mm] 0, & \text{if } x \in V \setminus W. \end{cases}$$

Hence, as we have mentioned above, φ is infinitely differentiable in V and we complete the construction of the required integral T.

This completes the proof of the problem. ▨

7.4   Problems related to the AC of a Function

> **Problem 7.9**
>
> *Rudin Chapter 7 Exercise 9.*

Proof. The function f discussed in §7.16 is called a singular function (see [8, p. 161] or [34, Definition 7.1.44, p. 170]). Let $0 < \alpha < 1$. Pick t so that $t^\alpha = 2$. Let $\delta_n = (\frac{2}{t})^n = 2^n t^{-n}$ for all $n \in \mathbb{N}$.

We remark that comprehensive materials about the relations between singular functions and Cantor sets can be found in [33, Chap. 4] and [34, Chap. 8]. In fact, our proof here relies heavily on the known results there.

We fix some notations first. Let $n \geq 0$. At the nth step, we have

$$E_n = \bigcup_{k=1}^{2^n} I_n(k), \tag{7.31}$$

where $I_n(1), I_n(2), \ldots, I_n(2^n)$ are the 2^n disjoint closed intervals, each of length $2^{-n}\delta_n = t^{-n}$. Next the deleted open segment between the intervals $I_n(k)$ and $I_n(k+1)$ is denoted by $J(k \cdot 2^{-n})$, where $k = 1, 2, \ldots, 2^n - 1$. Suppose that

$$D = \{m \cdot 2^{-n} \,|\, 0 \leq m \leq 2^n \text{ and } n \geq 0\}$$

which is the set of dyadic rationals in $[0, 1]$.

We claim that the function f discussed in §7.16 satisfies

$$f(x) = k \cdot 2^{-n} \tag{7.32}$$

for all $x \in J(k \cdot 2^{-n})$ and $k \cdot 2^{-n} \in D$. To this end, the hypothesis $x \in J(k \cdot 2^{-n})$ implies that there are k closed disjoint intervals $I_n(1), I_n(2), \ldots, I_n(k)$ on its left-hand side, so we derive from this and the fact [51, Eqn. (5), p. 145] that

$$f_n(x) = \int_0^x g_n(t)\,\mathrm{d}t = \sum_{m=1}^{k} \int_{I_n(m)} g_n(t)\,\mathrm{d}t = k \cdot 2^{-n}. \tag{7.33}$$

Since $x \notin E_n$ by the definition (7.31), we see from [51, Eqn. (6), p. 145] and the values (7.33) that

$$f(x) = \lim_{n \to \infty} f_n(x) = k \cdot 2^{-n}.$$

This proves our claim. By [51, Eqn. (3), p. 145], we have $E^c = \bigcup\limits_{n=1}^{\infty} E_n^c$. Since $x \notin E_n$, we have $x \in E_n^c \subseteq E^c$. Therefore, the f is well-defined on E^c by the values (7.32).

By [33, p. 84] or [34, p. 202], E is *nowhere dense* in $[0,1]$ so that E^c is a dense subset of $[0,1]$. Next, since f is uniformly continuous on $[0,1]$ and particularly on E^c, it follows from [49, Exercise 13, pp. 99, 100] that f has a unique continuous extension on $[0,1]$ which agrees with f on E^c. Hence the uniqueness shows that our f considered in §7.16 is the so-called **Lebesgue function associated with the Cantor-like set** E. Let $h : [0,1] \to \mathbb{R}$ be defined by

$$h(x) = x^{\alpha}.$$

Since $\alpha \in (0,1)$, it is immediate from basic calculus that h is strictly concave on $[0,1]$. Clearly, we know that

$$h(2^{-n}\delta_n) = (2^{-n}\delta_n)^{\alpha} = \frac{1}{t^{n\alpha}} = 2^{-n}, \tag{7.34}$$

where $n = 0, 1, 2, \ldots$. Thus it gives

$$h\big(2^{-(n-1)}\delta_{n-1}\big) = 2^{-(n-1)} = 2 \cdot 2^{-n} = 2h(2^{-n}\delta_n) > h(2^{-n+1}\delta_n)$$

for all $n = 1, 2, \ldots$. Since h is strictly increasing on $[0,1]$, we obtain $2^{-(n-1)}\delta_{n-1} > 2^{-n+1}\delta_n$ which is equivalent to $\delta_{n-1} > \delta_n$ for each $n = 1, 2, \ldots$. By the formula (7.34), we know that $h(\delta_0) = 1$ which implies $\delta_0 = 1$. In other words, the Cantor-like set E in §7.16 is constructed by means of our fixed concave function h. By [33, p. 97] or [34, Theorem 8.5.12, p. 205], it concludes that

$$|f(x) - f(y)| \le h\big(|x - y|\big) = |x - y|^{\alpha}$$

for all $x, y \in [0,1]$. Hence f belongs to $\operatorname{Lip}\alpha$ on $[0,1]$, completing the proof of the problem. ▨

Problem 7.10

Rudin Chapter 7 Exercise 10.

Proof. By Problem 5.11, there exists a constant $M > 0$ such that

$$|f(s) - f(t)| \le M|s - t| \tag{7.35}$$

for all $s, t \in [a, b]$. For every $\epsilon > 0$, if we take $\delta = \frac{\epsilon}{M}$, then for any n and any disjoint collection of segments $(\alpha_1, \beta_1), \ldots, (\alpha_n, \beta_n)$ in $[a, b]$ with

$$\sum_{i=1}^{n}(\beta_i - \alpha_i) < \frac{\epsilon}{M},$$

the inequality (7.35) implies that

$$\sum_{i=1}^{n} |f(\beta_i) - f(\alpha_i)| \le M \sum_{i=1}^{n} |\beta_i - \alpha_i| = M \sum_{i=1}^{n}(\beta_i - \alpha_i) < \epsilon.$$

By Definition 7.17, f is AC on $[a, b]$.

By Theorem 7.20, f' exists a.e. on $[a, b]$. Let f be differentiable at $x \in [a, b]$. Then we follow from the inequality (7.35) that

$$|f'(x)| = \lim_{h \to 0} \left| \frac{f(x + h) - f(x)}{h} \right| \le \lim_{h \to 0} M \left| \frac{(x + h) - x}{h} \right| = M,$$

i.e., $|f'| \le M$ a.e. on $[a, b]$. Hence $f' \in L^{\infty}$ and we have completed the proof of the problem. ▨

Problem 7.11

Rudin Chapter 7 Exercise 11.

Proof. Since f is AC on $[a, b]$, we get from Theorem 7.20 that

$$f(x) - f(a) = \int_a^x f'(t)\,dt \quad \text{and} \quad f(y) - f(a) = \int_a^y f'(t)\,dt$$

for any $a \le y \le x \le b$. Now their difference gives

$$f(x) - f(y) = \int_y^x f'(t)\,dt.$$

By Theorem 7.20 again, we have $f' \in L^1(m)$ so that Theorem 1.33 implies

$$|f(x) - f(y)| = \left| \int_y^x f'(t)\,dt \right| \le \int_y^x |f'(t)|\,dt. \tag{7.36}$$

Since $f' \in L^p$, we apply Theorem 3.5 (Hölder's Inequality) to the right-hand side of the inequality (7.36) to obtain

$$|f(x) - f(y)| \le \left\{ \int_y^x |f'(t)|^p\,dt \right\}^{\frac{1}{p}} \left\{ \int_y^x dt \right\}^{\frac{1}{q}} \le \|f'\|_p \cdot |x - y|^\alpha$$

and thus

$$\sup_{\substack{x,y \in [a,b] \\ x \ne y}} \frac{|f(x) - f(y)|}{|x - y|^\alpha} \le \|f'\|_p < \infty.$$

By the definition (see Problem 5.11), we conclude that $f \in \text{Lip}\,\alpha$, completing the proof of the problem. ∎

Problem 7.12

Rudin Chapter 7 Exercise 12.

Proof.

(a) Since φ is nondecreasing, we follow from [49, Theorems 4.29 & 4.30, pp. 95, 96] that $\varphi(x+)$ and $\varphi(x-)$ exist for every point x of $[a, b]$ and *the set of discontinuities of φ is at most countable*. Let this set be $A = \{x_1, x_2, \ldots\}$ and $x_1 < x_2 < \cdots$. We define $f : [a, b] \to \mathbb{R}$ by $f(x) = \sup_{t < x} \varphi(t)$. In fact, we have

$$f(x) = \begin{cases} \varphi(x), & \text{if } x \in [a, b] \setminus A; \\[2ex] \lim_{\substack{t \to 0 \\ t > 0}} \varphi(x_k - t) = \varphi(x_k-), & \text{if } x = x_k \text{ for some } k \in \mathbb{N}. \end{cases} \tag{7.37}$$

Let $x < y$ and $x, y \in [a, b]$. Then the nondecreasing property of φ ensures that

$$f(x) = \sup_{t < x} \varphi(t) \le \sup_{t < y} \varphi(t) = f(y)$$

which implies that f is nondecreasing on $[a, b]$.

Clearly, $\{x \in [a,b] \mid f(x) \neq \varphi(x)\} \subseteq A$ is at most countable. Given $x \in (a,b)$ and $\epsilon > 0$. By the definition (7.37), if $x = x_k$, then there is a $\delta > 0$ such that

$$|f(x_k) - \varphi(x_k - t)| < \epsilon \tag{7.38}$$

for all $t \in (0,\delta)$. By the construction of A, the number δ can be chosen so small such that $x_k - t \notin A$ for all $t \in (0,\delta)$, i.e., $x_k - t \in [a,b] \setminus A$ for all $t \in (0,\delta)$. Then $\varphi(x)$ is continuous at every $x_k - t$ and $\varphi(x_k - t) = f(x_k - t)$. Thus the inequality (7.38) reduces to

$$|f(x_k) - f(x_k - t)| < \epsilon$$

for all $t \in (0,\delta)$. In other words, f is a left-continuous function.

(b) Consider the function $g : [a,b] \to \mathbb{R}$ defined by

$$g(t) = \begin{cases} \varphi(t+) - \varphi(t-), & \text{if } t \in (a,b); \\[2mm] 0, & \text{if } t = a; \\[2mm] 0, & \text{if } t = b. \end{cases} \tag{7.39}$$

Since φ is nondecreasing, we have $g(t) \geq 0$ on $[a,b]$.

Firstly, we claim that

$$\sum_{t \in [a,b]} g(t) \leq \varphi(b) - \varphi(a). \tag{7.40}$$

Since $g(t) = 0$ on $[a,b] \setminus A$, we have

$$\sum_{t \in [a,b]} g(t) = \sum_{t \in A} g(t).$$

Let $A_N = \{x_1, x_2, \ldots, x_N\} \subseteq A$ for a positive integer N. Insert $\{y_0, y_1, y_2, \ldots, y_N\}$ into A_N in such a way that

$$a \leq y_0 \leq x_1 < y_1 < x_2 < \cdots < y_{N-1} < x_N \leq y_N \leq b. \tag{7.41}$$

Since φ is nondecreasing, we have

$$\varphi(x_k+) \leq \varphi(y_k) \leq \varphi(x_{k+1}-)$$

for $k = 1, 2, \ldots, N - 1$. For the end points x_1 and x_N, if $a \leq y_0 < x_1$, then $\varphi(x_1-) \geq \varphi(y_0) \geq \varphi(a)$; if $a = y_0 = x_1$, then we have obviously $\varphi(x_1) = \varphi(y_0) = \varphi(a)$. Similarly, if $x_N < y_N \leq b$, then $\varphi(x_N+) \leq \varphi(y_N) \leq \varphi(b)$; if $x_N = y_N = b$, then we certainly have $\varphi(x_N) = \varphi(y_N) = \varphi(b)$. Hence this analysis implies that

$$\varphi(b) - \varphi(a) \geq \varphi(y_N) - \varphi(y_0) \geq \sum_{k=1}^{N} [\varphi(x_k+) - \varphi(x_k-)] = \sum_{k=1}^{N} g(x_k). \tag{7.42}$$

Now our claim (7.40) follows directly by taking limit $N \to \infty$ in the inequality (7.42).

Secondly, we define $h : [a,b] \to \mathbb{R}$ by

$$h(x) = \sum_{t \in A} g(t)\chi_{(t,b]}(x) = \sum_{k=1}^{\infty} g(x_k)\chi_{(x_k,b]}(x). \tag{7.43}$$

By the claim (7.40) (or the second inequality in (7.42)) and the Weierstrass M-test, we know that the sum (7.43) actually converges uniformly on $[a, b]$. If $a \leq x < y \leq b$, then we have $\chi_{(t,b]}(x) \leq \chi_{(t,b]}(y)$ and this shows that

$$h(x) = \sum_{t \in A} g(t)\chi_{(t,b]}(x) \leq \sum_{t \in A} g(t)\chi_{(t,b]}(y) = h(y),$$

i.e., h is nondecreasing on $[a, b]$.

Thirdly, we need two lemmas about the properties of the function $F : [a, b] \to \mathbb{R}$ defined by $F(x) = f(x) - h(x)$.

Lemma 7.2

The function F is continuous on $[a, b]$.

Proof of Lemma 7.2. If $x \notin A$, then $f = \varphi$ is continuous at x. Furthermore, if $x_m < x < x_{m+1}$ for some $m \in \mathbb{N}$, then for all small enough δ, we have $(x - \delta, x + \delta) \subset (x_m, x_{m+1})$. In this case, we obtain

$$\lim_{\delta \to 0} h(x \pm \delta) = \lim_{\delta \to 0} \sum_{k=1}^{\infty} g(x_k)\chi_{(x_k,b]}(x \pm \delta) = \lim_{\delta \to 0} \sum_{k=1}^{m} g(x_k)\chi_{(x_k,b]}(x \pm \delta) = h(x),$$

i.e., h is continuous at x. By the definition, F is also continuous at x.

If $x \in A$, then $x = x_m$ for some $m \in \mathbb{N}$. On the one hand, the definition (7.37) gives

$$f(x_m+) = \lim_{\substack{\delta \to 0 \\ \delta > 0}} f(x_m + \delta) = \varphi(x_m+) \quad \text{and} \quad f(x_m-) = \varphi(x_m-).$$

On the other hand, the definitions (7.39) and (7.43) show

$$h(x_m+) = \lim_{\substack{\delta \to 0 \\ \delta > 0}} h(x_m + \delta) = \lim_{\substack{\delta \to 0 \\ \delta > 0}} \sum_{k=1}^{\infty} g(x_k)\chi_{(x_k,b]}(x_m + \delta) = \sum_{k=1}^{m} [\varphi(x_k+) - \varphi(x_k-)]$$

and

$$h(x_m-) = \lim_{\substack{\delta \to 0 \\ \delta < 0}} h(x_m + \delta) = \lim_{\substack{\delta \to 0 \\ \delta < 0}} \sum_{k=1}^{\infty} g(x_k)\chi_{(x_k,b]}(x_m + \delta) = \sum_{k=1}^{m-1} [\varphi(x_k+) - \varphi(x_k-)].$$

Consequently, they imply that

$$F(x_m+) = f(x_m+) - h(x_m+)$$
$$= \varphi(x_m+) - \sum_{k=1}^{m} [\varphi(x_k+) - \varphi(x_k-)]$$
$$= \varphi(x_m-) - \sum_{k=1}^{m-1} [\varphi(x_k+) - \varphi(x_k-)]$$
$$= f(x_m-) - h(x_m-)$$
$$= F(x_m-),$$

i.e., F is continuous at x_m and this ends the proof of the lemma. ∎

> **Lemma 7.3**
>
> The function F is nondecreasing on $[a, b]$.

Proof of Lemma 7.3. Let $x, y \in [a, b]$ be $x < y$. Suppose that $x_m < x \leq x_{m+1}$ for some $m \in \mathbb{N}$. If $x_m < x < y \leq x_{m+1}$, then the definition (7.43) gives

$$h(y) - h(x) = \sum_{k=1}^{\infty} g(x_k)\chi_{(x_k, b]}(y) - \sum_{k=1}^{\infty} g(x_k)\chi_{(x_k, b]}(x)$$
$$= \sum_{k=1}^{m} g(x_k)\chi_{(x_k, b]}(y) - \sum_{k=1}^{m} g(x_k)\chi_{(x_k, b]}(x)$$
$$= 0$$
$$\leq f(y) - f(x).$$

Suppose that

$$x_m < x \leq x_{m+1} < \cdots < x_{m+p} \leq y \tag{7.44}$$

for some $p \in \mathbb{N}$, then we get from the definition (7.39) that

$$h(y) - h(x) = \sum_{k=1}^{\infty} g(x_k)\chi_{(x_k, b]}(y) - \sum_{k=1}^{\infty} g(x_k)\chi_{(x_k, b]}(x)$$
$$= \sum_{k=1}^{m} g(x_k)\chi_{(x_k, b]}(y) + \sum_{k=m+1}^{m+p} g(x_k)\chi_{(x_k, b]}(y) - \sum_{k=1}^{m} g(x_k)\chi_{(x_k, b]}(x)$$
$$= \sum_{k=m+1}^{m+p} g(x_k)\chi_{(x_k, b]}(y)$$
$$\leq \sum_{k=m+1}^{m+p} [\varphi(x_k+) - \varphi(x_k-)]. \tag{7.45}$$

If we apply the same trick (the sequence (7.41) is replaced by the sequence (7.44) with a and b are replaced by x and y respectively) as in proving the inequality (7.42), we can easily deduce from the inequality (7.45) that $h(y) - h(x) \leq f(y) - f(x)$.

In conclusion, we have shown that

$$h(y) - h(x) \leq f(y) - f(x)$$

for all $x, y \in [a, b]$ with $x < y$. Equivalently, it means that

$$F(x) = f(x) - h(x) \leq f(y) - h(y) \leq F(y)$$

and so F is nondecreasing, as desired. ∎

Fourthly, we express F in terms of a finite and positive Borel measure. By Lemmas 7.2 and 7.3, we know that the function

$$g(x) = x + F(x)$$

is continuous and strictly increasing on $[a, b]$. Then g is one-to-one on $[a, b]$ so that its inverse function $g^{-1} : [g(a), g(b)] \to [a, b]$ is a continuous and bijective function on $[g(a), g(b)]$

by [49, Theorem 4.17, p. 90]. For every $E \subseteq [a,b]$ and $E \in \mathscr{B}$, since g^{-1} is continuous on $[g(a), g(b)]$, Definition 1.2(c) tells us that $g(E) = (g^{-1})^{-1}(E) \in \mathscr{B}$ so that we are able to define

$$\nu(E) = m\big(g(E)\big). \tag{7.46}$$

We have to show that ν is a positive Borel measure, but it follows exactly the same reasons as in [51, p. 147]. Furthermore, since m is a finite measure, the construction (7.46) implies that ν is also a finite measure. By Theorem 6.10 (The Lebesgue-Radon-Nikodym Theorem), there exists a unique pair of positive and finite Borel measures ν_{ac} and ν_{sc} such that

$$\nu_{\mathrm{ac}} \ll m, \quad \nu_{\mathrm{sc}} \perp m \quad \text{and} \quad \nu = \nu_{\mathrm{ac}} + \nu_{\mathrm{sc}}.$$

Now the function F and the measures ν and m are linked up in the following lemma:

> **Lemma 7.4**
>
> Let $\mu' = \nu - m$. Then μ' is finite and positive Borel measure. Furthermore, we have
>
> $$\mu'\big([a,x)\big) = F(x) - F(a). \tag{7.47}$$

Proof of Lemma 7.4. Clearly, the measure μ' is a complex Borel measure. The finiteness of μ' is also clear. If $x \in [a,b]$, then we deduce from the definition (7.46) that

$$\mu'\big([a,x)\big) = m\big(g[a,x)\big) - m([a,x)) = g(x) - g(a) - x + a = F(x) - F(a).$$

It remains to verify that μ' is positive. To this end, it is obvious from the definition and Lemma 7.3 that for $x, y \in [a,b]$ with $x < y$, we have

$$\nu\big((x,y)\big) = m\big(g\big((x,y)\big)\big) = g(y) - g(x) = y + F(y) - x - F(x) \geq y - x = m\big((x,y)\big).$$

Suppose that V is an open set in $[a,b]$. By [49, Exercise 29, p. 45], V can be expressed as a union of an at most countable collection of *disjoint* segments $\{V_i\}$. Since μ and m are measures, we have

$$\nu(V) = \nu\Big(\bigcup_{i=1}^{\infty} V_i\Big) = \bigcup_{i=1}^{\infty} \nu(V_i) \geq \bigcup_{i=1}^{\infty} m(V_i) = m\Big(\bigcup_{i=1}^{\infty} V_i\Big) = m(V). \tag{7.48}$$

Next, it is well-known that m is regular. Since $[a,b]$ is compact, it is σ-compact. Recall that ν is a finite positive Borel measure, so $\nu(K) < \infty$ for every compact set $K \subseteq [a,b]$. Therefore, ν is regular by Theorem 2.18. Thus if $E \in \mathscr{B}$ and $E \subseteq [a,b]$, then Theorem 2.14(c) implies that given $\epsilon > 0$, there exists an open set V containing E such that

$$\nu(V) < \nu(E) + \epsilon. \tag{7.49}$$

Combining the inequalities (7.48) and (7.49), we obtain

$$\nu(E) + \epsilon > \nu(V) \geq m(V) \geq m(E).$$

Since ϵ is arbitrary, we establish the fact that $\nu(E) \geq m(E)$ and then $\mu'(E) \geq 0$ for all $E \in \mathscr{B}$ and $E \subseteq [a,b]$. This proves the positivity of the measure μ' and ends the proof of Lemma 7.4.

We come to the final step of the proof of part (b). Recall that a complex Borel measure μ'' is called a **discrete measure** if there is a countable set $\{x_j\} \subset \mathbb{R}$ and complex numbers c_j such that $\sum |c_j| < \infty$ and $\mu'' = \sum c_j \delta_{x_j}$, where δ_x is the point mass at x, see [22, p. 106] or [51, Example 1.20(b), p. 17]. Now if we let

$$\mu''(E) = \sum_{k=1}^{\infty} g(x_k) \delta_{x_k}(E) \geq 0$$

for any $E \subseteq [a, b]$ and $E \in \mathscr{B}$, then it follows from the fact (7.40) that μ'' is finite. Besides, we deduce from this and the definition (7.43) easily that

$$\mu''\big([a, x)\big) = \sum_{k=1}^{\infty} g(x_k) \delta_{x_k}\big([a, x)\big) = \sum_{k=1}^{\infty} g(x_k) \chi_{(x_k, b]}(x) = h(x) - h(a). \tag{7.50}$$

Therefore, if we further define $\mu = \mu' + \mu''$, then μ must be a finite positive Borel measure on $[a, b]$ and it yields from the representations (7.47) and (7.50) that

$$\mu([a, x)) = \mu'([a, x)) + \mu''([a, x)) = F(x) - F(a) + h(x) - h(a) = f(x) - f(a). \tag{7.51}$$

This proves that μ is the desired measure.

(c) By Theorem 6.10 (The Lebesgue-Radon-Nikodym Theorem), we have

$$\mu = \mu_{\text{ac}} + \mu_{\text{sc}},$$

where $\mu_{\text{ac}} \ll m$ and $\mu_{\text{sc}} \perp m$. Suppose that ω is the Radon-Nikodym derivative of μ_{ac} with respect to m. Since $\omega \in L^1(m)$ on $[a, b]$, almost every $x \in [a, b]$ is a Lebesgue point of ω by Theorem 7.7.

Let $x \in (a, b)$ be a Lebesgue point of ω. Define $r = \min(x - a, b - x)$ and we can choose $\{r_i\} \subseteq (0, r)$ such that $r_i \to 0$ as $i \to \infty$. Then the two sequences of sets $\{E_i\}$ and $\{E_i'\}$ given by

$$E_i = [x, x + r_i) \quad \text{and} \quad E_i' = [x - r_i, x)$$

satisfy the inequalities

$$m(E_i) = r_i \geq \frac{1}{2} \cdot 2r_i = \frac{1}{2} m\big((x - r_i, x + r_i)\big) \quad \text{and} \quad m(E_i') \geq \frac{1}{2} m\big((x - r_i, x + r_i)\big)$$

for all $i = 1, 2, \ldots$. In other words, they mean that $\{E_i\}$ and $\{E_i'\}$ shrink to x nicely. Since μ is a complex Borel measure on $[a, b]$ and $d\mu = \omega \, dm + d\mu_s$, Theorem 7.14 and the formula (7.51) imply that

$$f_+'(x) = \lim_{i \to \infty} \frac{f(x + r_i) - f(x)}{r_i} = \lim_{i \to \infty} \frac{\mu\big([x, x + r_i)\big)}{m\big([x, x + r_i)\big)} = \omega(x) \tag{7.52}$$

a.e. $[m]$ on $[a, b]$. Similarly, we have

$$f_-'(x) = \lim_{i \to \infty} \frac{f(x - r_i) - f(x)}{-r_i} = \lim_{i \to \infty} \frac{\mu\big([x - r_i, x)\big)}{m\big([x - r_i, x)\big)} = \omega(x) \tag{7.53}$$

a.e. $[m]$ on $[a, b]$. Now the results (7.52) and (7.53) definitely give $f'(x)$ exists a.e. $[m]$ on $[a, b]$ and $f' = \omega \in L^1(m)$ on $[a, b]$.

Finally, since $d\mu_{\text{ac}} = \omega \, dm$, we have

$$\mu_{\text{ac}}\big([a, x)\big) = \int_a^x \omega \, dm = \int_a^x f' \, dm = \int_a^x f'(t) \, dt$$

so that

$$f(x) - f(a) = \mu\big([a,x)\big) = \mu_{\mathrm{ac}}\big([a,x)\big) + \mu_{\mathrm{sc}}\big([a,x)\big) = \int_a^x f'(t)\,\mathrm{d}t + s(x),$$

where $s(x) = \mu_{\mathrm{sc}}\big([a,x)\big)$ for all $x \in [a,b]$. Now $\mu_{\mathrm{sc}}\big([a,x)\big)$ is positive finite because μ is positive finite, so s is nondecreasing on $[a,b]$. Since $f' \in L^1(m)$ on $[a,b]$, Theorem 7.11 shows that

$$s'(x) = [f(x) - f(a)]' - \left(\int_a^x f'(t)\,\mathrm{d}t \right)' = f'(x) - f'(x) = 0$$

a.e. $[m]$ on (a,b).

(d) By Theorem 6.8(e), $\mu \perp m$ if and only if $\mu_{\mathrm{ac}} = 0$ if and only if

$$\int_E f'\,\mathrm{d}m = 0 \tag{7.54}$$

for every $E \in \mathscr{B}$ and $E \subseteq [a,b]$. Recall that $f' \in L^1(m)$. By Theorem 1.39(b), the result (7.54) is equivalent to $f'(x) = 0$ a.e. $[m]$ on $[a,b]$.

Similarly, $\mu \ll m$ if and only if $\mu_{\mathrm{sc}} = 0$ if and only if $s = 0$ on $[a,b]$ if and only if

$$f(x) - f(a) = \int_a^x f'\,\mathrm{d}m \tag{7.55}$$

for all $x \in [a,b]$. As Rudin pointed out in the discussion preceding Definition 7.17 (on p. 145), one may apply Theorem 6.11 to show that f is AC on $[a,b]$ if the formula (7.55) holds. Conversely, if f is AC on $[a,b]$, then f is automatically continuous on $[a,b]$ and since f is nondecreasing, Theorem 7.18 implies that the formula (7.55) holds.

(e) Suppose that $f'(p)$ exists at $p \in (a,b) \setminus A$. Then f is continuous at p and the definition (7.37) implies the existence of a $\delta > 0$ such that

$$f(x) = \varphi(x)$$

for all $x \in (p - \delta, p + \delta)$. If $h \in (0,\delta)$, then $p + h \in (p - \delta, p + \delta)$ and so

$$\left| \frac{\varphi(p+h) - \varphi(p)}{h} - f'(p) \right| = \left| \frac{f(p+h) - f(p)}{h} - f'(p) \right|. \tag{7.56}$$

By taking $h \to 0+$ in the expression (7.56), we get

$$|\varphi'_+(p) - f'(p)| = |f'_+(p) - f'(p)| = 0,$$

i.e., $\varphi'_+(p) = f'(p)$. The other side $\varphi'_-(p) = f'(p)$ can be done similarly. Thus we conclude that $\varphi'(p) = f'(p)$. Since $f'(x)$ exists a.e. $[m]$ on $[a,b]$, we have $\varphi'(x) = f'(x)$ a.e. $[m]$ on $[a,b]$.

Hence we have completed the proof of the problem. ∎

> **Remark 7.3**
>
> By the result of part (b) in Problem 7.12, we have
>
> $$\mu = \mu' + \mu'' = v - m + \mu'' = \underbrace{(\nu_a - m)}_{\substack{\text{absolutely}\\ \text{continuous}}} + \overbrace{\nu_s}^{\substack{\text{singular}\\ \text{continuous}}} + \underbrace{\mu''}_{\text{discrete}}.$$
>
> See also the decomposition in [22, Theorem 3.35, p. 106].

Problem 7.13

Rudin Chapter 7 Exercise 13.

Proof. By the definition,

$$BV = \left\{ f : [a,b] \to \mathbb{C} \;\middle|\; F(b) = \sup \sum_{i=1}^{N} |f(t_i) - f(t_{i-1})| < \infty \right\},$$

where the supremum is taken over all N and over all choices of $\{t_i\}$ such that

$$a = t_0 < t_1 < \cdots < t_N = b. \tag{7.57}$$

(a) Let $f : [a,b] \to \mathbb{C}$ be a bounded nondecreasing function. Then we have

$$\sum_{i=1}^{N} |f(t_i) - f(t_{i-1})| = \sum_{i=1}^{N} [f(t_i) - f(t_{i-1})] = f(b) - f(a) < \infty,$$

where $\{t_i\}$ satisfies the requirement (7.57). Hence we have $f \in BV$. If f is a bounded nonincreasing function, then we consider $-f$ which is a bounded nondecreasing function.

(b) Let $f \in BV$ be real. Define

$$f_1(x) = \frac{F(x) + f(x)}{2} \quad \text{and} \quad f_2(x) = \frac{F(x) - f(x)}{2}, \tag{7.58}$$

where F is given by [51, Eqn. (1), p. 147]. It is easy to see that $f = f_1 - f_2$, so it remains to prove that they are bounded and monotonic.

Suppose first that f is nondecreasing. By Theorem 7.19, $F + f$ and $F - f$ are nondecreasing,[e] so are f_1 and f_2. By the definition of the space BV, we have $0 \le F(x) \le F(b) < \infty$ on $[a,b]$. If $x \in [a,b]$, then the triangle inequality gives

$$|f(x)| \le |f(a)| + |f(x) - f(a)| \le |f(a)| + F(b).$$

Thus f is bounded on $[a,b]$. By the definition (7.59), f_1 and f_2 are also bounded on $[a,b]$. If f is nonincreasing, then the function $-f$ is nondecreasing but the functions $f_1 = \frac{F+f}{2} = \frac{F-(-f)}{2}$ and $f_2 = \frac{F-f}{2} = \frac{F+(-f)}{2}$ remain nondecreasing.[f]

(c) Suppose that $f \in BV$ and it is left-continuous. It suffices to prove that F is left-continuous at p, i.e., if $a < p \le b$ and $\epsilon > 0$, then there is a $\delta > 0$ so that

$$|F(p) - F(p - h)| < \epsilon \tag{7.59}$$

whenever $0 < h < \delta$. To start with, we define

$$F(x - h; x) = \sup \sum_{i=1}^{N} |f(t_i) - f(t_{i-1})|,$$

where $x - h = t_0 < t_1 < \cdots < t_N = x$. We claim that

$$F(x) - F(x - h) = F(x - h; x)$$

---

[e] The conclusion of this assertion *does not* depend on the assumption that f is AC on $[a,b]$.
[f] If we write $f = (-f_2) - (-f_1)$, then $-f_1$ and $-f_2$ are nonincreasing.

for every small $h > 0$ such that $x - h \in [a, b]$. By the definition, if $\{t_i\}$ and $\{s_j\}$ are partitions of $[a, x - h]$ and $[x - h, x]$ respectively, then $\{t_i\} \cup \{s_j\}$ is certainly a partition of $[a, b]$ and so

$$F(x) \geq F(x - h) + F(x - h; x). \tag{7.60}$$

On the other hand, let $\{t_0, t_1, \ldots, t_N\}$ be a partition of $[a, x]$. If $t_M < x - h < t_{M+1}$, then we may insert a number t' such that

$$a = t_0 < t_1 < \cdots \leq t_M < t' = x - h < t_{M+1} < \cdots < t_N = x.$$

Clearly, the triangle inequality implies

$$\sum_{i=1}^{N} |f(t_i) - f(t_{i-1})| \leq \left[\sum_{i=1}^{M} |f(t_i) - f(t_{i-1})| + |f(t') - f(t_M)| \right]$$

$$+ \left[|f(t_{M+1}) - f(t')| + \sum_{i=M+2}^{N} |f(t_i) - f(t_{i-1})| \right]$$

$$\leq F(x - h) + F(x - h; x). \tag{7.61}$$

Since the partition $\{t_0, t_1, \ldots, t_N\}$ is *arbitrary*, the inequality (7.61) asserts that

$$F(x) \leq F(x - h) + F(x - h; x). \tag{7.62}$$

Hence our desired claim follows immediately from the inequalities (7.60) and (7.62).

Since f is left-continuous at $p \in (a, b]$, given $\epsilon > 0$, there exists a $\delta > 0$ such that $h \in (0, \delta)$ implies

$$|f(p) - f(p - h)| < \frac{\epsilon}{2}.$$

For this same ϵ, we can find a partition $\{t_i\}$ of $[a, p]$, say

$$a = t_0 < t_1 < \cdots < t_N = p,$$

such that

$$F(a; p) - \frac{\epsilon}{2} < \sum_{i=1}^{N} |f(t_i) - f(t_{i-1})| \tag{7.63}$$

Since adding more points to the partition $\{t_i\}$ results in an increase of the summation (7.63), we may assume that $t_N - t_{N-1} = h \in (0, \delta)$ and this implies that

$$|f(t_N) - f(t_{N-1})| = |f(t_N) - f(t_N - h)| = |f(p) - f(p - h)| < \frac{\epsilon}{2}$$

and the inequality (7.63) becomes

$$F(a; p) - \frac{\epsilon}{2} < \frac{\epsilon}{2} + \sum_{i=1}^{N-1} |f(t_i) - f(t_{i-1})| \leq \frac{\epsilon}{2} + F(a; p - h)$$

or equivalently,

$$F(a; p) - F(a; p - h) < \epsilon.$$

However, we note that

$$0 \leq F(p) - F(p - h) = F(p - h; p) = F(a; p) - F(a; p - h) < \epsilon$$

which is exactly the expected result (7.59). Hence F and then f_1 as well as f_2 are left-continuous on $[a, b]$.

(d) By parts (b) and (c), $f = f_1 - f_2$, where f_1 and f_2 are left-continuous and nondecreasing on $[a, b]$. Now Problem 7.12(b) shows that there exist positive and finite Borel measures λ and ν on $[a, b]$ such that

$$f_1(x) - f_1(a) = \lambda\big([a, x)\big) \quad \text{and} \quad f_2(x) - f_2(a) = \nu\big([a, x)\big),$$

where $x \in [a, b]$. Then the difference $\mu = \lambda - \nu$ is easily seen to be a Borel measure on $[a, b]$ satisfying the equation in the question. In fact, the finiteness of λ and ν imply the finiteness of μ. (However, μ may *not* be positive.)

As Theorem 6.10 (The Lebesgue-Radon-Nikodym Theorem) tells us that there exists a unique pair of positive and finite measures $\mu_a \ll m$ and $\mu_s \perp m$ such that $\mu = \mu_a + \mu_s$. Furthermore, there is a unique $h \in L^1(m)$ on $[a, b]$ such that

$$\mu_a(E) = \int_E h \, dm \tag{7.64}$$

for every $E \in \mathscr{B}$.

If $\mu \ll m$, then Proposition 6.8(e) implies that $\mu_s = 0$ and thus the μ_a in the representation (7.64) can be replaced by μ. Now we put $E = [a, x)$ into the representation and use the result $f(x) - f(a) = \mu([a, x))$ to obtain

$$f(x) - f(a) = \mu([a, x)) = \int_{[a,x)} h \, dm = \int_a^x h \, dm.$$

Since $h \in L^1(m)$ on $[a, b]$, we follow from [9, Theorem 5.3.6, p. 339] or [22, Theorem 3.35, p. 106] that f is AC on $[a, b]$.[g] Conversely, suppose that f is AC on $[a, b]$. By Theorem 7.19, F, f_1 and f_2 are AC and nondecreasing on $[a, b]$. By Problem 7.12(d), we conclude that

$$\lambda \ll m \quad \text{and} \quad \nu \ll m.$$

Since $\mu = \lambda - \nu$, Proposition 6.8(c) ensures that $\mu \ll m$.

(e) By part (b), we have $f = f_1 - f_2$, where f_1 and f_2 are nondecreasing. By Problem 7.12(e), f_1' and f_2' exist a.e. $[m]$ on $[a, b]$. By Problem 7.12(c), we further know that $f_1', f_2' \in L^1(m)$ on $[a, b]$. Hence we conclude that f' exists a.e. $[m]$ on $[a, b]$ and $f' \in L^1(m)$.

We complete the proof of the problem. ▨

Problem 7.14

Rudin Chapter 7 Exercise 14.

Proof. Suppose that f and g are AC on $[a, b]$. By Definition 7.17, they are continuous on $[a, b]$, so the Extreme Value Theorem ([49, Theorem 4.16, p. 89]) ensures that we may assume that there exists a $M > 0$ such that

$$|f(x)| \le M \quad \text{and} \quad |g(x)| \le M \tag{7.65}$$

on $[a, b]$. Given $\epsilon > 0$, there corresponds a $\delta > 0$ such that

$$\sum_{k=1}^n |f(\beta_k) - f(\alpha_k)| < \frac{\epsilon}{2M} \quad \text{and} \quad \sum_{k=1}^n |g(\beta_k) - g(\alpha_k)| < \frac{\epsilon}{2M} \tag{7.66}$$

---

[g]These theorems *do not* need the assumptions of Theorem 7.18: f is continuous and nondecreasing on $[a, b]$.

for every n and any disjoint collection of segments $(\alpha_1, \beta_1), \ldots, (\alpha_n, \beta_n)$ in $[a, b]$ with

$$\sum_{k=1}^{n} (\beta_k - \alpha_k) < \delta.$$

It is clear that

$$\sum_{k=1}^{n} |f(\beta_k)g(\beta_k) - f(\alpha_k)g(\alpha_k)| = \sum_{k=1}^{n} |f(\beta_k)g(\beta_k) - f(\beta_k)g(\alpha_k) + f(\beta_k)g(\alpha_k) - f(\alpha_k)g(\alpha_k)|$$

$$\le \sum_{k=1}^{n} |f(\beta_k)g(\beta_k) - f(\beta_k)g(\alpha_k)| + \sum_{k=1}^{n} |f(\beta_k)g(\alpha_k) - f(\alpha_k)g(\alpha_k)|$$

$$\le M \sum_{k=1}^{n} |g(\beta_k) - g(\alpha_k)| + M \sum_{k=1}^{n} |f(\beta_k) - f(\alpha_k)|. \tag{7.67}$$

By substituting the inequalities (7.66) into the inequality (7.67), we gain

$$\sum_{k=1}^{n} |f(\beta_k)g(\beta_k) - f(\alpha_k)g(\alpha_k)| < \epsilon$$

which implies that fg is AC on $[a, b]$.

Now Theorem 7.20 shows that f, g and fg are differentiable a.e. on $[a, b]$ and $f', g', (fg)' \in L^1$ on $[a, b]$. Apply Theorem 7.20 to fg, we have

$$f(b)g(b) - f(a)g(a) = \int_a^b \big(f(x)g(x)\big)' \, \mathrm{d}x. \tag{7.68}$$

By the bounds (7.65), we see that $f'g, fg' \in L^1$ on $[a, b]$. Furthermore, since $(fg)' = f'g + fg'$ a.e. on $[a, b]$, we deduce from the formula (7.68) that

$$\int_a^b f'(x)g(x) \, \mathrm{d}x = f(b)g(b) - f(a)g(a) - \int_a^b f(x)g'(x) \, \mathrm{d}x,$$

completing the proof of the problem. ∎

Problem 7.15

Rudin Chapter 7 Exercise 15.

Proof. It is well-known [49, Example 5.6(b), p. 106] that the function $f : \mathbb{R} \to \mathbb{R}$ defined by

$$f(x) = \begin{cases} x^2 \sin \dfrac{1}{x}, & \text{if } x \neq 0; \\ \\ 0, & \text{otherwise} \end{cases}$$

has derivative given by

$$f'(x) = \begin{cases} 2x \sin \dfrac{1}{x} - \cos \dfrac{1}{x}, & \text{if } x \neq 0; \\ \\ 0, & \text{otherwise.} \end{cases}$$

Thus f is differentiable at all points but f' is *not* a continuous function. However, such function oscillates around 0 so that it is not monotonic and we have to "modify" it. To do this, we notice that since

$$\lim_{x \to \pm\infty} x \sin\frac{1}{x} = \lim_{x \to \pm\infty} \frac{\sin(\frac{1}{x})}{\frac{1}{x}} = 1,$$

$f'(x)$ is bounded on $\mathbb{R}$. Let $M > 0$ be a bound of f' on $\mathbb{R}$, i.e.,

$$-M \leq f'(x) \leq M \tag{7.69}$$

for all $x \in \mathbb{R}$. We consider the function $F : \mathbb{R} \to \mathbb{R}$ given by

$$F(x) = f(x) + Mx.$$

Now $F'(x) = f'(x) + M$ which exists *finitely* on $\mathbb{R}$. In addition, the inequalities (7.69) implies that

$$F'(x) \geq -M + M \geq 0$$

for all $x \in \mathbb{R}$. Hence [49, Theorem 5.11, p. 108] ensures that F is monotonically increasing on $\mathbb{R}$. Since f' is not continuous at 0, F' is also discontinuous at 0 and thus F is our desired function. This completes the proof of the problem. ∎

> **Problem 7.16**
>
> *Rudin Chapter 7 Exercise 16.*

Proof. We follow the hint given by Rudin. Since m is outer regular (see Theorem 2.14(c) and Theorem 2.20(b)) and $m(E) = 0$, we can find a sequence $\{V_n\}$ of open subsets of $\mathbb{R}$ such that $E \subseteq \cdots \subseteq V_2 \subseteq V_1$ and $m(V_n) < 2^{-n}$. This construction gives

$$E \subseteq \bigcap_{n=1}^{\infty} V_n. \tag{7.70}$$

Consider $F : [a, b] \to \mathbb{R}$ defined by

$$F(x) = \sum_{n=1}^{\infty} \chi_{V_n}(x) \geq 0.$$

Since each χ_{V_n} is measurable, it follows from Theorem 1.27 that

$$0 \leq \int_a^b F(x)\,\mathrm{d}x = \sum_{n=1}^{\infty} \int_a^b \chi_{V_n}(x)\,\mathrm{d}x = \sum_{n=1}^{\infty} \int_{V_n} \mathrm{d}x = \sum_{n=1}^{\infty} m(V_n) < \sum_{n=1}^{\infty} 2^{-n} < \infty.$$

In other words, $F \in L^1([a, b])$. Next we define $f : [a, b] \to \mathbb{R}$ by

$$f(x) = \int_a^x F(t)\,\mathrm{d}t \geq 0 \tag{7.71}$$

which is clearly nondecreasing. By Theorem 7.11, f is differentiable a.e. on $[a, b]$ and then we follow from Theorem 7.18 that f is AC on $[a, b]$.

Now it remains to show that $f'(x) = \infty$ on E. To see this, if $x \in E$, then the set relation (7.70) tells us that $x \in V_n$ for all $n \in \mathbb{N}$. Since each V_n is open, there exists a $\delta_n > 0$ such that

$$(x - \delta_n, x + \delta_n) \subseteq V_n$$

for all $n \in \mathbb{N}$. In fact, we may assume that $\{\delta_n\}$ is a decreasing sequence of positive numbers. Pick $y \in (x - \delta_n, x + \delta_n)$ so that

$$|f(y) - f(x)| = \left| \int_x^y F(t)\, \mathrm{d}t \right| \geq \left| \int_x^y \sum_{k=1}^n [\chi_{V_1}(t) + \cdots + \chi_{V_n}(t)]\, \mathrm{d}t \right| \tag{7.72}$$

Since $V_n \subseteq V_{n-1} \subseteq \cdots \subseteq V_2 \subseteq V_1$ and $t \in [x, y] \subseteq (x - \delta_n, x + \delta_n) \subseteq V_n,$[h] the inequality (7.72) becomes

$$|f(y) - f(x)| \geq n|y - x|.$$

Notice that $y \to x$ if and only if $n \to \infty$, so we have

$$|f'(x)| = \lim_{y \to x} \left| \frac{f(y) - f(x)}{y - x} \right| = \infty.$$

By the definition (7.71), we conclude that $f'(x) = \infty$ on E. We have completed the proof of the problem. ∎

7.5 Miscellaneous Problems on Differentiation

Problem 7.17

Rudin Chapter 7 Exercise 17.

Proof. We prove the assertions one by one:

- **μ is a Borel measure.** Suppose that $\{E_i\}$ is a collection of mutually disjoint Borel subsets of $\mathbb{R}^k$. Let $E = \bigcup_{i=1}^{\infty} E_i$. Since $\mu_n(E_i) \geq 0$ for all $i \in \mathbb{N}$, we know from [49, Exericse 3, p. 196] that

$$\mu(E) = \sum_{n=1}^{\infty} \mu_n(E) = \sum_{n=1}^{\infty} \sum_{i=1}^{\infty} \mu_n(E_i) = \sum_{i=1}^{\infty} \sum_{n=1}^{\infty} \mu_n(E_i) = \sum_{i=1}^{\infty} \mu(E_i).$$

  Clearly, we have $\mu(E) \geq 0$ for all $E \in \mathscr{B}$. By Definition 1.18(a), μ is a positive Borel measure on $\mathbb{R}^k$.

- **The relation between the Lebesgue decompositions of the μ_n and that of μ.** Since m is a positive σ-finite measure on $\mathscr{B}$, it follows from Theorem 6.10 (The Lebesgue-Radon-Nikodym Theorem) that

$$\mu = \mu_a + \mu_s \quad \text{and} \quad \mu_n = \mu_{n,a} + \mu_{n,s},$$

  where

$$\mu_a, \mu_{n,a} \ll m \quad \text{and} \quad \mu_s, \mu_{n,s} \perp m \tag{7.73}$$

  for all $n \in \mathbb{N}$. Let

$$\lambda_a = \sum_{n=1}^{\infty} \mu_{n,a} \quad \text{and} \quad \lambda_s = \sum_{n=1}^{\infty} \mu_{n,s}.$$

---

[h] Or $t \in [y, x]$.

Since μ and μ_n are positive, so are $\mu_a, \mu_{n,a}, \mu_s$ and $\mu_{n,s}$ and it deduces from the first assertion that both λ_a and λ_s are positive Borel measures of $\mathbb{R}^k$. Furthermore, we have

$$\mu = \sum_{n=1}^{\infty} \mu_n = \sum_{n=1}^{\infty} (\mu_{n,a} + \mu_{n,s}) = \sum_{n=1}^{\infty} \mu_{n,a} + \sum_{n=1}^{\infty} \mu_{n,s} = \lambda_a + \lambda_s.$$

By Theorem 6.8 and the relations (7.73), we assert that $\lambda_a \ll m$ and $\lambda_s \perp m$. Recall that the Lebesgue decomposition of μ relative to m is *unique*, it must be true that

$$\mu_a = \lambda_a = \sum_{n=1}^{\infty} \mu_{n,a} \quad \text{and} \quad \mu_s = \lambda_s = \sum_{n=1}^{\infty} \mu_{n,s}.$$

- **The truth of the formula.** By Theorem 6.10 (The Lebesgue-Radon-Nikodym Theorem) again, there exists a unique $h \in L^1(m)$ such that

$$\mu_a(E) = \int_E h \, dm \tag{7.74}$$

for all $E \in \mathscr{B}$. Similarly, for each $n \in \mathbb{N}$, there exists a unique $h_n \in L^1(m)$ such that

$$\mu_{n,a}(E) = \int_E h_n \, dm \tag{7.75}$$

for all $E \in \mathscr{B}$. Since $\mu_a, \mu_{n,a} \ll m$ for all $n \in \mathbb{N}$, we observe from Theorem 7.8 that

$$D\mu_a = h \quad \text{and} \quad D\mu_{n,a} = h_n \tag{7.76}$$

a.e. $[m]$ for all $n \in \mathbb{N}$. Since $\mu_s \perp m$ and $\mu_{n,s} \perp m$ for all $n \in \mathbb{N}$, Theorem 7.14 ensures that

$$D\mu_s = 0 \quad \text{and} \quad D\mu_{n,s} = 0 \tag{7.77}$$

a.e. $[m]$ for all $n \in \mathbb{N}$. Hence the two results (7.76) and (7.77) say that

$$D\mu = h \quad \text{and} \quad D\mu_n = h_n$$

a.e. $[m]$ for all $n \in \mathbb{N}$.

Recall from the second assertion and the application of the integrals (7.74) and (7.75) that

$$\int_E h \, dm = \mu_a(E) = \sum_{n=1}^{\infty} \mu_{n,a}(E) = \sum_{n=1}^{\infty} \int_E h_n \, dm \tag{7.78}$$

for all $E \in \mathscr{B}$. Since $h_n \in L^1(m)$, each h_n is measurable and Theorem 1.27 implies that the expression (7.78) can be further reduced to

$$\int_E h \, dm = \int_E \sum_{n=1}^{\infty} h_n \, dm$$

for all $E \in \mathscr{B}$, or equivalently,

$$\int_E \left(h - \sum_{n=1}^{\infty} h_n \right) dm = 0 \tag{7.79}$$

for all $E \in \mathscr{B}$.

Since $\mu_a(E) \geq \sum_{n=1}^{N} \mu_{n,a}(E)$ for every $N \in \mathbb{N}$, the integrals (7.74) and (7.75) show that

$$\int_E \left(h - \sum_{n=1}^{N} h_n \right) dm \geq 0 \tag{7.80}$$

for every $E \in \mathscr{B}$. Assume that $h < \sum_{n=1}^{N} h_n$ on E' with $m(E') > 0$. For this Borel set E', we derive from Proposition 1.24(a) that

$$\int_{E'} \left(h - \sum_{n=1}^{N} h_n \right) dm < 0$$

which contradicts the result (7.80). Consequently, we always have $h \geq \sum_{n=1}^{N} h_n$ a.e. $[m]$ on $\mathbb{R}^k$ and since N is arbitrary, this implies that

$$h - \sum_{n=1}^{\infty} h_n \geq 0$$

a.e. $[m]$ on $\mathbb{R}^k$. Now we apply Theorem 1.39(a) to the integral (7.79) to conclude that

$$h(\mathbf{x}) = \sum_{n=1}^{\infty} h_n(\mathbf{x})$$

a.e. $[m]$ on $\mathbb{R}^k$, then this shows the formula

$$D\mu(\mathbf{x}) = \sum_{n=1}^{\infty} (D\mu_n)(\mathbf{x})$$

a.e. $[m]$ on $\mathbb{R}^k$.

- **A corresponding theorem for a sequence $\{f_n\}$ of positive nondecreasing functions on $\mathbb{R}$.** Suppose that $f(x) = \sum_{n=1}^{\infty} f_n(x)$ converges on $\mathbb{R}$. We claim that $f'(x)$ exists and

$$f'(x) = \sum_{n=1}^{\infty} f_n'(x) \tag{7.81}$$

a.e. $[m]$ on $\mathbb{R}$. By the definition, it is trivial that f is *nondecreasing* and $0 < f(x) < \infty$ on $\mathbb{R}$. Let $[a, b]$ be a compact interval of $\mathbb{R}$. Suppose that

$$E = \{x \in (a, b) \,|\, f'(x) \text{ and } f_n'(x) \text{ exist}\}.$$

By Problem 7.12(e), we see that f and each f_n are differentiable a.e. $[m]$ on $[a, b]$ so that $[a, b] \setminus E$ is of measure zero.

Let $x \in E$ and $h > 0$ be sufficiently small so that $x + h \in [a, b]$. Then for any $N \in \mathbb{N}$, we have

$$\frac{f(x+h) - f(x)}{h} \geq \sum_{n=1}^{N} \frac{f_n(x+h) - f_n(x)}{h}$$

so if we let $h \to 0$, then we have

$$f'(x) \geq \sum_{n=1}^{N} f'_n(x). \tag{7.82}$$

Now we take $N \to \infty$ in the inequality (7.82) to conclude that

$$f'(x) \geq \sum_{n=1}^{\infty} f'_n(x) \tag{7.83}$$

a.e. $[m]$ on $[a, b]$. Recall that $x \in E$ so that the inequality (7.83) ensures that the series of its right-hand side is convergent a.e. $[m]$ on $[a, b]$. Consequently, it implies that

$$\lim_{n \to \infty} f'_n(x) = 0 \tag{7.84}$$

a.e. $[m]$ on $[a, b]$.

Next we define

$$s_N(x) = \sum_{n=1}^{N} f_n(x).$$

Since $s_N(b) \to f(b)$ as $N \to \infty$, there is a sequence $\{N_k\} \subseteq \mathbb{N}$ such that

$$0 \leq f(b) - s_{N_k}(b) < \frac{1}{2^k}. \tag{7.85}$$

Obviously, we have

$$g_k(x) = f(x) - s_{N_k}(x) = \sum_{n=N_k+1}^{\infty} f_n(x),$$

so each g_k is a *positive nondecreasing* function on $[a, b]$. To apply the first part of the proof, we have to show that the series $\sum_{k=1}^{\infty} g_k(x)$ converges on $[a, b]$. Combining this fact and the inequality (7.85), we obtain

$$0 \leq g_k(x) < \frac{1}{2^k}$$

on $[a, b]$. As a result, the series

$$\sum_{k=1}^{\infty} g_k = \sum_{k=1}^{\infty} (f - s_{N_k})$$

converges (uniformly) to $g(x)$ on $[a, b]$ by the Weierstrass M-test. Now we are able to apply the result (7.84) to the sequence $\{g_k\}$ of positive nondecreasing functions and conclude that

$$\lim_{k \to \infty} g_k(x) = \lim_{k \to \infty} [f'(x) - s'_{N_k}(x)] = 0$$

a.e. $[m]$ on $[a, b]$ or equivalently,

$$\lim_{k \to \infty} \sum_{n=1}^{N_k} f'_n(x) = f'(x) \tag{7.86}$$

a.e. $[m]$ on $[a, b]$. Since each f_n is nondecreasing on $[a, b]$, $f_n'(x) \geq 0$ for all $x \in E$. Hence the limit (7.86) implies that

$$f'(x) = \sum_{n=1}^{\infty} f_n'(x)$$

a.e. $[m]$ on $[a, b]$. Since any real number is contained in some compact interval $[a, b]$, our desired result (7.81) follows from this.

It completes the proof of the problem. ∎

Remark 7.4

The last assertion in Problem 7.17 is called **Fubini's Theorem on Differentiation**. See [33, Sec. C, pp. 527 – 529].

Problem 7.18

Rudin Chapter 7 Exercise 18.

Proof. First of all, we show that $\{\varphi_n\}$ is orthonormal on $[0, 1]$. Suppose that $m, n \in \mathbb{N}$ and $m \geq n$. If $m = n$, then $\varphi_n(t)\varphi_n(t) = \varphi_n^2(t) = 1$ and so

$$\int_0^1 \varphi_n(t)\varphi_m(t) \, \mathrm{d}t = 1.$$

If $m > n$, then we deduce from Theorem 7.26 (The Change-of-variables Theorem) that

$$\int_0^1 \varphi_n(t)\varphi_m(t) \, \mathrm{d}t = \int_0^1 \varphi_0(2^n t)\varphi_0(2^m t) \, \mathrm{d}t$$

$$= \frac{1}{2^n} \int_0^{2^n} \varphi_0(y)\varphi_0(2^{m-n} y) \, \mathrm{d}y$$

$$= \frac{1}{2^n} \sum_{i=1}^{2^{n-1}} \int_{2(i-1)}^{2i} \varphi_0(y)\varphi_0(2^{m-n} y) \, \mathrm{d}y. \tag{7.87}$$

Since $\varphi_0(y)$ is periodic with period 2, we have $\varphi_0(y) = 1$ on $[2i - 2, 2i - 1)$ and $\varphi_0(y) = -1$ on $[2i - 1, 2i)$. Therefore, we can reduce the expression (7.87) to

$$\int_0^1 \varphi_n(t)\varphi_m(t) \, \mathrm{d}t = \frac{1}{2^n} \sum_{i=1}^{2^{n-1}} \Big[\int_{2(i-1)}^{2i-1} \varphi_0(2^{m-n} y) \, \mathrm{d}y - \int_{2i-1}^{2i} \varphi_0(2^{m-n} y) \, \mathrm{d}y \Big]. \tag{7.88}$$

Note that the interval $[2^{m-n+1} i - 2^{m-n}, 2^{m-n+1} i)$ is of length 2^{m-n}, so the integrals on the right-hand side of the equation (7.88) are 0 and this gives

$$\int_0^1 \varphi_n(t)\varphi_m(t) \, \mathrm{d}t = 0$$

if $m \neq n$. By Definition 4.16, $\{\varphi_n\}$ is orthonormal on $[0, 1]$.

Since $\sum |c_n|^2 < \infty$, it yields from the Riesz-Fischer Theorem [49, Theorem 11.43, p. 330] that

$$f \sim \sum_{n=1}^{\infty} c_n \varphi_n \tag{7.89}$$

for some $f \in L^2([0,1])$, i.e., the series in (7.89) is the Fourier series of f.

Consider $a = j \cdot 2^{-N}$, $b = (j+1) \cdot 2^{-N}$, $a < t < b$ and $s_N = c_1\varphi_1 + \cdots + c_N\varphi_N$, where $j = 0, 1, 2, \ldots$. On the one hand, if $1 \le k \le N$, then $a \le t \le b$ implies that

$$j \cdot 2^{k-N} \le 2^k t \le (j+1)2^{k-N} \quad \text{and} \quad m\big([2^k a, 2^k b]\big) = 2^{k-N} \le 1$$

so that each $\varphi_k(t) = \varphi_0(2^k t)$ (of course, it is 1 or -1) is constant in $[a, b]$. This implies that

$$\frac{1}{b-a} \int_a^b s_N(t)\,\mathrm{d}t = s_N(t) \cdot \frac{1}{b-a} \int_a^b \mathrm{d}t = s_N(t).$$

On the other hand, if $k > N$, then $2^{k-N} > 1$ and we apply Theorem 7.26 (The Change-of-variables Theorem) to get

$$\int_a^b \varphi_k(t)\,\mathrm{d}t = \int_a^b \varphi_0(2^k t)\,\mathrm{d}t = \frac{1}{2^k}\int_{2^k a}^{2^k b} \varphi_0(y)\,\mathrm{d}y = \frac{1}{2^k}\int_{j \cdot 2^{k-N}}^{(j+1)\cdot 2^{k-N}} \varphi_0(y)\,\mathrm{d}y.$$

Using the same argument as in evaluating the integrals in the equation (7.88), we can show that

$$\frac{1}{2^k}\int_{j \cdot 2^{k-N}}^{(j+1)\cdot 2^{k-N}} \varphi_0(y)\,\mathrm{d}y = 0.$$

In conclusion, we arrive at

$$s_N(t) = \frac{1}{b-a}\int_a^b s_N\,\mathrm{d}m = \frac{1}{b-a}\int_a^b s_n\,\mathrm{d}m \tag{7.90}$$

for every $n = 1, 2, \ldots$. By the relation (7.89), $s_n \to f$ in L^2 on $[0,1]$. By Theorem 3.5 (Hölder's Inequality), we obtain

$$\int_a^b |s_n - f|\,\mathrm{d}m \le \left\{\int_a^b |s_n - f|^2\,\mathrm{d}m\right\}^{\frac{1}{2}}\left\{\int_a^b \mathrm{d}m\right\}^{\frac{1}{2}} = (b-a)^{\frac{1}{2}}\|s_n - f\|_2 \to 0$$

as $n \to \infty$. Thus this implies that $s_n - f \in L^1$ on $[0,1]$ for large enough n and so we derive from Theorem 1.33 that

$$\lim_{n\to\infty}\int_a^b s_n\,\mathrm{d}m = \int_a^b f\,\mathrm{d}m. \tag{7.91}$$

By combining the representation (7.90) and the limit (7.91), we establish

$$s_N(t) = \frac{1}{b-a}\int_a^b f\,\mathrm{d}m. \tag{7.92}$$

Since $f \in L^2$ on $[0,1]$, we have $f \in L^1$ on $[0,1]$ by Theorem 3.8. By Theorem 7.7, almost every $x \in [0,1]$ is a Lebesgue point of f. Let $p \in [0,1]$ be such a point and for each $N \in \mathbb{N}$, we define $E_N(p) = [2^{-N}, 2^{-N+1}]$. Clearly, we have

$$m\big(E_N(p)\big) = 2^{-N} \quad \text{and} \quad \sum_{N=1}^{\infty} 2^{-N} = 1.$$

Therefore, one can find a $N \in \mathbb{N}$ such that $p \in [2^{-N}, 2^{-N+1}]$. Furthermore, it is easy to check that the sequence $\{E_N(p)\}$ shrinks to p nicely because

$$E_N(p) \subseteq B(p, 2^{-N+1}) \quad \text{and} \quad m\big(E_N(p)\big) = \frac{1}{4}m\big(B(p, 2^{-N+1})\big)$$

for every $N \in \mathbb{N}$. Using Theorem 7.10 (Lebesgue Differentiation Theorem) and then the representation (7.92), we find that

$$f(p) = \lim_{N \to \infty} \frac{1}{m\big([2^{-N}, 2^{-N+1}]\big)} \int_{[2^{-N}, 2^{-N+1}]} f \, dm = \lim_{N \to \infty} s_N(p) = \sum_{N=1}^{\infty} c_N \varphi_N(p).$$

In other words, the series $\displaystyle\sum_{n=1}^{\infty} c_n \varphi_n$ converges to f a.e. in $[0,1]$.

Finally, we know that

$$\varphi_n(t+1) = \varphi_0\big(2^n(t+1)\big) = \varphi_0(2^n t + 2^n) = \varphi_0(2^n t) = \varphi_n(t)$$

for every $n \in \mathbb{N}$, the series and f are functions with period 1. Now for every $t \in \mathbb{R}$, we can write $t = t' + m$ with $t' \in [0,1]$ and $m \in \mathbb{Z}$. Furthermore, if t is a Lebesgue point of f, then it can be shown from the definition and the periodicity of f that t' is also a Lebesgue point of f. By this fact (for the third equality below), we obtain

$$\sum_{n=1}^{\infty} c_n \varphi_n(t) = \sum_{n=1}^{\infty} c_n \varphi_n(t' + m) = \sum_{n=1}^{\infty} c_n \varphi_n(t') = f(t') = f(t' + m) = f(t),$$

completing the proof of the problem. $\blacksquare$

Problem 7.19

Rudin Chapter 7 Exercise 19.

Proof. Suppose that $c \in \mathbb{R}$. Since f is continuous on $\mathbb{R}$, it is Boreal measurable. Thus the function $f_n(x) = n^c f(nx)$ is also Borel measurable for every $n \in \mathbb{N}$. By Theorem 1.14, $h_c(x)$ is Borel measurable.

(a) By the hypotheses, it is true that

$$\max_{x \in \mathbb{R}} f(x) = \max_{x \in [0,1]} f(x) = \max_{x \in (0,1)} f(x) \tag{7.93}$$

exists and it is positive. Let this number be M. If $x \leq 0$, then $nx \leq 0$ for all $n \in \mathbb{N}$. Similarly, if $x \geq 1$, then $nx \geq 1$ for all $n \in \mathbb{N}$. In other words, we have

$$h_c(x) = 0$$

for all $x \in \mathbb{R} \setminus (0,1)$.

Fix $x \in (0,1)$, then $0 < nx < 1$ if and only if

$$0 < n < \frac{1}{x}. \tag{7.94}$$

In this case, there are only *finitely many* n satisfying the inequalities (7.94), so we have

$$h_c(x) = \sup\{n^c f(nx) \mid n = 1, 2, 3, \ldots\} = \max\left\{ n^c f(np) \,\Big|\, n \leq \left[\tfrac{1}{x}\right] \right\} \leq \frac{M}{x^c}. \tag{7.95}$$

Since $c \in (0,1)$, direct computation shows that

$$\int_0^1 \frac{M}{x^c} \, dx = M x^{1-c}\big|_0^1 = M$$

and then

$$0 \le \int_{\mathbb{R}} h_c(x)\, dx = \int_0^1 h_c(x)\, dx \le M.$$

In conclusion, we have $h_c \in L^1(\mathbb{R})$ for $c \in (0,1)$.

(b) By the inequality (7.95), we see that

$$h_1(x) \le \frac{M}{x} \qquad (7.96)$$

for $x \in (0,1)$ and $h_1(x) = 0$ otherwise. Take $\lambda > 0$. If the real number x satisfies $|h_1(x)| = h_1(x) > \lambda$, then we must have $x \in (0,1)$ and furthermore $0 < x < \frac{M}{\lambda}$ by the inequality (7.96). In other words, we have

$$\{x \in \mathbb{R} \mid h_1(x) > \lambda\} \subseteq \left\{x \in \mathbb{R} \,\Big|\, 0 < x < \frac{M}{\lambda}\right\}$$

and so

$$\lambda \cdot m\{|h_1| > \lambda\} \le \lambda \cdot m\left\{x \in \mathbb{R} \,\Big|\, 0 < x < \frac{M}{\lambda}\right\} = \lambda \cdot \frac{M}{\lambda} = M.$$

By the definition in §7.5, we see that h_1 is in weak L^1.

Recall the fact (7.93), so the Extreme Value Theorem [49, Theorem 4.16, p. 89] ensures that $M = f(p)$ for some $p \in (0,1)$. Given $\epsilon > 0$. Consider the function $g : \mathbb{R} \to \mathbb{R}$ defined by

$$g(x) = f(x) - M + \epsilon$$

which is continuous on $\mathbb{R}$. Since $g(p) = f(p) - M + \epsilon = \epsilon > 0$, the sign-preserving property for continuous functions [64, Problem 7.15, p. 112] says that there exists a $\delta > 0$ small enough such that $g(x) > 0$ on $(p-\delta, p+\delta) \subset (0,1)$, i.e., $f(x) > M-\epsilon$ on $(p-\delta, p+\delta) \subset (0,1)$. Particularly, we may take $\epsilon = \frac{M}{2}$ so that

$$f(x) > \frac{M}{2} \qquad (7.97)$$

on $(p - \delta, p + \delta)$. Fix this δ. If $0 < y < 2\delta$, then $\frac{2\delta}{y} > 1$. Suppose that $N_y = [\frac{p-\delta}{y}] + 1$. Then it is easy to check that

$$\frac{p-\delta}{y} < N_y \le \frac{p-\delta}{y} + 1 = \frac{p-\delta+y}{y} < \frac{p+\delta}{y} \qquad (7.98)$$

which is equivalent to saying that $N_y y \in (p-\delta, p+\delta)$. Hence it follows from the inequalities (7.97) and (7.98) that

$$h_1(y) \ge N_y f(N_y y) > \frac{N_y M}{2} > \frac{(p-\delta)M}{2y}.$$

However, we know that $\int_0^{2\delta} \frac{dt}{t} = \infty$, so we conclude that $h_1 \notin L^1(\mathbb{R})$.

(c) Let $c > 1$ and $\delta > 0$ be the number such that the inequality (7.97) holds on the segment $(p - \delta, p + \delta) \subset (0,1)$. Now for each $k \in \mathbb{N}$, if $x \in (\frac{p-\delta}{k}, \frac{p+\delta}{k})$, then $kx \in (p - \delta, p + \delta)$ and thus

$$h_c(x) \ge k^c f(kx) > \frac{Mk^c}{2}$$

which means $\left(\frac{p-\delta}{k}, \frac{p+\delta}{k}\right) \subset \{x \in \mathbb{R} \mid h_c(x) > \frac{Mk^c}{2}\}$. Consequently, if we take $\lambda_k = \frac{Mk^c}{2}$, then we obtain

$$\lambda_k \cdot m\{h_c > \lambda_k\} > \lambda_k \cdot m\left(\left(\frac{p-\delta}{k}, \frac{p+\delta}{k}\right)\right) = \frac{Mk^c}{2} \cdot \frac{2\delta}{k} = M\delta k^{c-1}. \tag{7.99}$$

As M and δ are fixed as well as $c > 1$, the inequality (7.99) implies that

$$\lambda_k \cdot m\{h_c > \lambda_k\} \to \infty$$

as $k \to \infty$. Then we have proved that h_c is *not* in weak L^1 if $c > 1$.

We have ended the proof of the problem.

Problem 7.20

Rudin Chapter 7 Exercise 20.

Proof. By the definition, we have $\partial E = \overline{E} \setminus E^\circ$.

(a) If $m(\partial E) = 0$, then we have $m(\overline{E} \setminus E^\circ) = 0$. Since E° is the union of all open sets contained in E and $\overline{E}$ is the intersection of all closed sets containing E, we have $E^\circ \subseteq E \subseteq \overline{E}$ and $E^\circ, \overline{E} \in \mathscr{B}$. In fact, we have

$$m(E) = m(E^\circ)$$

by Theorem 1.36.[i]　Furthermore, since m is a complete measure, E is also Lebesgue measurable.[j]

(b) To prove this part, we need stronger versions of Theorems 7.7 and 7.10 (The Lebesgue Differentiation Theorem). Roughly speaking, the hypothesis $f \in L^1(\mathbb{R}^k)$ can be replaced by $f \in L^1_{\mathrm{loc}}(\mathbb{R}^k)$, where $L^1_{\mathrm{loc}}(\mathbb{R}^k)$ is the space of all **locally integrable functions**:

$$\int_K |f| \, dm < \infty$$

for every bounded measurable set $K \subseteq \mathbb{R}^k$, see [22, p. 95]. For convenience, we state the stronger versions of these theorems in a single lemma:

Lemma 7.5

(a) If $f \in L^1_{\mathrm{loc}}(\mathbb{R}^k)$, then almost every $\mathbf{x} \in \mathbb{R}^k$ is a Lebesgue point of f.

(b) Associate to each $\mathbf{x} \in \mathbb{R}^k$ a sequence $\{E_i(\mathbf{x})\}$ that shrinks to $\mathbf{x}$ nicely, and let $f \in L^1_{\mathrm{loc}}(\mathbb{R}^k)$. Then

$$f(\mathbf{x}) = \lim_{i \to \infty} \frac{1}{m\big(E_i(\mathbf{x})\big)} \int_{E_i(\mathbf{x})} f \, dm$$

at every Lebesgue point of f, hence a.e. $[m]$ on $\mathbb{R}^k$.

---

[i]See also [49, Theorem 2.27, p. 35; Exercise 9, p. 43].

[j]Notice the difference between a Borel measurable set and a Lebesgue measurable set: *Not* every subset of a Borel set with measure 0 is also Borel measurable, but Lebesgue measure is obtained by enlarging $\mathscr{B}$ to include all subsets of sets of Borel measure 0.

Let A be a (possibly uncountable) set. Suppose that

$$E = \bigcup_{\alpha \in A} \overline{B(\mathbf{x}_\alpha, r_\alpha)}, \tag{7.100}$$

where $\overline{B(x_\alpha, r_\alpha)}$ is a closed disc in $\mathbb{R}^2$ and $r_\alpha \geq 1$.

Assume that $m(\partial E) > 0$. Clearly, E° is an open subset of $\mathbb{R}^2$, so it is a Borel subset in $\mathbb{R}^2$. As in §7.12, we consider the characteristic function $f = \chi_{E^\circ}$. Denote $D_A(\mathbf{x})$ to be the density of the Lebesgue measurable set A at $\mathbf{x}$. Since $f \in L^1_{\mathrm{loc}}(\mathbb{R}^2)$, we apply Lemma 7.5 to conclude that

$$D_{E^\circ}(\mathbf{x}) = \lim_{r \to 0} \frac{m\big(E^\circ \cap B(\mathbf{x}, r)\big)}{m\big(B(\mathbf{x}, r)\big)} = \lim_{r \to 0} \frac{1}{m\big(B(\mathbf{x}, r)\big)} \int_{B(\mathbf{x}, r)} \chi_{E^\circ}\, \mathrm{d}m = \chi_{E^\circ}(\mathbf{x}) = 0$$

for almost every point $\mathbf{x} \in \partial E \subseteq (E^\circ)^c$. In particular, we choose a $\mathbf{p} \in \partial E$ such that

$$D_{E^\circ}(\mathbf{p}) = 0. \tag{7.101}$$

To continue our proof, we need the following result:

Lemma 7.6

There is a sequence $\{E_i(\mathbf{p})\}$ that shrinks to $\mathbf{p}$ nicely.

Proof of Lemma 7.6. For every $i \in \mathbb{N}$, we have $E \cap B(\mathbf{p}, \frac{1}{i}) \neq \varnothing$. (In fact, this is an equivalent definition of a boundary point of E, see [3, Definition 3.40, p. 64].) Pick $\{\mathbf{q}_i\}$ to be an *arbitrary* sequence temporarily such that $\mathbf{q}_i \in E \cap B(\mathbf{p}, \frac{1}{i})$. By the definition (7.100), we know that $\mathbf{q}_i \in \overline{B(\mathbf{x}_{\alpha_i}, r_{\alpha_i})}$ for some $\alpha_i \in A$. If $|\mathbf{p} - \mathbf{q}_i| \leq \frac{1}{2i}$, then we have

$$B\Big(\mathbf{q}_i, \frac{1}{2i}\Big) \subset B\Big(\mathbf{p}, \frac{1}{i}\Big) \tag{7.102}$$

and we define $E_i(\mathbf{p}) = B(\mathbf{q}_i, \frac{1}{2i})$. Otherwise, we consider the point

$$\mathbf{s}_{2i} \in \overline{B(\mathbf{x}_{\beta_{2i}}, r_{\beta_{2i}})} \cap B\Big(\mathbf{p}, \frac{1}{2i}\Big)$$

which satisfies $|\mathbf{p} - \mathbf{s}_{2i}| < \frac{1}{2i}$ so that if we let $\mathbf{q}_i = \mathbf{s}_{2i}$ and $\alpha_i = \beta_{2i}$, then we gain the set inclusion (7.102) in this case. (See Figure 7.2 below.) Now this process can be done continually and then we obtain a sequence $\{\mathbf{q}_i\}$ satisfying

$$\mathbf{q}_i \in E \cap B\Big(\mathbf{p}, \frac{1}{i}\Big) \quad \text{and} \quad E_i(\mathbf{p}) = B\Big(\mathbf{q}_i, \frac{1}{2i}\Big) \subset B\Big(\mathbf{p}, \frac{1}{i}\Big).$$

Clearly, we have

$$m(E_i(\mathbf{p})) = \pi \times \Big(\frac{1}{2i}\Big)^2 = \frac{1}{4} m\Big(B\Big(\mathbf{p}, \frac{1}{i}\Big)\Big) \tag{7.103}$$

for every $i \in \mathbb{N}$, therefore $\{E_i(\mathbf{p})\}$ shrinks to $\mathbf{p}$ nicely by the definition in §7.9, completing the proof of the lemma. $\blacksquare$

We go back to the proof of part (b). Recall that $\mathbf{q}_i \in \overline{B(\mathbf{x}_{\alpha_i}, r_{\alpha_i})}$. There are two cases:

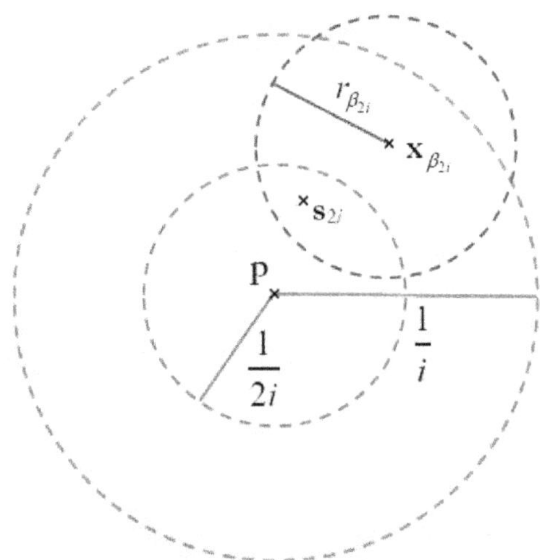

Figure 7.2: Construction of the sequence $E_i(\mathbf{p})$.

– **Case (i):** $\mathbf{q}_i \in \partial B(\mathbf{x}_{\alpha_i}, r_{\alpha_i})$. Geometrically, the area of $B(\mathbf{q}_i, r) \cap B(\mathbf{x}_{\alpha_i}, r_{\alpha_i})$ is always greater than that of $\frac{1}{4}B(\mathbf{q}_i, r)$ whenever $0 < r < r_{\alpha_i}$ so that

$$m\big(B(\mathbf{q}_i, r) \cap B(\mathbf{x}_{\alpha_i}, r_{\alpha_i})\big) \geq \frac{1}{4}m\big(B(\mathbf{q}_i, r)\big). \tag{7.104}$$

Since $B(\alpha, r_\alpha) \subseteq E$ for all $\alpha \in A$, we have $B(\alpha, r_\alpha) \subseteq E^\circ$ for all $\alpha \in A$. Furthermore, since $r_\alpha \geq 1$ for all $\alpha \in A$, we certainly have $\frac{1}{2i} < r_{\alpha_i}$ for all i. Hence these observations allow us to apply the equality (7.103) and the inequality (7.104) to get

$$\begin{aligned}
D_{E^\circ}(\mathbf{p}) &= \lim_{i \to \infty} \frac{m\big(E^\circ \cap B(\mathbf{p}, \frac{1}{i})\big)}{m\big(B(\mathbf{p}, \frac{1}{i})\big)} \\[2mm]
&= \lim_{i \to \infty} \frac{m\big(E^\circ \cap B(\mathbf{p}, \frac{1}{i})\big)}{4m\big(E_i(\mathbf{p})\big)} \\[2mm]
&\geq \frac{1}{4} \cdot \lim_{i \to \infty} \frac{m\big(B(\mathbf{x}_{\alpha_i}, r_{\alpha_i}) \cap B(\mathbf{q}_i, \frac{1}{2i})\big)}{m\big(B(\mathbf{q}_i, \frac{1}{2i})\big)} \\[2mm]
&\geq \frac{1}{16}
\end{aligned}$$

which contradicts the result (7.101).

– **Case (ii):** $\mathbf{q}_i \in B(\mathbf{x}_{\alpha_i}, r_{\alpha_i})$. Since $|\mathbf{p} - \mathbf{q}_i| < \frac{1}{2i}$ and $\mathbf{q}_i \in B(\mathbf{x}_{\alpha_i}, r_{\alpha_i}) \cap B(\mathbf{p}, \frac{1}{i})$, the geometry tells us that $r_{\alpha_i} > \frac{1}{2i}$ so that the inequality (7.104) holds in this case. Finally, we have $D_{E^\circ}(\mathbf{p}) \geq \frac{1}{16}$, a contradiction again.

In conclusion, we have proved that $m(\partial E) = 0$.

(c) The crux of the analysis in part (b) is the inequalities $r_{\alpha_i} > \frac{1}{2i}$. For unrestricted radii r_α, we can pick a sequence $\{r_{\alpha_i}\}$ satisfying this condition for *sufficiently large* i and thus $D_{E^\circ}(\mathbf{p}) \geq \frac{1}{16}$

(d) By the comments in §2.21, there exists a non-Borel subset of $\mathbb{R}$. Let this set be A. Suppose that

$$E = \bigcup_{x \in A} \overline{B\big((x, 0), 1\big)}.$$

Assume that E was Borel in $\mathbb{R}^2$. If we can show that $\mathbb{R} \times \{1\}$ is Borel in $\mathbb{R}^2$, then since

$$E \cap (\mathbb{R} \times \{1\}) = A \times \{1\},$$

A is also Borel in $\mathbb{R}$, a contradiction. Hence E must be non-Borel in $\mathbb{R}^2$. Now it remains to show that $\mathbb{R} \times \{1\}$ is Borel in $\mathbb{R}^2$ and this is the content of the following lemma:

> **Lemma 7.7**
>
> Let A and B be Boreal subsets of $\mathbb{R}$. Then $A \times B$ is also Borel in $\mathbb{R}^2$.

Proof of Lemma 7.7. Suppose that $\mathfrak{M}_1 = \{A \subseteq \mathbb{R} \mid A \times \mathbb{R} \text{ is Borel in } \mathbb{R}^2\}$. It is obvious that $\mathbb{R} \in \mathfrak{M}_1$. Let $A \in \mathfrak{M}_1$, i.e., $A \times \mathbb{R}$ is Borel in $\mathbb{R}^2$. Since $A^c \times \mathbb{R} = \mathbb{R}^2 \setminus (A \times \mathbb{R})$, $A^c \times \mathbb{R}$ is Borel in $\mathbb{R}^2$ by Comment 1.6(d). Suppose that $A_n \in \mathfrak{M}_1$ for all $n = 1, 2, \ldots$, i.e., each $A_n \times \mathbb{R}$ is Borel in $\mathbb{R}^2$ so that $\bigcup_{n=1}^{\infty} (A_n \times \mathbb{R})$ is Borel in $\mathbb{R}^2$. Since

$$\left(\bigcup_{n=1}^{\infty} A_n \right) \times \mathbb{R} = \left\{ (a,y) \,\middle|\, a \in \bigcup_{n=1}^{\infty} A_n, y \in \mathbb{R} \right\}$$

$$= \{(a_1,y) \mid a_1 \in A_1, y \in \mathbb{R}\} \cup \{(a_2,y) \mid a_2 \in A_1, y \in \mathbb{R}\} \cup \cdots$$

$$= (A_1 \times \mathbb{R}) \cup (A_2 \times \mathbb{R}) \cup \cdots$$

$$= \bigcup_{n=1}^{\infty} (A_n \times \mathbb{R}),$$

we establish that

$$\bigcup_{n=1}^{\infty} A_n \in \mathfrak{M}_1.$$

By Definition 1.3(a), $\mathfrak{M}_1$ is a σ-algebra in $\mathbb{R}$. Next, let V be open in $\mathbb{R}$. Since $V \times \mathbb{R}$ is open in $\mathbb{R}^2$ (see [42, p. 86]), it is Borel in $\mathbb{R}^2$ and thus $V \in \mathfrak{M}_1$. Hence $\mathscr{B} \subseteq \mathfrak{M}_1$.

Similarly, if we define $\mathfrak{M}_2 = \{B \subseteq \mathbb{R} \mid \mathbb{R} \times B \text{ is Borel in } \mathbb{R}^2\}$, then in the same way as above we can show that $\mathfrak{M}_2$ is also a σ-algebra in $\mathbb{R}$ containing $\mathscr{B}$. Finally, if $A, B \in \mathscr{B}$, then $A \times \mathbb{R}$ and $\mathbb{R} \times B$ are Borel in $\mathbb{R}^2$. Since $(A \times \mathbb{R}) \cap (\mathbb{R} \times B) = A \times B$, our desired result follows. ∎

(e) As Balcerzak and Kharazishvili [6, p. 205] pointed out that the union of an arbitrary family of convex bodies (compact convex sets with non-empty interiors) in $\mathbb{R}^2$ is Lebesgue measurable. Therefore, the closed discs can be replaced by arbitrary closed polygons provided that they are convex.[k]

Hence we have completed the proof of the problem. ∎

> **Problem 7.21**
>
> *Rudin Chapter 7 Exercise 21.*

---

[k] More generally, it is known that if $E = \bigcup X_\alpha$ and for each X_α, there exists an open ball B_α such that $B_\alpha \subseteq X_\alpha \subseteq \overline{B_\alpha}$, then E is Lebesgue measurable. See [33, Problem 22, p. 482].

Proof. Notice that the triangle inequality implies that $f, g \in BV$ on $[a, b]$ if and only if $f + g \in BV$ on $[a, b]$. Clearly, the function $g(t) = t$ belongs to BV on $[0, 1]$, so $\gamma \in BV$ on $[0, 1]$ if and only if $f \in BV$ on $[0, 1]$.

By the definition, the length of the graph of f is the total variation of γ on $[0, 1]$, say $V_\gamma(1)$. We want to show that

$$V_\gamma(1) = \int_0^1 \sqrt{1 + [f'(t)]^2}\, dt. \tag{7.105}$$

Suppose that f is AC on $[0, 1]$. Then γ is also AC on $[0, 1]$. Without loss of generality, we may assume that $f(0) = 0$; otherwise, we consider the function $\widetilde{f} : [0, 1] \to \mathbb{R}$ defined by $\widetilde{f} = f - f(0)$. This assumption implies that $\gamma(0) = 0$. By Theorem 7.18,[1] we know that $\gamma' \in L^1$ a.e. $[m]$ on $[0, 1]$ and

$$\gamma(x) = \int_0^x \gamma'(t)\, dt.$$

Therefore, if $x, y \in [0, 1]$ and $x < y$, then the above fact and Theorem 1.33 establish

$$|\gamma(y) - \gamma(x)| = \left| \int_x^y \gamma'(t)\, dt \right| \le \int_x^y |\gamma'(t)|\, dt = \int_x^y \sqrt{1 + f'(t)}\, dt. \tag{7.106}$$

Finally, it follows from the definition and the inequality (7.106) that if $0 = t_0 < t_1 < \cdots < t_N = 1$, then we have

$$V_\gamma(1) = \sup \sum_{i=1}^N |\gamma(t_i) - \gamma(t_{i-1})| \le \sup \sum_{i=1}^N \int_{t_{i-1}}^{t_i} \sqrt{1 + f'(t)}\, dt = \int_0^1 \sqrt{1 + f'(t)}\, dt.$$

This proves one direction.

For the other direction, since γ is AC on $[0, 1]$, V_γ is also AC on $[0, 1]$ by Theorem 7.19. Thus Theorem 7.18 and the fact $V_\gamma(0) = 0$ imply that

$$\int_0^1 V_\gamma'(t)\, dt = V_\gamma(1). \tag{7.107}$$

Employing the notation in the proof of Problem 7.13(c), if $0 \le x < y \le 1$, then we have

$$V_\gamma(y) - V_\gamma(x) = V_\gamma(x; y) = \sup \sum_{i=1}^N |\gamma(t_i) - \gamma(t_{i-1})|, \tag{7.108}$$

where the supremum is taken over all positive integer N and over all partitions $\{t_i\}$ such that $x = t_0 < t_1 < \cdots < t_N = y$. By repeated use of the triangle inequality, we certainly have

$$\sqrt{(y - x)^2 + [f(y) - f(x)]^2} = |\gamma(y) - \gamma(x)| \le \sum_{i=1}^N |\gamma(t_i) - \gamma(t_{i-1})| \le V_\gamma(x; y). \tag{7.109}$$

Therefore, it follows from the expressions (7.108) and (7.109) that

$$\sqrt{(y - x)^2 + [f(y) - f(x)]^2} \le V_\gamma(y) - V_\gamma(x)$$

and then

$$\sqrt{1 + \left[\frac{f(y) - f(x)}{y - x}\right]^2} \le \frac{V_\gamma(y) - V_\gamma(x)}{y - x}. \tag{7.110}$$

---

[1]Monotonicity of γ is *not* necessary for the implication (a) $\to$ (c).

Recall that f and V_γ are AC on $[0,1]$, f' and V'_γ exist a.e. $[m]$ on $[0,1]$ by Theorem 7.18, so the inequality (7.110) ensures that

$$\sqrt{1+[f'(x)]^2} \le V'_\gamma(x)$$

a.e. $[m]$ on $[0,1]$. Substituting this into the expression (7.107), we get

$$\int_0^1 \sqrt{1+[f'(t)]^2}\, dt \le V_\gamma(1)$$

which proves the other direction. Hence we have completed the proof of the problem.

Problem 7.22

Rudin Chapter 7 Exercise 22.

Proof.

(a) Assume that $f \ne 0$ on a measurable set $E \subseteq \mathbb{R}^k$ of positive measure. We claim that there corresponds a constant $c = c(f) > 0$ such that

$$(Mf)(\mathbf{x}) \ge c|\mathbf{x}|^{-k}$$

for sufficiently large $|\mathbf{x}|$. If it was not the case, then for each $n \in \mathbb{N}$, there exists a $\mathbf{x}_n \in \mathbb{R}^k$ with $|\mathbf{x}_n| > n$ such that

$$(Mf)(\mathbf{x}_n) < \frac{|\mathbf{x}_n|^{-k}}{n}.$$

Recall the basic fact that the Lebesgue measure of the ball $B(\mathbf{x}, r)$ in $\mathbb{R}^k$ is given by

$$\frac{\pi^{\frac{k}{2}}}{\Gamma(\frac{k}{2}+1)} r^k.$$

Thus it follows from the definition [51, Eqn. (4), p. 138] and this fact that

$$\int_{B(\mathbf{x}_n,r)} |f|\, dm \le m(B_r) \times (Mf)(\mathbf{x}_n) < \frac{\pi^{\frac{k}{2}}}{\Gamma(\frac{k}{2}+1)} r^k \times \frac{|\mathbf{x}_n|^{-k}}{n} \tag{7.111}$$

for all $r > 0$. Particularly, we pick the sequence $\{r_n\}$, where $r_n = \sqrt[2k]{n} \cdot |\mathbf{x}_n|$, and the inequality (7.111) becomes

$$\int_{\mathbb{R}^k} \chi_{B(\mathbf{x}_n,r_n)} |f|\, dm = \int_{B(\mathbf{x}_n,r_n)} |f|\, dm < \frac{\pi^{\frac{k}{2}}}{\Gamma(\frac{k}{2}+1)} \times \frac{1}{\sqrt{n}}. \tag{7.112}$$

By the Archimedean Property [49, Theorem 1.20(a), p. 9], for every $\mathbf{x} \in \mathbb{R}^k$, there is a positive integer n such that $n \ge |\mathbf{x}|$, so we have $|\mathbf{x}| < n < |\mathbf{x}_n|$ and then

$$|\mathbf{x} - \mathbf{x}_n| \le |\mathbf{x}| + |\mathbf{x}_n| < 2|\mathbf{x}_n| < \sqrt[2k]{n} \cdot |\mathbf{x}_n|$$

for large enough n (remember that k is *fixed*). In other words, this means that $\mathbf{x} \in B(\mathbf{x}_n, \sqrt[2k]{n} \cdot |\mathbf{x}_n|)$ and thus

$$\lim_{n\to\infty} \chi_{B(\mathbf{x}_n,\, \sqrt[2k]{n}|\mathbf{x}_n|)}(\mathbf{x}) = 1. \tag{7.113}$$

Let $|f_n| = \chi_{B(\mathbf{x}_n,r_n)}|f|$ for all $n \in \mathbb{N}$. Then each $|f_n|$ is measurable, $|f_n| \le |f|$ and the limit (7.113) implies

$$|f|(\mathbf{x}) = \lim_{n\to\infty} |f_n|(\mathbf{x})$$

for every $\mathbf{x} \in \mathbb{R}^k$. Since $f \in L^1(\mathbb{R}^k)$, the sequence $\{|f_n|\}$ satisfies the hypotheses of Theorem 1.34 (Lebesgue's Dominated Convergence Theorem). Thus we have

$$\lim_{n \to \infty} \int_{B(\mathbf{x}_n, r_n)} |f| \, \mathrm{d}m = \lim_{n \to \infty} \int_{\mathbb{R}^k} |f_n| \, \mathrm{d}m = \int_{\mathbb{R}^k} |f| \, \mathrm{d}m. \qquad (7.114)$$

Now we combine the inequality (7.112) and the result (7.114) to get

$$\int_{\mathbb{R}^k} |f| \, \mathrm{d}m \le \lim_{n \to \infty} \frac{\pi^{\frac{k}{2}}}{\Gamma(\frac{k}{2} + 1)} \times \frac{1}{\sqrt{n}} = 0.$$

By Theorem 1.39(a), $f = 0$ a.e. on $\mathbb{R}^k$, a contradiction. Hence there exists a constant $c = c(f) > 0$ and $r > 0$ such that for all $|\mathbf{x}| \ge r$, we have

$$(Mf)(\mathbf{x}) \ge c|\mathbf{x}|^{-k}$$

which implies that $\|Mf\|_1 = \infty$, a contradiction. Hence we must have $f = 0$ a.e. on $\mathbb{R}^k$.

(b) For every $x \in (0, \frac{1}{4})$, if $r \in (0, x]$, then we have $x - r \ge 0$ and $x + r < \frac{1}{2}$. Particularly, we take $r = x$ so that

$$\frac{1}{m(B_r)} \int_{B(x,r)} |f| \, \mathrm{d}m = \frac{1}{2r} \int_{x-r}^{x+r} |f(t)| \, \mathrm{d}t = \frac{1}{2x} \int_0^{2x} \frac{1}{t(\log t)^2} \, \mathrm{d}t. \qquad (7.115)$$

Since $t(\log t)^2 > 0$ on $(0, \frac{1}{2})$ and $t^{-1}(\log t)^{-2} \in \mathscr{R}$ on $[\epsilon, 2x]$, we apply [49, Theorem 11.33, p. 323] to obtain

$$\int_0^{2x} \frac{1}{t(\log t)^2} \, \mathrm{d}t \ge \int_\epsilon^{2x} \frac{1}{t(\log t)^2} \, \mathrm{d}t = \mathscr{R} \int_\epsilon^{2x} \frac{1}{t(\log t)^2} \, \mathrm{d}t = -\frac{1}{\log t}\Big|_\epsilon^{2x} = -\frac{1}{\log(2x)} + \frac{1}{\log \epsilon}$$

for every $\epsilon > 0$. As $\epsilon \to 0+$, we know that $\frac{1}{\log \epsilon} \to 0-$ and so

$$\int_0^{2x} \frac{1}{t(\log t)^2} \, \mathrm{d}t \ge -\frac{1}{\log(2x)}. \qquad (7.116)$$

Since $2x \in (0, \frac{1}{2})$, we have $-\frac{1}{\log(2x)} > 0$ and then we substitute the inequality (7.116) into the expression (7.115) to obtain

$$(Mf)(x) \ge \sup_{0 < r < \infty} \frac{1}{m(B_r)} \int_{B(x,r)} |f| \, \mathrm{d}m \ge -\frac{1}{2x \log(2x)} = \left| \frac{1}{2x \log(2x)} \right|,$$

as required.

Furthermore, direct computation gives

$$\int_0^{\frac{1}{4}} \frac{\mathrm{d}x}{|2x \log(2x)|} = \infty.$$

Recall the basic fact that $(Mf)(x) \ge 0$, so it implies that

$$\int_0^1 (Mf)(x) \, \mathrm{d}x \ge \int_0^{\frac{1}{4}} \frac{\mathrm{d}x}{|2x \log(2x)|} = \infty,$$

as desired.

Hence we have completed the proof of the problem. ∎

> **Problem 7.23**
>
> *Rudin Chapter 7 Exercise 23.*

Proof. We first show that $(SF)(\mathbf{x})$ is well-defined. To this end, let α_1 and α_2 be two complex numbers such that

$$\lim_{r \to 0+} \frac{1}{m(B_r)} \int_{B(\mathbf{x},r)} |f_1 - \alpha_1| \, dm = \lim_{r \to 0+} \frac{1}{m(B_r)} \int_{B(\mathbf{x},r)} |f_2 - \alpha_2| \, dm = 0 \tag{7.117}$$

for some $f_1, f_2 \in F$. Then we see that

$$\begin{aligned}
|\alpha_1 - \alpha_2| &= \frac{1}{m(B_r)} \int_{B(\mathbf{x},r)} |\alpha_1 - \alpha_2| \, dm \\
&\leq \frac{1}{m(B_r)} \int_{B(\mathbf{x},r)} (|\alpha_1 - f| + |f - g| + |g - \alpha_2|) \, dm \\
&= \frac{1}{m(B_r)} \int_{B(\mathbf{x},r)} |\alpha_1 - f| \, dm + \frac{1}{m(B_r)} \int_{B(\mathbf{x},r)} |g - \alpha_2| \, dm \\
&\quad + \frac{1}{m(B_r)} \int_{B(\mathbf{x},r)} |f - g| \, dm.
\end{aligned} \tag{7.118}$$

Since f and g belong to the *same* equivalence class F, $f = g$ a.e. on $\mathbb{R}^k$ and the last integral in the inequality (7.118) is actually 0. By taking $r \to 0+$ to the integrals in the inequality (7.118) and using the hypotheses (7.117), we obtain $\alpha_1 = \alpha_2$, i.e., $(SF)(\mathbf{x})$ is unique if $\mathbf{x}$ is a Lebesgue point of F.

Next, if $\mathbf{x}$ is a Lebesgue point of f and $f \in F$, then Definition 7.6 gives

$$\lim_{r \to 0+} \frac{1}{m(B_r)} \int_{B(\mathbf{x},r)} |f(\mathbf{y}) - f(\mathbf{x})| \, dm(\mathbf{y}) = 0$$

so that $\mathbf{x}$ is also a Lebesgue point of F and the uniqueness of Lebesgue points of F in the previous paragraph shows immediately that $(SF)(\mathbf{x}) = f(\mathbf{x})$. This completes the proof of the problem. $\blacksquare$

Integration on Product Spaces

8.1 Monotone Classes and Ordinate Sets of Functions

Problem 8.1

Rudin Chapter 8 Exercise 1.

Proof. Suppose that

$$\mathfrak{M} = \{\varnothing, \mathbb{R}, (-\infty, a), (-\infty, a], (b, \infty), [b, \infty) \mid a, b \in \mathbb{R}\}.$$

We claim that $\mathfrak{M}$ is a monotone class. To see this, let $A_i \in \mathfrak{M}$, $A_i \subset A_{i+1}$ and $A = \bigcup_{i=1}^{\infty} A_i$. We have several cases to consider.

- **Case (i):** $A_i = (-\infty, a_i)$. Then the condition $A_i \subset A_{i+1}$ forces that A_{i+1} is in one of the following forms:

$$(-\infty, a_i], \quad (-\infty, a_{i+1}) \quad \text{and} \quad (-\infty, a_{i+1}] \tag{8.1}$$

  for some $a_{i+1} > a_i$. Thus A is in one of the following forms:

$$(-\infty, a), \quad (-\infty, a] \quad \text{and} \quad \mathbb{R} \tag{8.2}$$

  for some $a \in \mathbb{R}$. Both cases say that $A \in \mathfrak{M}$.

- **Case (ii):** $A_i = (-\infty, a_i]$. Now A_{i+1} is still in one of the forms (8.1), so similar analysis shows that A is in one of the forms (8.2). Therefore, $A \in \mathfrak{M}$.

- **Case (iii):** $A_i = (b_i, \infty)$. In this case, A_{i+1} is in one of the forms:

$$[b_i, \infty), \quad (b_{i+1}, \infty) \quad \text{and} \quad [b_{i+1}, \infty) \tag{8.3}$$

  for some $b_{i+1} < b_i$. Then it can be shown easily that A is in one of the following forms:

$$(b, \infty), \quad [b, \infty) \quad \text{and} \quad \mathbb{R} \tag{8.4}$$

  for some $b \in \mathbb{R}$. Consequently, we have $A \in \mathfrak{M}$.

- **Case (iv):** $A_i = [b, \infty)$. The A_{i+1} is expressed in one of the forms (8.3) so that A is in one of the forms (8.4). In conclusion, we have $A \in \mathfrak{M}$.

Next, let $B_i \in \mathfrak{M}$, $B_i \supset B_{i+1}$ and $B = \bigcup_{i=1}^{\infty} B_i$. Similar to the sequence of sets $\{A_i\}$, there are also four cases for $\{B_i\}$. Since the analysis of these cases is very similar to those above, so we omit the details here. By Definition 8.1, we see that $\mathfrak{M}$ is a monotone class.

Clearly, $\mathbb{R} \in \mathfrak{M}$. Since $\mathbb{R} \setminus (-\infty, a) = [a, \infty)$, $\mathbb{R} \setminus (-\infty, a] = (a, \infty)$, $\mathbb{R} \setminus (b, \infty) = (-\infty, b]$ and $\mathbb{R} \setminus [b, \infty) = (-\infty, b)$, we immediately have $\mathbb{R} \setminus A \in \mathfrak{M}$ for every $A \in \mathfrak{M}$. However, $\mathfrak{M}$ is *not* a σ-algebra because $(-\infty, 0), (1, \infty) \in \mathfrak{M}$ but $(-\infty, 0) \cup (1, \infty) \notin \mathfrak{M}$ which violates Definition 1.3(a). This completes the proof of the problem. ■

Problem 8.2

Rudin Chapter 8 Exercise 2.

Proof. We have $f : \mathbb{R} \to [0, \infty)$ and

$$A(f) = \{(x, y) \in \mathbb{R}^2 \mid 0 < y < f(x)\}. \tag{8.5}$$

Now all the answers are affirmative.

(a) Since f is Lebesgue measurable, Theorem 1.17 (The Simple Function Approximation Theorem) ensures the existence of simple measurable functions s_n on $\mathbb{R}$ such that $0 \le s_1 \le s_2 \le \cdots \le f$ and $s_n(x) \to f(x)$ as $n \to \infty$ for every $x \in \mathbb{R}$.

Suppose that s is a nonnegative simple function on $\mathbb{R}$. Then, by Definition 1.6, we have

$$s = \sum_{i=1}^{n} \alpha_i \chi_{A_i},$$

where $A_1, A_2, \ldots, A_n$ can be assumed to be *mutually disjoint* Lebesgue measurable sets of $\mathbb{R}$. Therefore, we obtain

$$\begin{aligned} A(s) &= \{(x, y) \in \mathbb{R}^2 \mid 0 < y < s(x)\} \\ &= \bigcup_{i=1}^{n} \{(x, y) \in \mathbb{R}^2 \mid x \in A_i \text{ and } 0 < y < \alpha_i\} \\ &= \bigcup_{i=1}^{n} [A_i \times (0, \alpha_i)]. \end{aligned}$$

By Definition 8.1, $A(s)$ must be a Lebesgue measurable subset of $\mathbb{R}^2$. It is easy to check from the definition of the ordinate set of a function that the following lemma holds:

Lemma 8.1

If f and g are functions on $\mathbb{R}$ such that $f(x) \le g(x)$ for every $x \in \mathbb{R}$, then $A(f) \subseteq A(g)$.

Now we claim that

$$A(f) = \bigcup_{n=1}^{\infty} A(s_n). \tag{8.6}$$

The direction

$$\bigcup_{n=1}^{\infty} A(s_n) \subseteq A(f)$$

holds trivially because of the fact that $s_n \leq f$ for every $n = 1, 2, \ldots$ and Lemma 8.1. For the other direction, let $(x, y) \in A(f)$. This implies that $x \in \mathbb{R}$ and $0 < y < f(x)$. Since $s_n(x) \to f(x)$ as $n \to \infty$, there exists a $N \in \mathbb{N}$ such that $0 < y < s_N(x) \leq f(x)$ which means $(x, y) \in A(s_N)$ so that

$$A(f) \subseteq \bigcup_{n=1}^{\infty} A(s_n)$$

also holds. Hence we have established the claim (8.6). Finally, since each $A(s_n)$ is a Lebesgue measurable subset of $\mathbb{R}^2$, the relation (8.6) implies that $A(f)$ is also a Lebesgue measurable subset of $\mathbb{R}^2$.

(b) By the definition (8.5), the x-section of $A(f)$ is given by

$$[A(f)]_x = \{y \in \mathbb{R} \mid (x, y) \in A(f)\} = \{y \in \mathbb{R} \mid 0 < y < f(x)\} = (0, f(x)).$$

Since $\mathbb{R}$ is obviously σ-finite, it follows from Theorem 8.6 (see also Definition 8.7) that

$$m_2\big(A(f)\big) = (m \times m)\big(A(f)\big) = \int_{\mathbb{R}} m\big([A(f)]_x\big)\,\mathrm{d}x = \int_{\mathbb{R}} f(x)\,\mathrm{d}x.$$

(c) Recall that the graph of f, namely G, is given by

$$G = \{(x, y) \in \mathbb{R}^2 \mid y = f(x)\}.$$

For each $i \in \mathbb{N}$, the set $A(f + \frac{1}{i})$ is Lebesgue measurable in $\mathbb{R}^2$ by part (a). Then the set

$$A = \bigcap_{i=1}^{\infty} A\Big(f + \frac{1}{i}\Big) = \{(x, y) \in \mathbb{R}^2 \mid 0 < y \leq f(x)\}$$

is also Lebesgue measurable by Comment 1.6(c). Therefore, the set

$$A \setminus A(f) = \{(x, y) \in \mathbb{R}^2 \mid y = f(x) \text{ but } y > 0\}$$

is Lebesgue measurable in $\mathbb{R}^2$. Obviously, we have

$$\{(x, 0) \in \mathbb{R}^2 \mid 0 = f(x)\} = \{x \in \mathbb{R} \mid f(x) = 0\} \times \{0\} = f^{-1}(0) \times \{0\}.$$

Since f is Lebesgue measurable, $f^{-1}(0)$ is a Lebesgue measurable set in $\mathbb{R}$ by Definition 1.3(c).[a] Since $\{0\}$ is a Lebesgue measurable set in $\mathbb{R}$, we use Definition 8.1 to conclude that $f^{-1}(0) \times \{0\}$ is Lebesgue measurable in $\mathbb{R}^2$. Since

$$G = \big(A \setminus A(f)\big) \cup [f^{-1}(0) \times \{0\}],$$

the preceding analysis shows that G is a Lebesgue measurable subset of $\mathbb{R}^2$.

(d) For each *fixed* $x \in \mathbb{R}$, we consider the x-section of G:

$$G_x = \{y \in \mathbb{R} \mid (x, y) \in G\} = \{y \in \mathbb{R} \mid y = f(x)\} = \{f(x)\}.$$

By a similar argument as in part (b), we conclude easily from this that

$$m_2(G) = (m \times m)(G) = \int_{\mathbb{R}} m(G_x)\,\mathrm{d}x = 0$$

as required.

This completes the proof of the problem.

---

[a] Of course, we replace "for every open set V in Y" by "for every closed set V in Y" here.

8.2   Applications of the Fubini Theorem

Problem 8.3

Rudin Chapter 8 Exercise 3.

Proof. Note that

$$\varphi(x) = \int_0^1 f(x,y)\,dy$$

for each $x \in (0,1)$. Suppose that

$$f(x,y) = y^{(x-\frac{1}{2})^2-1}$$

in $(0,1) \times (0,1)$. It is clear that f is a positive continuous function in $(0,1) \times (0,1)$. If $x = \frac{1}{2}$, then $f(\frac{1}{2}, y) = y^{-1}$ so that

$$\varphi\left(\frac{1}{2}\right) = \int_0^1 y^{-1}\,dy = \ln y\big|_0^1 = \infty.$$

If $x \neq \frac{1}{2}$, then we have

$$\varphi(x) = \int_0^1 y^{(x-\frac{1}{2})^2-1}\,dy = \frac{1}{(x-\frac{1}{2})^2}$$

and this implies that

$$\int_0^1 \varphi(x)\,dx = \int_0^1 \frac{1}{(x-\frac{1}{2})^2}\,dx = -\left(x-\frac{1}{2}\right)^{-1}\Big|_0^1 = -4.$$

This is a required example, completing the proof of the problem. $\blacksquare$

Problem 8.4

Rudin Chapter 8 Exercise 4.

Proof.

(a) As Rudin pointed out, we may assume that f and g are Borel functions on $\mathbb{R}$. If $p = 1$, then it is in fact Theorem 8.14. If $p = \infty$, then $\|g\|_\infty < \infty$ and so

$$\begin{aligned}
|(f * g)(x)| &= \left|\int_{-\infty}^\infty f(x-t)g(t)\,dt\right| \\
&\leq \int_{-\infty}^\infty |f(x-t)| \cdot |g(t)|\,dt \\
&\leq \|g\|_\infty \int_{-\infty}^\infty |f(x-t)|\,dt \\
&= \|g\|_\infty \int_{-\infty}^\infty |f(t)|\,dt \\
&= \|g\|_\infty \cdot \|f\|_1.
\end{aligned}$$

Thus we have $\|f * g\|_\infty \leq \|f\|_1 \cdot \|g\|_\infty$.

Suppose that $1 < p < \infty$. Since f and g are measurable on $\mathbb{R}$, the functions $|f(x-t)|^{\frac{1}{q}}$ and $|f(x-t)|^{\frac{1}{p}} \cdot |g(y)|$ are also measurable on $\mathbb{R}$ by Theorem 1.7, where q is the conjugate exponent of p. By Theorem 3.5 (Hölder's Inequality), we derive that

$$
\begin{aligned}
\int_{-\infty}^{\infty} |f(x-y)g(y)|\,\mathrm{d}y &= \int_{-\infty}^{\infty} |f(x-y)|^{\frac{1}{q}} \cdot \left(|f(x-y)|^{\frac{1}{p}}|g(y)|\right)\,\mathrm{d}y \\
&\le \left\{ \int_{-\infty}^{\infty} |f(x-y)|\,\mathrm{d}y \right\}^{\frac{1}{q}} \times \left\{ \int_{-\infty}^{\infty} |f(x-y)| \cdot |g(y)|^p\,\mathrm{d}y \right\}^{\frac{1}{p}} \\
&= \|f\|_1^{\frac{1}{q}} \cdot \left\{ \int_{-\infty}^{\infty} |f(x-y)| \cdot |g(y)|^p\,\mathrm{d}y \right\}^{\frac{1}{p}}
\end{aligned}
$$

which gives

$$
\left\{ \int_{-\infty}^{\infty} |f(x-y)g(y)|\,\mathrm{d}y \right\}^{p} \le \|f\|_1^{\frac{p}{q}} \int_{-\infty}^{\infty} |f(x-y)| \cdot |g(y)|^p\,\mathrm{d}y. \tag{8.7}
$$

Define $F : \mathbb{R}^2 \to \mathbb{C}$ by

$$
F(x,y) = f(x-y)g(y).
$$

Using similar argument as in the proof of Theorem 8.14, we are able to show that the function $F \in L^1(\mathbb{R}^2)$. Since $\mathbb{R}$ is a σ-finite measure space, we get from Theorem 8.8 (The Fubini Theorem) that

$$
\varphi(x) = \int_{-\infty}^{\infty} F_x(y)\,\mathrm{d}y = \int_{-\infty}^{\infty} f(x-y)g(y)\,\mathrm{d}y = (f * g)(x)
$$

a.e. is in $L^1(\mathbb{R})$ so that the integral defining $(f * g)(x)$ exists for almost all x. This proves the first assertion.

To verify the second and the third assertions, we recall that $|f|, |g|^p \in L^1(\mathbb{R})$, so Theorem 8.14 ensures that the function $h : \mathbb{R} \to [0, \infty]$ given by

$$
h(x) = \int_{-\infty}^{\infty} |f(x-y)| \cdot |g(y)|^p\,\mathrm{d}y
$$

belongs to $L^1(\mathbb{R})$ which means that h is a measurable function on $\mathbb{R}$. By Theorem 1.33, we observe from the inequality (8.7) that

$$
\begin{aligned}
|(f * g)(x)|^p &= \left| \int_{-\infty}^{\infty} f(x-y)g(y)\,\mathrm{d}y \right|^{p} \\
&\le \left\{ \int_{-\infty}^{\infty} |f(x-y)g(y)|\,\mathrm{d}y \right\}^{p} \\
&\le \|f\|_1^{\frac{p}{q}} \int_{-\infty}^{\infty} |f(x-y)| \cdot |g(y)|^p\,\mathrm{d}y. \tag{8.8}
\end{aligned}
$$

Now we apply Theorem 8.8 (The Fubini Theorem) to the right-hand side of the inequality (8.8) to get

$$
\begin{aligned}
\int_{-\infty}^{\infty} |(f * g)(x)|^p\,\mathrm{d}x &\le \|f\|_1^{\frac{p}{q}} \int_{-\infty}^{\infty} \int_{-\infty}^{\infty} |f(x-y)| \cdot |g(y)|^p\,\mathrm{d}y\,\mathrm{d}x \\
&= \|f\|_1^{\frac{p}{q}} \int_{-\infty}^{\infty} \int_{-\infty}^{\infty} |f(x-y)| \cdot |g(y)|^p\,\mathrm{d}x\,\mathrm{d}y \\
&= \|f\|_1^{\frac{p}{q}} \cdot \|f\|_1 \int_{-\infty}^{\infty} |g(y)|^p\,\mathrm{d}y \\
&= \|f\|_1^{p} \cdot \|g\|_p^{p}
\end{aligned}
$$

which implies that $\|f * g\|_p \le \|f\|_1 \cdot \|g\|_p < \infty$.

(b) Let $p = 1$. If $f \geq 0$ and $g \geq 0$, then the definition gives

$$h(x) = (f * g)(x) = \int_{-\infty}^{\infty} f(x-y)g(y)\,\mathrm{d}y \geq 0. \tag{8.9}$$

Therefore we deduce from Theorem 8.8 (The Fubini Theorem) that

$$\|f * g\|_1 = \|h\|_1$$
$$= \int_{-\infty}^{\infty} h(x)\,\mathrm{d}x$$
$$= \int_{-\infty}^{\infty}\int_{-\infty}^{\infty} f(x-y)g(y)\,\mathrm{d}y\,\mathrm{d}x$$
$$= \int_{-\infty}^{\infty} g(y)\,\mathrm{d}y \int_{-\infty}^{\infty} f(x-y)\,\mathrm{d}x$$
$$= \|g\|_1 \cdot \|f\|_1$$

as required.

Let $p = \infty$. If $f \geq 0$ and $g = c \geq 0$, then the inequality (8.9) holds in this case. Recall that $f \in L^1(\mathbb{R})$ and $g \in L^\infty(\mathbb{R})$, so we have

$$|h(x)| = \left| \int_{-\infty}^{\infty} f(x-y)c\,\mathrm{d}y \right| = c\|f\|_1 < \infty$$

for almost all $x \in \mathbb{R}$. By Definition 3.7, we obtain

$$\|f * g\|_\infty = \|h\|_\infty = \|f\|_1 \times c = \|f\|_1 \cdot \|g\|_\infty.$$

(c) Suppose that $1 < p < \infty$ and $\|f * g\|_p = \|f\|_1 \cdot \|g\|_p$. By the definition of $f * g$ and the application of part (a), we observe that

$$\|f * g\|_p \leq \||f| * |g|\|_p \leq \||f|\|_1 \cdot \||g|\|_p = \|f\|_1 \cdot \|g\|_p = \|f * g\|_p$$

which forces $\||f| * |g|\|_p = \||f|\|_1 \cdot \||g|\|_p$. Thus we may assume that $f, g \geq 0$.

Recall that the inequality $\|f * g\|_p \leq \|f\|_1 \cdot \|g\|_p$ derives from (see the deduction of the inequality (8.8))

$$(f * g)(x)^p = \left\{ \int_{-\infty}^{\infty} f(x-y)g(y)\,\mathrm{d}y \right\}^p$$
$$= \left\{ \int_{-\infty}^{\infty} f(x-y)^{\frac{1}{q}} \cdot [f(x-y)^{\frac{1}{p}} g(y)]\,\mathrm{d}y \right\}^p$$
$$\leq \left\{ \int_{-\infty}^{\infty} f(x-y)\,\mathrm{d}y \right\}^{\frac{p}{q}} \times \left\{ \int_{-\infty}^{\infty} f(x-y)g(y)^p\,\mathrm{d}y \right\} \tag{8.10}$$
$$= \|f\|_1^{\frac{p}{q}} \times \left\{ \int_{-\infty}^{\infty} f(x-y)g(y)^p\,\mathrm{d}y \right\}$$

and so the equality $\|f * g\|_p = \|f\|_1 \cdot \|g\|_p$ forces that the equality (8.10) holds, but this means that the equality of Theorem 3.5 (Hölder's Inequality) holds there i.e., there are constants (with respect to y) $\alpha(x)$ and $\beta(x)$, not both zero, such that

$$\alpha(x)f(x-y) = \alpha(x)f(x-y)^{\frac{q}{q}} = \beta(x)[f(x-y)^{\frac{1}{p}}g(y)]^p = \beta(x)f(x-y)g(y)^p \tag{8.11}$$

for almost all $y \in \mathbb{R}$. Let $E = \{z \in \mathbb{R} \mid f(-z) > 0\}$. If $\|f\|_1 = 0$, then since $m(\mathbb{R}) = \infty$, $f = 0$ a.e. on $\mathbb{R}$ and we are done. Therefore, without loss of generality, we may assume that $\|f\|_1 = 1$ which gives $m(E) > 0$.

On the one hand, if $\beta(x) = 0$ *for some* $x \in \mathbb{R}$, then $\alpha(x) \neq 0$ and so the equation (8.11) implies that $f(x - y) = 0$ for almost all $y \in \mathbb{R}$ which is equivalent to $f = 0$ a.e. on $\mathbb{R}$. On the other hand, if $\alpha(x) = 0$ *for some* $x \in \mathbb{R}$, then $\beta(x) \neq 0$ and so

$$f(x - y)g(y)^p = 0 \tag{8.12}$$

for almost all $y \in \mathbb{R}$. Now if $y \in E_x = x + E = \{x + z \mid z \in E\}$, then $y = x + z$ or $-z = x - y$ for some $z \in E$ so that

$$f(x - y) > 0$$

for all $y \in E_x$. Therefore, it follows from the equation (8.12) that $g(y) = 0$ for almost all $y \in E_x$. Assume that $m(E_x) < \infty$. Then $m(E_x^c) = \infty$. Furthermore, we have $g(y) > 0$ for all $y \in E_x^c$ and this implies that

$$\|g\|_p^p = \int_{\mathbb{R}} g(y)^p \, dy = \int_{E_x^c} g(y)^p \, dy = \infty$$

which contradicts $g \in L^p(\mathbb{R})$.

(d) Suppose that $p = \infty$. We simply take $f(x) = \chi_{[0,1]}$ and $g(x) = 1$ on $\mathbb{R}$. Then we have $\|f\|_1 = \|g\|_\infty = 1$. Since $f \geq 0$ and $g \geq 0$ on $\mathbb{R}$, we deduce from Theorem 7.26 (The Change-of-variables Theorem) that

$$(f * g)(x) = \int_{-\infty}^{\infty} f(y)g(x - y) \, dy = \int_0^1 dy = 1.$$

Hence we always have

$$\|f * g\|_\infty = 1 > (1 - \epsilon)\|f\|_1 \cdot \|g\|_\infty.$$

Next, we suppose that $p < \infty$. Let $0 < \alpha < 1 - (1 - \epsilon)^p < 1$ and define

$$f(x) = \frac{1}{2\alpha}\chi_{[-\alpha,\alpha]}(x) \quad \text{and} \quad g(x) = \frac{1}{2^p}\chi_{[-1,1]}(x).$$

It is easy to see that

$$\|f\|_1 = \int_{-\infty}^{\infty} |f(x)| \, dx = \int_{-\alpha}^{\alpha} \frac{1}{2\alpha}\chi_{[-\alpha,\alpha]}(x) \, dx = 1$$

and

$$\|g\|_p = \int_{-\infty}^{\infty} |g(x)|^p \, dx = \int_{-1}^{1} \frac{1}{2}\chi_{[-1,1]}(x) \, dx = 1.$$

Since $f \geq 0$ and $g \geq 0$ on $\mathbb{R}$, we deduce from Theorem 7.26 (The Change-of-variables Theorem) that

$$(f * g)(x) = \int_{-\infty}^{\infty} f(y)g(x - y) \, dy = \frac{1}{2\alpha \cdot 2^{\frac{1}{p}}} \int_{-\alpha}^{\alpha} \chi_{[-1,1]}(x - y) \, dy. \tag{8.13}$$

Since $-1 \leq x - y \leq 1$ if and only if $x - 1 \leq y \leq x + 1$, we follow from the expression (8.13) that

$$(f * g)(x) = \frac{1}{2\alpha \cdot 2^{\frac{1}{p}}} \int_{-\alpha}^{\alpha} \chi_{[x-1,x+1]}(y) \, dy$$

$$= \frac{1}{2\alpha \cdot 2^{\frac{1}{p}}} m\big([-\alpha, \alpha] \cap [x-1, x+1]\big)$$

$$= \begin{cases} 0, & \text{if } x \in (-\infty, -1-\alpha]; \\[2ex] \dfrac{x+1+\alpha}{2\alpha \cdot 2^{\frac{1}{p}}} & \text{if } x \in (-1-\alpha, -1+\alpha]; \\[2ex] \dfrac{2\alpha}{2\alpha \cdot 2^{\frac{1}{p}}} & \text{if } x \in (-1+\alpha, 1-\alpha]; \\[2ex] \dfrac{1+\alpha-x}{2\alpha \cdot 2^{\frac{1}{p}}} & \text{if } x \in (1-\alpha, 1+\alpha]; \\[2ex] 0, & \text{if } x \in (1+\alpha, \infty). \end{cases}$$

Hence direct computation shows that

$$\begin{aligned} \|f * g\|_p^p &= \int_{-\infty}^{\infty} |(f * g)(x)|^p \, dx \\ &= \int_{-1-\alpha}^{1+\alpha} |(f * g)(x)|^p \, dx \\ &= \int_{-1-\alpha}^{-1+\alpha} \frac{1}{2(2\alpha)^p} (x+1+\alpha)^p \, dx + \int_{-1+\alpha}^{1-\alpha} \frac{(2\alpha)^p}{2(2\alpha)^p} \, dx \\ &\quad + \int_{1-\alpha}^{1+\alpha} \frac{1}{2(2\alpha)^p} (1+\alpha-x)^p \, dx \\ &= 2 \int_{-1-\alpha}^{-1+\alpha} \frac{1}{2(2\alpha)^p} (x+1+\alpha)^p \, dx + (1-\alpha). \end{aligned} \tag{8.14}$$

Since $(x+1+\alpha)^p \geq 0$ on $[-1-\alpha, -1+\alpha]$, the integral in the expression (8.14) is nonnegative so that

$$\|f * g\|_p^p \geq 1 - \alpha > (1-\epsilon)^p$$

and this is equivalent to saying that $\|f * g\|_p > (1-\epsilon)\|f\|_1 \cdot \|g\|_p$.

This completes the proof of the problem. ∎

Remark 8.1

The result in Problem 8.4(a) is sometimes called Minkowski's Inequality which is a special case of the Young's Convolution Inequality.

Problem 8.5

Rudin Chapter 8 Exercise 5.

Proof.

(a) Clearly, the function $\alpha : \mathbb{R}^2 \to \mathbb{R}$ defined by $\alpha(x, y) = x + y$ is continuous. We want to show the following result first:

> **Lemma 8.2**
>
> It is true that $\alpha^{-1}(E) \in \mathscr{B}_2$ if $E \in \mathscr{B}_1$. (See Problem 8.11 for the meaning of the notation $\mathscr{B}_k$ for any $k \in \mathbb{N}$.)

Proof of Lemma 8.2. Let $\mathfrak{M} = \{E \subseteq \mathbb{R} \mid \alpha^{-1}(E) \in \mathscr{B}_2\}$. Since α is continuous, $\alpha^{-1}(V) \in \mathscr{B}_2$ for every open set V in $\mathbb{R}$. In other words, $\mathfrak{M}$ contains the *standard topology* of $\mathbb{R}$. It is easy to check that

- $\mathbb{R} \in \mathfrak{M}$.

- If $E \in \mathfrak{M}$, then $E^c \in \mathfrak{M}$ because $\alpha^{-1}(E^c) = \alpha^{-1}(\mathbb{R}) \backslash \alpha^{-1}(E) = \mathbb{R}^2 \backslash \alpha^{-1}(E) \in \mathscr{B}_2$.

- If $E_n \in \mathfrak{M}$ so that $\alpha^{-1}(E_n) \in \mathscr{B}_2$ for every $n \in \mathbb{N}$, then we have

$$\alpha^{-1}\Big(\bigcup_{n=1}^{\infty} E_n \Big) = \bigcup_{n=1}^{\infty} \alpha^{-1}(E_n) \in \mathscr{B}_2.$$

By Definition 1.3(a), $\mathfrak{M}$ is a σ-algebra and so $\mathfrak{M} = \mathscr{B}_1$ which completes the proof of Lemma 8.2 $\blacksquare$

Let's return to the proof of the problem. We want to check Definition 6.1:

$$(\mu * \lambda)(E) = \sum_{i=1}^{\infty} (\mu * \lambda)(E^i) \tag{8.15}$$

for every partition $\{E^i\} \subseteq \mathscr{B}_1$ of $E \in \mathscr{B}_1$. For each $i = 1, 2, \ldots$, we denote

$$E_2^i = \{(x,y) \in \mathbb{R}^2 \mid x + y \in E^i\} \subseteq \mathbb{R}^2 \quad \text{and} \quad E_2 = \{(x,y) \in \mathbb{R}^2 \mid x + y \in E\} \subseteq \mathbb{R}^2.$$

Then we have

$$E_2 = \bigcup_{i=1}^{\infty} E_2^i.$$

Next, if $(x_0, y_0) \in E_2^i \cap E_2^j$ for $i \neq j$, then $x_0 + y_0 \in E^i \cap E^j$ but $E^i \cap E^j = \varnothing$, a contradiction. Thus the set $\{E_2^1, E_2^2, \ldots\}$ must be a partition of E_2 and since $\mu \times \lambda$ is a measure by Definition 8.7, we get

$$(\mu * \lambda)(E) = (\mu \times \lambda)(E_2) = \sum_{i=1}^{\infty} (\mu \times \lambda)(E_2^i) = \sum_{i=1}^{\infty} (\mu * \lambda)(E^i), \tag{8.16}$$

i.e., $\mu * \nu \in M$ and this proves the first assertion.

For the second assertion, since μ and λ are complex measures on $\mathbb{R}$, Theorem 6.12 implies that

$$d\mu = h_1 \, d|\mu| \quad \text{and} \quad d\lambda = h_2 \, d|\lambda|,$$

where $|h_1| = |h_2| = 1$ on $\mathbb{R}$. For any $A \in \mathscr{B}(\mathbb{R}^2)$, we observe from Definition 8.7 that

$$\begin{aligned}
(\mu \times \lambda)(A) &= \int_X \lambda(A_x) \, d\mu(x) \\
&= \int_X \lambda(A_x) h_1(x) \, d|\mu|(x)
\end{aligned}$$

$$= \int_X \left\{ \int_Y \chi_{A_x}(y)\, \mathrm{d}\lambda(y) \right\} h_1(x)\, \mathrm{d}|\mu|(x)$$

$$= \int_X \left\{ \int_Y \chi_A(x,y) h_2(y)\, \mathrm{d}|\lambda|(y) \right\} h_1(x)\, \mathrm{d}|\mu|(x). \tag{8.17}$$

Since $|\mu|$ and $|\lambda|$ are finite by Theorem 6.4, we obtain

$$\varphi^*(x) = \int_Y |g|_x\, \mathrm{d}|\lambda| = |\lambda|(Y) \quad \text{and} \quad \int_X \varphi^*\, \mathrm{d}|\mu| = |\lambda|(Y)|\mu|(X) < \infty.$$

Thus, by Theorem 8.8 (The Fubini Theorem), the integral (8.17) can be reduced to

$$(\mu \times \lambda)(A) = \int_X \int_Y \chi_A(x,y) h_2(y) h_1(x)\, \mathrm{d}|\lambda|(y)\, \mathrm{d}|\mu|(x)$$

$$= \int_{X \times Y} \chi_A(x,y) h_2(y) h_1(x)\, \mathrm{d}(|\mu| \times |\nu|)$$

$$= \int_A f(x,y)\, \mathrm{d}(|\mu| \times |\nu|), \tag{8.18}$$

where $f(x,y) = h_2(y) h_1(x)$. Obviously, Theorem 8.8 (The Fubini Theorem) also says that $f(x,y) \in L^1(|\mu| \times |\lambda|)$ so that Theorem 6.13 may be applied to the result (8.18) to get

$$|\mu \times \lambda|(A) = \int_A |f(x,y)|\, \mathrm{d}(|\mu| \times |\lambda|) = \int_A \mathrm{d}(|\mu| \times |\lambda|) = (|\mu| \times |\lambda|)(A), \tag{8.19}$$

i.e., the total variation of $\mu \times \lambda$ is $|\mu| \times |\lambda|$.

Since $\mu * \lambda \in M$, the series (8.16) converges absolutely and then the fact (8.16) implies

$$|(\mu * \lambda)|(\mathbb{R}) \le \sum_{i=1}^{\infty} |\mu * \lambda|(E^i) \le \sum_{i=1}^{\infty} |(\mu \times \lambda)|(E_2^i) \tag{8.20}$$

where $\{E^i\}$ is a partition of $\mathbb{R}$. By Theorem 6.2, $|\mu \times \lambda|$ is a positive measure. Recall that $\{E_2^i\}$ is a partition of E_2, so if we apply these facts and the result (8.19) to the inequality (8.20), then it becomes

$$|(\mu * \lambda)|(\mathbb{R}) \le |(\mu \times \lambda)|(E_2) = (|\mu| \times |\lambda|)(E_2) \le (|\mu| \times |\lambda|)(\mathbb{R}^2).$$

Hence we follow from the definition of the norm in M and Definition 8.7 ($\mathbb{R}$ is σ-finite) that

$$\|\mu * \lambda\| = |\mu * \lambda|(\mathbb{R}) \le (|\mu| \times |\lambda|)(\mathbb{R} \times \mathbb{R}) = |\mu|(\mathbb{R}) \times |\lambda|(\mathbb{R}) = \|\mu\| \cdot \|\lambda\|. \tag{8.21}$$

(b) Firstly, for $E \in \mathscr{B}_1$, we know from the definition of $\mu * \lambda$, Definition 8.7 and then [51, Eqn. (3), p. 163] that

$$\int_{\mathbb{R}} \chi_E(x)\, \mathrm{d}(\mu * \lambda) = (\mu * \lambda)(E)$$

$$= (\mu \times \lambda)(E_2)$$

$$= \int_{\mathbb{R}} \lambda(E_2^y)\, \mathrm{d}\mu(x)$$

$$= \int_{\mathbb{R}} \int_{\mathbb{R}} \chi_E(x+y)\, \mathrm{d}\mu(x)\, \mathrm{d}\lambda(y).$$

Thus the formula holds for χ_E and then it also holds for any simple function $s = \sum_{i=1}^{n} \alpha_i \chi_{A_i}$, where each A_i is a Borel set in $\mathbb{R}$, i.e.,

$$\int_{\mathbb{R}} s(x)\,\mathrm{d}(\mu * \lambda) = \int_{\mathbb{R}} \int_{\mathbb{R}} s(x+y)\,\mathrm{d}\mu(x)\,\mathrm{d}\lambda(y). \tag{8.22}$$

Next, we let $f \in C_0(\mathbb{R})$ so that $f : \mathbb{R} \to \mathbb{C}$ is continuous. By Definition 3.16, f is bounded by a positive constant M on $\mathbb{R}$ and Borel measurable. By [22, Theorem 2.10(b), p. 47], there exist simple Borel measurable functions $s_n : \mathbb{R} \to \mathbb{C}$ such that

$$0 \le |s_1| \le |s_2| \le \cdots \le |f| \tag{8.23}$$

and $s_n(x) \to f(x)$ as $n \to \infty$ for every $x \in \mathbb{R}$. Now we replace s by s_n in the formula (8.22) to get

$$\int_{\mathbb{R}} s_n(x)\,\mathrm{d}(\mu * \lambda) = \int_{\mathbb{R}} \int_{\mathbb{R}} s_n(x+y)\,\mathrm{d}\mu(x)\,\mathrm{d}\lambda(y). \tag{8.24}$$

By Theorem 6.12, there is a Borel measurable function h such that $|h(x)| = 1$ on $\mathbb{R}$ and $\mathrm{d}(\mu * \lambda) = h\,\mathrm{d}|\mu * \lambda|$. If we consider the sequence $\{s_n h\}$ of Borel measurable functions, then the inequalities (8.23) are still valid. Furthermore, we have $s_n(x)h(x) \to f(x)h(x)$ as $n \to \infty$ for every $x \in \mathbb{R}$. By the inequality (8.21), we see that

$$\|fh\|_1 = \int_{\mathbb{R}} |f(x)h(x)|\,\mathrm{d}|\mu * \lambda| \le M|(\mu * \lambda)|(\mathbb{R}) \le M|\mu|(\mathbb{R})| \cdot \lambda|(\mathbb{R}) < \infty.$$

Therefore, it means $fh \in L^1(\mu * \lambda)$ and we apply Theorem 1.34 (Lebesgue's Dominated Convergence Theorem) to the left-hand side of the formula (8.24) to obtain

$$\lim_{n \to \infty} \int_{\mathbb{R}} s_n(x)\,\mathrm{d}(\mu * \lambda) = \lim_{n \to \infty} \int_{\mathbb{R}} s_n(x)h(x)\,\mathrm{d}|\mu * \lambda| = \int_{\mathbb{R}} f(x)h(x)\,\mathrm{d}|\mu * \lambda| = \int_{\mathbb{R}} f\,\mathrm{d}(\mu * \lambda).$$

Now we may apply similar analysis to show that the limit of the right-hand side of the formula (8.24) is exactly

$$\int_{\mathbb{R}} \int_{\mathbb{R}} f(x+y)\,\mathrm{d}\mu(x)\,\mathrm{d}\lambda(y).$$

Hence we conclude that

$$\int_{\mathbb{R}} f\,\mathrm{d}(\mu * \lambda) = \int_{\mathbb{R}} \int_{\mathbb{R}} f(x+y)\,\mathrm{d}\mu(x)\,\mathrm{d}\lambda(y). \tag{8.25}$$

Since $\mu * \lambda \in M$, §6.18 on p. 130 indicates that the mapping

$$f \mapsto \int_{\mathbb{R}} f\,\mathrm{d}(\mu * \lambda)$$

is a *bounded linear functional* on $C_0(\mathbb{R})$. Consequently, Theorem 6.19 (The Riesz Representation Theorem) ensures that $\mu * \nu$ is the *unique* complex Borel measure satisfying the formula (8.25).

(c) Since $\mathbb{R}$ is commutative, $x + y \in E$ if and only if $y + x \in E$. Thus it follows from the formula (8.25) and Theorem 8.8 (The Fubini Theorem) that

$$\int f\,\mathrm{d}(\mu * \lambda) = \int \int f(x+y)\,\mathrm{d}\mu(x)\,\mathrm{d}\lambda(y) = \int \int f(y+x)\,\mathrm{d}\lambda(y)\,\mathrm{d}\mu(x) = \int f\,\mathrm{d}(\lambda * \mu).$$

Therefore, the uniqueness property in part (b) implies that $\mu * \lambda = \lambda * \mu$.

On the one hand, it is clear from the formula (8.25) and Theorem 8.8 (The Fubini Theorem) that

$$\int_{\mathbb{R}} f \, \mathrm{d}[(\mu * \lambda) * \nu] = \int_{\mathbb{R}} \int_{\mathbb{R}} \int_{\mathbb{R}} f(x + y + z) \, \mathrm{d}\mu(x) \, \mathrm{d}\lambda(y) \, \mathrm{d}\nu(z). \tag{8.26}$$

On the other hand, we also have

$$\int_{\mathbb{R}} f \, \mathrm{d}[\mu * (\lambda * \nu)] = \int_{\mathbb{R}} \underbrace{\left\{ \int_{\mathbb{R}} f(x + Y) \, \mathrm{d}\mu(x) \right\}}_{F} \mathrm{d}(\lambda * \nu)(Y)$$

$$= \int_{\mathbb{R}} \int_{\mathbb{R}} \int_{\mathbb{R}} f(x + y + z) \, \mathrm{d}\mu(z) \, \mathrm{d}\lambda(y) \, \mathrm{d}\nu(z). \tag{8.27}$$

Thus the associativity of the convolution in M follows immediately from the formulas (8.26) and (8.27), i.e., $(\mu * \lambda) * \nu = \mu * (\lambda * \nu)$.

By the formula (8.25) again, we see that

$$\int_{\mathbb{R}} f \, \mathrm{d}\big(\mu * (\lambda + \nu)\big) = \int_{\mathbb{R}} \int_{\mathbb{R}} f(x + y) \, \mathrm{d}\mu(x) \, \mathrm{d}[\lambda(y) + \nu(y)]$$

$$= \int_{\mathbb{R}} \int_{\mathbb{R}} f(x + y) \, \mathrm{d}\mu(x) \, \mathrm{d}\lambda(y) + \int_{\mathbb{R}} \int_{\mathbb{R}} f(x + y) \, \mathrm{d}\mu(x) \, \mathrm{d}\nu(y)$$

$$= \int_{\mathbb{R}} f \, \mathrm{d}(\mu * \lambda) + \int_{\mathbb{R}} f \, \mathrm{d}(\mu * \nu)$$

$$= \int_{\mathbb{R}} f \, \mathrm{d}(\mu * \nu + \mu * \nu)$$

which shows immediately that the convolution in M is distributive with respect to addition, i.e., $\mu * (\lambda + \nu) = \mu * \lambda + \mu * \nu$.

(d) By Definition 8.7, we have

$$(\mu \times \lambda)(Q) = \int_{\mathbb{R}} \mu(Q^y) \, \mathrm{d}\lambda(y) = \int \left\{ \int_{Q^y} \mathrm{d}\mu(x) \right\} \mathrm{d}\lambda(y), \tag{8.28}$$

where $Q^y = \{x \in \mathbb{R} \mid (x, y) \in Q\}$. Let $E \in \mathscr{B}_1$ and $y \in \mathbb{R}$. Then we have

$$E_2^y = \{x \in \mathbb{R} \mid (x, y) \in E_2\} = \{x \in \mathbb{R} \mid x + y \in E\} = \{x - y \in \mathbb{R} \mid x \in E\} = E - y.$$

By the formula (8.28), we have

$$(\mu \times \lambda)(E_2) = \int \left\{ \int_{E_2^y} \mathrm{d}\mu(x) \right\} \mathrm{d}\lambda(y) = \int \left\{ \int_{E-y} \mathrm{d}\mu(x) \right\} \mathrm{d}\lambda(y) = \int \mu(E - y) \, \mathrm{d}\lambda(y)$$

as required.

(e) Suppose that μ and λ are concentrated on the countable sets A and B respectively, i.e.,

$$\mu(E) = \lambda(F) = 0$$

whenever $E \cap A = \varnothing$ and $F \cap B = \varnothing$, where $A, B \in \mathscr{B}_1$. Let $Q \in \mathscr{B}_2$ and $Q \cap (A \times B) = \varnothing$. Then Theorem 8.2 implies that $Q_x \in \mathscr{B}_1$ for every $x \in \mathbb{R}$. Assume that $y_0 \in Q_x \cap B$. Then $(x, y_0) \in Q$ for every $x \in \mathbb{R}$. However, this means that $(x_0, y_0) \in Q \cap (A \times B)$ for some

$x_0 \in \mathbb{R}$, a contradiction. As a result, it must be true that $Q_x \cap B = \varnothing$. Similarly, we can show that $Q^y \cap A = \varnothing$. Therefore, we have

$$\lambda(Q_x) = 0 \quad \text{and} \quad \mu(Q^y) = 0$$

which definitely give

$$(\mu \times \lambda)(Q) = \int_{\mathbb{R}} \lambda(Q_x)\,\mathrm{d}\mu(x) = \int_{\mathbb{R}} \mu(Q^y)\,\mathrm{d}\lambda(y) = 0 \tag{8.29}$$

by Definition 8.7. In other words, $\mu \times \lambda$ is concentrated on $A \times B$.[b]

To continue the proof, we need the following lemma:

Lemma 8.3

$E_2 \cap (A \times B) = \varnothing$ if and only if $E \cap (A + B) = \varnothing$.

Proof of Lemma 8.3. It is easy to see that $x = a + b \in E \cap (A + B)$ if and only if $(a, b) \in E_2 \cap (A \times B)$, so our expected result follows. ∎

By the definition, if $E \cap (A + B) = \varnothing$, then Lemma 8.3 implies that $E_2 \cap (A \times B) = \varnothing$ and the analysis in the preceding paragraph shows that

$$(\mu * \lambda)(E) = (\mu \times \lambda)(E_2) = 0.$$

Since $A + B$ is countable, we have established the result that $\mu * \lambda$ is discrete.

Next, suppose that μ is continuous and $\lambda \in M$. By part (d), if $x \in \mathbb{R}$, then we have

$$(\mu * \lambda)(\{x\}) = \int_{\mathbb{R}} \mu(\{x\} - t)\,\mathrm{d}\lambda(t) = \int_{\mathbb{R}} \mu(\{x - t\})\,\mathrm{d}\lambda(t) = \int_{\mathbb{R}} 0\,\mathrm{d}\lambda(t) = 0.$$

Thus $\mu * \lambda$ is continuous.

Finally, let $\mu \ll m$ and $E \in \mathscr{B}_1$ satisfy $m(E) = 0$. By Theorem 2.20(c), we have $m(E - t) = m(E) = 0$ for every $t \in \mathbb{R}$ and then $\mu(E - t) = 0$ for every $t \in \mathbb{R}$. By this fact and part (d), we achieve

$$(\mu * \lambda)(E) = \int_{\mathbb{R}} \mu(E - t)\,\mathrm{d}\lambda(t) = \int_{\mathbb{R}} 0\,\mathrm{d}\lambda(t) = 0.$$

By Definition 6.7, we conclude that $\mu * \lambda \ll m$.

(f) By part (b) and the assumptions, we note that

$$\int_{\mathbb{R}} \mathrm{d}(\mu * \lambda) = \int_{\mathbb{R}} \int_{\mathbb{R}} \mathrm{d}\mu(x)\,\mathrm{d}\lambda(y) = \int_{\mathbb{R}} \int_{\mathbb{R}} f(x)g(y)\,\mathrm{d}m\,\mathrm{d}m. \tag{8.30}$$

By Theorem 8.14, since

$$f * g = \int_{\mathbb{R}} f(x - y)g(y)\,\mathrm{d}m,$$

we may further reduce the expression (8.30) to

$$\int_{\mathbb{R}} \mathrm{d}(\mu * \lambda) = \int_{\mathbb{R}} \int_{\mathbb{R}} f(x - y)g(y)\,\mathrm{d}m\,\mathrm{d}m. = \int_{\mathbb{R}} (f * g)\,\mathrm{d}m$$

and this is equivalent to $\mathrm{d}(\mu * \lambda) = (f * g)\,\mathrm{d}m$.

---

[b]It seems that it is enough to assume either μ or λ is discrete for the validity of the result (8.29). However, as Definition 8.7 indicates, $(\mu \times \lambda)(Q)$ can be computed by using two different integrals, so both μ and λ must be assumed to be discrete in order that the two integrals there are equal and thus $(\mu \times \lambda)(Q)$ makes sense.

(g) To being with, suppose that $X_1 = \{\mu \in M \mid \mu \text{ is discrete}\}$, $X_2 = \{\mu \in M \mid \mu \text{ is continuous}\}$ and $X_3 = \{\mu \in M \mid \mu \ll m\}$.

Recall from [40, Definition 3.3.1, p. 305] that a set X with binary operations $+$ and $*$ from $X \times X$ to X, and that $\cdot$ is a binary operation from $\mathbb{F} \times X$ to X is called an **algebra** if

- $(X, +, \cdot)$ is a vector space;
- for all $x, y, z \in X$ and every scalar $\alpha \in \mathbb{F}$, we have

  Condition (1). $x * (y * z) = (x * y) * z$;

  Condition (2). $x * (y + z) = (x * y) + (x * z)$ and $(x + y) * z = (x * z) + (y * z)$;

  Condition (3). $\alpha \cdot (x * y) = (\alpha \cdot x) * y = x * (\alpha \cdot y)$.

A subset of X is called a **subalgebra** if it is an algebra.

Now let's prove the assertions one by one:

- **Case(1):** $(X_1, +, *, \cdot)$ **is an algebra.** If μ and λ are concentrated on countable (Borel) sets A and B respectively, then $\mu(E) = 0$ and $\lambda(F) = 0$ whenever $E \cap A = \varnothing$ and $F \cap B = \varnothing$. Let $C = A \cup B$. If $E \cap (A \cup B) = \varnothing$, then we must have $E \cap A = \varnothing$ and $E \cap B = \varnothing$ which imply immediately that

$$(\alpha\mu + \beta\lambda)(E) = \alpha\mu(E) + \beta\lambda(E) = 0,$$

  where α and β are real. Thus $(X_1, +, \cdot)$ is a vector space. Next, by part (e), we know that the map $X_1 \times X_1 \to X_1$ given by

$$(\mu, \lambda) \mapsto \mu * \lambda$$

  is well-defined. Thus it can be seen that **Conditions (1)** to **(3)** come directly from the results of part (c). By the definition, $(X_1, +, *, \cdot)$ is an algebra.

- **Case (2):** X_2 **is an ideal in** M. Recall that Y is called an **ideal** of an algebra X if Y is a vector subspace of X and $xY, Yx \subseteq Y$ for every $x \in X$. Here

$$xY = \{x * y \mid y \in Y\} \quad \text{and} \quad Yx = \{y * x \mid y \in Y\}.$$

  Let $\mu, \lambda \in X_2$ so that $\mu(\{x\}) = \lambda(\{x\}) = 0$ for every $x \in \mathbb{R}$, but these imply trivially that

$$(\alpha\mu + \beta\lambda)(\{x\}) = \alpha\mu(\{x\}) + \beta\lambda(\{x\}) = 0$$

  for every real α and β. In other words, $(X_2, +, \cdot)$ is a vector space. Since M is commutative by part (c), it suffices to prove $X_2\lambda \subseteq X_2$ for every $\lambda \in M$, but this fact follows immediately from part (e). Hence we may conclude that X_2 is an ideal in M.

- **Case(3):** X_3 **is an ideal in** M. Let $\mu, \lambda \in X_3$. Then $\mu(E) = \lambda(E) = 0$ whenever $m(E) = 0$. For every $\alpha, \beta \in \mathbb{R}$, if $m(E) = 0$, then

$$(\alpha\mu + \beta\lambda)(E) = \alpha\mu(E) + \beta\lambda(E) = 0.$$

  Therefore, it means that $(\alpha\mu + \beta\lambda) \ll m$ and so $(X_3, +, \cdot)$ is a vector space. Next, for any $\mu \in X_3$ and $\lambda \in M$, $\mu * \lambda \ll m$ by part (e), so we have $\mu\lambda \in X_3$ and X_3 is an ideal in M by the definition.

  Recall that m is a positive σ-finite measure, thus for every $\mu \in X_3$, we have $\mu \ll m$ so that Theorem 6.10 (The Lebesgue-Radon-Nikodym Theorem) ensures that there is a unique $f \in L^1(\mathbb{R})$ such that $\mathrm{d}\mu = f\,\mathrm{d}m$. Therefore, it is reasonable to define $\Phi : (X_3, *) \to \big(L^1(\mathbb{R}), *\big)$ by

$$\Phi(\mu) = f.$$

We claim that Φ is in fact an isomorphism. Let λ be another element in X_3. Then there is a unique $g \in L^1(\mathbb{R})$ such that $\mathrm{d}\lambda = g\,\mathrm{d}m$. By part (f), we have

$$\mathrm{d}(\mu * \lambda) = (f * g)\,\mathrm{d}m. \tag{8.31}$$

By Theorem 6.10 (The Lebesgue-Radon-Nikodym Theorem) again, the $h \in L^1(\mathbb{R})$ satisfying the formula (8.31) must be unique. Hence we have

$$\Phi(\mu * \nu) = f * g,$$

i.e., Φ is a homomorphism. Next, if $\Phi(\mu) = \Phi(\lambda)$, then $\mathrm{d}\mu = \mathrm{d}\lambda$ which implies exactly $\mu = \lambda$. Thus Φ is injective. Next, suppose that $f \in L^1(\mathbb{R})$. By Definition 1.31, if $f = u + iv$, where u and v are real measurable functions, then we have

$$\mu(E) = \int_E f\,\mathrm{d}m = \int_E u^+\,\mathrm{d}\mu - \int_E u^-\,\mathrm{d}m + i\int_E v^+\,\mathrm{d}m - i\int_E v^-\,\mathrm{d}m. \tag{8.32}$$

where $E \in \mathscr{B}_1$. Since $u^\pm$ and $v^\pm$ are measurable and nonnegative, Theorem 1.29 guarantees that every integral in the expression (8.32) is a measure on $\mathscr{B}_1$. Thus, if $m(E) = 0$, then Proposition 1.24(e) tells us that $\mu(E) = 0$. By Definition 6.7, $\mu \ll m$ or equivalently $\mu \in X_3$ and Φ is surjective.

In conclusion, Φ is an isomorphism, as required.

(h) Consider the Dirac measure $\delta_0 : \mathscr{B}_1 \to [0, \infty]$, that is

$$\delta_0(E) = \begin{cases} 1, & \text{if } 0 \in E; \\ 0, & \text{otherwise.} \end{cases}$$

It is clear that $\delta_0 \in M$ and part (d) shows that

$$(\delta_0 * \mu)(E) = \int_{\mathbb{R}} \mu(E - t)\,\mathrm{d}\delta_0(t) = \mu(E)$$

for all $\mu \in M$. Hence δ_0 is a unit of M.

(i) Suppose that M is defined to be the Banach space of all complex Borel measures on $\mathbb{R}^k$ with norm $\|\mu\| = |\mu|(\mathbb{R}^k)$ and associate to each Borel set $E \subseteq \mathbb{R}^k$ the set

$$E_2 = \{(\mathbf{x}, \mathbf{y}) \mid \mathbf{x} + \mathbf{y} \in E\} \subseteq \mathbb{R}^k \times \mathbb{R}^k.$$

If $\mu, \lambda \in M$, we define

$$(\mu * \lambda)(E) = (\mu \times \lambda)(E_2)$$

for every Borel set $E \subseteq \mathbb{R}^k$. Then all the assertions from parts (a) to (h) also hold when we replace $\mathbb{R}$ and $\mathscr{B}_1$ by $\mathbb{R}^k$ and $\mathscr{B}_k$ respectively because $\mathbb{R}^k$ is always a commutative group and m_k is a translation invariant Borel measure on $\mathbb{R}^k$. The proofs go exactly the same, so we omit the details here.

For the k-dimensional torus T^k, we need to seek a measure similar to the Lebesgue measure m_k on $\mathbb{R}^k$. Note that T^k is a locally compact group with the σ-algebra $\mathscr{B}(T^k)$, so the **Haar measure** denoted by σ_k is what we want. (For the existence, uniqueness and basic properties of such a measure, the reader is suggested to read, for examples, [16,

Chap. 9], [22, §11.1, pp. 339 – 348] and [48, pp. 128 – 132, 211].) Once we have such a measure, we can follow the same line as above to obtain the corresponding results. Again, we omit the details here.

This completes the analysis of the problem.   ■

> **Problem 8.6**
>
> *Rudin Chapter 8 Exercise 6.*

Proof. Let $\mathbf{x} \in \mathbb{R}^k \setminus \{\mathbf{0}\}$ and $r = |\mathbf{x}|$. Then we have $r > 0$ so that it is meaningful to define $\mathbf{u} = \frac{\mathbf{x}}{r}$ which certainly gives the representation

$$\mathbf{x} = r\mathbf{u} \quad \text{and} \quad |\mathbf{u}| = 1. \tag{8.33}$$

Suppose that $\mathbf{x} = r'\mathbf{u}'$, where $r' > 0$ and $|\mathbf{u}'| = 1$. Then the equality $r\mathbf{u} = r'\mathbf{u}'$ implies that

$$r = |r\mathbf{u}| = |r'\mathbf{u}'| = r'$$

and so $\mathbf{u} = \mathbf{u}'$. In other words, the representation (8.33) is unique. Therefore, we may "identify" $\mathbb{R}^k \setminus \{\mathbf{0}\}$ as $(0, \infty) \times S_{k-1}$, i.e., the map

$$\varphi : \mathbb{R}^k \setminus \{\mathbf{0}\} \to (0, \infty) \times S_{k-1}$$

is a homeomorphism with the continuous inverse

$$\varphi^{-1} : (0, \infty) \times S_{k-1} \to \mathbb{R}^k \setminus \{\mathbf{0}\}$$

and they are given by

$$\varphi(\mathbf{x}) = \left(|\mathbf{x}|, \frac{\mathbf{x}}{|\mathbf{x}|} \right) \quad \text{and} \quad \varphi^{-1}(r, \mathbf{u}) = r\mathbf{u} \tag{8.34}$$

respectively. To proceed further, we need the following result:[c]

> **Lemma 8.4**
>
> Let $\mathfrak{M}_X$ and $\mathfrak{M}_Y$ be σ-algebras in X and Y respectively. Suppose that $\varphi : X \to Y$ is measurable and $\mu : \mathfrak{M}_X \to [0, \infty]$ is a measure. Define $\nu = \varphi_* \mu$ by
>
> $$\nu(E) = \mu\big(\varphi^{-1}(E)\big)$$
>
> for $E \in \mathfrak{M}_Y$. Then the set function ν is a measure on $\mathfrak{M}_Y$ and for every measurable function $g : Y \to [0, \infty]$, we have
>
> $$\int_Y g \, d\nu = \int_X (g \circ \varphi) \, d\mu.$$

Let ρ be the measure on $(0, \infty)$ given by $d\rho = r^{k-1} \, dr$. Now it suffices to prove that

$$\nu(E) = m_k\big(\varphi^{-1}(E)\big) = (\rho \times \sigma_{k-1})(E) \tag{8.35}$$

for every $E \in \mathscr{B}\big((0, \infty) \times S_{k-1}\big)$, where σ_{k-1} is the measure defined on S_{k-1} in the question. This is because if the formula (8.35) holds, then for every nonnegative Borel function f on $\mathbb{R}^k$,

---

[c]The measure ν in Lemma 8.4 is called the **push-forward measure** of μ. See [9, §3.6].

we deduce immediately from Lemma 8.4 (with $X = \mathbb{R}^k \setminus \{\mathbf{0}\}$, $Y = (0, \infty) \times S_{k-1}$, φ is given by the formula (8.34), $\mu = m_k$ and $g = f \circ \varphi^{-1}$) and Theorem 8.8 (The Fubini Theorem) that

$$
\begin{aligned}
\int_{\mathbb{R}^k \setminus \{\mathbf{0}\}} f \, \mathrm{d}m_k &= \int_{(0,\infty) \times S_{k-1}} (f \circ \varphi^{-1}) \, \mathrm{d}(\rho \times \sigma_{k-1}) \\
&= \int_{(0,\infty) \times S_{k-1}} f\big(\varphi^{-1}(r, \mathbf{u})\big) \, \mathrm{d}\sigma_{k-1} \, \mathrm{d}\rho \\
&= \int_0^\infty \int_{S_{k-1}} r^{k-1} f(r\mathbf{u}) \, \mathrm{d}\sigma_{k-1} \, \mathrm{d}r.
\end{aligned}
\tag{8.36}
$$

Since $\{\mathbf{0}\}$ is a Borel set in $\mathbb{R}^k$ and $m_k(\mathbf{0}) = 0$, the integral on the left-hand side of the equation (8.36) can be replaced by

$$
\int_{\mathbb{R}^k} f \, \mathrm{d}m_k
$$

and we have what we want. To prove the formula (8.35), we have several steps:

- **Step 1: The formula (8.35) holds for $(r_1, r_2) \times A$, where $A \in \mathscr{B}(S_{k-1})$.** On the one hand, it is clear from Definition 8.7 that

$$
\begin{aligned}
(\rho \times \sigma_{k-1})\big((r_1, r_2) \times A\big) &= \rho\big((r_1, r_2)\big) \sigma_{k-1}(A) \\
&= \sigma_{k-1}(A) \int_{r_1}^{r_2} r^{k-1} \, \mathrm{d}r \\
&= \frac{r_2^k - r_1^k}{k} \sigma_{k-1}(A).
\end{aligned}
\tag{8.37}
$$

On the other hand, we see from the definition (8.34) that

$$
\varphi\big((r_1, r_2) \times A\big) = \{r\mathbf{u} \mid 0 < r_1 < r < r_2 \text{ and } \mathbf{u} \in A\} = (r_2 \widetilde{A}) \setminus [(r_1 \widetilde{A}) \cup \{r_1 \mathbf{u} \mid \mathbf{u} \in A\}]
$$

so that

$$
\begin{aligned}
\nu\big((r_1, r_2) \times A\big) &= m_k\big(\varphi\big((r_1, r_2) \times A\big)\big) \\
&= m_k\big((r_2 \widetilde{A}) \setminus [(r_1 \widetilde{A}) \cup \{r_1 \mathbf{u} \mid \mathbf{u} \in A\}]\big) \\
&= m_k(r_2 \widetilde{A}) - m_k(r_1 \widetilde{A}) - 0 \\
&= (r_2^k - r_1^k) m_k(\widetilde{A}) \\
&= \frac{r_2^k - r_1^k}{k} \sigma_{k-1}(A).
\end{aligned}
\tag{8.38}
$$

Hence our claim follows by combining the two expressions (8.37) and (8.38).

- **Step 2: The formula (8.35) holds for $B \times A$, where $A \in \mathscr{B}(S_{k-1})$ and $B \in \mathscr{B}((0, \infty))$.** For any open V in $(0, \infty)$, we know from [49, Exercise 29, Chapter 2] that it can be represented as an at most countable union of disjoint segments in the form (a, b). Since both ν (by Lemma 8.4) and $\rho \times \sigma_{k-1}$ are measures on $\mathscr{B}((0, \infty) \times S_{k-1})$, **Step 1** ensures that the formula (8.35) also holds for $V \times A$ and hence for all $K \times A$, where K is closed in $(0, \infty)$.

  Next, let U be a bounded Borel set in $(0, \infty)$. The proof of Theorem 2.17(c) says that you can find an increasing sequence of closed sets K_n and a decreasing sequence of bounded open sets V_n such that

$$
K_1 \subseteq K_2 \subseteq \cdots \subset U \subset \cdots \subset V_2 \subset V_1
\tag{8.39}
$$

and $m(V_n \setminus K_n) \to 0$ as $n \to \infty$. Clearly, it is true that $V_1 \setminus K_1 \supseteq V_2 \setminus K_2 \supseteq \cdots$ and $V_1 \setminus K_1$ is a bounded set. Let $r^{k-1} \le M$ on $V_1 \setminus K_1$ for some positive constant M. Then we have

$$\rho(V_n \setminus K_n) = \int_{V_n \setminus K_n} r^{k-1} \, dm \le M \int_{V_n \setminus K_n} dm = M \cdot m(V_n \setminus K_n) \to 0 \qquad (8.40)$$

as $n \to \infty$. Therefore, we apply the set relations (8.39), then the fact that φ^{-1} sends Borel sets in $(0, \infty) \times S_{k-1}$ to Borel sets in $\mathbb{R}^k \setminus \{\mathbf{0}\}$ and finally Definition 8.7 to get

$$\begin{aligned}
\nu\big((V_n \setminus U) \times A\big) &\le \nu\big((V_n \setminus K_n) \times A\big) \\
&= (\rho \times \sigma_{k-1})\big((V_n \setminus K_n) \times A\big) \\
&= \rho(V_n \setminus K_n)\sigma_{k-1}(A). \qquad (8.41)
\end{aligned}$$

By using the inequality (8.40), the inequality (8.41) shows that $\nu(V_n \setminus U) \times A) \to 0$ as $n \to \infty$. In other words, it means that

$$\nu(V_n \times A) \to \nu(U \times A) \qquad (8.42)$$

as $n \to \infty$. Furthermore, if we suppose

$$K = \bigcup_{n=1}^{\infty} K_n \quad \text{and} \quad V = \bigcap_{n=1}^{\infty} V_n$$

which are an F_σ set and a G_δ set respectively, then Theorem 1.19(d) and (e) imply that

$$\rho(V) - \rho(K) = \lim_{n \to \infty} [\rho(V_n) - \rho(K_n)] = \lim_{n \to \infty} \rho(V_n \setminus K_n) = 0$$

so that $\rho(V) = \rho(K)$. Since $K, V \in \mathscr{B}\big((0, \infty)\big)$ and $K \subset U \subset V$, Theorem 1.36 asserts that ρ is complete and then

$$\rho(U) = \rho(K) = \rho(U) = \lim_{n \to \infty} \rho(V_n). \qquad (8.43)$$

Thus we establish from Definition 8.7 and the limits (8.42) and (8.43) that

$$\begin{aligned}
(\rho \times \sigma_{k-1})(U \times A) &= \rho(U)\sigma_{k-1}(A) \\
&= \lim_{n \to \infty} [\rho(V_n)\sigma_{k-1}(A)] \\
&= \lim_{n \to \infty} (\rho \times \sigma_{k-1})(V_n \times A) \\
&= \lim_{n \to \infty} \nu(V_n \times A) \\
&= \nu(U \times A).
\end{aligned}$$

Finally, if U is any bounded or unbounded Borel set in $(0, \infty)$, then we consider the sets

$$U_n = U \cap (n - 1, n]$$

for $n \in \mathbb{N}$. It is trivial that each U_n is a bounded Borel set, so the preceding paragraph gives

$$(\rho \times \sigma_{k-1})(U_n \times A) = \nu(U_n \times A) \qquad (8.44)$$

for each $n = 1, 2, \ldots$. Since $U_i \cap U_j = \varnothing$ if $i \ne j$, $(U_i \times A) \cap (U_j \times A) = \varnothing$ for all $i \ne j$ and then the result (8.44) shows that

$$(\rho \times \sigma_{k-1})(U \times A) = (\rho \times \sigma_{k-1})\bigg(\bigg(\bigcup_{n=1}^{\infty} U_n\bigg) \times A\bigg)$$

$$= (\rho \times \sigma_{k-1})\Big(\bigcup_{n=1}^{\infty} (U_n \times A) \Big)$$

$$= \sum_{n=1}^{\infty} (\rho \times \sigma_{k-1})(U_n \times A)$$

$$= \sum_{n=1}^{\infty} \nu(U_n \times A)$$

$$= \nu\Big(\bigcup_{n=1}^{\infty} U_n \times A \Big)$$

$$= \nu(U \times A),$$

i.e., the formula (8.35) holds for every $U \times A$, where $U \in \mathscr{B}\big((0,\infty)\big)$ and $A \in \mathscr{B}(S_{k-1})$.

- **Step 3: The formula (8.35) holds for $E \in \mathscr{B}\big((0,\infty) \times S_{k-1}\big)$.** Suppose that

$$\mathfrak{M} = \{E \in \mathscr{B}\big((0,\infty) \times S_{k-1}\big) \mid \text{the formula (8.35) holds for } E\} \subseteq \mathscr{B}\big((0,\infty) \times S_{k-1}\big). \tag{8.45}$$

By using similar argument as in the proof of Problem 8.11, we can prove that

$$\mathscr{B}\big((0,\infty)\big) \times \mathscr{B}(S_{k-1}) = \mathscr{B}\big((0,\infty) \times S_{k-1}\big).$$

By **Step 2**, we see that $\mathscr{B}\big((0,\infty)\big) \times \mathscr{B}(S_{k-1}) \subseteq \mathfrak{M}$. Thus it concludes from the set inclusion (8.45) that

$$\mathfrak{M} = \mathscr{B}\big((0,\infty) \times S_{k-1}\big).$$

When $k = 2$, for a nonnegative Borel function f on $\mathbb{R}^2$, the formula becomes

$$\int_{\mathbb{R}^2} f(\mathbf{x})\,\mathrm{d}m_2 = \int_0^{\infty} r\Big\{ \int_0^{2\pi} f(r\cos\theta, r\sin\theta)\,\mathrm{d}\theta \Big\}\,\mathrm{d}r.$$

We have to show that

$$\int_{S_1} f(r\mathbf{u})\,\mathrm{d}\sigma_1(\mathbf{u}) = \int_0^{2\pi} f(r\cos\theta, r\sin\theta)\,\mathrm{d}\theta \tag{8.46}$$

for every $r \in (0,\infty)$. To this end, let $\mathbf{u} \in S_1$. Then $\mathbf{u}$ is a point on the unit circle so that $\mathbf{u} = \big(\frac{x}{\sqrt{x^2+y^2}}, \frac{y}{\sqrt{x^2+y^2}}\big)$. We define $F : B(\mathbf{0},1) \setminus \{\mathbf{0}\} \to [0,\infty]$ by

$$F(\mathbf{u}) = f(r\mathbf{u}) = f\Big(\frac{rx}{\sqrt{x^2+y^2}}, \frac{ry}{\sqrt{x^2+y^2}} \Big) = f(r\cos\theta, r\sin\theta),$$

where $\theta \in [0, 2\pi)$. Recall the definition of σ_1 so that

$$\int_{S_1} f(r\mathbf{u})\,\mathrm{d}\sigma_1 = 2 \int_{B(\mathbf{0},1)\setminus\{\mathbf{0}\}} F(\mathbf{u})\,\mathrm{d}m_2$$

$$= 2 \int_0^1 a \int_0^{2\pi} f(r\cos\theta, r\sin\theta)\,\mathrm{d}\theta\,\mathrm{d}a$$

$$= \int_0^{2\pi} f(r\cos\theta, r\sin\theta)\,\mathrm{d}\theta$$

which is exactly the formula (8.46). When $k = 3$, the formula becomes

$$\int_{\mathbb{R}^3} f(\mathbf{x})\,\mathrm{d}m_3 = \int_0^{\infty} \int_0^{2\pi} \int_0^{\pi} r^2 f(r\cos\varphi\sin\theta, r\sin\varphi\sin\theta, r\cos\theta)\sin\theta\,\mathrm{d}\theta\,\mathrm{d}\varphi\,\mathrm{d}r,$$

where $\theta \in [0, \pi)$ and $\varphi \in [0, 2\pi)$. This completes the proof of the problem.

> **Remark 8.2**
>
> We note that it can be further shown that the Borel measure σ_{k-1} on S_{k-1} is unique, see [22, Theorem 2.49, pp. 78, 79] for a proof of this. For similar or other proofs of Problem 8.6, you are suggested to read [36, pp. 172, 173, 322], [54, Theorem 15.13, pp. 154, 155] or [58, §3.2, pp. 279 – 281].

8.3 The Product Measure Theorem and Sections of a Function

> **Problem 8.7**
>
> *Rudin Chapter 8 Exercise 7.*

Proof. Since X and Y are σ-finite, X is the union of countably many disjoint sets X_n with $\mu(X_n) < \infty$, and that Y is the union of countably many disjoint sets Y_m with $\lambda(Y_m) < \infty$. For each $X_n \times Y_m$, we have

$$\psi(X_m \times Y_m) = \mu(X_n)\lambda(Y_m) < \infty. \tag{8.47}$$

Suppose that Ω is the class of all $E \in \mathscr{S} \times \mathscr{T}$ for which the conclusion of the problem holds, i.e.,

$$\Omega = \{E \in \mathscr{S} \times \mathscr{T} \mid \psi(E) = (\mu \times \lambda)(E)\}.$$

If $E = A \times B$ for some $A \in \mathscr{S}$ and $B \in \mathscr{T}$, then $E_x = B$ if $x \in A$ and $E_x = \varnothing$ if $x \in B$. Thus we deduce from Definition 8.7 that

$$(\mu \times \lambda)(E) = \int_X \lambda(E_x)\, d\mu(x) = \int_A \lambda(B)\, d\mu(x) = \mu(A)\lambda(B) = \psi(A \times B) = \psi(E). \tag{8.48}$$

In other words, $A \times B \in \Omega$. Besides, if $A_i \in \Omega$, $A_i \cap A_j = \varnothing$ for $i \neq j$ and $A = \bigcup_{i=1}^{\infty} A_i$, then since ψ and $\mu \times \mu$ are measures,

$$\psi(A) = \psi\left(\bigcup_{i=1}^{\infty} A_i\right) = \sum_{i=1}^{\infty} \psi(A_i) = \sum_{i=1}^{\infty}(\mu \times \lambda)(A_i) = (\mu \times \lambda)\left(\bigcup_{i=1}^{\infty} A_i\right) = (\mu \times \lambda)(A). \tag{8.49}$$

Suppose that

$$\mathfrak{M} = \{E \in \mathscr{S} \times \mathscr{T} \mid E \cap (X_n \times Y_m) \in \Omega \text{ for all } n, m \in \mathbb{N}\} \subseteq \mathscr{S} \times \mathscr{T}. \tag{8.50}$$

Our goal is to show that $\mathfrak{M}$ is a monotone class containing $\mathscr{E}$ so that Theorem 8.3 (The Monotone Class Theorem) can be applied to obtain the reverse direction of the set inclusion (8.50).

We claim that $\mathfrak{M}$ is a monotone class. To see this, let $A_i \in \mathfrak{M}$, $A_i \subseteq A_{i+1}$ and $A = \bigcup_{i=1}^{\infty} A_i$. By the definition (8.50), we know that

$$\psi\big(A_i \cap (X_n \times Y_m)\big) = (\mu \times \lambda)\big(A_i \cap (X_n \times Y_m)\big) \tag{8.51}$$

for all $n, m \in \mathbb{N}$. Since $A_i \cap (X_n \times Y_m) \in \Omega$ and $\mathscr{S} \times \mathscr{T}$ is a σ-algebra in $X \times Y$, we have

$$A \cap (X_n \times Y_m) = \bigcup_{i=1}^{\infty}[A_i \cap (X_n \times Y_m)] \in \mathscr{S} \times \mathscr{T}.$$

Furthermore, since $A_i \cap (X_n \times Y_m) \subseteq A_{i+1} \cap (X_n \times Y_m)$ for all $n, m \in \mathbb{N}$, Theorem 1.19(d) and the equality (8.51) imply that

$$
\begin{aligned}
\psi\big(A \cap (X_n \times Y_m)\big) &= \psi\Big(\bigcup_{i=1}^{\infty}[A_i \cap (X_n \times Y_m)]\Big) \\
&= \lim_{i \to \infty} \psi\big(A_i \cap (X_n \times Y_m)\big) \\
&= \lim_{i \to \infty} (\mu \times \lambda)\big(A_i \cap (X_n \times Y_m)\big) \\
&= (\mu \times \lambda)\Big(\bigcup_{i=1}^{\infty}[A_i \cap (X_n \times Y_m)]\Big) \\
&= (\mu \times \lambda)\big(A \cap (X_n \times Y_m)\big).
\end{aligned}
$$

In other words, $A \cap (X_n \times Y_m) \in \Omega$ so that $A \in \mathfrak{M}$. Next, if $B_i \in \mathfrak{M}$, $B_i \supseteq B_{i+1}$ and $B = \bigcap_{i=1}^{\infty} B_i$. By the definition (8.50) again, we have

$$
\psi\big(B_i \cap (X_n \times Y_m)\big) = (\mu \times \lambda)\big(B_i \cap (X_n \times Y_m)\big) \tag{8.52}
$$

for all $n, m \in \mathbb{N}$. Since $B_i \cap (X_n \times Y_m) \in \Omega$ and $\mathscr{S} \times \mathscr{T}$ is a σ-algebra in $X \times Y$, we have

$$
B \cap (X_n \times Y_m) = \bigcap_{i=1}^{\infty}[B_i \cap (X_n \times Y_m)] \in \mathscr{S} \times \mathscr{T}
$$

by Comment 1.6(c). In addition, we follow from the inequality (8.47) that

$$
\psi\big(B_1 \cap (X_n \times Y_m)\big) \leq \psi(X_n \times Y_m) < \infty
$$

and

$$
(\mu \times \lambda)\big(B_1 \cap (X_n \times Y_m)\big) \leq (\mu \times \lambda)(X_n \times Y_m) < \infty.
$$

Since $B_i \cap (X_n \times Y_m) \supseteq B_{i+1} \cap (X_n \times Y_m)$ for all $n, m \in \mathbb{N}$, Theorem 1.19(e) and the equality (8.52) imply that

$$
\begin{aligned}
\psi\big(B \cap (X_n \times Y_m)\big) &= \psi\Big(\bigcap_{i=1}^{\infty}[B_i \cap (X_n \times Y_m)]\Big) \\
&= \lim_{i \to \infty} \psi\big(B_i \cap (X_n \times Y_m)\big) \\
&= \lim_{i \to \infty} (\mu \times \lambda)\big(B_i \cap (X_n \times Y_m)\big) \\
&= (\mu \times \lambda)\Big(\bigcap_{i=1}^{\infty}[B_i \cap (X_n \times Y_m)]\Big) \\
&= (\mu \times \lambda)\big(B \cap (X_n \times Y_m)\big).
\end{aligned}
$$

In other words, $B \cap (X_n \times Y_m) \in \Omega$ so that $B \in \mathfrak{M}$. By Definition 8.1, $\mathfrak{M}$ is a monotone class as claimed.

We claim that $\mathscr{E} \subseteq \mathfrak{M}$. Denote $Q = R_1 \cup R_2 \cup \cdots \cup R_n$, where each R_i is a measurable rectangle and $R_i \cap R_j = \varnothing$ for $i \neq j$. Then each R_i is in the form $A_i \times B_i$ for some $A_i \in \mathscr{S}$ and $B_i \in \mathscr{T}$, so it is clear from the equality (8.48) that $R_i \in \Omega$. Since

$$
R_i \cap (X_n \times Y_m) = (A_i \cap X_n) \times (B_i \cap Y_m),
$$

we have $R_i \cap (X_n \times Y_m) \in \Omega$ and then $R_i \in \mathfrak{M}$. Since $R_i \neq R_j$ for $i \neq j$, we observe from the fact (8.49) that $Q \in \mathfrak{M}$. Hence we have our claim that $\mathscr{E} \subseteq \mathfrak{M}$.

Now we have proven our goal that $\mathfrak{M}$ is a monotone class containing $\mathscr{E}$. These fact imply that

$$\mathscr{S} \times \mathscr{T} \subseteq \mathfrak{M}. \tag{8.53}$$

Consequently, our proof is completed if we compare the set inclusions (8.50) and (8.53). ■

> **Remark 8.3**
>
> Problem 8.7 is sometimes called the **Product Measure Theorem**.

> **Problem 8.8**
>
> *Rudin Chapter 8 Exercise 8.*

Proof. We follow the hint given by Rudin.

(a) Given $x \in \mathbb{R}$. We take

$$\alpha_n(x) = \frac{[nx]}{n},$$

where $[x]$ denotes the greatest integer function. Since $x \in [\frac{i}{n}, \frac{i+1}{n})$ if and only if $nx \in [i, i+1)$ so that

$$\alpha_n(x_0) = \frac{i}{n}$$

if $x_0 \in [\frac{i}{n}, \frac{i+1}{n})$. Thus this implies that

$$0 \leq x - \alpha_n(x) < \frac{1}{n}$$

for every $n \in \mathbb{N}$ and $\alpha_n(x) \to x$ as $n \to \infty$. Therefore, it is illegal to say that there are $n \in \mathbb{N}$ and $i(n) \in \mathbb{Z}$ such that

$$\frac{i(n) - 1}{n} = a_{i(n)-1} \leq x \leq a_{i(n)} = \frac{i(n)}{n}$$

and $a_{i(n)} \to x$ as $n \to \infty$. Now we simply write $a_i = a_{i(n)}$ and define $f_n : \mathbb{R}^2 \to \mathbb{R}$ by

$$f_n(x, y) = \frac{a_i - x}{a_i - a_{i-1}} f(a_{i-1}, y) + \frac{x - a_{i-1}}{a_i - a_{i-1}} f(a_i, y) \tag{8.54}$$

$$= \frac{a_i - x}{a_i - a_{i-1}} f_{a_{i-1}}(y) + \frac{x - a_{i-1}}{a_i - a_{i-1}} f_{a_i}(y).$$

By the hypotheses, $f_{a_{i-1}}(y)$ and $f_{a_i}(y)$ are Borel measurable. Obviously, both

$$\frac{a_i - x}{a_i - a_{i-1}} \quad \text{and} \quad \frac{x - a_{i-1}}{a_i - a_{i-1}}$$

are continuous in x so that they are Borel measurable. By Proposition 1.19(c), each $f_n(x, y)$ is Borel measurable. Furthermore, it satisfies

$$|f_n(x, y) - f(x, y)| = \left| \frac{a_i - x}{a_i - a_{i-1}} f(a_{i-1}, y) + \frac{x - a_{i-1}}{a_i - a_{i-1}} f(a_i, y) - \frac{a_i - x}{a_i - a_{i-1}} f(x, y) \right.$$

$$-\frac{x - a_{i-1}}{a_i - a_{i-1}} f(x, y) \Big|$$
$$\le |f(a_{i-1}, y) - f(x, y)| + |f(a_i, y) - f(x, y)|$$
$$= |f^y(a_{i-1}) - f^y(x)| + |f^y(a_i) - f^y(x)|, \tag{8.55}$$

where $y \in \mathbb{R}$. Since a_i and a_{i-1} tend to x as $n \to \infty$ and f^y is continuous for all $y \in \mathbb{R}$, we deduce from the inequality (8.55) that

$$f(x, y) = \lim_{n \to \infty} f_n(x, y).$$

By Theorem 1.14, f is also Borel measurable in $\mathbb{R}^2$.

(b) We prove it by induction on k. The case for $k = 1$ is obviously true. Assume that the statement is also true for $k = m$, i.e., if $g(x_1, x_2, \ldots, x_m)$ is continuous in each of the m variables separately, then g is a Borel function in $\mathbb{R}^m$. For $k = m+1$, $g(x_1, x_2, \ldots, x_{m+1})$ is continuous in each of the $m + 1$ variables separately. Now for each $x_{m+1} \in \mathbb{R}$, the function

$$g_{x_{m+1}}(x_1, x_2, \ldots, x_m)$$

is continuous separately in $x_1, \ldots, x_m$. Thus we get from our assumption that every $g_{x_{m+1}}$ is a Borel function in $\mathbb{R}^m$. By our hypothesis of the question, *for every* choice $y = (x_2, x_3, \ldots, x_{m+1})$, the function

$$g^y(x_1) = g(x_1, x_2, \ldots, x_{m+1})$$

is continuous in x_1.

Now if we define $\frac{i-1}{n} = a_{i-1} \le x_1 \le a_i = \frac{i}{n}$ and $g_n : \mathbb{R}^{m+1} \to \mathbb{R}$ by

$$g_n(x, y) = \frac{a_i - x_1}{a_i - a_{i-1}} g(a_{i-1}, y) + \frac{x_1 - a_{i-1}}{a_i - a_{i-1}} g(a_i, y),$$

then we use a similar argument as part (a) to conclude that each g_n is Borel measurable in $\mathbb{R}^{m+1}$ and the limit

$$g(x, y) = \lim_{n \to \infty} g_n(x, y)$$

implies that g is a Borel function in $\mathbb{R}^{m+1}$. Hence our desired result follows from induction.

This completes the proof of the problem. ◼

Problem 8.9

Rudin Chapter 8 Exercise 9.

Proof. For each $x \in \mathbb{R}$, let $\{p_n\} \subseteq E$ such that $p_n \to x$. By the hypotheses, let N be the Lebesgue measurable set such that $m(N) = 0$ and f^y is continuous on $\mathbb{R}$ for all $y \in \mathbb{R} \setminus N$. Now for every $y \in \mathbb{R} \setminus N$, these facts imply that

$$\lim_{n \to \infty} f_{p_n}(y) = \lim_{n \to \infty} f(p_n, y) = \lim_{n \to \infty} f^y(p_n) = f^y(x) = f(x, y) = f_x(y).$$

Since each f_{p_n} is Lebesgue measurable, Theorem 1.14 ensures that f_x is also Lebesgue measurable. In other words, what we have shown is that for each $x \in \mathbb{R}$, the function f_x is Lebesgue measurable on $\mathbb{R} \setminus N$.

Using the same idea of the proof of Problem 8.8, we define $f_n : \mathbb{R} \times (\mathbb{R} \setminus N) \to \mathbb{R}$ by

$$f_n(x, y) = \frac{a_i - x}{a_i - a_{i-1}} f(a_{i-1}, y) + \frac{x - a_{i-1}}{a_i - a_{i-1}} f(a_i, y),$$

where

$$\frac{i(n) - 1}{n} = a_{i(n)-1} \leq x \leq a_{i(n)} = \frac{i(n)}{n}.$$

Imitate Problem 8.8's proof, we conclude that each $f_n(x, y)$ is Lebesgue measurable and the inequality (8.55) holds for every $y \in \mathbb{R} \setminus N$ which further imply that

$$f(x, y) = \lim_{n \to \infty} f_n(x, y)$$

for every $y \in \mathbb{R} \setminus N$. By Theorem 1.14, the function $f : \mathbb{R} \times \mathbb{R} \setminus N \to \mathbb{R}$ is clearly Lebesgue measurable. Eventually, the completeness of m_2 guarantees that $f : \mathbb{R}^2 \to \mathbb{R}$ is Lebesgue measurable which ends the analysis of the proof. $\blacksquare$

> ### Problem 8.10
>
> *Rudin Chapter 8 Exercise 10.*

Proof. Define each $f_n : \mathbb{R}^2 \to \mathbb{R}$ by the representation (8.54), put[d]

$$h_n(y) = f_n\big(g(y), y\big) = \frac{a_i - g(y)}{a_i - a_{i-1}} f(a_{i-1}, y) + \frac{g(y) - a_{i-1}}{a_i - a_{i-1}} f(a_i, y)$$

if $g(y) \in [a_{i-1}, a_i]$. Similar to the proof of Problem 8.8,

$$f_{a_{i-1}}(y) = f(a_{i-1}, y) \quad \text{and} \quad f_{a_i}(y) = f(a_i, y)$$

are Lebesgue measurable on $\mathbb{R}$ by the hypotheses. Furthermore, since g is Lebesgue measurable on $\mathbb{R}$, both

$$\frac{a_i - g(y)}{a_i - a_{i-1}} \quad \text{and} \quad \frac{g(y) - a_{i-1}}{a_i - a_{i-1}}$$

are also Lebesgue measurable on $\mathbb{R}$. By Proposition 1.19(c), each $h_n(y)$ is Lebesgue measurable on $\mathbb{R}$.

Now our h_n and h satisfy

$$
\begin{aligned}
|h_n(y) - h(y)| &= \left| \frac{a_i - g(y)}{a_i - a_{i-1}} f(a_{i-1}, y) + \frac{g(y) - a_{i-1}}{a_i - a_{i-1}} f(a_i, y) - \frac{a_i - g(y)}{a_i - a_{i-1}} f\big(g(y), y\big) \right. \\
&\quad \left. - \frac{g(y) - a_{i-1}}{a_i - a_{i-1}} f\big(g(y), y\big) \right| \\
&\leq |f(a_{i-1}, y) - f\big(g(y), y\big)| + |f(a_i, y) - f\big(g(y), y\big)| \\
&= |f^y(a_{i-1}) - f^y\big(g(y)\big)| + |f^y(a_i) - f^y\big(g(y)\big)|.
\end{aligned}
$$

By using similar argument as in the proof of Problem 8.8, we can show that

$$h(y) = \lim_{n \to \infty} h_n(y)$$

which implies that h is Lebesgue measurable on $\mathbb{R}$ by Theorem 1.14, completing the proof of the problem. $\blacksquare$

---

[d]Remember that i actually depends on n.

8.4 Miscellaneous Problems

> **Problem 8.11**
>
> *Rudin Chapter 8 Exercise 11.*

Proof. Recall from the definition on [42, p. 86] that the product topology on $\mathscr{T}_{m+n} = \mathbb{R}^{m+n}$ is $\mathbb{R}^m \times \mathbb{R}^n$ which is the topology having as basis the collection

$$\mathcal{B}_{m+n} = \{U \times V \mid U \in \mathscr{T}_m \text{ and } V \in \mathscr{T}_n\}.$$

Furthermore, the result [42, Theorem 15.1, p. 86] shows that the basis $\mathcal{B}_{m+n}$ can be replaced by the basis

$$\mathcal{B}'_{m+n} = \{U \times V \mid U \in \mathcal{B}_m \text{ and } V \in \mathcal{B}_n\}.$$

For every $k \geq 1$, since $\mathbb{R}^k$ is second-countable, it has a countable basis.[e] Thus we may assume that $\mathcal{B}_m$ and $\mathcal{B}_n$ are countable and consequently $\mathcal{B}'_{m+n}$ is also countable. Obviously, we have $U \in \mathcal{B}_m \subseteq \mathscr{T}_m \subseteq \mathscr{B}_m$ and $V \in \mathcal{B}_n \subseteq \mathscr{T}_n \subseteq \mathscr{B}_n$ so that

$$U \times V \in \mathcal{B}_m \times \mathcal{B}_n \subseteq \mathscr{T}_m \times \mathscr{T}_n \subseteq \mathscr{B}_m \times \mathscr{B}_n. \tag{8.56}$$

If $W \in \mathscr{T}^{m+n}$, then it can be expressed as a union of basis elements of $\mathcal{B}'_{m+n}$. Since $\mathcal{B}'_{m+n}$ is countable, this union must be countable too. Therefore, it follows from this fact, the fact $\mathscr{B}_m \times \mathscr{B}_n$ is an σ-algebra and the set relation (8.56) that $\mathscr{T}_{m+n} \subseteq \mathscr{B}_m \times \mathscr{B}_n$ and this means that

$$\mathscr{B}_{m+n} \subseteq \mathscr{B}_m \times \mathscr{B}_n. \tag{8.57}$$

For the other direction, the projections $\pi_1 : \mathbb{R}^m \times \mathbb{R}^n \to \mathbb{R}^m$ and $\pi_2 : \mathbb{R}^m \times \mathbb{R}^n \to \mathbb{R}^n$ are continuous mappings because $\pi_1^{-1}(U) = U \times \mathbb{R}^n$ and $\pi_2^{-1}(V) = \mathbb{R}^m \times V$ which are open in $\mathbb{R}^m \times \mathbb{R}^n = \mathbb{R}^{m+n}$ if U and V are open in $\mathbb{R}^m$ and $\mathbb{R}^n$ respectively. If $A \in \mathscr{B}_m$, then $A \times \mathbb{R}^n$ is a Borel set in $\mathbb{R}^m \times \mathbb{R}^n = \mathbb{R}^{m+n}$. Similarly, if $B \in \mathscr{B}_n$, then $\mathbb{R}^m \times B$ ia a Borel set in $\mathbb{R}^{m+n}$. Therefore, we get $A \times \mathbb{R}^n, \mathbb{R}^m \times B \in \mathscr{B}_{m+n}$ and these imply that

$$A \times B = (A \times \mathbb{R}^n) \cap (\mathbb{R}^m \times B) \in \mathscr{B}_{m+n}. \tag{8.58}$$

We recall from Definition 8.1 that $\mathscr{B}_m \times \mathscr{B}_n$ is the *smallest* σ-algebra in $\mathbb{R}^m \times \mathbb{R}^n = \mathbb{R}^{m+n}$ containing every measurable rectangle $A \times B$, where $A \in \mathscr{B}_m$ and $B \in \mathscr{B}_n$. Then we deduce from the set relation (8.58) that

$$\mathscr{B}_m \times \mathscr{B}_n \subseteq \mathscr{B}_{m+n}. \tag{8.59}$$

Hence our desired result follows if we combine the set relations (8.57) and (8.59), completing the proof of the problem. $\blacksquare$

> **Problem 8.12**
>
> *Rudin Chapter 8 Exercise 12.*

Proof. Let $f(x, y) = (\sin x)e^{-xt}$. Since f is continuous in $\mathbb{R}^2$, f is a m_2-measurable function on $\mathbb{R}^2$. It is trivial that $(0, A)$ and $(0, \infty)$ are σ-finite, where $A > 0$. Furthermore, we have

$$\int_0^A \mathrm{d}x \int_0^\infty |f(x,t)|\,\mathrm{d}t = \int_0^A |\sin x|\,\mathrm{d}x \int_0^\infty e^{-xt}\,\mathrm{d}t = \int_0^A \frac{|\sin x|}{x}\,\mathrm{d}x. \tag{8.60}$$

---

[e] Read [42, §30] for details.

By [49, Exercise 7, p. 197], we have $\frac{2}{\pi} < \frac{\sin x}{x} < 1$ on $(0, \frac{\pi}{2})$. Therefore, the integral (8.60) becomes

$$\left| \int_0^A dx \int_0^\infty |f(x,t)|\, dt \right| = \left| \int_0^A \frac{|\sin x|}{x}\, dx \right|$$

$$\leq \left| \int_0^{\frac{\pi}{2}} \frac{\sin x}{x}\, dx \right| + \left| \int_{\frac{\pi}{2}}^A \frac{|\sin x|}{x}\, dx \right|$$

$$< \frac{\pi}{2} + \int_{\frac{\pi}{2}}^A \frac{dx}{x}$$

$$= \frac{\pi}{2} + \ln A - \ln \frac{\pi}{2}$$

$$< \infty$$

for large $A > 0$.

By Theorem 8.8 (The Fubini Theorem)[f] and the given relation, we see that

$$\int_0^A \frac{\sin x}{x}\, dx = \int_0^A \sin x\, dx \left(\int_0^\infty e^{-xt}\, dt \right)$$

$$= \int_0^A dx \int_0^\infty (\sin x) e^{-xt}\, dt$$

$$= \int_0^\infty dt \int_0^A (\sin x) e^{-xt}\, dx$$

$$= \int_0^\infty \left[\frac{1}{1+t^2} - \frac{e^{-At}}{1+t^2}(\cos A + t \sin A) \right] dt$$

$$= \frac{\pi}{2} - \int_0^\infty \frac{e^{-At}}{1+t^2}(\cos A + t \sin A)\, dt. \tag{8.61}$$

For $A \geq 1$, we notice that

$$\left| \frac{e^{-At}}{1+t^2}(\cos A + t \sin A) \right| \leq \left| \frac{1}{1+t^2} \right| e^{-At} + \left| \frac{t}{1+t^2} \right| e^{-At} \leq 2e^{-At} \leq 2e^{-t}. \tag{8.62}$$

Then the given relation implies that $e^{-t} \in L^1(m)$ on $(0, \infty)$. Obviously, for $t \geq 0$, we note from the second inequality in (8.62) that

$$\lim_{A \to \infty} \frac{e^{-At}}{1+t^2}(\cos A + t \sin A) = 0,$$

so we deduce from this and Theorem 1.34 (Lebesgue's Dominated Convergence Theorem) that

$$\lim_{A \to \infty} \int_0^\infty \frac{e^{-At}}{1+t^2}(\cos A + t \sin A)\, dt = 0. \tag{8.63}$$

Hence our desired result follows from the representation (8.61) and the limit (8.63), completing the proof of the problem. ∎

> **Remark 8.4**
>
> The improper integral in Problem 8.12 is known as **Dirichlet integral**.

---

[f]See the *Notes* on [51, p. 165]

Problem 8.13

Rudin Chapter 8 Exercise 13.

Proof. We follow Rudin's hint. Since μ is a complex measure on a σ-algebra $\mathfrak{M}$ in X, Theorem 6.12 implies that *there is* a measurable function h such that $|h(x)| = 1$ on X and $d\mu = h\, d|\mu|$. This asserts that there exists a real measurable function θ such that $h = e^{i\theta}$ and so

$$d\mu = e^{i\theta}\, d|\mu|.$$

Given $E \in \mathfrak{M}$ and we define A_α to be the subset of E where $\cos(\theta - \alpha) > 0$. Then we have

$$\mu(A_\alpha) = \int_{A_\alpha} d\mu = \int_{A_\alpha} e^{i\theta}\, d|\mu| = \int_E \chi_{A_\alpha} e^{i\theta}\, d|\mu|$$

and this gives

$$|\mu(A_\alpha)| = |e^{-i\alpha}\mu(A_\alpha)| \geq \operatorname{Re}\left[e^{-i\alpha}\mu(A_\alpha)\right] = \operatorname{Re}\left(\int_E \chi_{A_\alpha} e^{i(\theta-\alpha)}\, d|\mu|\right). \tag{8.64}$$

Since $\chi_{A_\alpha}\cos(\theta - \alpha) = \cos^+(\theta - \alpha)$ by the Corollary following Theorem 1.14, the inequality (8.64) becomes

$$|\mu(A_\alpha)| \geq \int_E \chi_{A_\alpha}\cos(\theta - \alpha)\, d|\mu| = \int_E \cos^+(\theta - \alpha)\, d|\mu|. \tag{8.65}$$

As in the proof of Lemma 6.3 from Rudin, we may choose α_0 so as to maximize the integral in the inequality (8.65). Since this maximum is at least as large as the average of the sum over $[-\pi, \pi]$, we obtain

$$|\mu(A_{\alpha_0})| \geq \frac{1}{2\pi}\int_{-\pi}^{\pi}\int_E \cos^+(\theta - \alpha)\, d|\mu|\, d\theta.$$

Since $|\mu|(E) < \infty$ by Theorem 6.4, E is σ-finite. It is clear that $[-\pi, \pi]$ is also σ-finite. Thus we deduce from Theorem 8.8 (The Fubini Theorem) and the integral in Lemma 6.3 that

$$|\mu(A_{\alpha_0})| \geq \frac{1}{2\pi}\int_{-\pi}^{\pi}\int_E \cos^+(\theta - \alpha)\, d|\mu|\, d\theta = \int_E \left[\frac{1}{2\pi}\int_{-\pi}^{\pi}\cos^+(\theta - \alpha)\, d\theta\right]d|\mu| = \frac{1}{\pi}|\mu|(E).$$

By [16, Example 9.2.1(c), p. 285], if T is the circle group and $f : [0, 2\pi] \to T$ is given by $f(t) = (\cos t, \sin t)$, then the Haar measure μ on T is given by

$$\mu(S) = \frac{1}{2\pi}m\big(f^{-1}(S)\big). \tag{8.66}$$

In particular, we have $\mu(T) = 1$. Since f is continuous and $2 < 2\pi$, if we consider $f([0, 2]) = A$, then we see from the definition (8.66) that

$$\mu(A) = \frac{1}{2\pi}m\big(f^{-1}(A)\big) = \frac{1}{2\pi}m([0, 2]) = \frac{1}{\pi} = \frac{1}{\pi}\mu(S).$$

Since μ is a positive measure, we have $|\mu| = \mu$ so that the Haar measure considered here gives an example showing that $\frac{1}{\pi}$ is the beast constant. We have completed the proof of the problem. ∎

Problem 8.14

Rudin Chapter 8 Exercise 14.

Proof. Since $f(t)t^\alpha$ and $t^{-\alpha}$ satisfy the hypotheses of Theorem 3.5 (Hölder's Inequality), we have

$$xF(x) = \int_0^x f(t)t^\alpha t^{-\alpha}\,dt \le \left\{\int_0^x f^p t^{\alpha p}\,dt\right\}^{\frac{1}{p}}\left\{\int_0^x t^{\alpha q}\,dt\right\}^{\frac{1}{q}} = \left\{\int_0^x f^p t^{\alpha p}\,dt\right\}^{\frac{1}{p}} \times \left(\frac{x^{-\alpha q+1}}{-\alpha q+1}\right)^{\frac{1}{q}}$$

which gives

$$F^p(x) \le (1-\alpha q)^{1-p}x^{-1-\alpha p}\int_0^x f^p t^{\alpha p}\,dt$$

and then

$$\int_0^\infty F^p(x)\,dx \le (1-\alpha q)^{1-p}\int_0^\infty x^{-1-\alpha p}\int_0^x f^p t^{\alpha p}\,dt\,dx \tag{8.67}$$

$$= (1-\alpha q)^{1-p}\int_0^\infty x^{-1-\alpha p}\int_0^\infty \chi_{(0,x)}(t)f^p t^{\alpha p}\,dt\,dx. \tag{8.68}$$

As the space $(0,\infty)$ is σ-finite, $\chi_{(0,x)}x^{-1-\alpha p}f^p t^{\alpha p} \ge 0$ and measurable on $(0,\infty) \times (0,\infty)$, Theorem 8.8 (The Fubini Theorem) allows the inequality (8.67) can be rewritten as

$$\int_0^\infty F^p(x)\,dx \le (1-\alpha q)^{1-p}\int_0^\infty f^p t^{\alpha p}\int_0^\infty \chi_{(0,x)}(t)x^{-1-\alpha p}\,dx\,dt$$

$$= (1-\alpha q)^{1-p}\int_0^\infty f^p t^{\alpha p}\int_t^\infty x^{-1-\alpha p}\,dx\,dt$$

$$= (1-\alpha q)^{1-p}\int_0^\infty f^p t^{\alpha p} \times \frac{t^{-\alpha p}}{\alpha p}\,dt$$

$$= (1-\alpha q)^{1-p}(\alpha p)^{-1}\int_0^\infty f^p\,dt.$$

If we define $g(\alpha) = (1-\alpha q)^{1-p}(\alpha p)^{-1}$ on $(0,\frac{1}{q})$, then we can apply basic calculus to verify that g will attain the absolute minimum at $\alpha = \frac{1}{pq}$ and its minimum value is

$$\left(\frac{p}{p-1}\right)^p.$$

Hence we have again proven the result of Problem 3.14 and this completes the proof of the problem. ∎

Problem 8.15

Rudin Chapter 8 Exercise 15.

Proof. Notice that

$$\varphi(x) = \begin{cases} 1 - \cos x, & \text{if } x \in [0, 2\pi]; \\ 0, & \text{otherwise.} \end{cases} \tag{8.69}$$

For all $x \in \mathbb{R}$, we have

$$(f * g)(x) = \int_{-\infty}^\infty f(x-y)g(y)\,dy = \int_{-\infty}^\infty \sin y\,dy = \int_0^{2\pi} \sin y\,dy = 0.$$

This proves **Statement (i)**.

Clearly, if $x \in \mathbb{R}$, then it is easy to see that

$$(g * h)(x) = \int_{-\infty}^{\infty} g(x - y)h(y)\,dy = \int_{-\infty}^{\infty} g(y)h(x - y)\,dy = \int_{0}^{2\pi} \varphi'(y)h(x - y)\,dy. \qquad (8.70)$$

Applying the Mean Value Theorem to $\varphi(t) = 1 - \cos t$, we get

$$|\varphi(x) - \varphi(y)| \leq |x - y|$$

for all $x, y \in [0, 2\pi]$. Next, for $x \in [0, 2\pi]$, we follow from the definition (8.69) that

$$h(x) = \int_{-\infty}^{x} \varphi(t)\,dt = \int_{0}^{x} (1 - \cos t)\,dt = x - \sin x.$$

Using the Mean Value Theorem to h, we obtain

$$|h(x) - h(y)| \leq 2|x - y|$$

for all $x, y \in [0, 2\pi]$. Hence we have

$$\varphi, h \in \mathrm{Lip}\,1$$

on $[0, 2\pi]$. By Problem 7.10, φ and h are AC on $[0, 2\pi]$ and then we may apply Problem 7.14 to the integral (8.70) to establish

$$\begin{aligned}
(g * h)(x) &= \int_{0}^{2\pi} \varphi'(y)h(x - y)\,dy \\
&= \varphi(2\pi)h(x - 2\pi) - \varphi(0)h(x) + \int_{0}^{2\pi} \varphi(y)h'(x - y)\,dy \\
&= \int_{0}^{2\pi} \varphi(y)\varphi(x - y)\,dy \\
&= \int_{-\infty}^{\infty} \varphi(y)\varphi(x - y)\,dy \\
&= (\varphi * \varphi)(x)
\end{aligned}$$

for all $x \in \mathbb{R}$. Suppose that $x \in (0, 2\pi]$. Then we have

$$(\varphi * \varphi)(x) = \int_{0}^{x} \varphi(y)\varphi(x - y)\,dy.$$

It is obvious that $\varphi(y) = 1 - \cos y > 0$ on $(0, x)$. Similarly, if $y \in (0, x)$, then $x - y \in (0, x)$ and so $\varphi(x - y) = 1 - \cos(x - y) > 0$ for all $y \in (0, x)$. Consequently, we deduce that

$$(\varphi * \varphi)(x) > 0 \qquad (8.71)$$

on $(0, 2\pi]$. Suppose that $x \in (2\pi, 4\pi)$. Then the definition (8.69) implies that

$$(\varphi * \varphi)(x) = \int_{0}^{x} \varphi(y)\varphi(x - y)\,dy = \int_{0}^{x - 2\pi} \varphi(y)\varphi(x - y)\,dy.$$

Obviously, $\varphi(y) = 1 - \cos y > 0$ on $(0, x - 2\pi)$. If $0 < y < x - 2\pi$, then $2\pi < x - y < x$ and so $\varphi(x - y) = 1 - \cos(x - y) > 0$ for all $y \in (0, x - 2\pi)$. Therefore, the inequality (8.71) is till valid whenever $x \in (2\pi, 4\pi)$. In conclusion, we establish that

$$(g * h)(x) = (\varphi * \varphi)(x) > 0$$

on $(0, 4\pi)$. This is **Statement (ii)**.

On the one hand, **Statement (i)** says that $f * g = 0$ which implies that $(f * g) * h = 0$. On the other hand, if $x \in \mathbb{R}$, since $f(x) = 1$ for all $x \in \mathbb{R}$, we have

$$[f * (g * h)](x) = \int_{-\infty}^{\infty} f(y)(g * h)(x - y)\, dy = \int_{-\infty}^{\infty} (g * h)(x - y)\, dy. \tag{8.72}$$

By Theorem 7.26 (The Change-of-variables Theorem) (or its special case on p. 156), we see that the last integral in the expression (8.72) becomes

$$\int_{-\infty}^{\infty} (g * h)(y)\, dy. \tag{8.73}$$

Recall from the definition (8.69) that $\varphi(y) = 0$ outside $[0, 2\pi]$ and $\varphi(x - y) = 0$ outside $[x - 2\pi, x]$. If $x \leq 0$ or $x \geq 4\pi$, then $[0, 2\pi] \cap [x - 2\pi, x]$ is of measure 0 so that $(\varphi * \varphi)(x) = 0$ outside $(0, 4\pi)$. Hence it follows from this fact and the expression (8.73) that

$$[f * (g * h)](x) = \int_{-\infty}^{\infty} (g * h)(y)\, dy = \int_{0}^{4\pi} (g * h)(y)\, dy = \int_{0}^{4\pi} (\varphi * \varphi)(y)\, dy > 0$$

by **Statement (ii)**.

Now the inconsistence in **Statement (iii)** is due to the fact that $f \notin L^1(\mathbb{R})$. This completes the proof of the problem. $\blacksquare$

Problem 8.16

Rudin Chapter 8 Exercise 16.

Proof. Suppose that $(X, \mathscr{L}, \mu)$ and $(Y, \mathscr{T}, \nu)$ are σ-finite measure spaces. Let f be an $(\mathscr{L} \times \mathscr{T})$-measurable function on $X \times Y$. If $0 \leq f \leq \infty$ and $1 \leq p < \infty$, then we have the analogy of Minkowski's inequality in the question.

If $p = 1$, then it is exactly Theorem 8.8 (The Fubini Theorem). Suppose that $1 < p < \infty$ and q is the conjugate exponent of p. Since the inequality holds trivially when

$$\int \left[\int f^p(x, y)\, d\mu(x) \right]^{\frac{1}{p}} d\lambda(y) = \infty,$$

so without loss of generality, we assume further that this integral is finite. Let $g \in L^q(\mu)$. By Theorem 8.8 (The Fubini Theorem) and Theorem 3.5 (Hölder's Inequality), we have

$$\int \left[\int f(x, y)\, d\lambda(y) \right] \cdot |g(x)|\, d\mu(x) = \int \int f(x, y) \cdot |g(x)|\, d\mu(x)\, d\lambda(y)$$

$$\leq \int \left\{ \int f^p(x, y)\, d\mu(x) \right\}^{\frac{1}{p}} \left\{ \int |g|^q\, d\mu(x) \right\}^{\frac{1}{q}} d\lambda(y)$$

$$= \|g\|_q \int \left[\int f^p(x, y)\, d\mu(x) \right]^{\frac{1}{p}} d\lambda(y). \tag{8.74}$$

Thus if we define $\Phi : L^q(\mu) \to [0, \infty]$ by

$$\Phi(g) = \int \left[\int f(x, y)\, d\lambda(y) \right] \cdot |g(x)|\, d\mu(x), \tag{8.75}$$

then the inequality (8.74) and the assumption imply that Φ is a bounded linear functional on $L^q(\mu)$. Applying Theorem 6.16 to Φ, there exists a unique $h \in L^p(\mu)$ such that

$$\Phi(g) = \int hg\, d\mu(x). \tag{8.76}$$

Since $\Phi(|g|) = \Phi(g)$, the uniqueness of h and the comparison of the representations (8.75) and (8.76) show that

$$h = \int f(x, y)\, d\lambda(y).$$

Furthermore, Theorem 6.16 gives that $\|\Phi\| = \|h\|_p$ and we deduce from this and the inequality (8.74) that

$$\left\{ \int \left[\int f(x, y)\, d\lambda(y) \right]^p d\mu(x) \right\}^{\frac{1}{p}} = \|h\|_p$$

$$= \|\Phi\|$$

$$= \sup\{|\Phi(g)| \,|\, g \in L^q(\mu) \text{ and } \|g\|_q = 1\}$$

$$\leq \int \left[\int f^p(x, y)\, d\mu(x) \right]^{\frac{1}{p}} d\lambda(y)$$

which completes the proof of the problem.

CHAPTER **9**

Fourier Transforms

9.1 Properties of The Fourier Transforms

Rudin Chapter 9 Exercise 1.

Proof. For every $y \in \mathbb{R}$, $|f(x)\mathrm{e}^{-ixy}| = |f(x)| = f(x)$ so that $f(x)\mathrm{e}^{-ixy} \in L^1$. By Definition 9.1 and Theorem 1.33, we have

$$|\widehat{f}(y)| = \frac{1}{\sqrt{2\pi}}\left|\int_{-\infty}^{\infty} f(x)\mathrm{e}^{-ixy}\,\mathrm{d}x\right| \le \frac{1}{\sqrt{2\pi}} \int_{-\infty}^{\infty} f(x)\,\mathrm{d}x = \widehat{f}(0)$$

for every $y \in \mathbb{R}$. Assume that $|\widehat{f}(y)| = \widehat{f}(0)$ for some $y \neq 0$, i.e.,

$$\left|\int_{-\infty}^{\infty} f(x)\mathrm{e}^{-ixy}\,\mathrm{d}x\right| = \int_{-\infty}^{\infty} f(x)\,\mathrm{d}x = \int_{-\infty}^{\infty} |f(x)\mathrm{e}^{-ixy}|\,\mathrm{d}x.$$

By Theorem 1.39(c), there is a constant α such that

$$\alpha f(x)\mathrm{e}^{-ixy} = |f(x)\mathrm{e}^{-ixy}| = |f(x)| = f(x) \tag{9.1}$$

a.e. on $\mathbb{R}$. Since $f(x) > 0$ on $\mathbb{R}$, the expression (9.1) implies that $\alpha \mathrm{e}^{-ixy} = 1$ a.e. on $\mathbb{R}$. Thus it gives $y = 0$, a contradiction. This completes the proof of the problem. $\blacksquare$

Rudin Chapter 9 Exercise 2.

Proof. We prove the assertions one by one.

- **The Fourier transform of** $\chi_{[a,b]}$**.** By Definition 9.1, we have

$$\widehat{\chi_{[a,b]}}(t) = \frac{1}{\sqrt{2\pi}} \int_{-\infty}^{\infty} \chi_{[a,b]}(x)\mathrm{e}^{-ixt}\,\mathrm{d}x$$

$$= \frac{1}{\sqrt{2\pi}} \int_{a}^{b} \mathrm{e}^{-ixt}\,\mathrm{d}x$$

$$
= \begin{cases} -\dfrac{(e^{-ibt} - e^{-iat})}{\sqrt{2\pi}\,it}, & \text{if } t \neq 0; \\[3ex] \dfrac{b - a}{\sqrt{2\pi}}, & \text{otherwise.} \end{cases} \tag{9.2}
$$

By similar argument, we see that $\widehat{\chi_{(a,b)}}$, $\widehat{\chi_{(a,b]}}$ and $\widehat{\chi_{[a,b)}}$ also give the same answer (9.2).

- **The expression of** $\chi_{[-n,n]} * \chi_{[-1,1]}$. We have $g_n = \chi_{[-n,n]}$ and $h = \chi_{[-1,1]}$. By Definition 9.1, we have

$$
\begin{aligned}
(g_n * h)(x) &= \frac{1}{\sqrt{2\pi}} \int_{-\infty}^{\infty} \chi_{[-n,n]}(x - y)\chi_{[-1,1]}(y)\,\mathrm{d}y \\
&= \frac{1}{\sqrt{2\pi}} \int_{-1}^{1} \chi_{[-n,n]}(x - y)\,\mathrm{d}y \\
&= \frac{1}{\sqrt{2\pi}} m\big([x - n, x + n] \cap [-1, 1]\big).
\end{aligned} \tag{9.3}
$$

If $x \leq -n - 1$, then $x + n \leq -1$ so that

$$
m\big([x - n, x + n] \cap [-1, 1]\big) = 0.
$$

If $-n - 1 \leq x \leq -n + 1$, then $-1 \leq x + n \leq 1$ and $x - n \leq -2n + 1 \leq -1$ so that

$$
m\big([x - n, x + n] \cap [-1, 1]\big) = x + n - (-1) = x + n + 1.
$$

If $-n + 1 \leq x \leq n - 1$, then $x + n \geq 1$ and $x - n \leq -1$ so that $[-1, 1] \subseteq [x - n, x + n]$ and

$$
m\big([x - n, x + n] \cap [-1, 1]\big) = 2.
$$

If $n - 1 \leq x \leq n + 1$, then $-1 \leq x - n \leq 1$ and $x + n \geq 2n - 1 \geq 1$ so that

$$
m\big([x - n, x + n] \cap [-1, 1]\big) = 1 - (x - n) = 1 - x + n.
$$

If $x \geq n + 1$, then $x - n \geq 1$ so that

$$
m\big([x - n, x + n] \cap [-1, 1]\big) = 0.
$$

- **Proof of** $\dfrac{2}{\pi}\widehat{f_n} = g_n * h$. We first claim that $f_n \in L^1$ for each $n = 1, 2, \ldots$. Recall the basic fact that $\dfrac{\sin x}{x} \to 1$ as $x \to 0$. Thus if we define $f_n(0) = n$, then f_n is continuous on $\mathbb{R}$. Particularly, f_n is bounded by a positive constant M_n on $[-1, 1]$. Since $|f_n(x)| \leq \frac{1}{x^2}$ if $|x| \geq 1$, we have

$$
\begin{aligned}
\int_{\mathbb{R}} |f_n(x)|\,\mathrm{d}x &= \int_{|x| \geq 1} |f_n(x)|\,\mathrm{d}x + \int_{|x| \leq 1} |f_n(x)|\,\mathrm{d}x \\
&\leq \int_{1}^{\infty} \frac{\mathrm{d}x}{x^2} + \int_{-\infty}^{-1} \frac{\mathrm{d}x}{x^2} + M_n \\
&= 2 + M_n.
\end{aligned}
$$

Therefore, our claim is true.

Next, since $g_n, h \in L^1$, Theorem 9.2(c) implies that $\widehat{g_n * h} = \widehat{g_n} \cdot \widehat{h}$. By the formulas (9.2), we see that

$$\widehat{g_n}(x) = \begin{cases} \sqrt{\dfrac{2}{\pi}} \cdot \dfrac{\sin nx}{x}, & \text{if } x \neq 0; \\[2ex] \sqrt{\dfrac{2}{\pi}}, & \text{otherwise} \end{cases} \quad \text{and} \quad \widehat{h}(x) = \begin{cases} \sqrt{\dfrac{2}{\pi}} \cdot \dfrac{\sin x}{x}, & \text{if } x \neq 0; \\[2ex] \sqrt{\dfrac{2}{\pi}}, & \text{otherwise.} \end{cases}$$

Therefore, we have

$$\widehat{(g_n * h)}(x) = \begin{cases} \dfrac{2}{\pi} \cdot \dfrac{\sin x \sin nx}{x^2}, & \text{if } x \neq 0; \\[2ex] \dfrac{2}{\pi}, & \text{otherwise} \end{cases}$$

and then

$$\widehat{(g_n * h)}(x) = \frac{2}{\pi} \cdot f_n(x) \tag{9.4}$$

for all $x \neq 0$. Now the equation (9.4) clearly gives

$$\frac{2}{\pi} \widehat{f_n}(-x) = \int_{-\infty}^{\infty} \underbrace{\frac{2}{\pi} f(t) e^{ixt} \, dm(x)}_{\text{It is } g \text{ in the theorem}} = \int_{-\infty}^{\infty} \underbrace{\widehat{(g_n * h)}(t) e^{ixt} \, dm(x)}_{\text{It is } f \text{ in the theorem}}$$

Furthermore, the relation (9.4) also gives $\widehat{(g_n * h)} \in L^1$. Recall the expression (9.3) ensures that $g_n * h \in L^1$. In other words, the function $g_n * h$ satisfies the hypotheses of Theorem 9.11 (The Inversion Theorem), so we conclude that

$$\frac{2}{\pi} \widehat{f_n}(-x) = (g_n * h)(x) \tag{9.5}$$

a.e. on $\mathbb{R}$. Since f_n is an odd function, the equation (9.5) reduces to

$$\frac{2}{\pi} \widehat{f_n}(x) = (g_n * h)(x)$$

a.e. on $\mathbb{R}$, as required.

- $\|f_n\|_1 \to \infty$ **as** $n \to \infty$**.** By [49, Exercise 7, p. 197], we have

$$\frac{\sin x}{x} > \frac{2}{\pi} \tag{9.6}$$

if $0 < x < \frac{\pi}{2}$. Again the fact $\frac{\sin x}{x} \to 1$ as $x \to 0$ implies that the inequality (9.6) also holds at $x = 0$. By the definition, we have

$$|f_n(x)| > \frac{2}{\pi} \cdot \frac{|\sin nx|}{x}$$

if $0 \le x < \frac{\pi}{2}$ and so

$$\|f_n\|_1 = \int_{-\infty}^{\infty} |f_n(x)| \, dm(x) \ge \frac{2}{\pi} \int_0^{\frac{\pi}{2}} \frac{|\sin nx|}{x} \, dm(x) = \frac{2}{\pi} \int_0^{\frac{n\pi}{2}} \frac{|\sin x|}{x} \, dm(x). \tag{9.7}$$

It is well-known that

$$\int_0^\infty \frac{|\sin x|}{x} \, dx \ge \sum_{k=0}^{n} \int_{k\pi}^{(k+1)\pi} \frac{|\sin x|}{x} \, dx \ge \sum_{k=0}^{n} \int_{k\pi}^{(k+1)\pi} \frac{|\sin x|}{(k+1)\pi} \, dx = \frac{2}{\pi} \sum_{k=0}^{n} \frac{1}{k+1} \to \infty$$

as $n \to \infty$, so the right-most integral in the inequality (9.7) tends to ∞ as $n \to \infty$.

- **The mapping $f \mapsto \widehat{f}$ maps L^1 into a proper subset of C_0.** Suppose that $\mathcal{F} : L^1 \to C_0$ is given by

$$\mathcal{F}(f) = \widehat{f}.$$

By Definition 9.1, $\mathcal{F}$ is linear. If $f \in L^1$ and $\|f\|_1 = 1$, then Theorem 9.6 gives $\|\widehat{f}\|_\infty \le 1$ so that

$$\|\mathcal{F}\| - \sup\{\|\widehat{f}\|_\infty \mid f \in L^1 \text{ and } \|f\|_1 = 1\} \le 1,$$

i.e., $\mathcal{F}$ is bounded. If $\mathcal{F}(f) = \mathcal{F}(g)$, then the linearity of $\mathcal{F}$ gives

$$\widehat{f - g} = \mathcal{F}(f - g) = \mathcal{F}(f) - \mathcal{F}(g) = 0.$$

By Theorem 9.12 (The Uniqueness Theorem), we obtain $f(x) = g(x)$ for almost all $x \in \mathbb{R}$ which means that $\mathcal{F}$ is injective.

Assume that $\mathcal{F}$ was surjective. Since both L^1 and C_0 are Banach, we know from Theorem 5.10 that there is a $\delta > 0$ such that

$$\|\widehat{f}\|_\infty \ge \delta \|f\|_1 \tag{9.8}$$

for all $f \in L^1$. Recall the facts from the above assertions that

$$f_n \in L^1, \quad g_n * h \in L^1, \quad \frac{2}{\pi}\widehat{f_n} = g_n * h \quad \text{and} \quad \|\widehat{f_n}\|_\infty = \frac{\pi}{2}\|g_n * h\|_\infty = \frac{\pi}{2} \cdot \sqrt{\frac{2}{\pi}} = \sqrt{\frac{2}{\pi}},$$

so we get from the inequality (9.8) that

$$\|f_n\|_1 \le \frac{1}{\delta}\sqrt{\frac{2}{\pi}} < \infty$$

but this contradicts the fact $\|f_n\|_1 \to \infty$. Hence $\mathcal{F}$ is *not* surjective.

- **The set $\mathcal{B} = \{\widehat{f} \mid f \in L^1\}$ is dense in C_0.** To solve this part, we need the locally compact Hausdorff version of the **Stone-Weierstrass Theorem**[a]. In fact, it is

> **Lemma 9.1 (The Stone-Weierstrass Theorem)**
>
> If $\mathcal{A}$ is a subalgebra of C_0 that separates points on $\mathbb{R}$ and vanishes at no point of $\mathbb{R}$, then $\mathcal{A}$ is dense in C_0.

It is clear that $\mathcal{B}$ is a vector subspace of C_0. For every pair of $f, g \in L^1$, Theorem 8.14 says that L^1 is closed under convolution, i.e., $f * g \in L^1$. By Theorem 9.2(c), we get

$$\widehat{f * g} = \widehat{f} \cdot \widehat{g} \in \mathcal{B}.$$

By the definition, $\mathcal{B}$ is an algebra. Suppose that $p, q \in \mathbb{R}$ and we have two cases for consideration.

- **Case (i): $p < q$ and $p \neq -q$.** Consider the function

$$f(x) = \mathrm{e}^{-|x|}. \tag{9.9}$$

By [51, Eqn. (3), p. 183], since f is even in $\mathbb{R}$, we have

$$\widehat{f}(t) = \int_{-\infty}^{\infty} f(x)\mathrm{e}^{-ixt}\,\mathrm{d}m(x) = \int_{-\infty}^{\infty} f(x)\mathrm{e}^{ixt}\,\mathrm{d}m(x) = h_1(t) = \sqrt{\frac{2}{\pi}} \cdot \frac{1}{1 + t^2}. \tag{9.10}$$

Since $\widehat{f}(x) = \widehat{f}(y)$ if and only if $x = \pm y$, we obtain $\widehat{f}(p) \neq \widehat{f}(q)$ in our case.

---

[a] See [49, pp. 159 – 165] for the compact cases and [22, §4.7] for the locally compact Hausdorff cases

– **Case (ii):** $p < q$ **and** $p = -q$. We consider $g(x) = f(x)e^{ix}$, where f is the function (9.9). By Theorem 9.2(a), we have

$$\widehat{g}(t) = \widehat{f}(t-1) = \sqrt{\frac{2}{\pi}} \cdot \frac{1}{1+(t-1)^2}.$$

If $\widehat{g}(p) = \widehat{g}(-q) = \widehat{g}(q)$, then we have $(q-1)^2 = (q+1)^2$ which implies that $q = 0$, a contradiction. Hence we have $\widehat{g}(-q) \neq \widehat{g}(q)$.

In other words, $\mathcal{B}$ separates points on $\mathbb{R}$. Furthermore, we see from the expression (9.10) that $\widehat{f}(t) \neq 0$ for each $t \in \mathbb{R}$. Thus $\mathcal{B}$ vanishes at no point of $\mathbb{R}$. Hence our desired result follows immediately from Lemma 9.1.

This completes the proof of the problem.

Remark 9.1

The result in Problem 9.2 that $\mathcal{F}(L^1) \subset C_0$ is called the **Riemann-Lebesgue Lemma** ([58, Theorem 1.4, p. 80]). See also §5.14 from Rudin.

Problem 9.3

Rudin Chapter 9 Exercise 3.

Proof. By similar analysis as in Problem 9.2, if $g_\lambda = \chi_{[-\lambda,\lambda]}$, then for $t \neq 0$, we have

$$\widehat{g_\lambda} = \sqrt{\frac{2}{\pi}} \cdot \frac{\sin \lambda t}{t}.$$

Note that $\widehat{g_\lambda} \notin L^1$, so we can't apply any inversion theorems (Theorems 9.11 or 9.14) directly. To solve this problem, we apply Theorem 9.13.

It is clear that $g_\lambda, \widehat{g_\lambda} \in L^2$. Using the same notation as in Theorem 9.13(d), we have

$$\psi_A(x) = \int_{-A}^{A} \widehat{g_\lambda}(t)e^{ixt}\, dm(t) = \frac{1}{\sqrt{2\pi}} \cdot \sqrt{\frac{2}{\pi}} \int_{-A}^{A} \frac{\sin \lambda t}{t} e^{ixt}\, dt = \frac{1}{\pi} \int_{-A}^{A} \frac{\sin \lambda t}{t} e^{ixt}\, dt$$

By Theorem 9.13(d) again,

$$\|\psi_A - g_\lambda\|_2 \to 0 \tag{9.11}$$

as $A \to \infty$. If we restrict A to be a positive integer, then the result (9.11) means that $\{\psi_A\}$ is a Cauchy sequence in L^2 with limit g_λ.[b] Thus it follows from Theorem 3.12 that $\{\psi_A\}$ has a subsequence which converges pointwise almost everywhere to g_λ, i.e.,

$$\lim_{A \to \infty} \int_{-A}^{A} \frac{\sin \lambda t}{t} e^{ixt}\, dt = \lim_{k \to \infty} \int_{-A_k}^{A_k} \frac{\sin \lambda t}{t} e^{ixt}\, dt = \pi \chi_{[-\lambda,\lambda]}(x) = \begin{cases} \pi, & \text{if } x \in [-\lambda, \lambda]; \\ 0, & \text{otherwise} \end{cases} \tag{9.12}$$

for almost all real x, where $\lambda > 0$.

Since the result (9.12) only holds for almost all real x, if we want to evaluate the limit *for all* $x \in \mathbb{R}$, we need an inversion theorem which is applicable to g_λ. Fortunately, there is such a

---

[b] Here we use the basic fact [49, Theorem 3.11(a)].

theorem for **piecewise smooth functions**[c]. Indeed, we have (see [21, Theorem 7.6, p. 220] or [55, Theorem 6.6.2, p. 330])

> **Lemma 9.2**
>
> If f is a piecewise smooth function on $\mathbb{R}$ such that $f \in L^1$, then we have
>
> $$\lim_{A \to \infty} \int_{-A}^{A} \widehat{f}(t) e^{ixt}\, dm(t) = \frac{\pi}{2}[f(x-) + f(x+)]$$
>
> for every $x \in \mathbb{R}$.

Now the g_λ has discontinuities only at $x = \pm\lambda$, so it is piecewise smooth on $\mathbb{R}$ and $g_\lambda \in L^1$. Apply Lemma 9.2 to g_λ, we establish at once that

$$\lim_{A \to \infty} \int_{-A}^{A} \frac{\sin \lambda t}{t} e^{ixt}\, dt = \frac{\pi}{2}[g_\lambda(x-) + g_\lambda(x+)] = \begin{cases} \pi, & \text{if } x \in (-\lambda, \lambda); \\[2mm] \frac{\pi}{2} & \text{if } x = \pm\lambda; \\[2mm] 0, & \text{otherwise.} \end{cases}$$

Hence we have ended the analysis of the proof. ∎

> **Problem 9.4**
>
> *Rudin Chapter 9 Exercise 4.*

Proof. Suppose that there exists a $g \in L^1 \cap L^2$ such that $\widehat{g} \notin L^1$ and

$$f(x) = \widehat{g}(-x) \tag{9.13}$$

on $\mathbb{R}$. Then the hypotheses immediately give $f \notin L^1$. Since $g \in L^2$, Theorem 9.13 (The Plancherel Theorem) implies that

$$\|\widehat{g}\|_2 = \|g\|_2 < \infty.$$

Thus we have $\widehat{g} \in L^2$ and we deduce immediate from the hypothesis (9.13) that $f \in L^2$. Consequently, we must have $f \in L^2 \setminus L^1$. Now it remains to show that $\widehat{f} \in L^1$. To this end, we need the following property of Fourier transform ([55, Theorem 6.5.1, p. 324]):

> **Lemma 9.3**
>
> If $f \in L^2$, then we have $\widehat{\widehat{f}} = f^-$, where $\widehat{\widehat{f}}$ denotes the Fourier transform of $\widehat{f}$ and $f^-(x) = f(-x)$.

Now the expression (9.13) can be written as $f^- = \widehat{g}$. Since $g \in L^2$, this and Lemma 9.3 imply that

$$\widehat{f^-} = \widehat{\widehat{g}} = g^-.$$

Since $g \in L^1$, we have $g^- \in L^1$ and then $\widehat{f^-} \in L^1$. Recall from Definition 9.1 that

$$\widehat{f^-}(t) = \int_{-\infty}^{\infty} f^-(x) e^{-ixt}\, dm(x) = \int_{-\infty}^{\infty} f(-x) e^{-ixt}\, dm(x) = \int_{-\infty}^{\infty} f(x) e^{ixt}\, dm(x) = \overline{\widehat{f}(t)}.$$

---

[c] For the definition, see [55, Definition 1.4.2, p. 29].

Hence, we may conclude that $\widehat{f} \in L^1$.

Finally, we give examples of the problem. Consider $g_n = \chi_{[-n,n]}$ which is clearly an element of $L^1 \cap L^2$ for every $n \in \mathbb{N}$. Since

$$\widehat{g_n}(t) = \int_{-\infty}^{\infty} g_n(x) \mathrm{e}^{-ixt} \, \mathrm{d}m(x) = \frac{1}{\sqrt{2\pi}} \int_{-n}^{n} \mathrm{e}^{-ixt} \, \mathrm{d}x = \sqrt{\frac{2}{\pi}} \cdot \frac{2\sin nt}{t},$$

we have $\widehat{g_n} \notin L^1$. Therefore, the expression (9.13) ensures that each function

$$f_n(x) = \sqrt{\frac{2}{\pi}} \cdot \frac{2\sin nx}{x}$$

satisfies the hypotheses, completing the proof of the problem.

> ### Problem 9.5
>
> *Rudin Chapter 9 Exercise 5.*

Proof. By Theorem 9.6, we have $\widehat{f}$ is continuous on $\mathbb{R}$ which vanishes at infinity. Thus $\widehat{f}$ is bounded by a positive constant, say M, on $[-1, 1]$. In addition, it is obvious that

$$|\widehat{f}(t)| \le |t\widehat{f}(t)|$$

if $|t| > 1$ and so our hypothesis implies that

$$\int_{\mathbb{R}} |\widehat{f}(t)| \, \mathrm{d}t = \int_{|t| \le 1} |\widehat{f}(t)| \, \mathrm{d}t + \int_{|t| > 1} |\widehat{f}(t)| \, \mathrm{d}t \le M + \int_{|t| > 1} |t\widehat{f}(t)| \, \mathrm{d}t < \infty.$$

Therefore, $\widehat{f} \in L^1$ and Theorem 9.11 (The Inversion Theorem) ensures that

$$f(x) = g(x) = \int_{-\infty}^{\infty} \widehat{f}(t) \mathrm{e}^{ixt} \, \mathrm{d}m(t) \tag{9.14}$$

a.e. on $\mathbb{R}$.

It remains to show that g is differentiable with the mentioned derivative in the question. To this end, we define $F(t) = \widehat{f}(t)$ and $G(t) = -it\widehat{f}(t)$ so that

$$G(x) = -ixF(x)$$

for all $x \in \mathbb{R}$. Since $|G(t)| = |t\widehat{f}(t)|$, the hypotheses give $G \in L^1$. Thus we deduce from Theorem 9.2(f) that $\widehat{F}$ is differentiable and

$$\widehat{F}'(t) = \widehat{G}(t) \tag{9.15}$$

for all $t \in \mathbb{R}$. By the integral (9.14) and Definition 9.1, we see that

$$g(x) = \int_{-\infty}^{\infty} F(t) \mathrm{e}^{ixt} \, \mathrm{d}m(t) = \widehat{F}(-x).$$

Using the equation (9.15) and the definition of G, we conclude that

$$g'(x) = -\widehat{F}'(-x) = -\widehat{G}(-x) = -\int_{-\infty}^{\infty} G(t)\mathrm{e}^{ixt} \, \mathrm{d}m(t) = i\int_{-\infty}^{\infty} t\widehat{f}(t)\mathrm{e}^{ixt} \, \mathrm{d}m(t).$$

Hence we have completed the proof of the problem.

Problem 9.6

Rudin Chapter 9 Exercise 6.

Proof. Let $f = \chi_{[-1,1]}$. Clearly, $f \in L^1$ and $f' = 0$ a.e. on $\mathbb{R}$. Thus we have $f' \in L^1$ too. However, we see from Definition 9.1 that

$$\widehat{f'}(t) = \int_{-\infty}^{\infty} f'(x)\mathrm{e}^{-ixt}\,\mathrm{d}m(x) = 0$$

a.e. on $\mathbb{R}$ and

$$ti\widehat{f}(t) = ti\int_{-\infty}^{\infty} f(x)\mathrm{e}^{-ixt}\,\mathrm{d}m(x) = \frac{ti}{\sqrt{2\pi}}\int_{-1}^{1}\mathrm{e}^{-ixt}\,\mathrm{d}x = \frac{ti}{\sqrt{2\pi}}\cdot\frac{2\sin t}{t} = i\sqrt{\frac{2}{\pi}}\sin t.$$

Hence we conclude that $\widehat{(f')}(t) \neq ti\widehat{f}(t)$ which completes the proof of the problem. ∎

Problem 9.7

Rudin Chapter 9 Exercise 7.

Proof. Let $f \in S$. Denote $\mathcal{F}[f(x)](t) = \widehat{f}(t)$, or simply $\mathcal{F}[f(x)] = \widehat{f}$, to be the Fourier transform of f. We want to show that, for every m and $n = 0, 1, 2, \ldots$, there are positive numbers $A_{mn}(\mathcal{F}[f(x)]) < \infty$ such that

$$|x^n D^m \mathcal{F}[f(x)]| \le A_{mn}(\mathcal{F}[f(x)]).$$

Since $|x^2 f(x)| \le A_{02}(f)$ for every $x \in \mathbb{R}$, it means that

$$|f(x)| \le \frac{A_{02}}{x^2}$$

for every $x \neq 0$. By this, we have $f \in L^1$ and thus Theorem 9.6 gives

$$\big|\mathcal{F}[f(x)](t)\big| \le \big\|\mathcal{F}[f(x)]\big\|_{\infty} \le \|f\|_1 < \infty.$$

It is easy to prove that if $f \in S$, then $f'(x) \in S$ and $xf(x) \in S$. These facts show that $D^n\big((-ix)^m f(x)\big)$ belongs to S for every $m, n = 0, 1, 2, \ldots$ and so

$$\big|\mathcal{F}\big[D^n\big((-ix)^m f(x)\big)\big]\big| < \infty \tag{9.16}$$

for every $m, n = 0, 1, 2, \ldots$.

By Theorem 9.2(f), we observe that

$$\frac{\mathrm{d}}{\mathrm{d}t}\{\mathcal{F}[f(x)](t)\} = \mathcal{F}[-ixf(x)](t) \quad \text{and} \quad \frac{\mathrm{d}^2}{\mathrm{d}t^2}\{\mathcal{F}[f(x)](t)\} = \mathcal{F}[(-ix)^2 f(x)](t).$$

Therefore, we see that

$$D^m \mathcal{F}[f(x)](t) = \frac{\mathrm{d}^m}{\mathrm{d}t^m}\{\mathcal{F}[f(x)](t)\} = \mathcal{F}[(-ix)^m f(x)](t) \tag{9.17}$$

for every $m = 0, 1, 2, \ldots$. To continue the proof, we need the following lemma:

Lemma 9.4

Let $f \in S$. Then we have $\mathcal{F}[f'(x)](t) = it\mathcal{F}[f(x)](t)$.

Proof of Lemma 9.4. By Integration by Parts, we obtain

$$\int_{-N}^{N} f'(x)e^{-ixt}\,dx = f(x)e^{-ixt}\Big|_{-N}^{N} - \int_{-N}^{N} f(x)\,d(e^{-ixt})$$

$$= \left[f(N)e^{-iNt} - f(-N)e^{iNt}\right] + it\int_{-N}^{N} f(x)e^{-ixt}\,dx. \qquad (9.18)$$

Since $f \in S$, $f(x) \to 0$ as $x \to \pm\infty$. Thus we follow from the expression (9.18) by letting $N \to \infty$ that

$$\mathcal{F}[f'(x)](t) = it\mathcal{F}[f(x)](t),$$

completing the proof of the lemma.

Using Lemma 9.4 repeatedly, we establish the relation

$$\mathcal{F}[D^n f(x)](t) = (it)^n \mathcal{F}[f(x)](t) \qquad (9.19)$$

for every $n = 0, 1, 2, \ldots$. Particularly, by replacing $f(x)$ by $(-ix)^m f(x)$ in the formula (9.19), we gain

$$\mathcal{F}[D^n\left((-ix)^m f(x)\right)](t) = (it)^n \mathcal{F}[(-ix)^m f(x)](t)$$

or equivalently,

$$\mathcal{F}[(-ix)^m f(x)](t) = (it)^{-n} \mathcal{F}[D^n\left((-ix)^m f(x)\right)](t). \qquad (9.20)$$

Now we combine formulas (9.17) and (9.20) to conclude

$$(it)^n D^m \mathcal{F}[f(x)](t) = \mathcal{F}[D^n\left((-ix)^m f(x)\right)](t)$$

and we use the result (9.16) to get

$$\left|t^n D^m \mathcal{F}[f(x)](t)\right| = \left|\mathcal{F}[D^n\left((-ix)^m f(x)\right)]\right| < \infty$$

for every $m, n = 0, 1, 2, \ldots$. Hence we have proven that $\mathcal{F}[f(x)] = \widehat{f} \in S$.

The function $f(x) = e^{-x^2}$ is called a **Gaussian function** which is an element of S. Similarly, the function $f(x) = P(x)e^{-x^2}$ belongs to S for any polynomial $P(x)$. This completes the proof of the problem.

Remark 9.2

The class S in Problem 9.7 is called the **Schwartz space** or **space of rapidly decreasing functions on $\mathbb{R}$.**

Problem 9.8

Rudin Chapter 9 Exercise 8.

Proof. We are going to show the assertions one by one.

- $h = f * g$ **is uniformly continuous on $\mathbb{R}$.** We claim that $(f * g)_z = f_z * g$. To see this, we note from the definition that

$$(f*g)_z(x) = (f*g)(x-z) = \int_{-\infty}^{\infty} f(x-z-y)g(y)\,dy = \int_{-\infty}^{\infty} f_z(x-y)g(y)\,dy = (f_z*g)(x).$$

It is clear that either p or q must be finite. Suppose that $1 \le p < \infty$. We need the following result:

> **Lemma 9.5**
>
> Let $p, q \in [1, \infty]$ be conjugate exponents. If $f \in L^p$ and $g \in L^q$, then $f * g \in L^\infty$ and
> $$\|f * g\|_\infty \le \|f\|_p \cdot \|g\|_q. \tag{9.21}$$

Proof of Lemma 9.5. Suppose that $\sigma(f)(x) = f(-x)$. By Theorem 3.8 and the fact that $\sigma(f_{-x})(y) = f_{-x}(-y) = f(-y + x)$ for every $y \in \mathbb{R}$, if $x \in \mathbb{R}$, then

$$|(f * g)(x)| = \left| \int_{-\infty}^{\infty} f(x - y) g(y) \, dm(y) \right| = \|\sigma(f_{-x}) \times g\|_1 \le \|\sigma(f_{-x})\|_p \cdot \|g\|_q. \tag{9.22}$$

In addition, we have

$$\|\sigma(f_x)\|_p = \left\{ \int_{-\infty}^{\infty} |f(-y - x)|^p \, dy \right\}^{\frac{1}{p}} = \left\{ \int_{-\infty}^{\infty} |f(y)|^p \, dy \right\}^{\frac{1}{p}} = \|f\|_p. \tag{9.23}$$

Hence we derive (9.23) by simply putting the result (9.23) back into the inequality (9.22). This proves Lemma 9.5. $\blacksquare$

Now it is easily seen that the convolution of functions f and g is distributive, so for every $x \in \mathbb{R}$, it follows from Lemma 9.5 that

$$\begin{aligned}
|(f * g)(x - \delta) - (f * g)(x)| &= |(f * g)_\delta(x) - (f * g)(x)| \\
&\le \|(f * g)_\delta - f * g\|_\infty \\
&= \|(f_\delta - f) * g\|_\infty \\
&\le \|f_\delta - f\|_p \cdot \|g\|_q
\end{aligned}$$

which implies that $|(f * g)(x - \delta) - (f * g)(x)| \to 0$ as $\delta \to 0$. Hence $f * g$ is uniformly continuous on $\mathbb{R}$.

- $h \in C_0$ **if** $1 < p < \infty$. Recall from Theorem 3.14 that $C_c(\mathbb{R})$ is dense in L^p and L^q. Choose $\{f_n\} \subseteq C_c(\mathbb{R}) \subset L^p$ and $\{g_n\} \subseteq C_c(\mathbb{R}) \subset L^q$ such that $\|f_n - f\|_p \to 0$ and $\|g_n - g\|_q \to 0$ as $n \to \infty$.

  Let M be a positive constant such that $\operatorname{supp}(f_n), \operatorname{supp}(g_n) \subseteq [-M, M]$. Thus for every $|x| > 2M$ and every $y \in \mathbb{R}$, we have either $y \in \mathbb{R} \setminus [-M, M]$ or $x - y \in \mathbb{R} \setminus [-M, M]$. Otherwise, it will give the contradiction that $|x| = |x - y + y| \le 2M$. Thus this fact gives $f_n(x - y) g_n(y) = 0$ and then

  $$(f_n * g_n)(x) = 0$$

  for every x such that $|x| > 2M$, i.e., $\operatorname{supp}(f_n * g_n) \subseteq [-2M, 2M]$ which means that each $f_n * g_n$ is an element of $C_c(\mathbb{R})$.

  Next, by Lemma 9.5 and then Theorem 3.9, we obtain

  $$\begin{aligned}
  |(f_n * g_n)(x) - (f_m * g_m)(x)| &\le |(f_n * g_n)(x) - (f_n * g_m)(x)| \\
  &\quad + |(f_n * g_m)(x) - (f_m * g_m)(x)| \\
  &= \left| [f_n * (g_n - g_m)](x) \right| + \left| [(f_n - f_m) * g_m](x) \right| \\
  &\le \|f_n\|_p \cdot \|g_n - g_m\|_q + \|f_n - f_m\|_p \cdot \|g_m\|_q
  \end{aligned}$$

$$\leq (\|f_n - f\|_p + \|f\|_p) \times \|g_n - g_m\|_q$$
$$+ \|f_n - f_m\|_p \times (\|g_m - g\|_q + \|g\|_q). \tag{9.24}$$

Given $\epsilon > 0$. Now there exists a $N \in \mathbb{N}$ such that $n, m \geq N$ imply that

$$\|f_n - f\|_p \leq 1, \quad \|g_m - g\|_p \leq 1, \quad \|g_n - g_m\|_q < \frac{\epsilon}{2(1 + \|f\|_p)} \quad \text{and} \quad \|f_n - f_m\|_p < \frac{\epsilon}{2(1 + \|g\|_q)}.$$

Hence if $n, m \geq N$, then we deduce from the inequality (9.24) that

$$|(f_n * g_n)(x) - (f_m * g_m)(x)| < \epsilon$$

for every $x \in \mathbb{R}$. In other words, $\{f_n * g_n\}$ is Cauchy in $C_c(\mathbb{R})$ relative to the metric induced by the supremum norm.

Finally, using similar argument as in proving the inequality (9.24), we can show that one can find a positive integer N' such that $n \geq N'$ implies

$$|(f_n * g_n)(x) - (f * g)(x)| < \epsilon.$$

Consequently, it means that
$$\|f_n * g_n - f * g\|_\infty \to 0 \tag{9.25}$$
as $n \to \infty$. By Theorem 3.17, we know that $C_0(\mathbb{R}) = \overline{C_c(\mathbb{R})}$ so the limit (9.25) establishes the fact that $f * g \in C_0(\mathbb{R})$.

- **A counter-example.** Consider $f = \chi_{[-1,1]}$ and $g = 1$ so that $f \in L^1$ and $g \in L^\infty$. However, for any $x \in \mathbb{R}$, we note that

$$h(x) = \int_{-\infty}^{\infty} f(x - y)g(y)\,\mathrm{d}y = \int_{-\infty}^{\infty} \chi_{[x-1,x+1]}(y)\,\mathrm{d}y = m\big([x - 1, x + 1]\big) = 2.$$

Thus h *does not* vanish at ∞, i.e., $h \notin C_0(\mathbb{R})$.

This completes the proof of the problem.

Problem 9.9

Rudin Chapter 9 Exercise 9.

Proof. We have

$$g(x) = \int_{\mathbb{R}} \chi_{[x,x+1]}(t)f(t)\,\mathrm{d}t = \int_{\mathbb{R}} \chi_{[0,1]}(t - x)f(t)\,\mathrm{d}t. \tag{9.26}$$

By the change of variable formula [51, p. 156], the integral (9.26) reduces to

$$g(x) = \int_{\mathbb{R}} \chi_{[0,1]}(t)f(t + x)\,\mathrm{d}t = \int_{\mathbb{R}} \chi_{[0,1]}(t)f_{-x}(t)\,\mathrm{d}t.$$

For every $x \in \mathbb{R}$, by Theorem 9.5, the mapping $x \mapsto f_{-x}$ is continuous at x. In other words, given $\epsilon > 0$, there exists a $\delta > 0$ such that $\|f_{-x} - f_{-y}\|_p < \epsilon$ for all $y \in \mathbb{R}$ with $|x - y| < \delta$. Since $f_{-x} - f_{-y} \in L^p$ and $\chi_{[0,1]} \in L^q$, where q be the conjugate exponent of p, we deduce from Theorem 3.8 and then Theorem 1.33 that

$$|g(x) - g(y)| = \left| \int_{\mathbb{R}} \chi_{[0,1]}(t)[f_{-x}(t) - f_{-y}(t)]\,\mathrm{d}t \right|$$
$$= \big\| \chi_{[0,1]} \cdot (f_{-x} - f_{-y}) \big\|_1$$

$$\leq \|\chi_{[0,1]}\|_q \cdot \|f_{-x} - f_{-y}\|_p$$
$$< \epsilon$$

for all $y \in \mathbb{R}$ with $|x - y| < \delta$. Thus g is continuous at x.

Next, suppose that a and b are real numbers with $a < b$. Since $f \in L^1$, we have $f \in L^p([a,b])$ and then Theorem 3.5 (Hölder's Inequality) implies that

$$\int_a^b |f(t)|\, dt \leq \left\{ \int_a^b dt \right\}^{\frac{1}{q}} \left\{ \int_a^b |f(t)|^p\, dt \right\}^{\frac{1}{p}} = (b-a)^{\frac{1}{q}} \cdot \left\{ \int_a^b |f(t)|^p\, dt \right\}^{\frac{1}{p}}. \tag{9.27}$$

In particular, if $a = n$ and $b = n+1$, then we know from the inequality (9.27) that

$$\int_n^{n+1} |f(t)|\, dt \leq \left\{ \int_n^{n+1} |f(t)|^p\, dt \right\}^{\frac{1}{p}}. \tag{9.28}$$

Given $\epsilon > 0$. Again $f \in L^p$ gives

$$\int_{-\infty}^{\infty} |f(t)|^p\, dt = \sum_{n=-\infty}^{\infty} \int_{-n}^{n} |f(t)|^p\, dt < \infty.$$

Thus there is a positive integer N such that for all $n \in \mathbb{Z}$, $|n| \geq N$ implies that

$$\int_n^{n+1} |f(t)|^p\, dt < \epsilon^p. \tag{9.29}$$

Combining the inequalities (9.28) and (9.29), we get

$$\int_n^{n+1} |f(t)|^p\, dt < \epsilon$$

if $|n| \geq N$. In other words, we have

$$\lim_{|n| \to \infty} \int_n^{n+1} |f(t)|^p\, dt = 0$$

and equivalently,

$$\lim_{|x| \to \infty} \int_x^{x+1} |f(t)|^p\, dt = 0. \tag{9.30}$$

Finally, we use Theorem 1.33 and then apply the limit (9.30) to the inequality (9.27) with $a = x$ and $b = x+1$ to establish

$$\lim_{|x| \to \infty} |g(x)| \leq \lim_{|x| \to \infty} \int_x^{x+1} |f(t)|\, dt \leq \lim_{|x| \to \infty} \left\{ \int_x^{x+1} |f(t)|^p\, dt \right\}^{\frac{1}{p}} = 0.$$

Hence we conclude that g vanishes at infinity and then $g \in C_0(\mathbb{R})$.

Suppose that $f \in L^\infty$. Therefore, we can say that $|f(x)| \leq \|f\|_\infty < \infty$ for almost all $x \in \mathbb{R}$. Thus if $x < y$, then we have

$$|g(x) - g(y)| = \left| \int_x^{x+1} f(t)\, dt - \int_y^{y+1} f(t)\, dt \right|$$
$$= \left| \int_x^0 f(t)\, dt + \int_0^{x+1} f(t)\, dt - \int_y^0 f(t)\, dt - \int_0^{y+1} f(t)\, dt \right|$$
$$= \left| \int_x^y f(t)\, dt - \int_{x+1}^{y+1} f(t)\, dt \right|$$

$$\leq \int_x^y |f(t)|\,dt + \int_{x+1}^{y+1} |f(t)|\,dt$$

$$\leq 2(y-x)\|f\|_\infty.$$

Hence this means that g is uniformly continuous on $\mathbb{R}$ and furthermore,

$$|g(x)| = \left| \int_x^{x+1} f(t)\,dt \right| \leq \int_x^{x+1} |f(t)|\,dt \leq \int_x^{x+1} \|f\|_\infty\,dt = \|f\|_\infty$$

for almost every $x \in \mathbb{R}$ so that $\|g\|_\infty \leq \|f\|_\infty$. This completes the proof of the problem.

Problem 9.10

Rudin Chapter 9 Exercise 10.

Proof. Let's prove the results one by one.

- **C_c^∞ does not consist of 0 alone.** In Problem 7.8(d), we define the bump function $\psi : \mathbb{R} \to [0, \infty)$ by

$$\psi(x) = \begin{cases} \exp\left(-\dfrac{1}{1-x^2}\right), & \text{if } x \in (-1, 1); \\[2mm] 0, & \text{otherwise.} \end{cases} \tag{9.31}$$

This is an element of C^∞. Since $\operatorname{supp}(\psi) = [-1, 1]$, it must be true that $\psi \in C_c^\infty$.

- **If $f \in L^1_{\text{loc}}$ and $g \in C_c^\infty$, then $f * g \in C^\infty$.** To begin with, we have to show that $f * g$ is well-defined first. Since $\operatorname{supp}(g)$ is compact in $\mathbb{R}$, let M be a positive number such that $\operatorname{supp}(g) \subseteq [-M, M]$. Thus we have

$$(f * g)(x) = \int_{\mathbb{R}} f(x-y)g(y)\,dy = \int_{-M}^{M} f(x-y)g(y)\,dy. \tag{9.32}$$

As g is continuous on $[-M, M]$, the Heine-Borel Theorem says that it is bounded. Let a bound of g be K. Then $|g(y)| \leq K$ on $[-M, M]$. Therefore, the boundedness of g and the fact $f \in L^1_{\text{loc}}$ indicate that the integral in the equation (9.32) is finite. Consequently, $f * g$ is well-defined.

Next, we claim that $(f * g)' = f * g'$. In fact, if $x \in \mathbb{R}$ and $0 < \delta < 1$, then we write

$$\frac{(f * g)(x + \delta) - (f * g)(x)}{\delta}$$

$$= \frac{1}{\delta}\left[\int_{-\infty}^{\infty} f(x+\delta-y)g(y)\,dy - \int_{-\infty}^{\infty} f(x-y)g(y)\,dy \right]$$

$$= \frac{1}{\delta}\left[\int_{-M}^{M} f(x+\delta-y)g(y)\,dy - \int_{-M}^{M} f(x-y)g(y)\,dy \right]$$

$$= \frac{1}{\delta}\left[\int_{x-(M-\delta)}^{x+(M+\delta)} f(y)g(x+\delta-y)\,dy - \int_{x-M}^{x+M} f(y)g(x-y)\,dy \right]. \tag{9.33}$$

Since $x+\delta-[x+(M+\delta)] = -M$, $x+\delta-[x-(M+\delta)] = M+2\delta$ and $\operatorname{supp}(g) \subseteq [-M, M]$, the first integral in (9.33) can be replaced by

$$\int_{x-(M+1)}^{x+(M+1)} f(y)g(x+\delta-y)\,dy.$$

Similarly, we may extend the lower limit and the upper limit of the second integral in (9.33) to $x - (M + 1)$ and $x + (M + 1)$ respectively. Therefore, the expression (9.33) can be reduced to

$$\frac{(f * g)(x + \delta) - (f * g)(x)}{\delta} = \int_{x-(M+1)}^{x+(M+1)} \frac{g(x - y + \delta) - g(x - y)}{\delta} \cdot f(y)\, dy. \tag{9.34}$$

Now we *fix* $x \in \mathbb{R}$. Let $g = g_1 + ig_2$. Since $g \in C_c^\infty$, g_1 and g_2 are real differentiable functions. It follows from the Mean Value Theorem [49, Theorem 5.10, p. 108] that there exist $\xi, \eta \in (x - y, x - y + \delta)$ such that

$$g_1(x - y + \delta) - g_1(x - y) = \delta g_1'(\xi) \quad \text{and} \quad g_2(x - y + \delta) - g_2(x - y) = \delta g_2'(\eta).$$

Then they imply that $g(x - y + \delta) - g(x - y) = \delta[g_1'(\xi) + ig_2'(\eta)]$ which gives

$$\left| \frac{g(x - y + \delta) - g(x - y)}{\delta} \right| \le \sqrt{[g_1'(\xi)]^2 + [g_2'(\eta)]^2} \le \sqrt{|g'(\xi)|^2 + |g'(\eta)|^2}. \tag{9.35}$$

Since $g \in C_c^\infty$, g' is continuous and thus bounded on $[x - y, x - y + \delta]$. Recall that $y \in [x - (M + 1), x + (M + 1)]$ and $0 < \delta < 1$. Thus, there is an $A > 0$ such that $|g'(t)| \le A$ *for every* $t \in [-M - 2, M + 2]^{\mathrm{d}}$ and then we observe from the inequality (9.35) that

$$\left| \frac{g(x - y + \delta) - g(x - y)}{\delta} \cdot f(y) \right| \le \sqrt{2} A |f(y)|.$$

Since f is integrable on $[x - (M + 1), x + (M + 1)]$, Theorem 1.34 (Lebesgue's Dominated Convergence Theorem) ensures that the expression (9.34) gives

$$
\begin{aligned}
(f * g)'(x) &= \lim_{\delta \to 0} \frac{(f * g)(x + \delta) - (f * g)(x)}{\delta} \\
&= \lim_{\delta \to 0} \int_{x-(M+1)}^{x+(M+1)} \frac{g(x - y + \delta) - g(x - y)}{\delta} \cdot f(y)\, dy \\
&= \int_{x-(M+1)}^{x+(M+1)} \lim_{\delta \to 0} \frac{g(x - y + \delta) - g(x - y)}{\delta} \cdot f(y)\, dy \\
&= \int_{x-(M+1)}^{x+(M+1)} g'(x - y) \cdot f(y)\, dy \\
&= (g' * f)(x).
\end{aligned}
$$

Since x is arbitrary, we have $f * g \in C^1$ and $(f * g)' = g' * f$. Since g has compact support, $g^{(n)}$ also has compact support for every $n \in \mathbb{N}$. By induction, we have obtained

$$(f * g)^{(n)} = g^{(n)} * f.$$

- **Existence of a sequence $\{g_n\} \subseteq C_c^\infty$ such that $\|f * g_n - f\|_1 \to 0$ as $n \to \infty$ for every $f \in L^1$.** If $f \in L^1$, then $f \in L_{\mathrm{loc}}^1$ and so $f * g \in C^\infty$ for any $g \in C_c^\infty$. Let

$$c = \int_{-1}^{1} \psi(x)\, dx,$$

where ψ is the bump function (9.31). It is clear that c is a finite positive number because $\psi > 0$. If we define $\varphi : \mathbb{R} \to [0, \infty)$ by

$$\varphi(x) = c^{-1} \psi(x),$$

---

^dNotice that A *does not* depend on x, y or δ.

then φ satisfies the conditions that $\varphi \in C_c^\infty$, $\operatorname{supp}(\varphi) = [-1,1]$, $\varphi \in L^\infty$ and $\|\varphi\|_1 = 1$. Next, for $\epsilon > 0$, we define[e]

$$\varphi_\epsilon(x) = \epsilon^{-1}\varphi(\epsilon^{-1}x).$$

Then it is easily checked that

$$\varphi_\epsilon \in C_c^\infty, \quad \operatorname{supp}(\varphi_\epsilon) = [-\epsilon, \epsilon], \quad \varphi_\epsilon \in L^\infty \quad \text{and} \quad \|\varphi_\epsilon\|_1 = \|\varphi\|_1 = 1.$$

We observe that

$$
\begin{aligned}
f * \varphi_\epsilon(x) - f(x) &= \int_{\mathbb{R}} f(x-y)\varphi_\epsilon(y)\,\mathrm{d}y - \int_{\mathbb{R}} f(x)\varphi_\epsilon(y)\,\mathrm{d}y \\
&= \int_{\mathbb{R}} [f(x-y) - f(x)]\varphi_\epsilon(y)\,\mathrm{d}y \\
&= \int_{\mathbb{R}} [f(x-y) - f(x)]\varphi(\epsilon^{-1}y)\epsilon^{-1}\,\mathrm{d}y.
\end{aligned}
\tag{9.36}
$$

If we let $y = \epsilon z$, then the expression (9.36) becomes

$$|f * \varphi_\epsilon(x) - f(x)| = \left| \int_{\mathbb{R}} [f_{\epsilon z}(x) - f(x)]\varphi(z)\,\mathrm{d}z \right|. \tag{9.37}$$

Since $f \in L^1$ and $\varphi \in L^\infty$, Theorem 3.8 (with $p = 1$ and $q = \infty$) makes sure that $[f_{\epsilon z}(x) - f(x)]\varphi(z) \in L^1$ (with respect to z). By Theorem 1.33, the expression (9.37) gives

$$\|f * \varphi_\epsilon - f\|_1 = \int_{\mathbb{R}} |f * \varphi_\epsilon(x) - f(x)|\,\mathrm{d}x \le \int_{\mathbb{R}} \left[\int_{\mathbb{R}} \underbrace{|f_{\epsilon z}(x) - f(x)| \cdot |\varphi(z)|}_{\text{nonnegative integrand}}\,\mathrm{d}z \right]\mathrm{d}x. \tag{9.38}$$

Since the integrand in the expression (9.38) is nonnegative, we may apply Theorem 8.8 (The Fubini Theorem) to change the order of integration in the expression (9.38) to conclude that

$$\|f * \varphi_\epsilon - f\|_1 \le \int_{\mathbb{R}} |\varphi(z)| \cdot \|f_{\epsilon z} - f\|_1\,\mathrm{d}z. \tag{9.39}$$

By Theorem 9.5, we note that $\|f_{\epsilon z} - f\|_1 \to 0$ as $\epsilon \to 0$ for each $z \in \mathbb{R}$. Furthermore, since $\|f_{\epsilon z} - f\|_1 \le 2\|f\|_1$, we apply Theorem 1.34 (Lebesgue's Dominated Convergence Theorem) to the inequality (9.39) to obtain

$$\|f * \varphi_\epsilon - f\|_1 \to 0$$

as $\epsilon \to 0$. Now, if we let $\epsilon = \frac{1}{n}$ and $g_n = \varphi_{\frac{1}{n}}$, then we can obtain our desired sequence.

- **Existence of a sequence $\{g_n\} \subseteq C_c^\infty$ such that $(f * g_n)(x) \to f(x)$ a.e. for every** $f \in L_{\text{loc}}^1$. For each *fixed* $n \in \mathbb{N}$, let $K_n = [-\frac{1}{n}, \frac{1}{n}] \subset V_n = (-\frac{1}{n} - 2^{-n}, \frac{1}{n} + 2^{-n})$. By [49, Exercise 6, p. 289], there exists functions $\psi_1, \ldots, \psi_s \in C^\infty$ such that[f]

  - $0 \le \psi_i \le 1$ for $1 \le i \le s$;

  - $\operatorname{supp}(\psi_i) \subset V_n$;

  - $\psi_1(x) + \psi_2(x) + \cdots + \psi_s(x) = 1$ for every $x \in K_n$.

---

[e]The collection $\{\varphi_\epsilon\}$ is called an **approximate identity** or **a sequence of mollifiers on** $\mathbb{R}$, see [48, Definition 6.31, p. 173] or [11, p. 108].

[f]For a proof of this, please read [63, Problem 10.6, pp. 266, 267]. Of course, our s depends on n.

If we define $g_n : \mathbb{R} \to \mathbb{R}$ by

$$g_n = \frac{n}{2}(\psi_1 + \psi_2 + \cdots + \psi_s),$$

then the above conditions imply that $g_n \in C_c^\infty$, $g_n(x) = \frac{n}{2}$ on K_n, $g_n(x) = 0$ outside V_n and $0 \le g_n(x) \le \frac{n}{2}$ for all $x \in \mathbb{R}$.

Let $x \in \mathbb{R}$ and $f(x) \ge 0$ on $\mathbb{R}$. On the one hand, we have

$$(f * g_n)(x) \ge \frac{n}{2} \int_{-\frac{1}{n}}^{\frac{1}{n}} f(x-y)\,dy = \frac{1}{m\left(B\left(0, \frac{1}{n}\right)\right)} \int_{B\left(x, \frac{1}{n}\right)} f(y)\,dy. \tag{9.40}$$

Since $f \in L^1_{\text{loc}}$, the integral (9.40) is finite. On the other hand, we have

$$(f * g_n)(x) \le \frac{n}{2} \int_{-\frac{1}{n}-2^{-n}}^{\frac{1}{n}+2^{-n}} f(x-y)\,dy$$

$$= \left(1 + \frac{n}{2^n}\right) \times \frac{1}{m\left(B(0, \frac{1}{n} + 2^{-n})\right)} \int_{B(x, \frac{1}{n}+2^{-n})} f(y)\,dy. \tag{9.41}$$

Similarly, the integral (9.41) is finite because of the fact that $f \in L^1_{\text{loc}}$. Now if $x \in \mathbb{R}$ is a Lebesgue point of f, then we combine the inequalities (9.40) and (9.41) to get

$$\lim_{n \to \infty} (f * g_n)(x) = \lim_{n \to \infty} \frac{1}{m\left(B\left(0, \frac{1}{n}\right)\right)} \int_{B\left(x, \frac{1}{n}\right)} f(y)\,dy = f(x). \tag{9.42}$$

By Lemma 7.5, we see that almost every $x \in \mathbb{R}$ is a Lebesgue point of f. Hence we conclude from this that $(f * g_n)(x) \to f(x)$ a.e. on $\mathbb{R}$ for every $f \in L^1_{\text{loc}}$ and $f(x) \ge 0$ on $\mathbb{R}$.

For arbitrary $f \in L^1_{\text{loc}}$, we can write $f = f^+ - f^-$ and apply the above analysis to f^+ and f^- individually so that the limit (9.42) also holds when f is replaced by f^+ and f^-. It is clear from Lemma 7.5 that almost every $x \in \mathbb{R}$ is a *common* Lebesgue point of f, f^+ and f^-, so

$$\lim_{n \to \infty} (f * g_n)(x) = \lim_{n \to \infty} (f^+ * g_n)(x) - \lim_{n \to \infty} (f^- * g_n)(x) = f^+(x) - f^-(x) = f(x).$$

Consequently, the result (9.42) is also true in this general case.

- $(f * h_\lambda)(x) \to f(x)$ **a.e. if $f \in L^1$ as $\lambda \to 0$ and $f * h_\lambda \in C^\infty$.** First of all, we notice that the h_λ is in the form of the so-called **Poisson kernel** on $\mathbb{R}$, see [21, p. 228]. Next, it is easy to see that

$$h_\lambda(x) = \lambda^{-1} h_1(\lambda^{-1} x),$$

where $h_1(x) = \frac{1}{1+x^2}$. Furthermore, we obtain from the hypotheses of h_λ that h_1 is an even function such that it is decreasing on $(0, \infty)$ and $h_1 \in L^1$. With the aid of [56, Theorem 1.25, p. 13], we see that

$$\lim_{\lambda \to 0} (f * h_\lambda)(x) = f(x) \int_{\mathbb{R}} h_1(t)\,dm(t) = f(x)$$

for every Lebesgue point x of f. By Theorem 7.7, we obtain our desired result that $(f * h_\lambda)(x) \to f(x)$ a.e. on $\mathbb{R}$ if $f \in L^1$ as $\lambda \to 0$.

It remains to prove that $f * h_\lambda \in C^\infty$. To see this, the definition of h_λ shows clearly that $h_\lambda \in C^\infty$. In addition, direct computation gives easily that $h_\lambda^{(n)}$ is bounded on $\mathbb{R}$ for every positive integer n. As a consequence of [22, Proposition 8.10, p. 242], we may conclude immediately that $f * h_\lambda \in C^\infty$.

This completes the proof of the problem.

9.2 The Poisson Summation Formula and its Applications

> **Problem 9.11**
>
> *Rudin Chapter 9 Exercise 11.*

Proof. Suppose that f is a Schwartz function. We are going to show that the Poisson summation formula holds in this situation and consider the limiting case as $\alpha \to 0$ under the extra hypothesis that $\widehat{f} \in L^1$.

- **The validity of the Poisson summation formula.** We first assume that $f \in S$. Consider

$$F(x) = \sum_{k=-\infty}^{\infty} f(x + 2k\pi)$$

which is a periodic function of period 2π, with Fourier coefficients

$$\widehat{F}(n) = \frac{1}{2\pi} \int_{-\pi}^{\pi} F(x) e^{-inx}\, dx = \frac{1}{2\pi} \int_{-\pi}^{\pi} \Big(\sum_{k=-\infty}^{\infty} f(x + 2k\pi) e^{-inx} \Big) dx. \tag{9.43}$$

Since $f \in S$, there exists a positive constant M such that $|f(x)| \le M$ and $|x^2 f(x)| \le M$ for all $x \in \mathbb{R}$. In particular, if $k \ge 1$, then

$$|f(x + 2k\pi)| \le \frac{M}{(x + 2k\pi)^2} \le \frac{M}{(2k-1)^2 \pi^2}$$

for all $x \in [-\pi, \pi]$. By the Weierstrass M-test [49, Theorem 7.10, p. 148], we see that the series

$$\sum_{k=1}^{\infty} f(x + 2k\pi) e^{-inx}$$

converges uniformly and absolutely on $[-\pi, \pi]$. Similarly, the series

$$\sum_{k=1}^{\infty} f(x - 2k\pi) e^{-inx}$$

also converges uniformly and absolutely on $[-\pi, \pi]$. If $k = 0$, then

$$\left| \int_{-\pi}^{\pi} f(x) e^{-inx}\, dx \right| \le \int_{-\pi}^{\pi} |f(x)|\, dx \le 2\pi M.$$

Consequently, the series

$$\sum_{k=-\infty}^{\infty} f(x + 2k\pi) e^{-inx} \tag{9.44}$$

converges uniformly and absolutely on $[-\pi, \pi]$ so that the integral and summation in the expression (9.43) can be interchanged (by [49, Corollary, p. 152]) and we get

$$\widehat{F}(n) = \frac{1}{2\pi} \sum_{k=-\infty}^{\infty} \int_{-\pi}^{\pi} f(x + 2k\pi) e^{-inx}\, dx$$

$$= \frac{1}{2\pi} \sum_{k=-\infty}^{\infty} \int_{(2k-1)\pi}^{(2k+1)\pi} f(x) e^{-inx}\, dx$$

$$= \frac{1}{2\pi} \int_{-\infty}^{\infty} f(x) e^{-inx} \, dx$$

$$= \varphi(n)$$

for every $n \in \mathbb{Z}$. In other words, we have

$$\sum_{k=-\infty}^{\infty} f(x + 2k\pi) = F(x) = \sum_{n=-\infty}^{\infty} \varphi(n) e^{inx}. \tag{9.45}$$

If we put $x = 0$ in the formula (9.45), then we get

$$\sum_{k=-\infty}^{\infty} f(2k\pi) = \sum_{n=-\infty}^{\infty} \varphi(n). \tag{9.46}$$

Let $\gamma > 0$ and $g(x) = f(\gamma x)$. Since $g \in S$, the preceding formula (9.46) also gives

$$\sum_{k=-\infty}^{\infty} g(2k\pi) = \sum_{n=-\infty}^{\infty} \psi(n), \tag{9.47}$$

where

$$\psi(t) = \frac{1}{2\pi} \int_{-\infty}^{\infty} g(x) e^{-itx} \, dx.$$

Direct computation shows

$$\psi(t) = \frac{1}{2\pi} \int_{-\infty}^{\infty} f(\gamma x) e^{-itx} \, dx = \frac{1}{\gamma} \varphi\left(\frac{t}{\gamma}\right)$$

and after substituting this back into the formula (9.47) and then using the definition of g
, we obtain

$$\sum_{k=-\infty}^{\infty} f(2\gamma k\pi) = \sum_{n=-\infty}^{\infty} \frac{1}{\gamma} \varphi\left(\frac{n}{\gamma}\right). \tag{9.48}$$

Suppose that $\alpha = \frac{1}{\gamma} > 0$ and $\beta = \frac{2\pi}{\alpha} > 0$. Then we have $\alpha\beta = 2\pi$ and the formula (9.48)
becomes

$$\sum_{k=-\infty}^{\infty} f(k\beta) = \alpha \sum_{n=-\infty}^{\infty} \varphi(n\alpha). \tag{9.49}$$

- **The case when** $\alpha \to 0$. Now we suppose further that the inequality (9.51) in Remark
  9.3 below holds. On the one hand, we know that

$$\lim_{\alpha \to 0} \alpha \sum_{n=-\infty}^{\infty} \varphi(n\alpha) = \lim_{N \to \infty} \frac{1}{N} \sum_{n=-\infty}^{\infty} \varphi\left(\frac{n}{N}\right) = \int_{-\infty}^{\infty} \varphi(x) \, dx = \int_{-\infty}^{\infty} \widehat{f}(x) \, dm(x)$$

because $\varphi(n) = \frac{1}{\sqrt{2\pi}} \widehat{f}(n)$. On the other hand, we have

$$\lim_{\alpha \to 0} \sum_{k=-\infty}^{\infty} f(k\beta) = \lim_{\alpha \to 0} \sum_{k=-\infty}^{\infty} f\left(\frac{2\pi k}{\alpha}\right) = \lim_{N \to \infty} \sum_{k=-\infty}^{\infty} f(2\pi k N). \tag{9.50}$$

Again the fact $f \in S$ implies that

$$|f(x)| \le \frac{M}{|x|^2}$$

holds for all $x \neq 0$, where M is a positive constant. Thus this shows that

$$\left| \sum_{\substack{k=-\infty \\ k \neq 0}}^{\infty} f(2\pi k N) \right| \leq \sum_{\substack{k=-\infty \\ k \neq 0}}^{\infty} |f(2\pi k N)| \leq \sum_{\substack{k=-\infty \\ k \neq 0}}^{\infty} \frac{M}{|2\pi k N|^2} \to 0$$

as $N \to \infty$ and then the limit (9.50) becomes

$$\lim_{\alpha \to 0} \sum_{k=-\infty}^{\infty} f(k\beta) = f(0).$$

Hence we arrive at the result

$$f(0) = \int_{-\infty}^{\infty} \widehat{f}(x) \, dm(x)$$

which is in agreement with Theorem 9.11 (The Inversion Theorem).

We have completed the proof of the problem. $\blacksquare$

Remark 9.3

If there exists a constant M and $p > 1$ such that

$$|f(x)| \leq \frac{M}{|x|^p} \tag{9.51}$$

for all sufficiently large x, then the series (9.44) still converges uniformly and absolutely on $[-\pi, \pi]$ so that the Poisson summation formula (9.49) also holds in this case.

Problem 9.12

Rudin Chapter 9 Exercise 12.

Proof. We note that f is *not* a Schwartz function, but the result [49, Theorem 8.6(f)] ensures that f satisfies the inequality (9.51), so we may apply the first assertion of Problem 9.11. First of all, since $e^{-|x|} \cos tx$ and $e^{-|x|} \sin tx$ are even and odd functions in x respectively, we have

$$\varphi(t) = \frac{1}{2\pi} \int_{-\infty}^{\infty} e^{-|x|} e^{-itx} \, dx = \frac{1}{\pi} \int_{0}^{\infty} e^{-x} \cos tx \, dx.$$

By [27, §2.662 Eqn. 2, p. 228], we have

$$\int e^{-x} \cos tx \, dx = \frac{(-\cos xt + t \sin xt) e^{-x}}{1 + t^2}$$

so that

$$\varphi(t) = \frac{1}{(1 + t^2)\pi}. \tag{9.52}$$

Substituting the expression (9.52) and $f(x) = e^{-|x|}$ into the formula (9.49), we obtain

$$\sum_{k=-\infty}^{\infty} e^{-|k|\beta} = \alpha \sum_{n=-\infty}^{\infty} \frac{1}{(1 + n^2\alpha^2)\pi}$$

$$\frac{1+e^{-\beta}}{1-e^{-\beta}} = \frac{1}{\pi}\sum_{n=-\infty}^{\infty}\frac{\alpha}{1+n^2\alpha^2}$$

$$\frac{e^{\beta}+1}{e^{\beta}-1} = \frac{1}{\pi}\sum_{n=-\infty}^{\infty}\frac{\alpha}{1+n^2\alpha^2}, \tag{9.53}$$

where $\alpha\beta = 2\pi$. If we take $\beta = 2\pi\gamma$, then $\alpha = \frac{1}{\gamma}$. Hence we deduce from the formula (9.53) that

$$\frac{e^{2\pi\gamma}+1}{e^{2\pi\gamma}-1} = \frac{1}{\pi}\sum_{n=-\infty}^{\infty}\frac{\gamma}{\gamma^2+n^2} \tag{9.54}$$

and the desired result follows if we replace γ by α in the formula (9.54). This completes the proof of the problem. ◼

Problem 9.13

Rudin Chapter 9 Exercise 13.

Proof.

(a) By Definition 9.1 and similar argument as in the proof of Problem 9.12, we have

$$\widehat{f_c}(t) = \int_{-\infty}^{\infty} f_c(x)e^{-ixt}\,dm(x) = \int_{-\infty}^{\infty} e^{-cx^2}e^{-ixt}\,dm(x) = \sqrt{\frac{2}{\pi}}\int_{0}^{\infty} e^{-cx^2}\cos tx\,dx. \tag{9.55}$$

By [27, §3.922, Eqn. 4, p. 494], we see that

$$\int_{0}^{\infty} e^{-cx^2}\cos tx\,dx = \frac{1}{2}\sqrt{\frac{\pi}{c}}\exp\left(-\frac{t^2}{4c}\right),$$

so the formula (9.55) becomes

$$\widehat{f_c}(t) = \frac{1}{\sqrt{2c}}\exp\left(-\frac{t^2}{4c}\right).$$

(b) By part (a), if $\widehat{f_c} = f_c$, then we have

$$\frac{1}{\sqrt{2c}}\exp\left(-\frac{x^2}{4c}\right) = \exp(-cx^2)$$

$$\exp\left(\frac{4c^2-1}{4c}x^2\right) = \sqrt{2c} \tag{9.56}$$

for every $x \in \mathbb{R}$. Since the right-hand side of the equation (9.56) is constant, $4c^2 - 1 = 0$ and then $c = \frac{1}{2}$. Direct checking immediately shows that $c = \frac{1}{2}$ satisfies the equation (9.56).

(c) Let $a, b \in (0, \infty)$. By the definition, we see that

$$(f_a * f_b)(x) = \int_{-\infty}^{\infty} f_a(x-y)f_b(y)\,dy$$

$$= \int_{-\infty}^{\infty} \exp[-a(x-y)^2 - by^2]\,dy$$

$$= \int_{-\infty}^{\infty} \exp\{-[(a+b)y^2 - 2axy + ax^2]\}\,dy. \tag{9.57}$$

Using [27, §2.33, Eqn. 1, p. 108] and the facts that[g] $\mathrm{erf}(\infty) = 1$ and $\mathrm{erf}(-\infty) = -1$, we are able to represent the integral (9.57) as

$$(f_a * f_b)(x) = \frac{1}{2}\sqrt{\frac{\pi}{a+b}}\,\exp\left(-\frac{ab}{a+b}x^2\right) \times \mathrm{erf}\left(\sqrt{a+b}\,y - \frac{ax}{\sqrt{a+b}}\right)\Big|_{y=-\infty}^{y=\infty}$$

$$= \sqrt{\frac{\pi}{a+b}}\,\exp\left(-\frac{ab}{a+b}x^2\right).$$

Therefore, we conclude that

$$\gamma = \sqrt{\frac{\pi}{a+b}} \quad\text{and}\quad c = \frac{ab}{a+b}.$$

(d) By Problem 9.7, we know that $f_c \in S$. Thus the definition and the formula [27, §2.33, Eqn. 1, p. 108] together give

$$\varphi(t) = \frac{1}{2\pi}\int_{-\infty}^{\infty} e^{-cx^2}e^{-itx}\,dx = \frac{1}{2\pi}\int_{-\infty}^{\infty}\exp[-(cx^2 + itx)]\,dx = \frac{1}{2\sqrt{\pi c}}\exp\left(-\frac{t^2}{4c}\right).$$

Hence we establish from the formula (9.49) that

$$\sum_{k=-\infty}^{\infty} e^{-ck^2\beta^2} = \frac{2\pi}{\beta}\sum_{n=-\infty}^{\infty}\frac{1}{2\sqrt{\pi c}}\exp\left[-\frac{1}{4c}\times\left(\frac{2n\pi}{\beta}\right)^2\right]$$

$$= \frac{1}{\beta}\sqrt{\frac{\pi}{c}}\sum_{n=-\infty}^{\infty}\exp\left(-\frac{n^2\pi^2}{c\beta^2}\right).$$

We complete the proof of the problem.   ▨

9.3   Fourier Transforms on $\mathbb{R}^k$ and its Applications

> **Problem 9.14**
>
> *Rudin Chapter 9 Exercise 14.*

Proof. Recall that if $f \in L^1(\mathbb{R}^k)$, then

$$\widehat{f}(\mathbf{y}) = \int_{\mathbb{R}^k} f(\mathbf{x})\exp(-i\mathbf{x}\cdot\mathbf{y})\,dm_k(\mathbf{x}). \tag{9.58}$$

We first show the analogue of Theorem 9.6 that $\widehat{f} \in C_0(\mathbb{R}^k)$ and $\|\widehat{f}\|_\infty \le \|f\|_1$. To this end, we imitate the proof of Theorem 9.6. The required inequality is obvious from the definition (9.58) and the integrability of f.

    To prove $\widehat{f} \in C_0(\mathbb{R}^k)$, we need the analogue of Theorem 9.5 for $\mathbb{R}^k$:[h]

---

[g] Here $\mathrm{erf}(x)$ is the **error function**.

[h] Its proof is very similar to that of Theorem 9.5, so we omit the details here.

> **Lemma 9.6**
>
> For any function f on $\mathbb{R}^k$ and every $\mathbf{y} \in \mathbb{R}^k$, let $f_{\mathbf{y}}$ be the translate of f defined by
>
> $$f_{\mathbf{y}}(\mathbf{x}) = f(\mathbf{x} - \mathbf{y})$$
>
> for $\mathbf{x} \in \mathbb{R}^k$. If $1 \le p < \infty$ and $f \in L^p(\mathbb{R}^k)$, then the mapping
>
> $$\mathbf{y} \mapsto f_{\mathbf{y}}$$
>
> is a uniformly continuous mapping of $\mathbb{R}^k$ into $L^p(\mathbb{R}^k)$.

Let $\mathbf{t} \in \mathbb{R}^k$. If $\mathbf{t}_n \to \mathbf{t}$, then

$$|\widehat{f}(\mathbf{t}_n) - \widehat{f}(\mathbf{t})| \le \int_{\mathbb{R}^k} |f(\mathbf{x})| \cdot |\exp(-i\mathbf{t}_n \cdot \mathbf{x}) - \exp(-i\mathbf{t} \cdot \mathbf{x})| \, dm_k(\mathbf{x}). \qquad (9.59)$$

Since the integrand of the integral (9.59) is bounded by $2|f(\mathbf{x})| \in L^1(\mathbb{R}^k)$, Theorem 1.34 (Lebesgue's Dominated Convergence Theorem) implies that

$$\widehat{f}(\mathbf{t}_n) \to \widehat{f}(\mathbf{t})$$

as $n \to \infty$. In other words, $\widehat{f}$ is continuous on $\mathbb{R}^k$.

Next we have to prove that f vanishes at infinity. Suppose that $\mathbf{y} = (\eta_1, \ldots, \eta_k)$ and $\eta_1 \neq 0$. Take

$$\mathbf{y}_1 = \left(\frac{\pi}{\eta_1}, 0, \ldots, 0\right).$$

Since $e^{-\pi i} = -1$, we have

$$\exp\left[-i\mathbf{y} \cdot \left(\mathbf{x} + \mathbf{y}_1\right)\right] = \exp\left\{-i\left[\eta_1\left(\xi_1 + \frac{\pi}{\eta_1}\right) + \sum_{j=2}^{k} \eta_j \xi_j\right]\right\}$$

$$= \exp\left(-i\sum_{j=1}^{k} \xi_j \eta_j\right) \times e^{-i\pi}$$

$$= -\exp(-i\mathbf{x} \cdot \mathbf{y})$$

so that the definition (9.58) implies that

$$\widehat{f}(\mathbf{y}) = -\int_{\mathbb{R}^k} f(\mathbf{x})\exp\left[-i\mathbf{y} \cdot \left(\mathbf{x} + \mathbf{y}_1\right)\right] dm_k(\mathbf{x}) = -\int_{\mathbb{R}^k} f(\mathbf{x} - \mathbf{y}_1)\exp(-i\mathbf{y} \cdot \mathbf{x}) \, dm_k(\mathbf{x}).$$

Thus we get from this that

$$2\widehat{f}(\mathbf{y}) = \int_{\mathbb{R}^k} [f(\mathbf{x}) - f(\mathbf{x} - \mathbf{y}_1)]\exp(-i\mathbf{y} \cdot \mathbf{x}) \, dm_k(\mathbf{x})$$

$$= \int_{\mathbb{R}^k} [f(\mathbf{x}) - f_{\mathbf{y}_1}(\mathbf{x})]\exp(-i\mathbf{y} \cdot \mathbf{x}) \, dm_k(\mathbf{x}). \qquad (9.60)$$

Now we apply Lemma 9.6 to the integral (9.60), we assert that

$$2|\widehat{f}(\mathbf{y})| \le \|f - f_{\mathbf{y}_1}\|_1 = \|f_{\mathbf{0}} - f_{\mathbf{y}_1}\|_1 \qquad (9.61)$$

which tends to 0 as $|\mathbf{y}_1| \to 0$ or equivalently, as $\eta_1 \to \pm\infty$. Or we can say this way: Given $\epsilon > 0$, there exists a $\delta > 0$ such that

$$\|f_{\mathbf{0}} - f_{\mathbf{y}_1}\|_1 < 2\epsilon$$

for all $\mathbf{y}_1 \in \mathbb{R}^k$ for which $|\mathbf{y}_1| < \delta$ or $|\eta_1| > \frac{\pi}{\delta}$. Notice also that $\eta_1 \to \pm\infty$ means that $|\mathbf{y}| \to \infty$.

By similar argument as the previous part, we note that the result (9.61) also holds if $\mathbf{y}_1$ is replaced by any $\mathbf{y}_j = (0, \ldots, 0, \frac{\pi}{\eta_j}, 0, \ldots, 0)$ with $\eta_j \neq 0$, where $j = 2, 3, \ldots, k$. Write it explicitly, we have

$$\|f_{\mathbf{0}} - f_{\mathbf{y}_j}\|_1 < 2\epsilon \tag{9.62}$$

for all $\mathbf{y}_j \in \mathbb{R}^k$ for which $|\eta_j| > \frac{\pi}{\delta}$, where $j = 1, 2, \ldots, k$. Now for arbitrary $\mathbf{y} \in \mathbb{R}^k \setminus \{\mathbf{0}\}$, there is *at least* one j such that $|\eta_j| \geq \frac{|\mathbf{y}|}{k}$. Otherwise, we have $|\eta_j| < \frac{|\mathbf{y}|}{k}$ for every $j = 1, 2, \ldots, k$, but this means that $k^2 < 1$, a contradiction. Thus if we take

$$K = \left\{ \mathbf{y} \in \mathbb{R}^k \,\middle|\, |\mathbf{y}| \leq \frac{k\pi}{\delta} \right\},$$

then K is compact and for $\mathbf{y} \notin K$, we have

$$|\eta_j| \geq \frac{|\mathbf{y}|}{k} > \frac{\pi}{\delta}$$

for some $1 \leq j \leq k$. Hence the inequality (9.62) always holds for this j and then we conclude from the inequality (9.61) that

$$|\widehat{f}(\mathbf{y})| < \epsilon$$

if $\mathbf{y} \notin K$. By Definition 3.16, $\widehat{f}$ vanishes at infinity, i.e., $\widehat{f} \in C_0(\mathbb{R}^k)$.

Our next target is to obtain the analogue of Proposition 9.8. Similar to the work in §9.7, we put

$$H(\mathbf{t}) = \exp[-(|t_1| + |t_2| + \cdots + |t_k|)],$$

where $\mathbf{t} = (t_1, t_2, \ldots, t_k)$. Let $\lambda > 0$. Then $0 < H(\mathbf{t}) \leq 1$ and $H(\lambda \mathbf{t}) \to 1$ as $\lambda \to 0$. Define

$$h_\lambda(\mathbf{x}) = \int_{\mathbb{R}^k} H(\lambda \mathbf{t}) \exp(i\mathbf{t} \cdot \mathbf{x}) \, dm_k(\mathbf{t}).$$

A simple computation with repeated applications of Theorem 8.8 (The Fubini Theorem), it gives

$$h_\lambda(\mathbf{x}) = \prod_{j=1}^{k} \left(\frac{2}{\pi}\right)^{\frac{1}{2}} \frac{\lambda}{\lambda^2 + \xi_j^2}.$$

By Theorem 8.8 (The Fubini Theorem) and [51, Eqn. (4), p. 183], we get

$$\int_{\mathbb{R}^k} h_\lambda(\mathbf{x}) \, dm_k(\mathbf{x}) = \prod_{j=1}^{k} \left[\int_{\mathbb{R}} \left(\frac{2}{\pi}\right)^{\frac{1}{2}} \frac{\lambda}{\lambda^2 + \xi_j^2} \, dm(\xi_1) \right] = 1.$$

By this result, we establish the analogue of Proposition 9.8: If $f \in L^1(\mathbb{R}^k)$, then

$$(f * h_\lambda)(\mathbf{x}) = \int_{\mathbb{R}^k} H(\lambda \mathbf{t}) \widehat{f}(\mathbf{t}) \exp(i\mathbf{x} \cdot \mathbf{t}) \, dm_k(\mathbf{t}). \tag{9.63}$$

Similarly, we have the following analogue of Theorem 9.10: If $1 \leq p < \infty$ and $f \in L^p(\mathbb{R}^k)$, then

$$\lim_{\lambda \to 0} \|f * h_\lambda - f\|_p = 0. \tag{9.64}$$

Now it is time to prove the required results.

- **The Inversion Theorem for $\mathbb{R}^k$.** Suppose that $f, \widehat{f} \in L^1(\mathbb{R}^k)$. The integrand on the right-hand side of the formula (9.63) are bounded by $|\widehat{f}(\mathbf{t})|$. By Theorem 1.34 (Lebesgue's Dominated Convergence Theorem) and the fact that $H(\lambda \mathbf{t}) \to 1$ as $\lambda \to 0$, the right-hand side of the formula (9.63) converges to

$$g(\mathbf{x}) = \int_{\mathbb{R}^k} \widehat{f}(\mathbf{t}) \exp(i\mathbf{x} \cdot \mathbf{t}) \, dm_k(\mathbf{t}).$$

for all $\mathbf{x} \in \mathbb{R}^k$. By the limit (9.64) and Theorem 3.12, there is a sequence $\{\lambda_n\}$ such that $\lambda_n \to 0$ and

$$\lim_{n \to \infty} (f * h_{\lambda_n})(\mathbf{x}) = f(\mathbf{x})$$

a.e. on $\mathbb{R}^k$. Consequently, we have $f(\mathbf{x}) = g(\mathbf{x})$ a.e. on $\mathbb{R}^k$. Since $\widehat{f} \in L^1(\mathbb{R}^k)$ and $g(-\mathbf{x})$ is the Fourier transform of $\widehat{f}$, the analogue of Theorem 9.6 above implies that $g \in C_0(\mathbb{R}^k)$ as desired.

- **The Plancherel Theorem for $\mathbb{R}^k$.** Fix $f \in L^1(\mathbb{R}^k) \cap L^2(\mathbb{R}^k)$. Put $\widetilde{f}(\mathbf{x}) = \overline{f(-\mathbf{x})}$ and $g = f * \widetilde{f}$. Then we have

$$g(\mathbf{x}) = \int_{\mathbb{R}^k} f(\mathbf{x} - \mathbf{y})\overline{f(-\mathbf{y})} \, dm_k(\mathbf{y}) = \int_{\mathbb{R}^k} f(\mathbf{x} + \mathbf{y})\overline{f}(\mathbf{y}) \, dm_k(\mathbf{y}) = \langle f_{-\mathbf{x}}, f \rangle.$$

By the use of Lemma 9.6 and Theorem 4.6, it can be shown that g is a continuous function on $\mathbb{R}^k$ and thus Theorem 4.2 (The Schwarz Inequality) gives

$$|g(\mathbf{x})| \le \|f_{-\mathbf{x}}\|_2 \cdot \|f\|_2 = \|f\|_2^2$$

so that g is bounded on $\mathbb{R}^k$. Since $f \in L^1(\mathbb{R}^k)$, we have $\widetilde{f} \in L^1(\mathbb{R}^k)$ and then $g \in L^1(\mathbb{R}^k)$ by the analogue of Theorem 8.14 for $\mathbb{R}^k$.[i] By the formula (9.63), we have

$$(g * h_\lambda)(\mathbf{0}) = \int_{\mathbb{R}^k} H(\lambda \mathbf{t})\widehat{g}(\mathbf{t}) \, dm_k(\mathbf{t}). \tag{9.65}$$

Since g is continuous and bounded on $\mathbb{R}^k$, the analogue of Theorem 9.9[j] shows that

$$\lim_{\lambda \to 0} (g * h_\lambda)(\mathbf{0}) = g(\mathbf{0}) = \|f\|_2^2. \tag{9.66}$$

Now the analogues of Theorem 9.2(c) and (d) imply that

$$\widehat{g} = \widehat{f} \times \widehat{\widetilde{f}} = \widehat{f} \times \overline{\widehat{f}} = |\widehat{f}|^2 \ge 0.$$

Since $H(\lambda \mathbf{t})$ is measurable and increases to 1 as $\lambda \to 0$, Theorem 1.26 (Lebesgue's Monotone Convergence Theorem) implies that

$$\lim_{\lambda \to 0} \int_{\mathbb{R}^k} H(\lambda \mathbf{t})\widehat{g}(\mathbf{t}) \, dm_k(\mathbf{t}) = \int_{\mathbb{R}^k} |\widehat{f}(\mathbf{t})|^2 \, dm_k(\mathbf{t}). \tag{9.67}$$

Hence it follows from the expression (9.65) and the two limits (9.66) and (9.67) that

$$\|\widehat{f}\|_2^2 = \lim_{\lambda \to 0} \int_{\mathbb{R}^k} H(\lambda \mathbf{t})\widehat{g}(\mathbf{t}) \, dm_k(\mathbf{t}) = \lim_{\lambda \to 0} (g * h_\lambda)(\mathbf{0}) = \|f\|_2^2 < \infty. \tag{9.68}$$

Suppose that

$$Y = \{\widehat{f} \mid f \in L^1(\mathbb{R}^k) \cap L^1(\mathbb{R}^k)\}.$$

---

[i] The analogue of Theorem 8.14 is that $f, g \in L^1(\mathbb{R}^k)$ implies $f * g \in L^1(\mathbb{R}^k)$, but we won't give a proof here.
[j] If $g \in L^\infty(\mathbb{R}^k)$ and g is continuous at a point $\mathbf{x} \in \mathbb{R}^k$, then $(g * h_\lambda)(\mathbf{x}) \to g(\mathbf{x})$ as $\lambda \to 0$.

The equality (9.68) implies that Y is a subspace of $L^2(\mathbb{R}^k)$. We claim that Y is dense in $L^2(\mathbb{R}^k)$. To this end, we recall from Problem 4.1 that $(Y^\perp)^\perp = \overline{Y}$. Now we observe that $Y^\perp = \{0\}$ if and only if

$$\overline{Y} = \{0\}^\perp = \{f \in L^2(\mathbb{R}^k) \mid \langle f, 0 \rangle = 0\} = L^2(\mathbb{R}^k).$$

Thus it suffices to show that $\overline{Y} = \{0\}^\perp$. Let $\mathbf{a} \in \mathbb{R}^k$ and $\lambda > 0$. Denote the mapping $\mathbf{x} \mapsto \exp(i\mathbf{a} \cdot \mathbf{x}) H(\lambda \mathbf{x})$ by F. It is clear that $F(\mathbf{x}) \in L^1(\mathbb{R}^k) \cap L^2(\mathbb{R}^k)$ for all $\mathbf{x} \in \mathbb{R}^k$ and

$$\widehat{F}(\mathbf{t}) = \int_{\mathbb{R}^k} F(\mathbf{x}) \exp(-i\mathbf{x} \cdot \mathbf{t}) \, dm_k(\mathbf{t}) = \int_{\mathbb{R}^k} \exp\big(i(\mathbf{a} - \mathbf{t}) \cdot \mathbf{x}\big) H(\lambda \mathbf{x}) \, dm_k(\mathbf{x}) = h_\lambda(\mathbf{a} - \mathbf{t})$$

so that $h_\lambda \in Y$. If $w \in Y^\perp = \{f \in L^2(\mathbb{R}^k) \mid \langle f, g \rangle = 0 \text{ for all } g \in Y\}$, then it follows that

$$(h_\lambda * \overline{w})(\mathbf{a}) = \int_{\mathbb{R}^k} h_\lambda(\mathbf{a} - \mathbf{t}) \overline{w(\mathbf{t})} \, dm_k(\mathbf{t}) = 0$$

for all $\mathbf{a} \in \mathbb{R}^k$. By the limit (9.64), we see that $w = 0$, i.e., $Y^\perp = \{0\}$ and thus $\overline{Y} = L^2(\mathbb{R}^k)$.

Now we define $\Phi : L^1(\mathbb{R}^k) \cap L^2(\mathbb{R}^k) \to Y$ by

$$\Phi(f) = \widehat{f}.$$

Then the equality (9.68) certainly shows that Φ is an $L^2(\mathbb{R}^k)$-isometry. Since both $L^1(\mathbb{R}^k) \cap L^2(\mathbb{R}^k)$ and Y are *dense* in $L^2(\mathbb{R}^k)$, as the paragraph before [51, Eqn. (10), p. 187] indicates that Φ can be extended to an isometry

$$\widetilde{\Phi} : L^2(\mathbb{R}^k) \to L^2(\mathbb{R}^k).$$

If we denote $\widehat{f} = \widetilde{\Phi}(f)$ (of course, it is the Plancherel transform), then the equality (9.68) holds for every $f \in L^2(\mathbb{R}^k)$. Using a similar argument as in the proof of Theorem 4.18, we can prove the Parserval formula

$$\langle f, g \rangle = \int_{\mathbb{R}^k} f(\mathbf{x}) \overline{g(\mathbf{x})} \, dm_k(\mathbf{x}) = \int_{\mathbb{R}^k} \widehat{f}(\mathbf{t}) \overline{\widehat{g}(\mathbf{t})} \, dm_k(\mathbf{t}) = \big\langle \widehat{f}, \widehat{g} \big\rangle$$

for every $f, g \in L^2(\mathbb{R}^k)$. Finally, for the symmetric relation in Theorem 9.13(d), the φ_A and ψ_A should be replaced by

$$\widetilde{\varphi_A}(\mathbf{t}) = \int_{[-A,A]^k} f(\mathbf{x}) \exp(-i\mathbf{x} \cdot \mathbf{t}) \, dm_k(\mathbf{x}) \quad \text{and} \quad \widetilde{\psi_A}(\mathbf{x}) = \int_{[-A,A]^k} \widehat{f}(\mathbf{t}) \exp(i\mathbf{x} \cdot \mathbf{t}) \, dm_k(\mathbf{t})$$

respectively. Employing a similar argument as shown in [51, p. 187], it can be verified that

$$\|\widetilde{\varphi_A} - \widehat{f}\|_2 \to 0 \quad \text{and} \quad \|\widetilde{\psi_A} - f\|_2 \to 0$$

as $A \to \infty$.

- **The analogue of Theorem 9.23.** To every complex homomorphism φ on $L^1(\mathbb{R}^k)$ (except for $\varphi = 0$) there corresponds a unique $\mathbf{t} \in \mathbb{R}^k$ such that

$$\varphi(f) = \widehat{f}(\mathbf{t}).$$

Here it is seen that the content up to Eqn. (6) in §9.22 can be extended to $\mathbb{R}^k$ easily. If $\mathbf{x} = \xi\mathbf{e}_1 + \xi_2\mathbf{e}_2 + \cdots + \xi_k\mathbf{e}_k$, where $\{\mathbf{e}_1, \mathbf{e}_2, \ldots, \mathbf{e}_k\}$ is the usual basis of $\mathbb{R}^k$, then we have

$$\beta(\mathbf{x}) = \beta(\xi\mathbf{e}_1 + \xi_2\mathbf{e}_2 + \cdots + \xi_k\mathbf{e}_k) = \beta(\xi_1\mathbf{e}_1) \times \beta(\xi_2\mathbf{e}_2) \times \cdots \times \beta(\xi_1\mathbf{e}_1). \tag{9.69}$$

Let $\beta_j : \mathbb{R} \to \mathbb{R}$ be defined by $\beta_j(\xi_j) = \beta(\xi_j \mathbf{e}_j)$ for $1 \le j \le k$. Since β is continuous on $\mathbb{R}^k$, each β_j is continuous on $\mathbb{R}$. Since β is not identity 0, each β_j is not identically 0. Furthermore, each β_j satisfies

$$\beta_j(x + y) = \beta\big((x + y)\mathbf{e}_j\big) = \beta(x\mathbf{e}_j)\beta(y\mathbf{e}_j) = \beta_j(x)\beta_j(y).$$

Thus each β_j satisfies the hypotheses qon [51, p. 192] which implies $\beta_j(\xi_j) = \exp(-it_j\xi_j)$ for a unique $t_j \in \mathbb{R}$. By the formula (9.69), we obtain

$$\beta(\mathbf{x}) = \exp(-it_1\xi_1) \times \exp(-it_2\xi_2) \times \cdots \times \exp(-it_k\xi_k) = \exp(-i\mathbf{t} \cdot \mathbf{x}),$$

where $\mathbf{t} = (t_1, t_2, \ldots, t_k)$ and it is unique. Hence, we get from this that

$$\varphi(f) = \int_{\mathbb{R}^k} f(\mathbf{x})\beta(\mathbf{x}) \, dm_k(\mathbf{x}) = \int_{\mathbb{R}^k} f(\mathbf{x}) \exp(-i\mathbf{t} \cdot \mathbf{x}) \, dm_k(\mathbf{x}) = \widehat{f}(\mathbf{t}).$$

We have completed the proof of the problem.

> ### Problem 9.15
>
> *Rudin Chapter 9 Exercise 15.*

Proof. Let $\mathbf{A} : \mathbb{R}^k \to \mathbb{R}^k$ be a linear operator with $\det \mathbf{A} > 0$. Thus $\mathbf{A}^{-1}$ exists. Let $\mathbf{y} = \mathbf{A}^{-1}(\mathbf{x})$. Then we follow from Theorem 7.26 (The Change-of-variables Theorem) (with $X = Y = V = \mathbb{R}^k$, $T = \mathbf{A}^{-1}$ which is one-to-one and differentiable on $\mathbb{R}^k$, $T(\mathbb{R}^k) = \mathbf{A}^{-1}(\mathbb{R}^k) = \mathbb{R}^k$ and $\mathbf{y} = \mathbf{A}^{-1}(\mathbf{x})$) that

$$\widehat{g}(\mathbf{z}) = \widehat{f \circ A}(\mathbf{z})$$
$$= \underbrace{\int_{\mathbb{R}^k} f\big(\mathbf{A}(\mathbf{y})\big) \exp(-i\mathbf{y} \cdot \mathbf{z}) \, dm_k(\mathbf{y})}_{\text{This is } \int_Y f \, dm.}$$
$$= \int_{\mathbb{R}^k} \underbrace{f(\mathbf{x}) \exp[-i\mathbf{A}^{-1}(\mathbf{x}) \cdot \mathbf{z}] \times |\det \mathbf{A}^{-1}|}_{\text{This is } (f \circ T)(\mathbf{x}) = (f \circ \mathbf{A}^{-1})(\mathbf{x}).} \, dm_k(\mathbf{x}) \qquad (9.70)$$

for every $\mathbf{z} \in \mathbb{R}^k$. We need a result from linear algebra:

> ### Lemma 9.7
>
> If $\mathbf{A}$ is a $k \times k$ matrix, then for all $\mathbf{x}, \mathbf{y} \in \mathbb{C}^k$, we have $\langle \mathbf{A}(\mathbf{x}), \mathbf{y} \rangle = \langle \mathbf{x}, \mathbf{A}^T(\mathbf{y}) \rangle$, where $\mathbf{A}^T$ is the transpose of $\mathbf{A}$.

Proof of Lemma 9.7. Using [41, p. 109; Example 5.3.1, p. 286], it can be seen easily that

$$\langle \mathbf{A}(\mathbf{x}), \mathbf{y} \rangle = [\mathbf{A}(\mathbf{x})]^T \mathbf{y} = \mathbf{x}^T \mathbf{A}^T(\mathbf{y}) = \langle \mathbf{x}, \mathbf{A}^T(\mathbf{y}) \rangle,$$

completing the proof of Lemma 9.7.

Thus we apply Lemma 9.7 to the expression (9.70) to deduce that

$$\widehat{g}(\mathbf{z}) = |\det \mathbf{A}| \cdot \int_{\mathbb{R}^k} f(\mathbf{x}) \exp[-i\mathbf{x} \cdot \mathbf{A}^{-T}(\mathbf{z})] \, dm_k(\mathbf{x}) = |\det \mathbf{A}| \cdot \widehat{f}(\mathbf{A}^{-T}(\mathbf{z})).$$

Consequently, we conclude that

$$\widehat{f \circ \mathbf{A}} = |\det \mathbf{A}| \cdot (\widehat{f} \circ \mathbf{A}^{-T}). \tag{9.71}$$

Particularly, let $\mathbf{A}$ be a rotation matrix of $\mathbb{R}^k$. Then it preserves the distance of $\mathbf{x}$ from the origin, i.e., $\langle \mathbf{A}(\mathbf{x}), \mathbf{A}(\mathbf{x}) \rangle = \langle \mathbf{x}, \mathbf{x} \rangle$. By [4, Proposition 5.1.13, p. 134], $\mathbf{A}$ is orthogonal which means

$$\mathbf{A}^T = \mathbf{A}^{-1}.$$

Furthermore, this implies that $|\det \mathbf{A}| = 1$, so the formula (9.71) can be simplified to

$$\widehat{f \circ \mathbf{A}} = \widehat{f} \circ \mathbf{A}. \tag{9.72}$$

In other words, if f is invariant under rotations, then $f \circ \mathbf{A} = f$ and the expression (9.72) implies

$$\widehat{f} \circ \mathbf{A} = \widehat{f},$$

i.e., $\widehat{f}$ is also invariant under rotations. This completes the proof of the problem. $\blacksquare$

Problem 9.16

Rudin Chapter 9 Exercise 16.

Proof. We make the following assumption: Suppose that $f \in S(\mathbb{R}^k)$, where $S(\mathbb{R}^k)$ is the class of all functions $f : \mathbb{R}^k \to \mathbb{C}$ such that $f \in C^\infty$ and $x^\beta \partial^\alpha f$ is bounded for all multi-indices α and β.[k] Let $1 \le j \le k$, $\mathbf{x} = (\mathbf{x}_j, x_j)$ and $F(\mathbf{x}) = \dfrac{\partial f}{\partial x_j}$, where $\mathbf{x} = (x_1, \ldots, x_j, \ldots, x_k)$ and $\mathbf{x}_j = (x_1, \ldots, x_{j-1}, x_{j+1}, \ldots, x_k)$. We note that

$$\mathbf{x} \cdot \mathbf{y} = x_1 y_1 + x_2 y_2 + \cdots + x_k y_k = \mathbf{x}_j \cdot \mathbf{y}_j + x_j y_j.$$

For every $\mathbf{y} \in \mathbb{R}^k$, we follow from this, the definition in Problem 9.14 and Lemma 9.4 that

$$\widehat{F}(\mathbf{y}) = \int_{\mathbb{R}^k} \frac{\partial f}{\partial x_j}(\mathbf{x}) \exp(-i\mathbf{x} \cdot \mathbf{y}) \, dm_k(\mathbf{x})$$

$$= \int_{\mathbb{R}^{k-1}} \exp(-i\mathbf{x}_j \cdot \mathbf{y}_j) \, dm_{k-1}(\mathbf{x}_k) \underbrace{\left[\int_{\mathbb{R}} \frac{\partial f}{\partial x_j}(\mathbf{x}_j, x_j) \exp(-i x_j \cdot y_j) \, dm(x_j) \right]}_{\text{Apply Lemma 9.4 to this.}}$$

$$= \int_{\mathbb{R}^{k-1}} \exp(-i\mathbf{x}_j \cdot \mathbf{y}_j) \, dm_{k-1}(\mathbf{x}_k) \left[i y_j \int_{\mathbb{R}} f(\mathbf{x}) \exp(-i x_j \cdot y_j) \, dm(x_j) \right]$$

$$= i y_j \int_{\mathbb{R}^k} f(\mathbf{x}) \exp(-i\mathbf{x} \cdot \mathbf{y}) \, dm_k(\mathbf{x})$$

$$= i y_j \widehat{f}(\mathbf{y})$$

which implies that

$$\widehat{\frac{\partial^2 f}{\partial x_j^2}}(\mathbf{y}) = -y_j^2 \widehat{f}(\mathbf{y}).$$

Therefore, we have

$$\widehat{g}(\mathbf{x}) = -\sum_{j=1}^{k} x_j^2 \widehat{f}(\mathbf{x}) = -|\mathbf{x}|^2 \cdot \widehat{f}(\mathbf{x}) \tag{9.73}$$

---

[k]In fact, S is the Schwartz space on $\mathbb{R}^k$, see Remark 9.2.

for all $\mathbf{x} \in \mathbb{R}^k$.

Suppose that f has continuous second derivatives so that the Laplacian Δf is well-defined. Suppose further that $\text{supp}\,(f)$ is compact. In this case, $f \in L^1(\mathbb{R}^k)$ so that $\widehat{f}$ is well-defined on $\mathbb{R}^k$. Furthermore, we suppose that $\mathbf{A}$ is a rotation of $\mathbb{R}^k$ about the origin $\mathbf{0}$ and $g_{\mathbf{A}} = \Delta(f \circ \mathbf{A})$. Then we deduce from the formulas (9.72) and (9.73) that

$$\widehat{g_{\mathbf{A}}}(\mathbf{x}) = -|\mathbf{x}|^2 \cdot \widehat{f \circ \mathbf{A}}(\mathbf{x}) = -|\mathbf{A}\mathbf{x}|^2 \cdot (\widehat{f} \circ \mathbf{A})(\mathbf{x}) = -|\mathbf{A}\mathbf{x}|^2 \cdot \widehat{f}(\mathbf{A}\mathbf{x}) = \widehat{g}(\mathbf{A}\mathbf{x}) = \widehat{g \circ \mathbf{A}}(\mathbf{x}).$$

By the definition, we have $\Delta(\widehat{f \circ \mathbf{A}}) = (\widehat{\Delta f}) \circ \mathbf{A}$. To go further, we need the analogue of Theorem 9.12 (The Uniqueness Theorem):

> **Lemma 9.8**
>
> If $f \in L^1(\mathbb{R}^k)$ and $\widehat{f}(\mathbf{t}) = 0$ for all $\mathbf{t} \in \mathbb{R}^k$, then
>
> $$f(\mathbf{x}) = 0 \tag{9.74}$$
>
> a.e. on $\mathbb{R}^k$. Furthermore, if f is continuous on $\mathbb{R}^k$, then the expression (9.74) holds everywhere on $\mathbb{R}^k$.

Proof of Lemma 9.8. Since $\widehat{f} = 0$ on $\mathbb{R}^k$, we have $\widehat{f} \in L^1(\mathbb{R}^k)$ and the result (9.74) follows from the Inversion Theorem for $\mathbb{R}^k$ (Problem 9.14). If f is continuous on $\mathbb{R}^k$, the definition of continuity ensures that the relation (9.74) holds everywhere on $\mathbb{R}^k$. $\blacksquare$

Now we let $F = \Delta(f \circ \mathbf{A}) - (\Delta f) \circ \mathbf{A}$. It is clear that $f \circ \mathbf{A} \in S(\mathbb{R}^k)$ if $f \in S(\mathbb{R}^k)$. Similar to the proof of Problem 9.7, we can show that

$$\partial^\alpha f \in S(\mathbb{R}^k) \quad \text{and} \quad \partial^\alpha f \in L^1(\mathbb{R}^k).$$

Thus we have $\Delta(f \circ \mathbf{A}) \in L^1(\mathbb{R}^k)$ and $(\Delta f) \circ \mathbf{A} \in L^1(\mathbb{R}^k)$ so that $F \in L^1(\mathbb{R}^k)$. Since $\widehat{F}(\mathbf{t}) = 0$ for all $\mathbf{t} \in \mathbb{R}^k$ and F is continuous on $\mathbb{R}^k$, Lemma 9.8 implies that

$$\Delta(f \circ \mathbf{A}) = (\Delta f) \circ \mathbf{A} \tag{9.75}$$

holds everywhere on $\mathbb{R}^k$. It is well-known that any rotation about a point $\mathbf{p}$ is the composition of a rotation about the origin $\mathbf{0}$ and a translation. By this fact, the analysis preceding Lemma 9.8 and the hypothesis that Δ commutes with translations, we conclude that the expression (9.75) is still valid for every rotation about every point $\mathbf{p}$.

It is time to consider the general situation. Fix a $\mathbf{p} \in \mathbb{R}^k$. There exists a $r > 0$ such that $\mathbf{p} \in K_r = \overline{B(\mathbf{0}, r)} \subset V_r = B(\mathbf{0}, 2r)$. By [49, Exercise 6, p. 289], there exists functions $\psi_1, \ldots, \psi_s \in C^\infty(\mathbb{R}^k)$ such that[1]

- $0 \le \psi_i \le 1$ for $1 \le i \le s$;

- $\text{supp}\,(\psi_i) \subset V_r$;

- $\psi_1(\mathbf{x}) + \psi_2(\mathbf{x}) + \cdots + \psi_s(\mathbf{x}) = 1$ for every $\mathbf{x} \in K_r$.

If we define $\Psi : \mathbb{R}^k \to \mathbb{R}$ by

$$\Psi = \psi_1 + \psi_2 + \cdots + \psi_s,$$

---

[1]For a proof of this, please read [63, Problem 10.6, pp. 266, 267].

then the above conditions imply that $\Psi \in C_c^\infty(\mathbb{R}^k)$ and $\Psi(\mathbf{x}) = 1$ on K_r. Thus we have $f \cdot \Psi = f$ on K_r. Particularly, we have

$$(f \cdot \Psi)(\mathbf{p}) = f(\mathbf{p}) \quad \text{and} \quad \frac{\partial \Psi}{\partial x_j}(\mathbf{p}) = \frac{\partial^2 \Psi}{\partial x_j^2}(\mathbf{p}) = 0$$

for every $1 \le j \le k$. As a consequence, these imply that

$$\frac{\partial^2(f \cdot \Psi)}{\partial x_j^2}(\mathbf{p}) = \Psi(\mathbf{p}) \cdot \frac{\partial^2 f}{\partial x_j^2}(\mathbf{p}) + 2\frac{\partial f}{\partial x_j}(\mathbf{p}) \cdot \frac{\partial \Psi}{\partial x_j}(\mathbf{p}) + f \cdot \frac{\partial^2 \Psi}{\partial x_j^2}(\mathbf{p}) = \frac{\partial^2 f}{\partial x_j^2}(\mathbf{p}) \tag{9.76}$$

for $1 \le j \le k$. By the definition of the Laplacian and the result (9.76), we yield

$$(\Delta(f \cdot \Psi))(\mathbf{p}) = \sum_{j=1}^k \frac{\partial^2(f \cdot \Psi)}{\partial x_j^2}(\mathbf{p}) = \sum_{j=1}^k \frac{\partial^2 f}{\partial x_j^2}(\mathbf{p}) = (\Delta f)(\mathbf{p}). \tag{9.77}$$

Note that $f \cdot \Psi$ is compactly supported in this case. If $\mathbf{Ap} = \mathbf{q}$, then $\mathbf{q} \in K_r$. On the one hand, the expressions (9.75) (with f replaced by $f \cdot \Psi$) and (9.77) give

$$\big[\Delta(f \cdot \Psi) \circ \mathbf{A}\big](\mathbf{p}) = [(\Delta(f \cdot \Psi)) \circ \mathbf{A}](\mathbf{p}) = \big(\Delta(f \cdot \Psi)\big)(\mathbf{q}) = (\Delta f)(\mathbf{q}) = [(\Delta f) \circ \mathbf{A}](\mathbf{p}). \tag{9.78}$$

On the other hand, since

$$[(f \cdot \Psi) \circ \mathbf{A}](\mathbf{p}) = [(f \circ \mathbf{A})(\mathbf{p})] \cdot [(\Psi \circ \mathbf{A})(\mathbf{p})] = (f \cdot \Psi)(\mathbf{q}) = f(\mathbf{q}) = (f \circ \mathbf{A})(\mathbf{p})$$

on K_r, we have $(f \cdot \Psi) \circ \mathbf{A} = f \circ \mathbf{A}$ on K_r so that

$$\big[\Delta((f \cdot \Psi) \circ \mathbf{A})\big](\mathbf{p}) = \big[\Delta(f \circ \mathbf{A})\big](\mathbf{p}) \tag{9.79}$$

Finally, by combining the expressions (9.78) and (9.79) and using the fact that $\mathbf{p}$ is arbitrary, we may conclude that our desired formula (9.75) also holds everywhere on $\mathbb{R}^k$ in this general case. This completes the proof of the problem. ▟

> ### Problem 9.17
>
> *Rudin Chapter 9 Exercise 17.*

Proof. We prove the case for $\mathbb{R}^k$ directly. Suppose that $\varphi : \mathbb{R}^k \to \mathbb{C}$ is a Lebesgue measurable character. We define $\Phi : L^1(\mathbb{R}^k) \to \mathbb{C}$ by

$$\Phi(f) = \int_{\mathbb{R}^k} f(\mathbf{x})\varphi(\mathbf{x})\, dm_k(\mathbf{x}). \tag{9.80}$$

Now direct computation gives

$$\Phi(f * g) = \int_{\mathbb{R}^k} (f * g)(\mathbf{x})\varphi(\mathbf{x})\, dm_k(\mathbf{x}) = \int_{\mathbb{R}^k} \int_{\mathbb{R}^k} f(\mathbf{x} - \mathbf{y})g(\mathbf{y})\varphi(\mathbf{x})\, dm_k(\mathbf{y})\, dm_k(\mathbf{x}). \tag{9.81}$$

Since φ is a character, we have $\varphi(\mathbf{x}) = \varphi(\mathbf{x} - \mathbf{y})\varphi(\mathbf{y})$. Substituting this into the expression (9.81), we see that

$$\Phi(f * g) = \int_{\mathbb{R}^k} \int_{\mathbb{R}^k} f(\mathbf{x} - \mathbf{y})g(\mathbf{y})\varphi(\mathbf{x} - \mathbf{y})\varphi(\mathbf{y})\, dm_k(\mathbf{y})\, dm_k(\mathbf{x})$$

$$= \int_{\mathbb{R}^k} \Big[\int_{\mathbb{R}^k} f(\mathbf{x} - \mathbf{y})\varphi(\mathbf{x} - \mathbf{y})\, dm_k(\mathbf{x})\Big] g(\mathbf{y})\varphi(\mathbf{y})\, dm_k(\mathbf{y})$$

$$= \int_{\mathbb{R}^k} \Phi(f) g(\mathbf{y}) \varphi(\mathbf{y}) \, dm_k(\mathbf{y})$$

$$= \Phi(f) \cdot \Phi(g).$$

Therefore, Φ is a linear functional on $L^1(\mathbb{R}^k)$. In other words, Φ is a complex homomorphism of the Banach algebra $L^1(\mathbb{R}^k)$. Hence the analogue of Theorem 9.23 in Problem 9.14 shows that one may find a $\beta \in L^\infty(\mathbb{R}^k)$ and a unique $\mathbf{y} \in \mathbb{R}^k$ such that

$$\beta(\mathbf{x}) = \exp(-i\mathbf{y} \cdot \mathbf{x}) \quad \text{and} \quad \Phi(f) = \int_{\mathbb{R}^k} f(\mathbf{x}) \exp(-i\mathbf{y} \cdot \mathbf{x}) \, dm_k(\mathbf{x}) = \widehat{f}(\mathbf{y}). \tag{9.82}$$

Next, we deduce from the formulas (9.80) and (9.82) that

$$\int_{\mathbb{R}^k} f(\mathbf{x})[\varphi(\mathbf{x}) - \exp(-i\mathbf{y} \cdot \mathbf{x})] \, dm_k(\mathbf{x}) = 0 \tag{9.83}$$

holds for all $f \in L^1(\mathbb{R}^k)$. Since $f(\mathbf{x}) = e^{-|\mathbf{x}|}$ is in $L^1(\mathbb{R}^k)$ and $f(\mathbf{x}) \neq 0$ on $\mathbb{R}^k$, the result (9.83) implies that

$$\varphi(\mathbf{x}) = \exp(-i\mathbf{y} \cdot \mathbf{x}) \tag{9.84}$$

a.e. on $\mathbb{R}^k$. To proceed further, we need the following result which is a generalization of Problem 7.6 and its proof uses the **Steinhaus Theorem** in $\mathbb{R}^k$ (see Remark 7.2), but we don't present it here.

Lemma 9.9

Suppose that G is a subgroup of $\mathbb{R}^k$ (relative to addition), $G \neq \mathbb{R}^k$, and G is Lebesgue measurable. Then $m_k(G) = 0$.

To apply this lemma, we consider $h(\mathbf{x}) = \varphi(\mathbf{x}) - \exp(-i\mathbf{y} \cdot \mathbf{x})$ and $G = \{\mathbf{x} \in \mathbb{R}^k \,|\, h(\mathbf{x}) = 0\}$. For $\mathbf{a}, \mathbf{b} \in G$, if $\mathbf{a} + \mathbf{b} \notin G$, then we have $h(\mathbf{a} + \mathbf{b}) \neq 0$ and

$$\varphi(\mathbf{a})\varphi(\mathbf{b}) = \varphi(\mathbf{a} + \mathbf{b}) \neq \exp[-i\mathbf{y} \cdot (\mathbf{a} + \mathbf{b})] = \exp(-i\mathbf{y} \cdot \mathbf{a}) \exp(-i\mathbf{y} \cdot \mathbf{b}) = \varphi(\mathbf{a})\varphi(\mathbf{b})$$

which implies $1 \neq 1$, a contradiction. Thus $\mathbf{a} + \mathbf{b} \in G$ which means G is a subgroup of $\mathbb{R}^k$ (relative to vector addition). Besides, since h is obviously Lebesgue measurable by Proposition 1.9(c) and $G = h^{-1}(0)$, G is a Lebesgue measurable set. Since $m_k(G) > 0$, Lemma 9.9 forces that

$$G = \mathbb{R}^k$$

so that the equation (9.84) holds for all $\mathbf{x} \in \mathbb{R}^k$. Now the continuity of the exponential function implies that φ is continuous on $\mathbb{R}^k$. This completes the proof of the problem. ∎

9.4 Miscellaneous Problems

Problem 9.18

Rudin Chapter 9 Exercise 18.

Proof. First of all, the equation

$$f(x + y) = f(x) + f(y) \tag{9.85}$$

for all $x, y \in \mathbb{R}$ is called the **Cauchy functional equation**.

- **Existence of real discontinuous functions f satisfying the equation (9.85).** It is well-known that the Hausdorff Maximality Theorem, the Axiom of Choice and Zorn's Lemma are equivalent to each other. To prove this part, we would like to apply Zorn's Lemma which says that

> **Lemma 9.10 (Zorn's Lemma)**
>
> If X is a partially ordered set and every totally ordered subset of X has an upper bound, then X has a maximal element.

As a consequence, Zorn's Lemma can be used to prove that every vector space has a basis. In fact, if A is a linearly independent subset of a vector space V over a field K, then *there is* a basis $\mathcal{B}$ of V that contains A.[m] By this result, there exists a basis of $\mathbb{R}$ over $\mathbb{Q}$ with the usual addition and scalar multiplication.[n]

Let $x \in \mathbb{R}$ and H be a basis of $\mathbb{R}$ over $\mathbb{Q}$. Thus x has a unique representation

$$x = r_1 b_1 + r_2 b_2 + \cdots + r_n b_n,$$

where $b_1, b_2, \ldots, b_n \in H$ and $r_1, r_2, \ldots, r_n \in \mathbb{Q}$. Define $f : \mathbb{R} \to \mathbb{R}$ by

$$f(x) = r_1 + r_2 + \cdots + r_n.$$

It is obvious that f satisfies the equation (9.85). Assume that f was continuous on $\mathbb{R}$. Then $f(\mathbb{R})$ must be connected (see [49, Theorem 4.22, p. 93]), but $f(\mathbb{R}) = \mathbb{Q}$ which is a contradiction.

- **If f is Lebesgue measurable and satisfies the equation (9.85), then f is continuous.** We use the following special form of Theorem 2.24 (Lusin's Theorem):[o]

> **Lemma 9.11**
>
> Suppose that $f : [a, b] \to \mathbb{R}$ is Lebesgue measurable. For every $\epsilon > 0$, there is a compact set $K \subseteq [a, b]$ such that $m([a, b] \setminus K) < \epsilon$ and $f|_K$ is continuous.

By Lemma 9.11, there exists a compact set $K \subseteq [0, 1]$ such that

$$m(K) > \frac{2}{3} \tag{9.86}$$

and $f|_K$ is continuous. Since K is compact, f is actually uniformly continuous on K. Given $\epsilon > 0$. There is a $\delta > 0$ such that $|f(x) - f(y)| < \epsilon$ for every $x, y \in K$ with $|x - y| < \delta$. Without loss of generality, we may assume that $\delta < \frac{1}{3}$. Let $\eta \in (0, \delta)$ and $K - \eta = \{x - \eta \mid x \in K\}$. Then $K - \eta \subseteq [-\eta, 1 - \eta]$. Assume that $K \cap (K - \eta) = \varnothing$. Since $K, K - \eta \subseteq [-\eta, 1]$, we have

$$m(K) + m(K - \eta) = m\big(K \cup (K - \eta)\big) \leq m([-\eta, 1]) = 1 + \eta. \tag{9.87}$$

By Theorem 2.20(c), $m(K - \eta) = m(K)$, so we deduce from the inequalities (9.86) and (9.87) that $1 + \eta \geq 2m(K) > \frac{4}{3}$ and thus $\eta > \frac{1}{3}$, a contradiction. Hence we have

$$K \cap (K - \eta) \neq \varnothing.$$

---

[m] For a hint of this proof, please refer to [42, Exercise 8, p. 72]

[n] This kind of basis is called a **Hamel basis**.

[o] Read [22, Exercise 44, p. 64] or [47, p. 66]

Let $p \in K \cap (K - \eta)$. Then $p, p + \eta \in K$ and $|p + \eta - p| = \eta < \delta$ so that

$$|f(\eta) - f(0)| = |f(p) + f(\eta) - f(p)| = |f(p + \eta) - f(p)| < \epsilon. \tag{9.88}$$

In other words, f is continuous at 0. Now for any $x \in [0, 1]$, the inequality (9.88) implies clearly that

$$|f(x + \eta) - f(x)| = |f(x) + f(\eta) - f(x)| = |f(\eta) - f(0)| < \epsilon.$$

Hence f is continuous on $[0, 1]$.

Next, if $z \in [1, 2]$, then we have $z = x + y$ for some $x, y \in [0, 1]$. Since $f(z) = f(x) + f(y)$, we have

$$|f(z + \eta) - f(z)| = |f(x + \eta) - f(x)| < \epsilon$$

and the above paragraph shows that f is also continuous on $[1, 2]$. Now this kind of argument can be applied repeatedly and we conclude that f is continuous on $\mathbb{R}$.

- **If $\mathrm{graph}\,(f)$ is not dense in $\mathbb{R}^2$ and satisfies the equation (9.85), then f is continuous.** Suppose that f is discontinuous. Then f *cannot* be of the form $f(x) = cx$ for any constant c. We have

$$\mathrm{graph}\,(f) = \{(x, f(x)) \mid x \in \mathbb{R}\}.$$

Choose a nonzero real number x_1. Then there exists another nonzero real number x_2 such that

$$\frac{f(x_1)}{x_1} \neq \frac{f(x_2)}{x_2}$$

or equivalently,

$$\begin{vmatrix} x_1 & f(x_1) \\ x_2 & f(x_2) \end{vmatrix} \neq 0.$$

In other words, the vectors $\mathbf{u} = (x_1, f(x_1))$ and $\mathbf{v} = (x_2, f(x_2))$ are linearly independent and thus span the space $\mathbb{R}^2$. Let $\mathbf{p} \in \mathbb{R}^2$. Given $\epsilon > 0$. Since $\mathbb{Q}^2$ is dense in $\mathbb{R}^2$, one can find $q_1, q_2 \in \mathbb{Q}$ such that

$$|\mathbf{p} - q_1 \mathbf{u} - q_2 \mathbf{v}| < \epsilon.$$

Clearly, we have

$$q_1 \mathbf{u} + q_2 \mathbf{v} = (q_1 x_1 + q_2 x_2, f(q_1 x_1 + q_2 x_2)).$$

Therefore, the set

$$G = \{(x, f(x)) \in \mathbb{R}^2 \mid x = q_1 x_1 + q_2 x_2 \text{ and } q_1, q_2 \in \mathbb{Q}\}$$

is dense in $\mathbb{R}^2$. Since $G \subseteq \mathrm{graph}\,(f)$, $\mathrm{graph}\,(f)$ is dense in $\mathbb{R}^2$.

- **The form of all continuous functions satisfying the equation (9.85).** In fact, by a similar argument as in the proof used in [63, Problem 8.6, p. 178], it can be shown that the function f satisfies

$$f(r) = r f(1)$$

for every $r \in \mathbb{Q}$. Now for every $x \in \mathbb{R}$, since $\mathbb{Q}$ is dense in $\mathbb{R}$, we can find a sequence $\{r_n\} \subseteq \mathbb{Q}$ such that $r_n \to x$ as $n \to \infty$. Since f is continuous, we have

$$f(x) = \lim_{n \to \infty} f(r_n) = \lim_{n \to \infty} r_n f(1) = x f(1).$$

Hence every continuous solution of the equation (9.85) must be in the form

$$f(x) = f(1) \cdot x.$$

We have completed the proof of the problem.   ▨

> **Problem 9.19**
>
> *Rudin Chapter 9 Exercise 19.*

Proof. Let $f = \chi_A$ and $g = \chi_B$. Since $f, g \in L^1$, their convolution $h = f * g$ is well-defined by Theorem 8.14. Suppose that p and q are conjugate exponents. Direct computation gives

$$\|f\|_p = \int_{-\infty}^{\infty} |\chi_A(x)|^p \, dm(x) = \frac{1}{\sqrt{2\pi}} \int_A dx = \frac{m(A)}{\sqrt{2\pi}} < \infty$$

and

$$\|g\|_q = \int_{-\infty}^{\infty} |\chi_B(x)|^q \, dm(x) = \frac{1}{\sqrt{2\pi}} \int_B dx = \frac{m(B)}{\sqrt{2\pi}} < \infty.$$

Thus we have $f \in L^p$ and $g \in L^q$. By Problem 9.8, $h = \chi_A * \chi_B$ is (uniformly) continuous on $\mathbb{R}$.

We consider

$$\int_{\mathbb{R}} (f * g)(x) \, dm(x) = \int_{\mathbb{R}} \int_{\mathbb{R}} f(x - y) g(y) \, dm(y) \, dm(x). \tag{9.89}$$

Recall that $\mathbb{R}$ is σ-finite. If we set $F(x, y) = f(x - y)g(y) = \chi_A(x - y)\chi_B(y) \geq 0$ on $\mathbb{R}^2$, then F satisfies the hypotheses of Theorem 8.8 (The Fubini Theorem). We notice that

$$\psi(y) = \int_{\mathbb{R}} F^y \, dm(x) = \int_{\mathbb{R}} f(x - y)g(y) \, dm(x) = g(y) \int_{\mathbb{R}} f(x - y) \, dm(x) = g(y) \int_{\mathbb{R}} f(x) \, dm(x),$$

so the integral on the right-hand side in the expression (9.89) can be simplified to

$$\int_{\mathbb{R}} (f * g)(x) \, dm(x) = \int_{\mathbb{R}} \psi(y) \, dm(y)$$

$$= \int_{\mathbb{R}} \left[g(y) \int_{\mathbb{R}} f(x) \, dm(x) \right] dm(y)$$

$$= \left\{ \int_{\mathbb{R}} f(x) \, dm(x) \right\} \times \left\{ \int_{\mathbb{R}} g(y) \, dm(y) \right\}$$

$$= \frac{1}{2\pi} m(A) m(B) > 0.$$

Hence it is impossible that $\chi_A * \chi_B = 0$.

Recall the definition that

$$(\chi_A * \chi_B)(x) = \int_{-\infty}^{\infty} \chi_A(x - y) \chi_B(y) \, dm(y). \tag{9.90}$$

Since $y \in B$ if and only if $\chi_B(y) = 1$, the integral (9.90) reduces to

$$(\chi_A * \chi_B)(x) = \int_B \chi_A(x - y) \, dm(y). \tag{9.91}$$

Similarly, $x - y \in A$ if and only if $\chi_A(x - y) = 1$. Define $x - A = \{x - a \mid a \in A\}$. We notice that $x - y \in A$ implies that $y \in x - A$ and then $y \in B \cap (x - A)$. These facts show that the integral (9.91) can be further simplified to

$$(\chi_A * \chi_B)(x) = \int_{B \cap (x - A)} dm(y) = m\big(B \cap (x - A)\big) \geq 0. \tag{9.92}$$

By the previous paragraph, there exists a $x_0 \in \mathbb{R}$ such that $(\chi_A * \chi_B)(x_0) > 0$, so the expression (9.92) gives

$$m\big(B \cap (x_0 - A)\big) > 0.$$

In other words, we have $B \cap (x_0 - A) \neq \varnothing$ and we can find a $b \in B$ such that $x_0 = a + b$ for some $a \in A$. Consequently, we have

$$x_0 \in A + B. \tag{9.93}$$

Finally, since $\chi_A * \chi_B$ is continuous at x_0, there exists a neighborhood I around x_0 such that $(\chi_A * \chi_B)(x) > 0$ on I. Hence we conclude from the relation (9.93) that

$$I \subseteq A + B,$$

completing the proof of the problem.

Index

Bibliography

[1] C. D. Aliprantis and O. Burkinshaw, *Principles of Real Analysis*, 3rd ed., San Diego: Academic Press, 1998.

[2] T. M. Apostol, *Calculus Vol. 1: One-Variable Calculus, with an Introduction to Linear Algebra*, 2nd ed., John Wiley & Sons, Inc., 1967.

[3] T. M. Apostol, *Mathematical Analysis*, 2nd ed., Reading, Mass.: Addison-Wesley Pub. Co. , 1974.

[4] M. Artin, *Algebra*, 2nd ed., Boston: Pearson Prentice Hall, 2011.

[5] H. Bauer, *Measure and Integration Theory*, Berlin: W. de Gruyter, 2001.

[6] M. Balcerzak and A. Kharazishvili, On uncountable unions and intersections of measurable sets, *Georgian Math. J.* Vol. 6, No. 3, pp. 201 – 212, 1999.

[7] S. K. Berberian, *Measure and Integration*, New York: Macmillan, 1965.

[8] R. P. Boas, *A Primer of Real Functions*, 4th ed., Washington, D. C.: The Mathematical Association of America Inc., 1996.

[9] V. I. Bogachev, *Measure Theory Volume I*, Berlin; New York: Springer, 2007.

[10] V. I. Bogachev, *Measure Theory Volume II*, Berlin; New York: Springer, 2007.

[11] H. Brezis, *Functional Analysis, Sobolev Spaces and Partial Differential Equations*, New York: Springer, 2011.

[12] A. Browder, *Mathematical Analysis: An Introduction*, New York; Hong Kong: Springer, 1996.

[13] M. R. Burke, Weakly dense subsets of the measure algebra, *Proc. Amer. Math. Soc.*, Vol 106, No. 4, pp. 867 – 874, 1989.

[14] F. S. Cater, A partition of the unit interval, *Amer. Math. Monthly*, Vol. 91, No. 9, pp. 564 – 566, 1984.

[15] R. Cignoli and J. Hounie, Functions with arbitrarily small periods, *Amer. Math. Monthly*, Vol. 85, No. 7, pp. 582 – 584, 1978.

[16] D. L. Cohn, *Measure Theory*, 2nd ed., New York, NY: Birkhäuser, 2013.

[17] J. Dieudonné, Un exemple d'un espace normal non susceptible d'une structure uniforme d'espace complet, *C. R. Acad. Sci. Paris*, 209, pp. 145 – 147, 1939.

[18] C. Eckhardt, Free products and the lack of state-preserving approximations of nuclear C^*-algebras, *Proc. Amer. Math. Soc.*, Vol. 141, No. 8, pp. 2719 – 2727, 2013.

[19] J. Fabrykowski, On continuted fractions and a certain example of a sequence of continuous functions, *Amer. Math. Monthly*, Vol. 95, No. 6, pp. 537 – 539, 1988.

[20] J. Feldman, *Another Riesz Representation Theorem*, http://www.math.ubc.ca/~feldman/m511/rieszmarkov.pdf (updated October 9, 2018).

[21] G. B. Folland, *Fourier Analysis and its Applications*, Providence, R. I.: American Mathematical Society, 2009.

[22] G. B. Folland, *Real Analysis: Modern Techniques and Their Applications*, 2nd ed., New York: Wiley, 1999.

[23] J. B. Fraleigh, *A First Course in Abstract Algebra*, 7th ed., Boston: Addison-Wesley, 2003.

[24] P. M. Fitzpatrick, *Advanced Calculus*, 2nd ed., Belmont, Calif.: Thomson Brooks/Cole, 2006.

[25] X. Fu and C. K. Lai, Translational absolute continuity and Fourier frames on a sum of singular measures, *J. Funct. Anal.*, Vol. 274, No. 9, pp. 2477 – 2498, 2018.

[26] B. R. Gelbaum and J. M. H. Olmsted, *Counterexamples in Analysis*, Mineola, N.Y.: Dover Publications, 2003.

[27] I. S. Gradshteyn and I. M. Ryzhik, *Table of Integrals, Series, and Products*, 7th ed., Academic Press, 2007.

[28] L. M. Graves, *The Theory of Functions of Real Variables*, New York: Mc-Graw Hill Inc., 1946.

[29] G. H. Hardy, *A Course of Pure Mathematics*, 10th ed., Cambridge: Cambridge University Press, 2002.

[30] G. H. Hardy, *Divergent Series*, Oxford: Clarendon Press, 1949.

[31] G. H. Hardy, J. E. Littlewood and G. Polya, *Inequalities*, Cambridge: Cambridge University Press, 1934.

[32] A. Hatcher, *Algebraic Topology*, Cambridge University Press, 2002.

[33] F. Jones, *Lebesgue Integration on Euclidean Space*, Rev. ed., Boston: Jones and Bartlett, 2001.

[34] R. Kannan and C. K. Krueger, *Advanced Analysis on the Real Line*, New York: Springer, 1996.

[35] R. B. Kirk, Sets which split families of measurable sets, *Amer. Math. Monthly*, Vol. 79, No. 8, pp. 884 – 886, 1972.

[36] A. W. Knapp, *Basic Real Analysis*, Boston, MA: Birkhäuser Boston, 2005.

[37] J. W. Lewin, A Truly Elementary Approach to the Bounded Convergence Theorem, *Amer. Math. Monthly*, Vol. 93, No. 5, pp. 395 – 397, 1986.

[38] W. A. Luxemburg, Arzelà's Dominated Convergence Theorem for the Tiemann Integral, *Amer. Math. Monthly*, Vol. 78, No. 9, pp. 970 – 979, 1971.

[39] B. Magajna, C^*-convex sets and completely positive maps, *Integr. Equ. Oper. Theory*, Vol. 85, No. 1, pp. 37 – 62, 2016.

[40] R. E. Megginson, *An Introduction to Banach Space Theory*, New York; Hong Kong: Springer, 1998.

[41] C. D. Meyer, *Matrix Analysis and Applied Linear Algebra*, Philadelphia: SIAM, 2000.

[42] J. R. Munkres, *Topology*, 2nd ed., Upper Saddle River, N.J.: Prentice-Hall, 2000.

[43] G. Myerson, First-class Functions, *Amer. Math. Monthly*, Vol. 98, No. 3, pp. 237 – 240, 1991.

[44] P. N. Natarajan, *An Introduction to Ultrametric Summability Theory*, 2nd ed., New Delhi: Springer, 2015.

[45] W. P. Novinger, Mean convergence in L^p spaces, *Proc. Amer. Math. Soc.*, Vol. 34, No. 2, pp. 627 – 628, 1972.

[46] D. R. Pitts and V. Zarikian, Unique pseudo-expectations for C^*-inclusions, *Illinois J. Math.*, Vol. 59, No. 2, pp. 449 – 483, 2015.

[47] H. L. Royden and P. M. Fitzpatrick, *Real Analysis*, 4th ed., Boston: Prentice Hall, 2010.

[48] W. Rudin, *Functional Analysis*, 2nd ed., Mc-Graw Hill Inc., 1991.

[49] W. Rudin, *Principles of Mathematical Analysis*, 3rd ed., Mc-Graw Hill Inc., 1976.

[50] W. Rudin, Well-distributed measurable sets, *Amer. Math. Monthly*, Vol. 90, No. 1, pp. 41 – 42, 1983.

[51] W. Rudin, *Real and Complex Analysis*, 3rd ed., Mc-Graw Hill Inc., 1987.

[52] S. Saks, Integration in Abstract Metric Spaces, *Duke Math. J.*, Vol. 4, No. 2, pp. 408 – 411, 1938.

[53] Nadish de Silva, A Concise, Elementary Proof of Arzelà's Bounded Convergence Theorem, *Amer. Math. Monthly*, Vol. 117, No. 10, pp. 918 – 920, 2000.

[54] R. L. Schilling, *Measures, Integrals and Martingales*, Cambridge: Cambridge University Press, 2005.

[55] E. Stade, *Fourier Analysis*, Hoboken, N.J. : Wiley-Interscience, 2005.

[56] E. M. Stein and G. L. Weiss, *Introduction to Fourier Analysis on Euclidean Spaces*, Princeton, N. J.: Princeton University Press, 1971.

[57] E. M. Stein and R. Shakarchi, *Fourier Analysis: an Introduction*, Princeton, N. J.: Princeton University Press, 2003.

[58] E. M. Stein and R. Shakarchi, *Real Analysis: Measure Theorey, Integration and Hilbert Spaces*, Princeton, N. J.: Princeton University Press, 2005.

[59] R. S. Strichartz, *The Way of Analysis*, revised ed., Boston, Mass.: Jones and Bartlett, 2000.

[60] B. S. Thomson, Strong Derivatives and Integrals, *Real Anal. Exchange*, Vol. 39, No. 2, pp. 469 – 488, 2013/14.

[61] R. L. Wheeden and A. Zygmund, *Measure and Integral: An Introduction to Real Analysis*, 2nd ed., Boca Raton, FL: CRC Press, 2015.

[62] A. Villani, Another Note on the Inclusion of $L^p(\mu) \subset L^q(\mu)$, *Amer. Math. Monthly*, Vol. 92, No. 7, pp. 485 – 487, 1985.

[63] K. W. Yu, *A Complete Solution Guide to Principles of Mathematical Analysis*, Amazon.com, 2018.

[64] K. W. Yu, *Problems and Solutions for Undergraduate Real Analysis I*, Amazon.com, 2018.

[65] W. P. Ziemer, *Modern Real Analysis*, 2nd ed., Cham: Springer, 2017.

[66] V. A. Zorich, *Mathematical Analysis I*, Berlin: Springer, 2004.

[67] V. A. Zorich, *Mathematical Analysis II*, Berlin: Springer, 2004.

[68] A. Zygmund, *Trigonometric Series*, 2nd ed., Cambridge: Cambridge University Press, 1959.

www.ingramcontent.com/pod-product-compliance
Lightning Source LLC
Chambersburg PA
CBHW081249130726
47998CB00010B/2724